Advances in
ORGANOMETALLIC CHEMISTRY

VOLUME 55

Advances in Organometallic Chemistry

EDITED BY

ROBERT WEST

ORGANOSILICON
RESEARCH CENTER
DEPARTMENT OF CHEMISTRY,
UNIVERSITY OF WISCONSIN,
MADISON WI, USA

ANTHONY F. HILL

RESEARCH SCHOOL OF CHEMISTRY,
INSTITUTE OF ADVANCED STUDIES,
AUSTRALIAN NATIONAL UNIVERSITY
CANBERRA, ACT
AUSTRALIA

MARK J. FINK

DEPARTMENT OF CHEMISTRY
TULANE UNIVERSITY
NEW ORLEANS, LOUISIANA, USA

FOUNDING EDITOR

F. GORDON A. STONE

VOLUME 55

Amsterdam • Boston • Heidelberg • London • New York • Oxford • Paris
San Diego • San Francisco • Singapore • Sydney • Tokyo
Academic Press is an imprint of Elsevier

Academic Press is an imprint of Elsevier
84 Theobald's Road, London WC1X 8RR, UK
Radarweg 29, PO Box 211, 1000 AE Amsterdam, The Netherlands
Linacre House, Jordan Hill, Oxford OX2 8DP, UK
30 Corporate Drive, Suite 400, Burlington, MA 01803, USA
525 B Street, Suite 1900, San Diego, CA 92101-4495, USA

First edition 2008

ISBN: 978-0-12-373978-0
ISSN: 0065-3055

For information on all Academic Press publications
visit our website at books.elsevier.com

Printed and bound in USA

08 09 10 11 12 10 9 8 7 6 5 4 3 2 1

Contents

Pentadienyl Complexes of the Group 4 Transition Metals

LOTHAR STAHL and RICHARD D. ERNST

Mixed Metal Acetylide Complexes

PRADEEP MATHUR, SAURAV CHATTERJEE and VIDYA D. AVASARE

Transition Metal Organometallic Synthesis Utilising Diorganoiodine(III) Reagents

ALLAN J. CANTY, THOMAS RODEMANN and JOHN H. RYAN

Contributors

Numbers in parentheses indicated the pages on which the authors' contributions begin.

VIDYA D. AVASARE (205), Chemistry Department, Indian Institute of Technology-Bombay, Powai, Bombay, India

ALLAN J. CANTY (283), School of Chemistry, University of Tasmania, Hobart, Tasmania, Australia

SAURAV CHATTERJEE (205), Chemistry Department, Indian Institute of Technology-Bombay, Powai, Bombay, India

GULLIVER T. DALTON (63), Department of Chemistry, Australian National University, Canberra, ACT, Australia and Laser Physics Centre, Research School of Physical Sciences and Engineering, Australian National University, Canberra, ACT, Australia

RICHARD D. ERNST (139), Department of Chemistry, University of Utah, Salt Lake City, UT, USA

MARK G. HUMPHREY (63), Department of Chemistry, Australian National University, Canberra, ACT, Australia

PRADEEP MATHUR (205), Chemistry Department, Indian Institute of Technology-Bombay, Powai, Bombay, India

JOSEPH P. MORRALL (63), Department of Chemistry, Australian National University, Canberra, ACT, Australia and Laser Physics Centre, Research School of Physical Sciences and Engineering, Australian National University, Canberra, ACT, Australia

THOMAS RODEMANN (283), School of Chemistry, University of Tasmania, Hobart, Tasmania, Australia

AROOP K. ROY (1), Personal Care and Pharmaceuticals R&D, Noveon (A Subsidiary of The Lubrizol Corporation), Cleveland, OH, USA

Current Affiliation: Performance Additives – Personal Care, Wacker Chemical Corporation, Adrian, MI, USA.

JOHN H. RYAN (283), CSIRO Molecular and Health Technologies, Ian Wark Laboratory, Bayview Avenue, Clayton, Vic., Australia

MAREK SAMOC (63), Laser Physics Centre, Research School of Physical Sciences and Engineering, Australian National University, Canberra, ACT, Australia

LOTHAR STAHL (139), Department of Chemistry, University of North Dakota, Grand Forks, ND, USA

Preface

With this volume, Mark Fink officially joins us as Editor. Mark is Professor of Chemistry at Tulane University, where he leads a diversified program of research on multiply bonded and low-coordinate silicon and germanium compounds, palladium–silicon compounds in catalysis, and precursor molecules, especially of gallium, for CVD processing. He is a pioneer in the study of organosilicon compounds in molecular beams. His perspective on main group chemistry, and the interface with transition-metal chemistry and applications, ideally places him to maintain the synergy between main-group and transition-metal organometallic chemistry that *Advances in Organometallic Chemistry* has endeavored to chronicle. It is a pleasure to welcome Mark to our editorial team.

Anthony F. Hill and Robert West

A Review of Recent Progress in Catalyzed Homogeneous Hydrosilation (Hydrosilylation)

AROOP K. ROY*

Personal Care and Pharmaceuticals R&D, Noveon (A Subsidiary of The Lubrizol Corporation), 9911 Brecksville Road, Cleveland, OH 44141, USA

I
INTRODUCTION

Without a doubt, amongst silicon–carbon as well as other silicon-heteroelement bond-forming reactions, the catalyzed hydrosilation (or hydrosilylation) reaction is the most fascinating chemical transformation. Even for a specialized chemical reaction, the steady periodic reviews that have appeared over the last 30 years on the subject are clear testimony to the diverse and continuing utility of this reaction in organic, inorganic, bio-organic, organometallic, polymer and materials chemistry. The last extensive review of published literature in the area (covering research through about 1996) was provided as recently as 1998 by Ojima and coworkers.[1] This, together with Ojima's earlier review[2] and the treatise by Marciniec,[3] considered by many as the "Bible" on hydrosilation, chart the history of developments in the area in compelling detail, since Speier's,[4] Voronkov and coworkers'[5] and Harrod and Chalk's[6] descriptive accounts of the reaction covering the first two decades of explosive growth appeared in the late 1970s. An account that is concise yet which covers all of the important aspects of the reaction has been provided by Brook.[7] Marciniec also recently reviewed C=C and C≡C hydrosilation.[8]

The aim of this brief review, covering literature only since 1990 up through ca. mid-2005, is to update the readership and the practicing scientist on some of the significant advances that have been made in the field. These have occurred

*Corresponding author. Present affiliation: Performance Additives – Personal Care, Wacker Chemical Corporation, 3301 Sutton Road, Adrian, MI 49221, USA.
E-mail: nirakr2@gmail.com (A.K. Roy).

particularly in areas such as new ligands for catalysts, the range of transition metal species that effectively catalyze the reaction, asymmetric hydrosilation, significant growth in application to a plethora of new polymer and materials synthesis and, most notably, strides in unraveling the inner workings of the catalytic cycle. In this last context, the terms precatalyst and catalyst have been used interchangeably in this review to denote starting compounds or complexes that facilitate hydrosilation. The word combination "true catalyst" has only been used where the actual catalytic species has been positively identified.

To preview the progress in the field, completely new ligand classes such as diimines and N-heterocyclic carbenes have surfaced; several lanthanides are established as quite effective catalysts; asymmetric hydrosilation is now an extremely important tool in chiral synthesis; materials accessed *via* hydrosilation range from new block copolymers to POSS-based composites and dendrimers, and evidence is substantial that more than one type of catalytic cycle is operative for different transition metals or groups of transition metals and perhaps even for different oxidation states of some metals.

The author's association with industrial hydrosilation processes and research for a number of years has been particularly enlightening in connection with several commercially important reactions and their nuances, oddities and challenges. This has helped the development of certain perspectives on the reaction which will undoubtedly infiltrate parts of this review. It is hoped that these occasional views will enhance the value of this chapter to the readership. Also, because Ojima's second review was published fairly recently there will be overlap of coverage. However, the organization of this review is not as elaborate and it will only cover hydrosilation occurring, intended to occur or perceived to occur in homogeneous media, except for brief but important comparisons with and discussion on some new developments in the area of heterogeneous catalysis.

II

HYDROSILATION WITH IRON TRIAD CATALYSTS

Amongst the Fe-triad metals Fe, Ru and Os, only a few hydrosilations have been reported using the first and third row metal catalysts. Very recently, however, a new type of Fe(0) complex, containing a diimine-based pincer ligand, and also containing two coordinated dinitrogen molecules was shown to be an effective and, remarkably, regiospecific catalyst for the hydrosilation of a variety of terminal alkenes with $PhSiH_3$ or Ph_2SiH_2. For example, styrene which generally leads to both terminal and internal adducts, produced only the terminal hydrosilation product [Eq. (1)], as did all other terminal alkenes examined.[9] The novel iron complex was also highly successful in catalyzing the hydrosilation of diphenyl acetylene with phenylsilane to produce a monosilylated product. This unique activity for an iron complex, the few previous examples of which have been known to be poor hydrosilation catalysts, suggests that modulation of metal activity using newly designed ligands may allow the synthesis of newer and much more effective catalysts from transition metals that are less expensive than the traditional Pt or Rh-based complexes. The observed high lability of N_2 ligands in **1** and the isolation of both η^2-SiH and π-alkyne complexes *via* displacement of N_2 afford valuable insight on the catalytic activity of this unusual iron complex.

Styrene + $PhSiH_3$ —"Fe"→ $PhCH_2CH_2SiH_2Ph$

Exclusive β-adduct

"Fe" = **1** (pyridine bis(imine) Fe(N_2)$_2$ complex, N-Ar substituents); Ar = 2,6-^{i}Pr$_2$-C_6H_3 (1)

Ruthenium complexes are particularly effective for the hydrosilation of alkynes. A variety of terminal as well as internal alkynes have been hydrosilated in respectable to excellent yields using a number of catalyst types. With 5 mol% [$RuCl_2$(*p*-cymene)]$_2$ as catalyst, sterereoselectivities upwards of 95:5::*Z*:*E* and yields in the range 78–98% are achieved for many terminal alkynes, including several containing functional groups.[10] Interestingly, a regioselectivity reversal to high α-product is observed for alkynes with a hydroxyl group α- or β- to the internal alkyne carbon, but not for this substituent at the γ position or beyond or for a benzylated hydroxyl group. For complexes with phosphine ligands, basicity of the phosphine plays a dominant role in stereoselectivity and overall product distribution. This has been nicely demonstrated by Ozawa and coworkers.[11] With Ph_3P ligands and tertiary arylsilanes, >99:1 *E* selectivity was observed for a number of terminal alkynes, whereas a Ru–Si complex with Pr^i_3P ligands and a Ru-complex with Cy_3P in place of Ph_3P afforded *Z*-products, in most cases with selectivity >90:10 [Eq. (2)]. Further, overall products yields were mostly >96%. Such phosphine-substituent effects were also reported by Oro and coworkers.[12] Both groups propose mechanistic pathways to the products that incorporate similar intermediates, but Oro's group also postulates the involvement of polynuclear hydride-bridged Ru complexes as a key factor in the stark selectivity differences. It appears from the electronic character of the silanes used by the two groups that the silane substituents also exert some effect on the stereoselectivity. In fact, the effect of various silanes on product distribution (which includes enyne formation) with $RuCl_2(PPh_3)_3$ has been reported.[13] Nevertheless, since the product vinylsilanes can be of further utility in various organic transformations, simple phosphine ligand substitution offers a convenient way to arrive at ultimate products with different geometries or isomeric identities.

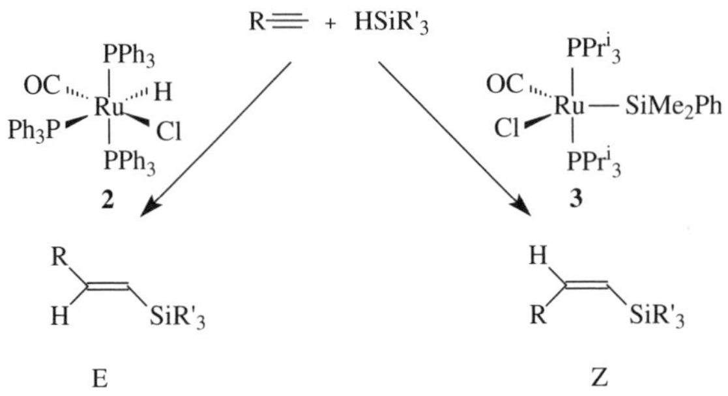

$$R{\equiv} + HSiR'_3 \quad (2)$$

Unlike the high anti-Markovnikov or β-selectivity exhibited by the above neutral phosphine complexes, cationic Ru complexes have been reported to provide high regioselectivity towards Markovnikov adducts (in good to high yields) for both terminal and internal alkynes.[14] More fascinating still is the use of Grubbs' first-generation olefin-metathesis catalyst [$Ru{=}CHPhCl_2(PCy_3)_2$] that yields adducts with regio- or stereoselectivity based on the nature of the alkyne and the silane.[15]

Although many metal-carbonyl clusters technically do not fall in the category of homogeneous catalysts, several substituted ruthenium-carbonyl clusters are soluble, and $Ru_3(CO)_{12}$ has been shown to behave in some cases as if lower nuclearity fragments derived from it are the catalytic species in hydrosilation reactions. The hydrosilation of allyl chloride is an important industrial reaction that is also wrought with side reactions. This will be addressed in a later Section XI but $Ru_3(CO)_{12}$ is one of a few catalysts that provide yields of useful 3-chloropropylsilanes in excess of 70% *via* hydrosilation of allyl chloride. Using a one to three fold molar excess of $HSi(OMe)_3$ over allyl chloride at 80 °C, >70% yield of chloropropyltrimethoxysilane was obtained by Tanaka and coworkers.[16] A μ-silane diruthenium complex **4** catalyzes the hydrosilation of imines and ketones with secondary silanes *via* cooperative action between the ruthenium centers.[17] A novel catalytic hydrosilation of esters using either $Ru_3(CO)_{12}$ or $[RuCl_2(CO)_3]_2$, in the presence of diethylamine and ethyl iodide as co-catalysts, leads to aldehydes in good to excellent yields following hydrolysis of the alkyl silyl acetals. Many other TM carbonyl clusters were inactive in this reaction.[18]

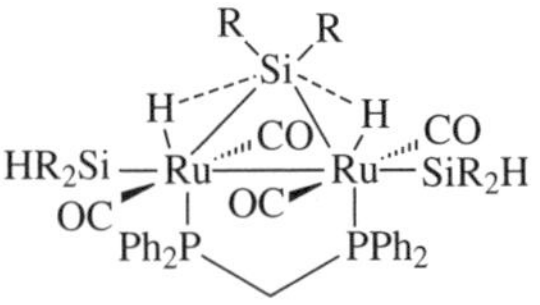

4

A few osmium complexes are active catalysts for the hydrosilation of phenylacetylene with Et_3SiH. Whereas [$RuHCl(CO)(PPr^i_3)_2$] is stereospecific in providing *Z*-selectivity as well as quantitative yield, the corresponding Os complex yields high *Z*-adduct or *E*-adduct depending on whether a stoichiometric amount or excess silane, respectively, is used.[19] Some other examples of Os complexes have also been reported for the hydrosilation of phenylacetylene with Et_3SiH.[20]

III
COBALT TRIAD CATALYSTS

Unlike the iron triad complexes, cobalt triad compounds, particularly those of Rh and Ir are much more active in hydrosilation catalysis, although the cobalt carbonyl $Co_2(CO)_8$ has played a very important role in furthering understanding of the catalytic cycle with transition metals. Recently, a simple cobalt salt, $CoBr_2$, in conjunction with added Bu^n_3P, ZnI_2 and Bu_4NBH_4 was shown to afford 1,4-hydrosilated isoprene (silicon at C-4) in 90% isolated yield, where most metal catalysts derived from Ru, Rh, Pd and Pt usually lead to a mixture of regioisomers.[21]

Mixed Co–Rh carbonyl clusters have been reported by Ojima and coworkers to catalyze hydrosilations of 1-hexyne, isoprene, cyclohexanone and cyclohexenone [Eq. (3)] and compared against catalysis by $Rh_4(CO)_{12}$, Wilkinson's catalyst and $Co_2(CO)_8$.[22] Stereoselectivity mostly in the range 85–95% towards the *Z*-adduct was obtained with 1-hexyne under various and often mild conditions, although some *E*-adduct and Markovnikov product were always present, except in the case of $Co_2(CO)_8$ which gave no *Z*-adduct. Lower levels of catalyst led to higher *cis/trans* product ratios. Silane substituents also affected product distribution, with alkyl- and alkyl/aryl silanes providing high *Z* product isomer, and alkyl/chloro and alkoxysilanes leading to *E*- and α-adduct only. For isoprene hydrosilation, product selectivity toward 1-silyl adduct was high for all clusters, except $Co_2(CO)_8$ which again differed from the rest in affording a high 4-silyl adduct. It is of interest to note that although the Rh catalysts are different, Ojima proposed a selectivity mechanism that incorporates an isomerization on the Rh complex that leads from an *E*-alkene to a *Z*-alkene, while Hiyama and coworkers recently suggested an isomerization catalyzed by $RhI(PPh_3)_3$ and heteroatom-substituted hydrosilane that converts *Z*-alkenylsilanes to the *E*-isomers.[23] The latter was not observed with $RhCl(PPh_3)_3$ in Ojima's study with Et_3SiH as the hydrosilane.

Catalyst
Hydrosilane
Substrate-, Silane-, Catalyst-dependent Hydrosilation Product:
stereo-, regioisomer (and silylenolether) (3)

Catalysts studied: $Rh_4(CO)_{12}$, $Co_2Rh_2(CO)_{12}$, $Co_3Rh(CO)_{12}$, $Co_2(CO)_8$, $RhCl(PPh_3)_3$.
Silanes examined: Et_3SiH, Et_2MeSiH, $EtMe_2SiH$, $PhMe_2SiH$, Ph_2SiH_2, $ClMe_2SiH$, Cl_2MeSiH, $(MeO)_3SiH$. Not all catalysts or silanes investigated for all substrates.

In the context of bimetallic catalysts, an unusual cationic Ti–Rh bimetallic complex **5** exhibits an interesting cooperative effect in catalyzing the hydrosilation of acetophenone with Ph_2SiH_2.[24] A weak bonding interaction between rhodium and a Cl on titanium apparently stabilizes a lower-valent Rh intermediate such that the bimetallic complex delivers a much higher hydrosilation yield than the corresponding Rh(COD) monometallic complex based on 1,2-bis(diphenylphosphino)benzene.

The Ti is thought not to be directly involved, since Cp_2TiCl_2 does not catalyze this reaction.

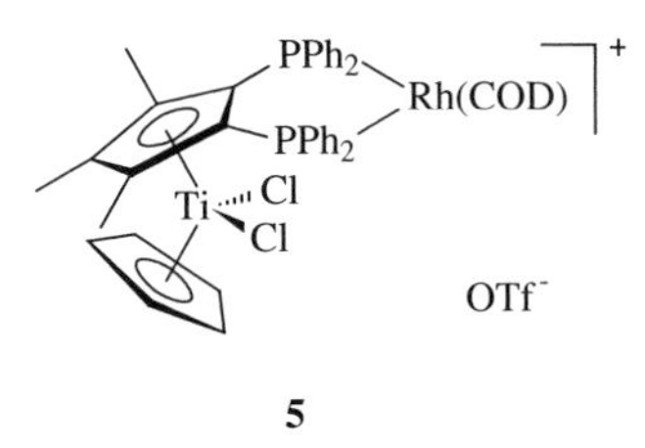

5

Takeuchi and coworkers discovered that stereoselectivity in 1-alkyne hydrosilation can be very effectively controlled *via* solvent and catalyst choice with Rh complexes, and have reported extensively on functional and non-functional 1-alkyne hydrosilations (in excellent yields) and the utility of adducts in further transformations.[25] For example, 1-hexyne hydrosilation with Et_3SiH in EtOH or DMF using $[RhCl(COD)]_2$ leads to high *Z*-selectivity, whereas in nitrile solvents, in the presence of added Ph_3P, the same Rh dimer precatalyst affords very high *E*-selectivity. The high *E*-selectivity is explained by invoking a "near-naked" cationic complex derived from $[Rh(COD)_2]BF_4/2PPh_3$, where only high cationic character leads to high *trans*-adduct (using BPh_4 as the counter anion gave the *Z*-isomer as the major product). In contrast to the findings above, Basato and coworkers report an unusual oxothioether and acetato ligated cationic dirhodium complex, with the same BF_4^- counterion, $[Rh_2(\mu\text{-}OAc)_3(RSCH_2Z)]BF_4$ {R = Me, Ph; Z = C(O)OEt, $CH_2C(O)OMe$}, leads to high *Z*-selectivity (62% conversion) for the hydrosilation of 1-hexyne with $PhMe_2SiH$.[26] Similarly, for the hydrosilation of phenylacetylene with triphenyl-, triethoxy- or triethylsilane, Faller and D'Alliesi report high *E* or *Z* selectivity and little or no α-adduct depending on whether the catalyst is $[RhCp^*(BINAP)](SbF_6)_2$ or $[RhCl_2Cp^*]_2$, respectively.[27] Such dramatic changes in product selectivity as reported by these groups underscore the effect of electronic and steric changes in a metal's coordination sphere on its catalytic behavior, although such effects are often not predictable. With Wilkinson's catalyst $[RhCl(PPh_3)_3]$, the reaction of styrene with Ph_2SiH_2 yields only the β-adduct and other side products, but no α-adduct: whereas $PhSiH_3$ is unreactive towards styrene at room temperature but yields the *E*-adduct exclusively with phenylacetylene as the substrate.[28]

Recently, zwitterionic Rh complexes **6** were reported to be interesting hydrosilation catalysts.[29] The complexes, similar to the cationic rhodium $[Rh(P\text{–}P)(solv)_2]^+$ of Osborne and Schrock, though less active than the corresponding cationic complexes, provide only the beta-adduct from the hydrosilation of styrene with Ph_2SiH_2. Further, unlike the cationic complexes the zwitterionic compounds are highly tolerant of coordinating solvents like acetonitrile and are appreciably soluble in non-polar hydrocarbons such as benzene.

R$_2$ E Rh L L E R$_2$ B

6

E = P, R = phenyl;
E = N, R = methyl

One of the newer developments in catalytic chemistry is performing multiple, distinct transformations in a "one-pot" fashion, using one or more catalysts. Tandem hydrosilation-cyclizations have been reported for a number of catalysts, including the relatively new class of catalysts based on NHC (*N*-heterocyclic carbene) ligands.[30–36] Of the metal compounds that have been utilized, Rh complexes are the catalysts of choice for these reactions. Reported research in the area covers a plethora of unsaturated substrates, including enynes, diynes, allenynes, enediynes and triynes, with many containing heteroatoms that allow the construction of heterocycles from acyclic starting materials. The number of publications in the area of generating carbocycles and heterocycles in the last 15 years, for "silylcarbocyclization" alone, are too numerous to cover in detail within the scope of this review. Hence, a select few are represented in [Eqs. (4)–(8)]. The use of hydrosilation of ynals as a synthetic tool for the generation of allenes has also been reported.[37]

O H CO_2Me —— 1% $RhCl(PPh_3)_3$ / Et_3SiH / toluene,50°C ——> $OSiEt_3$ CO_2Me + $OSiEt_3$ CO_2Me 42-81% (4)

Z R + $(MeO)_3SiH$ or $(EtO)_3SiH$ (3 equiv) —— $Rh(acac)(CO)_2$ (2/5 mol%) / toluene, 60 °C / CO (1 atm) ——> Z R H SiR'_3 (5)

Z = TsN, O, $(EtO_2C)_2C$; R = Me, n-Bu, Ph 46-82%

Z R —— N N Bu / Rh(COD)Cl / Et_3SiH or $PhMe_2SiH$ / CH_2Cl_2, reflux ——> Z R $SiEt_3(SiMe_2Ph)$ (6)

34-90%

Z = TsN, $(MeO_2C)_2C$, $(EtO_2C)_2C$, O, $(MeOC)_2C$, $Me_2C(OC)_2C$; R = Me, Bu, Ph

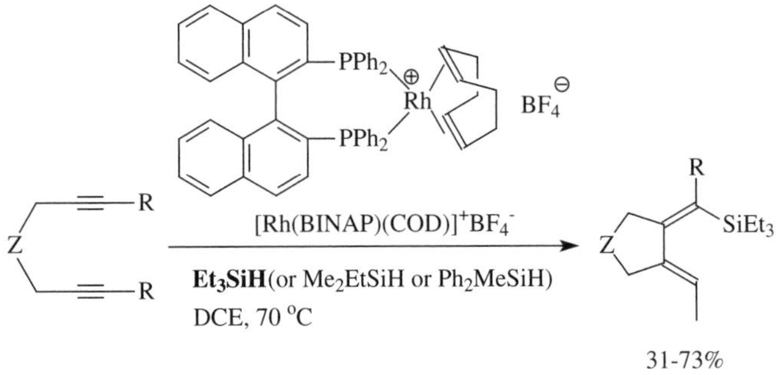

PPh$_2$ Rh PPh$_2$ $BF_4^{\ominus}$

Z, R, R; $[Rh(BINAP)(COD)]^+BF_4^-$ / Et_3SiH (or Me_2EtSiH or Ph_2MeSiH), DCE, 70 °C → Z, R, $SiEt_3$; 31-73% (7)

Z = TsN, O, $(MeO_2C)_2C$, $(R'O)_2C$; R = Me, other combinations; R' = Me, Bn, etc. Not all Z, R, R' combinations examined.

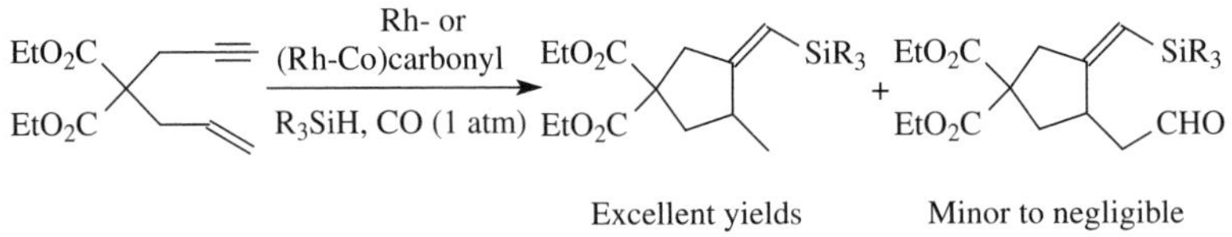

EtO_2C, EtO_2C; Rh- or (Rh-Co)carbonyl / R_3SiH, CO (1 atm) → EtO_2C, EtO_2C, SiR_3 (Excellent yields) + EtO_2C, EtO_2C, SiR_3, CHO (Minor to negligible) (8)

R_3 = Me_2Ph, Ph_2Me, $(EtO)_3$, $(EtO)_2Me$, $(MeO)_3$, Ph_3, $^tBuMe_2$. Not all silanes used with all catalysts.

For both acyclic and cyclic α,β-unsaturated ketones and in general for α,β-unsaturated carbonyl compounds such as aldehydes and esters, $[Rh(OH)(COD)]_2$ is a more efficient catalyst than other types of Rh(I) catalysts.[38] Products are obtained quickly in excellent yields under mild conditions (often at r.t.) and acyclic substrates afford *E*/*Z* ratios from 2:1 to 4:1. Crabtree and Rivera investigated the effect of hydrogen bonding on catalytic activity of specially designed NHC complexes of Rh, and report that a strategically placed amide group on the "wing tip" of the carbene ligand significantly alters selectivity in the hydrosilation of PhCH=CHCOMe with Et_3SiH (that can lead to four different products).[39] Methodically run elimination experiments lead to their proposal that H-bonding is involved in the transition state for hydrosilation between the substrate and the catalyst.

For application toward the construction of peptidomimetics, Sieburth and coworkers studied the hydrosilation of Boc-protected enamines with a variety of silanes, using $Rh_2(OAc)_4$.[40] Exclusive α-selectivity leads to the protected α-amino silanes which are readily deprotected with TFA. However, the yield is dependent on both silane and alkene substituents.

The use of a bowl-shaped phosphine, with three terphenyl groups that are 2,6-methyl-substituted on the outer rings, $P(tm\text{–}tp)_3$, as an added ligand to $[\{RhCl(C_2H_4)_2\}_2]$ dramatically increases the Rh dimer's catalytic activity toward the hydrosilation of cyclohexanone and other ketones, when compared to many other bulky and non-bulky phospines with tertiary silanes.[41] Marciniec and coworkers have reported monomeric and dimeric siloxy-rhodium complexes, $[Rh(OSiMe_3)(PCy_3)(cod)]$ and $[\{Rh(\mu\text{-}OSiMe_3)(diene)\}_2]$, that exihibit high activity in the hydrosilation of allyl glycidyl ether and 1-hexene respectively.[42] The dimeric SiO–Rh complex was found to be much superior to the corresponding

Cl–Rh dimer. Stoichiometric reaction of hydrosilanes with both catalysts indicated the formation of hydridorhodium intermediates and the absence or low-level formation of dehydrogenative silylation products is suggestive of a Chalk–Harrod mechanistic pathway for the catalysis.

Recently, a comparison of the catalytic efficacy of Rh(I), (II) and (III) complexes against Karstedt's catalyst was made, since Rh is the most effective metal following Pt for hydrosilation reactions but Pt is most often used for commercial processes.[43] Both a model reaction, hydrosilation of 1-octene with triethoxysilane, and a commercially useful reaction, hydrosilation of allyl alcohol ethoxylate with poly (dimethyl-*co*-methylsiloxane) were studied, with respect to rate of reaction and terminal vs. internal addition of SiH. Amongst the various phosphine-based and one non-phosphine-based Rh complexes examined, none fared well against Karstedt's catalyst over a 4- and 13-hour reaction period in refluxing toluene, except $[Rh(CO)(PPh_3)(Tp)]$ {Tp = tris(pyrazolyl)borate}, which was comparable in yield and selectivity. For the polymer reaction, again only a Tp–Rh complex, $[Rh(CO)_2(Tp)]$, compared favorably in rate but not in selectivity. Rh(II) and (III) complexes were totally outperformed by Karstedt's catalyst in the study.

Heterogeneous catalysts wield a particular superiority over the homogeneous type, especially in industrial processes – they are often easily separated from products and reused, although they also have disadvantages including lower activity compared to the "soluble" catalysts. Thus, chemists and chemical engineers alike are increasingly interested in harnessing this advantage for homogeneous catalysts which offer them versatility in many more ways. For hydrosilation, reports have recently appeared on progress in this area that takes advantage of the solubility properties of fluorocarbon-groups in solvents. The temperature-dependent miscibility of $CF_3C_6F_{11}$ with toluene allowed the hydrosilation of 2-cyclohexen-1-one with $PhMe_2SiH$ at RT under biphasic conditions, with 90–85% distilled yields and high retention of TON (turnover number) over three cycles using the same charge of catalyst $[RhCl\{P(CH_2CH_2(CF_2)_{n-1}CF_3)_3\}_3]$ ($n = 6$, 8).[44] Similar results were obtained in one-phase higher temperature reactions with hexane or diethyl ether, where the catalyst was separated by cooling to $-30\,^{\circ}C$, when the mixtures phase separate. The one-phase reactions were faster. Further optimizations were done with hydrosilation of cyclohexanone. In another report, improved activity of the fluorous Rh catalysts, making them comparable to $RhCl(PPh_3)_3$, is achieved *via* the use of 4-silylaryl phosphines that are substituted with fluoroalkyl groups on silicon.[45] With a clever reverse approach, recovery and reuse of the non-fluorous Wilkinson's catalyst was demonstrated for the hydrosilation of 1*H*,1*H*,2H-perfluoro-1-alkenes, using a fluorinated solvent to extract the product. In yet another novel approach van Koten's group communicates the use of a fluorous (imidazolium) ionic liquid to separate a fluorinated version of Wilkinson's catalyst that could be recycled up to 15 times without significant loss of activity.[46] Although tungsten is not a member of the Co-triad, a tungsten catalyst was found to precipitate out *via* an unusual "liquid clathrate" formation mechanism, following the consumption of reactants in the hydrosilation of carbonyl compounds (in a solvent-free medium.)[47] Using phase separation of a mixture of solvents that is homogeneous at hydrosilation temperature, to separate the catalyst for recycle has also been claimed for dried H_2PtCl_6.[48] In the author's laboratory, indications were obtained that a certain true Pt catalyst precipitated out at the end of hydrosilation due to high crystallinity and insolubility in the product, but was capable of redissolving under reaction conditions to catalyze a second charge of reactants without loss of activity and selectivity (see Section IV for catalyst description). Although the precipitated catalyst was not filtered out, the same behavior was noted for a different olefinic substrate and silane combination.[49]

Most new Ir catalysts that have been studied in the last 15 years for hydrosilation appear to be cationic species. However, Tanke and Crabtree reported a class of neutral Ir(I) complexes **7** with a very unusual ligand – tris(diphenyloxophosphoranyl)methanide, abbreviated as "triso."[50] Complexes **7** with cyclooctene (COE or coe) or ethylene as the ancillary ligand were found to catalyze only terminal alkyne hydrosilation, with very high $Z{:}E$ stereoselectivity and β-regiospecificity. Turnover rate and $Z{:}E$ ratio were found to be dependent on the steric and electronic character of the substrate 1-alkyne. In the presence of iminophosphines of the type 2-$Ph_2PC_6H_4N{=}CR^1R^2$, $[IrCl(COD)]_2$ has been shown to be a moderately active catalyst for the hydrosilation of acetophenone with Ph_2SiH_2.[51]

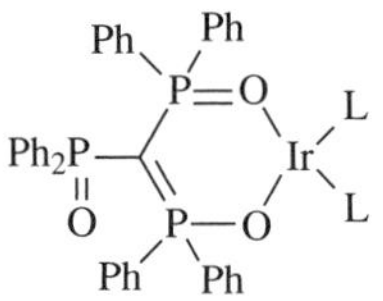

7

L = coe/C_2H_4/cod/CO

The cationic complexes $[Ir(\eta^2\text{-}{}^iPr_2PCH_2CH_2NMe_2)(\text{diolefin})][BF_4]$ and $[Ir(\eta^2\text{-}{}^iPr_2PCH_2CH_2OMe)(\text{diolefin})][BF_4]$, where diolefin is either 1,5-cyclooctadiene (COD or cod)) or tetrafluorobenzobarrelene (TFB), efficiently catalyze the hydrosilation of phenylacetylene with triethylsilane, although all possible isomeric products as well as styrene and $PhC{\equiv}CSiEt_3$ are formed.[52] Oro, Sola and coworkers have reported on the versatility of 1,5-cyclooctadiene as a ligand, *via* flexible hapticity, in phenylacetylene hydrosilation catalysis by the cationic complex $[Ir(NCCH_3)(PMe_3)(\eta\text{-}1,2,5,6\text{-cod})]BF_4$.[53] The cationic silyl and silylene complexes $[Cp^*Ir(H)(PMe_3)(\eta^2\text{-}SiPh_2C_6H_4)][B(C_6F_5)_4]$ and $[Cp^*Ir(H)(PMe_3)(SiPh_2)][B(C_6F_5)_4]$ have been shown by Tilley, Bergman and Klei to be effective precatalysts for the hydrosilation of acetophenone with Ph_3SiH and Ph_2SiH_2, respectively.[54] For these complexes, high ability of the anion to dissociate is critical to catalytic activity, as complexes with triflate anion showed no catalytic behavior. Finally, the certainty that well-designed, newer ligands will advance the boundaries of hydrosilation catalysis is beautifully demonstrated by a tandem hydroamination-hydrosilation reaction communicated by Messerle and coworkers.[55] Cationic Ir(I) complexes **8** and **9**, based on bis(pyrazolyl)- and bis(imidazolyl)methane ligands catalyze the conversion of 4-pentyne-1-amine to *N*-silyl-2-methylpyrrolidine, in a one-pot reaction and in nearly quantitative yields [Eq. (9)]. The internal hydroamination occurs first, but the silane can be added later or may be present in the starting reaction mixture – a remarkable feature of the two-step reaction.

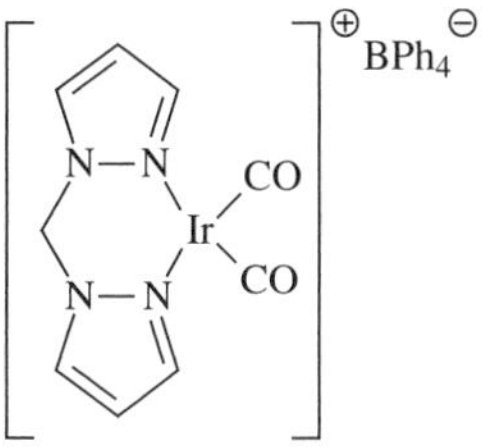

8 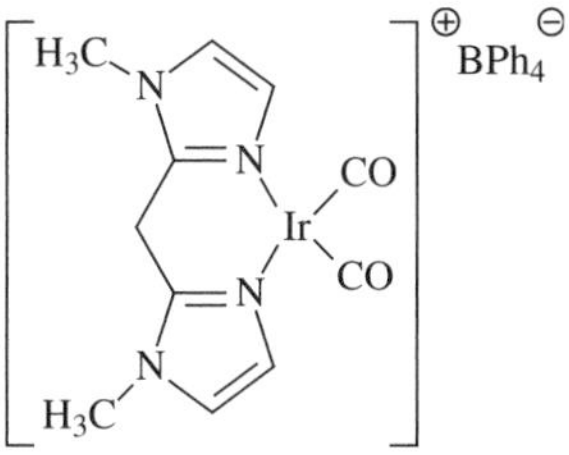 **9**

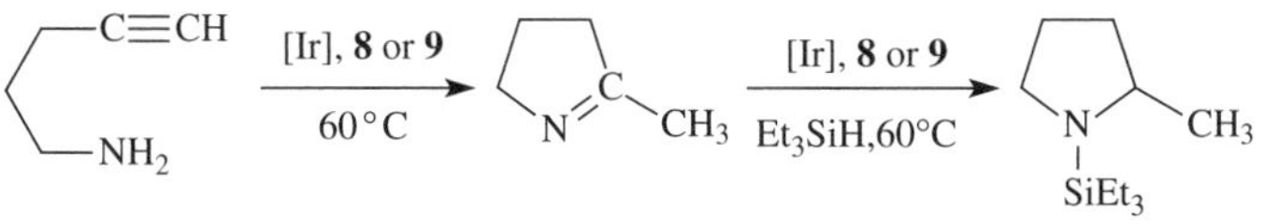

(9)

IV
NICKEL-TRIAD CATALYSTS

Like its other first-row preceding neighbors, Ni is known to be a rather poor hydrosilation-catalyst source-metal. Nevertheless, research effort in recent years has led to the development of Ni-based catalysts that show improved activity, and in some cases opposite regioselectivity compared to workhorse catalysts based on Pt or Rh. Of particular note is a series of catalysts **10a–c** where an indenyl ligand anchors the Ni.[56] Zargarian and coworkers systematically investigated the effect of a ring-tethered NMe_2 ligand on the activity of the Ni center containing various phosphine ligands, towards hydrosilation of styrene with $PhSiH_3$ and Ph_2SiH_2. The modulating action of the captive, nearby NMe_2 ligand on the cationic precatalyst center generated from **10a** was evident from the result that the catalyst species with PMe_3 ligand was the most active, *via* the balance struck between phosphine donor strength and steric bulk, when compared with PPh_3 and PCy_3.[56a] Yet, the same research group showed that for precatalyst **10b** with a ring substituent incapable of coordinating to the Ni center, the species containing a PPh_3 ligand is a more active catalyst than one with a PMe_3 ligand.[56b] In the latter case, although the silane/styrene ratio plays an important role, the ease of dissociation of the phosphine from the Ni center influences the activity of the catalyst. This is even more accentuated in **10c**; the rate of reaction is significantly increased even without cationic initiation when two Me_3Si are present at the 1,3-positions of the ring anchor.[56c] It is also noteworthy that all of these catalysts yield the internally or α-silylated product almost exclusively, with yields reaching up to 95% when excess $PhSiH_3$ is used relative to styrene.

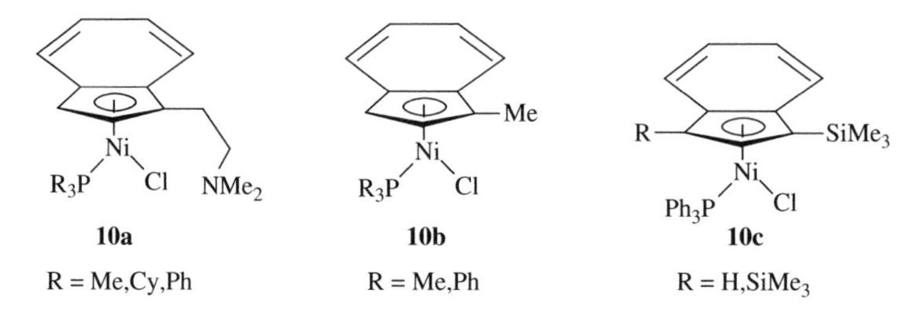

The use of Ni(0) catalysts for the hydrosilation of phenyl acetylene with Ph_3SiH and of various internal alkynes and conjugated diynes with Ph_2SiH_2, $PhMe_2SiH$ and Et_3SiH has also been reported.[57,58] It is of interest to note that although the Ni equivalent of Karstedt's catalyst has been prepared, dehydrogenative silylation of alkenes leading to unsaturated products is the dominant reaction and hydrosilation a secondary pathway with this Ni catalyst.[59]

The fact that Pd-based catalysts are not used in industrial hydrosilation processes (with perhaps rare exceptions), despite their significant commercial utility in the closely related hydrogenation and in myriad other organic transformations, is the result of generally low activity of palladium complexes in catalyzing olefin hydrosilations in particular. However, Pd compounds often provide selectivity that is different from that of Pt complexes. For example, Gulinski and James have reported high α-selectivity and α-specificity in the hydrosilations of styrene and vinyltrichlorosilane, respectively, with $HSiCl_3$, using various Pd complexes with bis(diphenylphospino)methane as a ligand.[60] On the other hand, a catalyst based on $Pd_2(dba)_3 \cdot CHCl_3$ and two moles Cy_3P (per mol Pd) provided high yield (0.5 mol% Pd based on silane), exclusive *E*-selectivity and very high β:α selectivity for room temperature hydrosilations of a variety of terminal alkynes with Ph_3SiH or Ph_2MeSiH.[61] An unusual rate enhancement was observed by the same research group for 1-octyne hydrosilation when water was used as a solvent.

The ability of Pd complexes to catalyze tandem cyclization–hydrosilation reactions, that could provide a synthetic route to carbocycles of interest in the fields of biochemistry and medicine, was recently demonstrated. Widenhoefer's group has developed phenanthroline-based complexes **11** and others that quite efficiently bring about this dual transformation for a wide variety of functionalized dienes, with high regio- and stereoselectivities [e.g., Eq. (10)].[62]

E, E — [Pd], $HSiMe_2Ph$, 25°C → E, E, $SiMe_2Ph$, Me (10)

E = CO_2Me — Single diastereomer, 91%

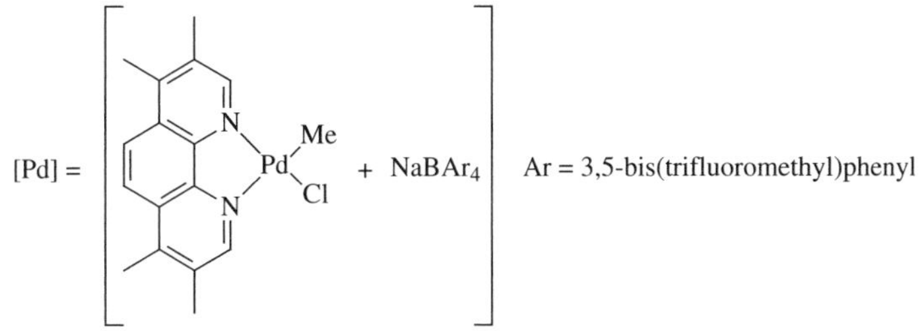

[Pd] = [N, N, Pd, Me, Cl + $NaBAr_4$] Ar = 3,5-bis(trifluoromethyl)phenyl

11

There have been a number of important advances in hydrosilation with platinum compounds and complexes as catalysts. Perhaps the most intriguing of these, for a rather simple compound of Pt, is that PtO_2 has been shown to facilitate hydrosilation of unsaturated amines with exceptional selectivity.[63] A number of primary, secondary and even tertiary allylic amines were hydrosilated with methyldiethoxysilane, providing γ/β ratios of 95/5 for the corresponding products in total yields >95% (although, distilled yield for 3-aminopropylmethyldimethoxysilane was only 60%). The question remains whether PtO_2 leads to an active homogeneous catalyst, especially since it is claimed to be filterable (no recyclability shown) at the end of reaction, and Pt/C (10% w/w Pt) yielded identical results. Nevertheless, the reported results are important from the perspective that allylic amines, whose hydrosilation is known to be problematic, are important substrates for the manufacture of a number of commercially useful functional silanes and silicones.

Amongst catalyst modifications that achieve purely rate effects, two reports are noteworthy. One is the hydrosilation rate enhancement in the presence of β-cyclodextrins, for Karstedt's catalyst [Pt_2(*sym*-tetramethyldivinyldisiloxane)$_3$] and Lamoreaux's catalyst.[64] Lamoreaux's catalyst (prepared from H_2PtCl_6 and 1-octanol), showed the most dramatic rate accelerations. The second instance of rate moderation is reported by Endo and collaborators.[65] They report the use of isocyanides in just 2:1 molar ratio with Pt, to dramatically suppress hydrosilation of vinylsilane (with Et_3SiH) or vinyl-terminated siloxane with methylhydrogensiloxane copolymer until 60–70 °C. Above this temperature, rapid hydrosilation occurs. This is in contrast with traditional cure inhibitors where a large excess of inhibitor (relative to Pt), typically alkynols or maleates, is needed to prevent premature cure.

An unusual bicyclic aminophosphine (proazaphosphatrane) ligand, appears to form a mononuclear Pt complex upon reaction of the bulky phosphine with a xylene solution of Karstedt's catalyst. Of several active members of a series of these catalysts systems, $[Pt\{P(^iBuNCH_2CH_2)_3N\}\{\eta^4\text{-}(CH_2{=}CHMe_2Si)_2O\}]$ **12** catalyzes hydrosilations of a variety of terminal alkynes (including ones containing heteroatom functionality) with Ph_3SiH or Et_3SiH, in a manner exhibiting high stereo- and regioselectivity. Thus, propargyl alcohol with triphenylsilane yielded a $\beta(E)$:α product distribution of 98:2, with no $\beta(Z)$ product evident [Eq. (11)]. Comparisons with Karstedt's catalyst alone showed significant to dramatic improvements for the phosphine-based system in most cases. For a similar, phosphine-modified Karstedt's catalyst, tandem Pt-catalyzed hydrosilation/Pd-catalyzed cross-coupling of terminal alkynes has also been reported.[66] Recently, the facilitating effect of a thioether group at the homoallylic position of olefins was demonstrated for H_2PtCl_6-catalyzed hydrosilations with $PhMe_2SiH$.[67]

CH≡CCH₂OH + Ph₃SiH —[Pt "complex" **12**, 1 mol%; THF, 20 °C, 20-30 min]→ HO–CH₂–CH=CH–SiPh₃ + HO–CH₂–C(=CH₂)–SiPh₃ (11)

β–(E), 98 α, 2

Over the last 8 or 9 years a series of papers have been published on platinum catalysts bearing electron-deficient olefins as activity modifier ligands.[68–71] Amongst several such catalysts derived by modifying Karstedt's complex, Osborn, Fisher and coworkers showed the 2-methylnaphthoquinone-ligated complex **13** to be the most active (compared to Karstedt's complex alone) and stable, particularly in the presence of excess 2-methylnaphthoquinone, for the model hydrosilation of trimethylvinylsilane with triethylsilane. Elsevier and collaborators also found various fumarate complexes **14** (among other similar complexes with maleate and quinones) to be more active and more stable compared to $Pt(norbornene)_3$ in the hydrosilation of styrene with triethylsilane. Using an interesting dinuclear complex with bridging η^2,η^2-olefin-bonded benzoquinone, $[Pt(\mu\text{-benzoquinone})(norbornene)]_2$ **15**, a 92% yield of the β-hydrosilation product was obtained by Itoh, Yamamoto and Ohno for the hydrosilation of styrene with $PhMe_2SiH$ [Eq. (12)].

Me₂Si, O, Me₂Si, Pt, O, O

13

RO, O, Pt, O, OR

14

R = H, Me, 2-butyl, phenyl, 1-naphthyl, 2-naphthyl

CH₂=CH-R + PhMe₂SiH —[cat. **15**; hexane, 30 °C, 1hour]→ PhMe₂Si–CH₂–CH₂–R (12)

R = phenyl, 92%; R = n-hexyl, 96%

N-heterocyclic-carbene complexes of Pt have also arrived on the scene of hydrosilation catalysis. Markó and collaborators report a catalyst where the bridging siloxane ligand has been replaced with an imidazolylidene ligand to provide a monomeric complex **16**.[72] These Pt–carbene catalysts, especially the cyclohexyl-substituted complex, are much superior to Karstedt's catalyst for olefin hydrosilation with respect to chemo- and regioselectivity, showing a high level of tolerance for various functional groups containing oxygen and providing good to excellent yields of product. That Pt–carbene complexes, as a new class of Pt-compounds, bring promise for improving various facets of hydrosilation catalysis is further borne out by Elsevier and coworkers' account of improved reactions with catalysts **17a**, **17b** and **18**.[73] For the hydrosilation of styrene with Et_3SiH, **17b** and **18** showed high activity, and although β/α selectivity was about 4/1, almost no dehydrogenative silation product formed.

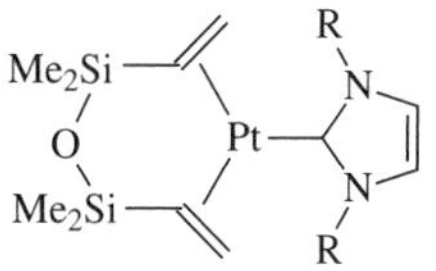

16
R = Me, cyclohexyl(cy), ^{t}Bu

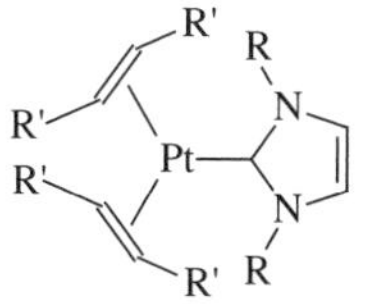

17a
R = mesityl, R' = CO_2Me

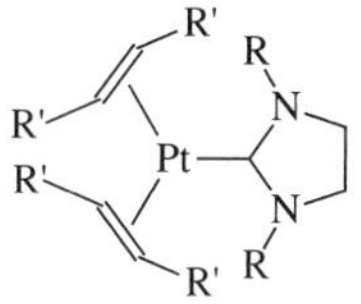

17b
R = mesityl, R' = CO_2Me

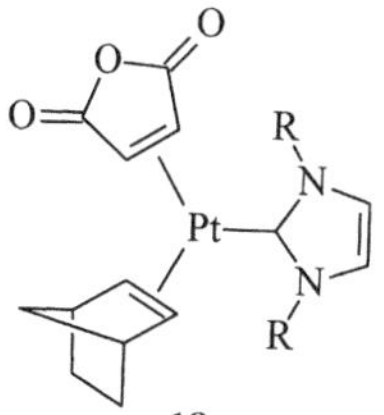

18
R = mesityl

Methylenecyclopropanes are useful and highly reactive intermediates in organic synthesis. Osakada and coworkers have reported the hydrosilation of methylenecyclopropanes with a variety of hydrosilanes using $PtI_2(PPh_3)_2$, $Pt(H)I(PPh_3)_2$ or $Pt(H)Cl(PPh_3)_2$ as catalyst, where the ring is preserved in the product and yields are high in many cases.[74] Hydrosilation of enynes, where the stereochemistry and nature of the product are controlled by the geometry (*cis* or *trans*) around the olefinic bond of the substrate as well as by the catalyst (Pt, Rh or Ru), has been described by Ozawa and coworkers.[75]

A series of simple alkene–Pt–silyl complexes that are the first true catalyst species ever to be observed in Pt-catalyzed hydrosilation was recently identified by Roy and Taylor.[76] Of these, **19a** was isolated and characterized by multinuclear NMR spectroscopy and X-ray crystal structure determination. Although the focused aim of the project was to elucidate the catalytic cycle with Pt catalysts (as discussed in more detail in Section X), **19a** as well as other members of the series were proven highly active, stable and recyclable catalysts in the hydrosilation of 1-hexene and styrene with chlorohydridosilanes.

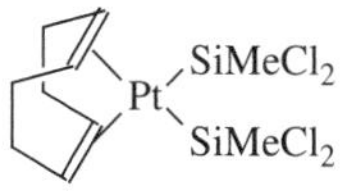

19a

Although their work does not strictly fall within the scope of this review, Hartwig and Tsukuda's report on dehydrogenative silation *via* arene-H and alkane-H activation deserves special mention.[77] They used a special class of Pt complexes, $[Pt(Me_2)(H)(Tp_2^{Me})]\{Tp_2^{Me}$ = hydridotris(3,5-dimethylpyrazolyl)borate} to demonstrate inter- or intramolecular C–H and Si–H irreversible dehydrogenative coupling without the need for a hydrogen acceptor. This new catalytic reaction certainly blazes a promising path in Si–C bond-forming chemistry.

V

GROUP IVB, GROUP IIIB, LANTHANIDES AND ACTINIDES

As oxophiles, Group IVB transition metals (Ti, Zr and Hf) are perhaps better suited for the hydrosilation of the C=O unsaturation (which will be addressed in greater detail in Section VII). Nevertheless, catalysis of SiH addition to carbon–carbon and carbon–nitrogen unsaturated compounds have also been investigated to a significant extent, using primarily metallocene complexes. During the course of their investigation of catalyzed silane oligomerization, Corey and Zhu discovered that using the catalyst system $Cp_2MCl_2/n\text{-}BuLi$ (M=Ti, Zr, Hf), both acyclic (terminal and internal) and cyclic olefins undergo hydrosilation with $PhMeSiH_2$.[78] The major course of the reaction and the nature of the silated product are dependent, however, on the type of olefin and the metal. Waymouth and Kesti studied the hydrosilation of styrene (in detail), as well as that of other olefins with several silanes (primary, secondary and tertiary), using the active species derived from $[Cp_2ZrCl_2/2\ n\text{-}BuLi]$.[79] Only β-silyl adducts were obtained, although styrene yielded a dehydrogenative silation product as well as ethylbenzene, and product distribution was dependent on reagent ratios. It was also found that the Zr-based catalyst gave higher yields of hydrosilation product than similar systems derived from Ti and Hf. Very recently, however, in a dramatic demonstration of the effect of changes in catalyst composition, Takahashi *et al.*, showed that when three equivalents of *n*-BuLi (instead of 2) per Cp_2ZrCl_2 are used, the regioselectivity of the hydrosilation of styrene with Ph_2SiH_2 is reversed, and only the α- or internal adduct is obtained.[80] This reversal of selectivity with increase in RLi/Zr ratio was also observed for *s*-BuLi, but PhLi gave only the α-adduct.

Takahashi's group has also reported hydrosilation of a number of alkynes with various silanes, using the $[Cp_2TiCl_2/2\ n\text{-}BuLi]$ combination, which leads to products with excellent regio- and *E*-selectivity [Eq. (13)].[81] It is of note that known zirconocene and hafnocene species do not catalyze the hydrosilation of alkynes.

$$R^1\text{—}\equiv\text{—}R^2 + H\text{-}[Si] \xrightarrow[\text{THF, 1 h}]{Cp_2TiCl_2/2\ n\text{-}BuLi} R^1\text{CH}{=}\text{C}(R^2)[Si] \quad 52\text{-}97\% \qquad (13)$$

R^1 = Et, n-Pr, n-Bu, n-pentyl, n-hexyl, n-octyl, $SiMe_3$
R^2 = H, Me, Et, n-Pr. [Si] = $SiHPh_2$, SiHMePh, SiH_2Ph
Not all combinations of R^1 and R^2 examined

Recently, some new alkylzirconium complexes, $[Zr(R)\{N(Ar)SiMe_2\}_3CH]$ (R = *n*-Bu, CH_2Ph; Ar = *p*-tolyl, *p*-fluorophenyl), based on a a tripodal (triamido) ligand, were found to catalyze alkene hydrosilation, although their catalytic activity is very poor.[82]

Harrod, Samuel and coworkers reported selective C=N hydrosilation (with concomitant partial hydrogenation in most cases) of the aromatic pyridine ring with $PhMeSiH_2$, using $[Cp_2TiMe_2]$ as catalyst [Eq. (14)].[83] A catalytic cycle for the reaction is proposed, with $[Cp_2TiH]$ envisioned as a key reactive intermediate in the catalytic cycle. Cp_2TiMe_2 also catalyzes the reduction of ethyl benzoate with $PhMeSiH_2$, as well as an interesting reduction-polymerization of both β- and γ-butyrolactone that leads to the alternating copolymers -$[OSiPhMeO(CH_2)_3CHMe]_n$- and -$[OSiPhMeO(CH_2)_3CH_2]_n$-, respectively.[84] The highly reactive $[Cp_2TiH]$ is also implicated as the "catalytic" species for the reduction-polymerization reaction.

$$\text{pyridine} + 2\,PhMeSiH_2 \xrightarrow[80\,^\circ C,\ 8\,h]{[Cp_2TiMe_2],\ 10\,mol\%} \text{N-(SiHMePh)-tetrahydropyridine}\ (94\%) + 1/n(PhMeSi)_n \qquad (14)$$

It is appropriate to consider Group IIIB, lanthanide and actinide metals together because of the preponderance of the +3 oxidation state in catalysts based on these metals. Like Group IVB metal complexes, those of Group IIIB and lanthanide metals are marked by characteristic Lewis acidic behavior. This is turn has mechanistic implications in hydrosilation, particularly with regard to potential olefin polymerization as a side reaction. Again, as with Group IVB metals, for Group IIIB and lanthanide/actinide metallocene-based complexes dominate the catalyst scene. Following discovery efforts by Voskoboynikov,[85] Tanaka[86] and their coworkers, the research groups of Marks and of Molander carried out some impressive catalyst development work in the area, and seminal publications from these groups have contributed enormously to catalyst refinement and understanding of *modus operandi* for catalysts based on this group of metals. (For a review of lanthanide catalysis in organic synthesis, and for additional references of lanthanide catalyzed hydrosilation, see Ref. [87].) A selected few works of the latter groups, along with others' are discussed below.

Of the Group IIIB metals, yttrium and lutetium based catalysts have been investigated in hydrosilation. Molander and Julius reported $[Y(\eta^5\text{-}C_5Me_5)_2\{CH(SiMe_3)_2\}]$ to be a selective and efficient precatalyst for hydrosilating mono- and 1,1-disubstituted olefins.[88] Appreciable selectivity and high to excellent yields are obtained at ambient temperatures with 3 mol% catalyst that take advantage of the catalyst's sensitivity to steric effects. For example, 3-vinylcyclohexene is hydrosilated by $PhSiH_3$ only at the exocyclic olefinic bond with 97% isolated yield of the β- or terminal adduct. For hydrosilation of alkynylsilanes, prepared to probe hydrosilation of terminal alkynes, which as unprotected substrates are too acidic for successful catalyzed reaction, Molander and coworkers clearly demonstrated the effect of catalyst structure and alkynylsilyl substituent on the substrate.[89] However, regioselectivity control is dominated by catalyst structure (e.g., **20**, **21** or **22**) where selectivity improvements are achieved by reducing Cp-substitution or by changing the metal used. It is noteworthy that one regio- and stereo-isomer is the major product [Eq. (15)]. Additionally, Molander and Retsch reported hydrosilation of a variety of internal alkynes (linear, branched and hetero-atom substituted) with $PhSiH_3$, using [Y(Cp′$_2$)(Me)(THF)] as precatalyst, where many alkynes afforded a single product with exclusive (*E*)-stereochemistry and addition of Si to the less hindered carbon.[90]

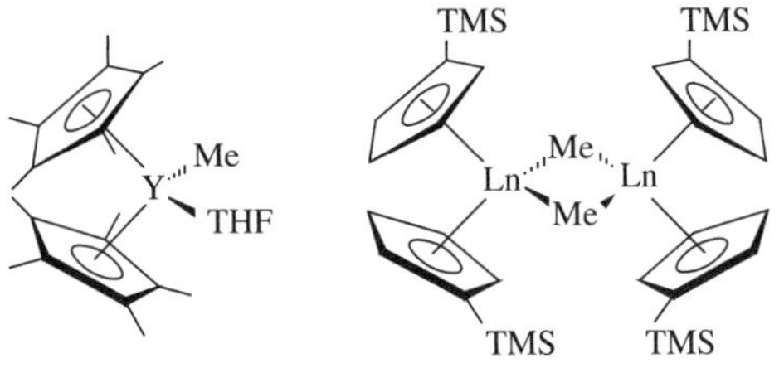
Y
Me
THF
TMS
TMS
Ln
Me
Me
Ln
TMS
TMS

20 Ln = Y, **21**; Ln = Lu, **22**; TMS = Me_3Si

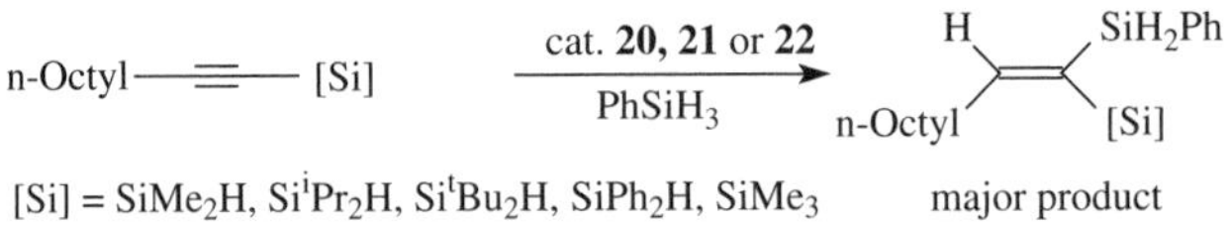
n-Octyl
[Si]
cat. 20, 21 or 22
PhSiH3
H
SiH2Ph
n-Octyl
[Si]
[Si] = SiMe2H, SiiPr2H, SitBu2H, SiPh2H, SiMe3
major product

(15)

In 1995, Marks and coworkers published a particularly insightful study on the hydrosilation of aralkenes, 1-alkenes and 1,1-disubstituted alkenes with $PhSiH_3$, using a number of lanthanide catalysts, but focusing on a series of Sm catalysts (e.g., **23**) with varied ligand structures. The work revealed that alleviation of steric congestion around the metal's coordination sphere, either *via* tying the Cp rings of the catalyst with a silicon bridge (which creates a cone effect, and blocks one side but widens the opposite side of the metal to ligand access) or by using a larger ionic radius metal center, significantly improves both activity and regioselectivity of the catalyst.[91] This is most dramatically demonstrated in the hydrosilation of styrene with $PhSiH_3$ [Eq. (16)]. Soon thereafter, the ability to use primary silanes as highly efficient chain terminators for olefin polymerization catalyzed by Y, Lu, La and Sm metallocene–hydride complexes was demonstrated by Marks' group.[92]

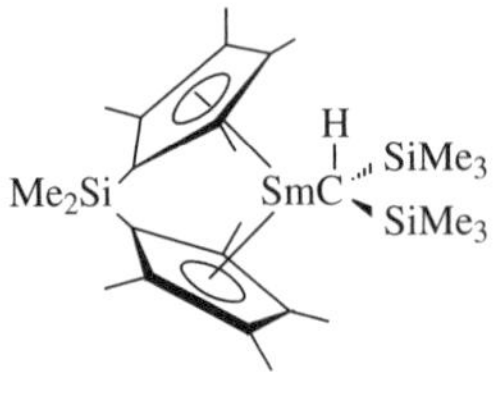
H
SiMe3
Me2Si
SmC
SiMe3

23

$$\text{PhCH=CH}_2 + \text{PhSiH}_3 \xrightarrow[\text{Me}_2\text{SiCp''}_2\text{SmR}(\mathbf{23}),\ 60\,^{\circ}\text{C}]{\text{Cp'}_2\text{LnR or}} \text{PhCH(CH}_3\text{)SiH}_2\text{Ph} \quad (16)$$

Cp' = η^5-Me_5C_5, Cp'' = η^5-Me_4C_5
R = $CH(SiMe_3)_2$

α- or internal adduct (yield)
Ln = Lu, 5.8 (5.5); Sm, 82 (78);
Nd, 90 (84); La, 96 (88)
Cp''$_2$Sm, >99 (98)

Hydrosilation of dienes using 1 mol% $[Nd(Cp'_2)\{CH(SiMe_3)_2\}]$ has been described by Tanaka *et al.*[93] With $PhSiH_3$, isoprene yields (*E*)-2-methyl-2-butenylphenylsilane as the major product, whereas 1,5-hexadiene and 1,6-heptadiene yield a cyclopentylmethylsilane as the primary product.

In the context of the role of monomer–dimer equilibrium in catalytic activity, Voskoboynikov *et al.* have shown that dissociation of a dimeric Ln-precatalyst to a monomeric species is not an absolute prerequisite for catalytic activity, as hydrosilation of 1-octene (to yield primarily the terminal adduct) with $PhMeSiH_2$ displayed identical selectivity, rate law and rate constant for a metal whether the dimer was $[LnCp^*_2(\mu\text{-}H)]_2$ or $[LnCp^*_2(\mu\text{-}H)(\mu\text{-}Me)]_2$.[94] A dimeric complex, $[Y(\mu\text{-}H)(\eta^5{:}\eta^1\text{-}C_5Me_4CH_2SiMe_2NCMe_3(THF)]_2$, with a two-atom bridge (CH_2SiMe_2) between the Cp ring and the ligating N-atom, was shown to be a more active catalyst than the one with just a $SiMe_2$ bridge, for the hydrosilation of 1-decene with $PhSiH_3$.[95] However, for styrene hydrosilation, this catalyst showed lower selectivity for internal adduct than achievable with the "constrained geometry" catalyst of Marks described above. Recently, the first constrained-geometry cyclopentadienylphosphido complexes of the type $[Ln(R)(\eta^5{:}\eta^1\text{-}C_5Me_4SiMe_2PR')]$ (Ln=Y, Yb, Lu; R'=Ph, Cy, 2,4,6-^{t}Bu-C_6H_2-; R=H, CH_2SiMe_3) in dimeric and other structural variations, were reported.[96] Nearly all the complexes were active towards highly *β*-selective hydrosilation of 1-decene with $PhSiH_3$ (rt, 5 mol% Ln), with the "CyP-LuCH_2SiMe_3" complex being the most active (100% yield in 10 min). This Lu complex was also highly active for other olefin hydrosilations.

Once in a while, as was related in the section on Pt catalysts (PtO_2), simple and readily available metal compounds and complexes turn out to be excellent catalysts. This is also the case with easily obtained $La[N(SiMe_3)_2]_3$. Livinghouse and Horino[97] recently described hydrosilation of a variety of olefins and dienes with $PhSiH_3$ using the tris-amido compound [Eq. (17)], where high activity and selectivity (at 3 mol% catalyst, 25 °C) nearly mirror those observed by Marks and coworkers with their metallocene-based Ln complexes. Yet, the analogous tris-amido complexes of Nd and Sm exhibited much lower activity, and several other simple lanthanum compounds as well as $Y[N(SiMe_3)_2]_3$ totally failed to catalyze the very same hydrosilation (that of 1-hexene with $PhSiH_3$)! It is this extraordinary difference between high activity and no activity amongst closely similar Group IIIB/lanthanide metal complexes that perhaps differentiates this metal group most from the main stream hydrosilation catalysts based on Pt and Rh. In a positive sense, the stark differences in activity are potentially exploitable in multi-step transformations or dormant catalyst applications.

$$\mathrm{R{\smallsetminus}\!\!=} \ + \ PhSiH_3 \xrightarrow[25\ ^\circ C,\ 5\text{-}40h]{[La\{N(SiMe_3)_2\}_3],\ 3\ mol\%} R\text{-}CH(SiH_2Ph)CH_3 \quad (17)$$

R = n-C_4H_9, C_6H_5, 4-(MeO)C_6H_4, C_6H_5(of alpha-methylstyrene)

yield : 98-99%, α-adduct: 96-99%

In contrast with lanthanide(III) catalysts, some divalent lanthanide complexes [Ln(η^2-PhNCPh$_2$(hmpa)$_n$)] and [Ln{N(CHPh$_2$)(Ph)}(NPh$_2$)(hmpa)$_n$)] (Ln = Yb, Sm, hmpa = hexamethylphosphoric triamide) show considerably different mediating activity.[98] For example, hydrosilation of styrene with $PhSiH_3$ (10 mol% catalyst, THF, rt) yields β- or terminal adduct as the predominant product, and dehydrogenative silation occurs with terminal alkynes. Additionally, imines, which are not amenable to hydrosilation with Ln(III) catalysts, are readily hydrosilated by the Ln(II) compounds in good yields. Further, both types of Yb catalysts yield a silacyclopentene derivative as a secondary product from reactions of conjugated dienes with $PhSiH_3$ (and also with Ph_2SiH_2 for isoprene). It is of interest to note that although several other metallocene lanthanide catalysts[91,93,96] (but not all[88]) produce a silylmethylcyclopentane *via* hydrosilation of α,ω-dienes, incorporation of silicon to afford a silacycle is typically not observed, except also for the tris-amido lanthanum complex which generates both cyclic products from 1,5-hexadiene and $PhSiH_3$.[97]

Although not particularly attractive from the perspective of general synthetic utility in hydrosilation, effective use of actinide catalysts for the hydrosilation of terminal alkynes (not viable with lanthanide catalysts) and alkenes has been demonstrated by Eisen and coworkers. Highly chemo- and regioselective hydrosilation of terminal alkynes (and enynes) with $PhSiH_3$ (affording β(E) adduct in excellent yield) is achieved with the ligation-facilitating *ansa*-$Me_2SiCp''_2Th(^nBu)_2$, where large rate increases (up to nearly 1000 fold) are observed over the corresponding Cp′-based complex (Cp″ = Me_4C_5, Cp′ = Me_5C_5).[99] Hydrosilation of alkenes also occurs with the *ansa*-catalyst to yield primarily β-adduct, but yields are moderate and considerable amounts of the corresponding alkanes are produced. The same group has also reported the use of Cp'_2UMe_2 to catalyze hydrosilation of terminal alkynes, but the reactions are influenced heavily by a number of factors, and multiple products are formed.[100]

VI

MANGANESE, RHENIUM AND COPPER

These metals, belonging to groups on either side of the main block of transition metals (Group VIIIB or 8, 9 and 10) commonly examined in hydrosilation catalysis, have only been employed very rarely to facilitate hydrosilation. Very few reports have emerged during the last 15 years on catalysts based on these metals and their use has generally been connected with special cases of hydrosilation.

Esters have been considered largely inert to hydrosilation, but in 1995 Cutler and coworkers showed that acetyl complexes of manganese, $[Mn\{C(O)CH_3\}(CO)_4(PPh_3)]$ and $[Mn\{C(O)CH_3\}(CO)_5]$, effectively catalyze hydrosilation of esters to successively produce silylacetal (with $PhSiH_3$, Ph_2SiH_2 or Me_2PhSiH) and then ether (with $PhSiH_3$ or Ph_2SiH_2 but not Me_2PhSiH) in moderate to good yields, using 1.5–3 mol% catalyst.[101] Cutler's group also reported hydrosilation of ketones with $PhMe_2SiH$ and Ph_2SiH_2 using $[Mn(Y)(CO)_4L]$ ($Y{=}CH_3$, $C(O)CH_3$, Br; $L{=}PPh_3$, CO) and compared activity of the phosphine–acetyl–carbonyl complex in particular with Wilkinson's catalyst $Rh(PPh_3)_3Cl$. In several cases, with acetone, acetophenone, cyclohexanone and 2-cyclohexenone as substrates, activity of the Mn-based catalyst was superior to that of the Rh complex.[102]

The first example of hydrosilation catalyzed by a rhenium complex, $[Re(O)_2I(PPh_3)_2]$, **24**, was recently communicated by Toste *et al.*,[103] and its implications to new avenues in reduction catalysis by high oxidation state transition-metal complexes has been explored by Thiel.[104] The most intriguing aspect of this catalysis is that reductions, including hydrosilation, have traditionally been catalyzed with electron-rich transition metals in their lower oxidation states. Bucking this wisdom and practice, the dioxorhenium(V) complex very effectively catalyzes the hydrosilation of aldehydes and ketones with tertiaryhydrosilanes at catalyst loading of around 2 mol% and at 60–70 °C, with yields of 60–95%. Furthermore, the catalyst is tolerant of several functional groups including cases of amino- and nitro substituents. Indication of any enantioselectivity, however, was not reported.

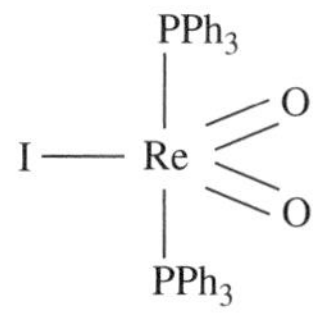

PPh3
I — Re (=O)(=O)
PPh3

24

The role of copper species in the "Direct Process" for the production of methylchlorosilanes from silicon metal and methyl chloride is widely known in silicon chemistry, but outside of this crucial yet largely mysterious catalytic role, the use of copper compounds in organosilicon synthesis is extremely rare. However, in 1993 Boudjouk's group reported the use of Cu_2O as well as several other Cu(I), Cu(II) compounds or metallic copper, in conjunction with TMEDA for the exclusive, high-yield ($>$90%), β-hydrosilation of methyl acrylate with Cl_3SiH or $MeCl_2SiH$ [Eq. (18)].[105] The optimal ratio for Cu/TMEDA/acrylate/silane was 7.0/10/20/30, with a reflux temperature of 80 °C, but addition of reactants to the Cu-TMEDA at 0 °C. Methyl methacrylate gave much lower yields, and triethylsilane or triethoxysilane failed to react. Despite this limited success, the catalysis represents an important step forward in the hydrosilation of acrylates, which had only yielded mixtures of α- and β-adducts with all previously tested catalysts.

$$CH_2{=}CH{-}C(O)OR + HSiR'Cl_2 \xrightarrow{[Cu],\ TMEDA} R'Cl_2SiCH_2{-}CH_2{-}C(O)OR \qquad (18)$$

R = Me, Et; R' = Cl, Me, Ph

[Cu] = Cu, Cu(I) and Cu(II) compounds

Exclusive β- or terminal adduct

In yet another example of the utility of N-heterocyclic carbene ligands, the use of Cu-NHC complexes (e.g., **25**) to catalyze the hydrosilation of hindered and functionalized ketones with Et_3SiH was very recently described by Nolan and coworkers [Eq. (19)].[106] Functionalities on the ketone substrate such as halide, ether and tertiary amine are well tolerated and excellent conversions are achievable using a number of variations of **25** (pre-synthesized or generated *in situ*) with different *N*-substitution on the ring.

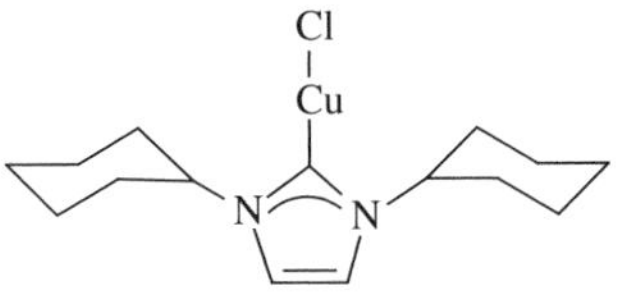

Cl–Cu (N,N'-dicyclohexylimidazol-2-ylidene)

25

$$R^1R^2C{=}O + R_3SiH \xrightarrow[\text{Toluene, 80°C}]{\text{3 mol\% (NHC) CuCl}} R^1R^2CH{-}OSiR_3 \qquad (19)$$

NHC = N-Heterocyclic Carbene

65-99%

VII
CATALYSIS BY ACIDS, BASES, OR FREE RADICALS

The use of Lewis acids as catalysts for the hydrosilation of unsaturated compounds, hydrocarbons in particular, has seen a surge of interest and activity over the last dozen years. This is a refreshing and bold development since Lewis acids are also excellent catalysts for polymerization of olefins and acetylenes.

In 1993, Otera and collaborators reported a comparative rate study on the reduction of aldehydes and ketones by Et_3SiH catalyzed by the Lewis acid Me_3SiClO_4 and the Brønsted acid $HClO_4$, in acetonitrile containing water.[107] Surprisingly, even in the presence of water, the catalytic reaction of Me_3SiClO_4 (generated from Ph_3CClO_4 and Me_3SiH) with the carbonyl compound was faster to afford alcohol products. Catalysis by the silyl perchlorate was at least two orders of magnitude faster than that by perchloric acid. Aluminum chloride and $EtAlCl_2$ were found to be very effective catalysts for the hydrosilation of alkynes and allenes by Et_3SiH.[108] While $HfCl_4$ gave low yields, $ZrCl_4$ gave only a trace of product, and Et_2AlCl failed to catalyze the reaction. A key feature of the catalysis is the high stereoselectivity in terminal or internal alkyne hydrosilation, leading to *cis*- or *Z*-vinylsilane *via trans*-addition of the Et_3SiH [Eq. (20)]. The order of addition of catalyst and reagents is critical since, in the absence of silane, the Lewis acidic catalyst can oligomerize the acetylenic substrate.

$$R^1{-}C{\equiv}C{-}R^2 + Et_3SiH \xrightarrow[\text{toluene or n-hexane, } 0\,^\circ\text{C}]{AlCl_3 \text{ or } EtAlCl_2} (R^1)(H)C{=}C(SiEt_3)(R^2) \tag{20}$$

R^1 = alkyl, aryl;
R^2 = mostly H, also alkyl, aryl

β-(Z)-product, exclusive or major

The first examples of the hydrosilation of linear and cyclic alkenes with trialkylsilanes in the presence of a range of Lewis acid catalysts have been reported by Jung and coworkers.[109] Like the alkyne and allene hydrosilations above, the alkene reactions are also highly regio- and stereoselective, producing the isomer with silyl addition to the less hindered carbon in moderate to good yields [Eq. (21)]. For cyclic substrates, such as 1-methylcyclohexene, only the *cis*-isomer is obtained. The activity of the catalysts examined follows the order $AlBr_3 > AlCl_3 > HfCl_4 > EtAlCl_2 > ZrCl_4 > TiCl_4$. Added trialkylchlorosilanes act as promoters, but may naturally be generated *in situ* from the trialkylsilane and trace HCl present in the catalyst. It must be said, however, that despite the surprising ability to mediate olefin and alkyne hydrosilation, the extreme reactivity of the Al–halogen bond coupled with the need to use quite high loadings of catalyst, limit the use of aluminum halide Lewis acids for broad application in hydrosilation.

$$(R^1)(H)C{=}C(R^2)(R^3) + Et_3SiH \xrightarrow[CH_2Cl_2,\ -20\,^\circ\text{C to rt}]{AlCl_3,\ 0.2\ \text{eqv}} (R^1)(Et_3Si)CH{-}C(R^2)(H)(R^3) \tag{21}$$

R^1, R^2 = mostly H;
R^3 = alkyl, aryl;
R^1, R^2, R^3 = Me

A familiar and synthetically important Lewis acid compound, $B(C_6F_5)_3$, that holds much promise of superiority over chloroaluminum species as a hydrosilation catalyst was recently very successfully utilized by Gevorgyan's group to add a number of alkyl- and arylhydrosilanes to a wide variety of linear and cycloaliphatic as well as aralkenes.[110] The products, with the same regio- and stereoselectivity as with $AlCl_3$-catalyzed reactions, were obtained in mostly excellent yields, using 5 mol% of the borane [Eq. (22)]. In several instances of identical olefin hydrosilation, the borane catalyst provided much higher yields of adduct. In a second instance of $B(C_6F_5)_3$-catalyzed hydrosilation under mild conditions and low borane loading, Rosenberg and collaborators very recently communicated an unusual reaction of silanes or disilanes with thiobenzophenone leading to new types of silyl thioether compounds.[111]

$$R^1(H)C{=}CR^2R^3 + (R^4)_3SiH \xrightarrow[CH_2Cl_2,\ rt]{B(C_6F_5)_3,\ 5\ mol\%} (R^4)_3Si{-}CH(R^1){-}CH(R^2)(R^3) \quad (22)$$

R^1, R^2 = H, alkyl;
R^3 = alkyl, aryl;
R^4 = alkyl, Ph

85-98%

Also very recently, Yamazaki and coworkers reported that regioselectivity in the hydrosilation of propiolate esters with $(Me_3Si)_3SiH$ is controlled by both substituents and the presence or absence of a Lewis acid such as $AlCl_3$, $EtAlCl_2$ or Et_2AlCl.[112] Methyl and ethyl esters gave α-adduct with $AlCl_3$, but β-adduct in the absence of Lewis acid or in the presence of either $EtAlCl_2$ or Et_2AlCl. On the other hand, for the trifluoroethyl ester all the Lewis acids gave α-adduct exclusively. The difference was explained using competitive free-radical and ionic mechanisms.

Fluoride-on-alumina promoted hydrosilation of benzaldehyde with Et_3SiH, affording benzyloxytriethylsilane quantitatively was described recently.[113] The formation of benzylbenzoate as a byproduct *via* self-condensation of benzaldehyde, is totally suppressed in DMF as a solvent, whereas in hexane or THF, it is the major product formed. Other basic species such as KNH_2/Al_2O_3, MgO or CaO showed much lower activity compared to KF/Al_2O_3.

Although efforts in free-radical catalyzed hydrosilation continue at a low level, no major advance has been reported or any particularly useful synthetic application developed in recent years and over the last decade. In 1992, Boardman reported results of hydrosilation of 1-octene with $Me_2HSiOSiMe_3$ facilitated by photo-activated $[Pt(Cp)Me_3]$.[114] With as little as 10 ppm Pt, and following an initial UV irradiation, rapid hydrosilation occurred (TON up to 10,000 per Pt per minute), with nearly quantitative yield of the β-adduct. Evidence indicated involvement of heterogeneous colloidal Pt as the active species. Lewis and Salvi studied the photo-initiated ($\lambda > 300$ nm) hydrosilation of triethylvinylsilane with triethyl silane using $[Pt(acac)_2]$ (acac = acetylacetonate) as the precatalyst.[115] No reaction was observed in the dark, but upon short (10 min) irradiation at 25 °C, ca. 85% conversion of reactants to $Et_3SiCH_2CH_2SiEt_3$ was obtained. Unlike Boardman's study, inhibition studies with mercury and DBCOT (dibenzocyclooctatetraene) indicated a homogeneous catalytic species during the fast initial stages of hydrosilation.

Rapid hydrosilation of dimethylvinylsilane with near-UV-activated $[Pt(acac)_2]$ has been reported.[116] Poly[(dimethyl)vinylsilane] with M_w ca. 5500 g/mol is the major product but oligomers and other side-products also form. Active species apparently remain "alive" for extended periods as molecular weight increase of the initial product to 12,300 g/mol in 6 months is observed. Brook and coworkers, on the other hand, report relatively poor hydrosilation–polymerization of diphenylvinylsilane *via* catalysis with radical generators such as AIBN, benzoyl peroxide and di-^{t}butyl peroxide.[117] Photo-hydrosilation of 1-octene with pentamethyldisiloxane and of a vinyl-silicone with a hydro-silicone, using four trimethyl(β-dicarbonyl) Pt(IV) complexes (OEt, Me, Ph, CF_3 groups, respectively on carbonyl carbons) was examined by Fouassier and coworkers.[118] Sonochemically-assisted hydrosilation of 2- and 4-substituted cyclohexanones, in the presence of $Rh(PPh_3)_3Cl$, has been described.[119]

VIII
ASYMMETRIC HYDROSILATION

It could be argued that catalytic asymmetric synthesis is one of the top fields of research currently in organic chemistry. The impetus for the wide and deep interest in this field is provided primarily by two fronts: the pressing need in the pharmaceutical industry to develop new drugs and the rapidly advancing scientific progress in establishing structure–function/structure–activity relationships and interrelationships of biologically active molecules. It is not too far from the truth to assert that together with C–C catenation, actually hand-in-hand with it, chirality of molecules defines the stuff of life. Because of the availability of a number of relatively simple methods to remove silyl groups from an organic compound to generate new derivatives, research in asymmetric hydrosilation has grown steadily in the last 15–20 years. Like hydrogenation, hydrosilation offers the potential to convert a sp^2 carbon to a sp^3 asymmetric center, and not surprisingly catalyst design in asymmetric hydrogenation has helped guide the development of catalysts for asymmetric hydrosilation. Several reviews on the subject have already appeared[120] and the summary below will attempt to capture some of the recent advances in this exciting area of silicon chemistry.

Asymmetric hydrosilation of olefins can be achieved *via* either Markovnikov or anti-Markovnikov addition of SiH depending on the olefin substituents, but for alkynes a double-hydrosilation is necessary to generate a new chiral center. Hayashi's group has been instrumental in the development of catalysts for these two types of unsaturated compounds. Of particular note is the design and development of monophosphine ($-PPh_2$) or MOP catalysts which were conceived based on the now-well-known asymmetric ligand template 1,1′-binaphthyl substituted at the 2 and 2′ positions with donor groups **26**. A large number of MOPs **27** have been synthesized[121] and Pd complexes of many of these are excellent catalysts for the regio- and enantioselective hydrosilation of various olefins as well as terminal alkynes. For terminal olefins, a catalyst prepared from $[PdCl(\eta^3\text{-}C_3H_5)]$ and (S)–MeO–MOP at P/Pd:2/1, e.g., affords the trichlorosilyl α-adduct in very high yields, and the corresponding alcohols obtained *via* oxidative removal of the silyl group exhibit enantiomeric excesses of 94–97% [Eq. (23)].[122] Conversion of bicycloalkenes such as norbornene to products in >90% ee was also accomplished with [Pd]/(R)–MeO–MOP.[123] Reaction of conjugated dienes (cyclic and acyclic) with $HSiCl_3$ produces adducts with 62–91% ee, using a Pd catalyst containing a MOP modified with 6,6′-binaphthyl substituents that help solubilize the catalyst.[124] For 1-aralkenes hydrosilated with $HSiCl_3$ and catalyzed by [Pd]/(R)–MeO–MOP, product yields and enantiomeric excesses range from 80 to 99% and 71 to 85%, respectively. Arylacetylenes undergo sequential double hydrosilation with $HSiCl_3$ that are catalyzed by $[PtCl_2(\eta^2\text{-}C_2H_4)_2]$ and a Pd-fluoroMOP, respectively, to optimize yield.[125] Although yield of the α-adduct is low, enantioselectivity of the vicinal diols following oxidative cleavage of C–Si is excellent [Eq. (24)].

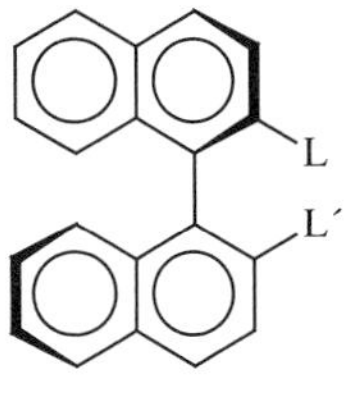

26

L, L′ = donor group

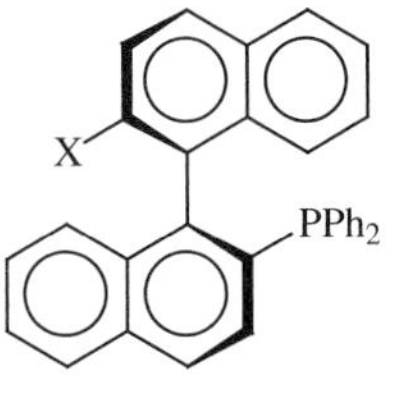

27

X = OMe, OCH_2Ph, OPr^i, Et, CN, CH_2NMe_2, CO_2Me, CO_2H, OH, H(R), C_6H_2-3,5-dimethyl-4-OMe, C_6H_2-3,5-dimethyl-4-OMe with bis(6,6′-^{n}octyl) substituents on binaphthyl

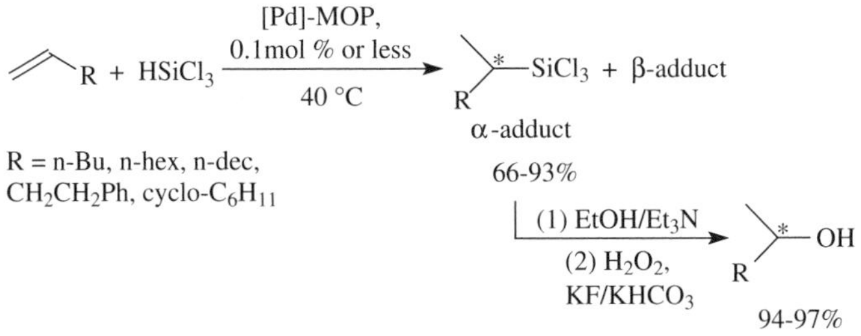

$$\text{CH}_2{=}\text{CH-R} + \text{HSiCl}_3 \xrightarrow[40\,^\circ\text{C}]{\text{[Pd]-MOP, 0.1mol \% or less}} \text{R-C*H(CH}_3\text{)-SiCl}_3\ (\alpha\text{-adduct, 66-93\%}) + \beta\text{-adduct} \xrightarrow[\text{(2) H}_2\text{O}_2\text{, KF/KHCO}_3]{\text{(1) EtOH/Et}_3\text{N}} \text{R-C*H(CH}_3\text{)-OH}\ (94\text{-}97\%) \quad (23)$$

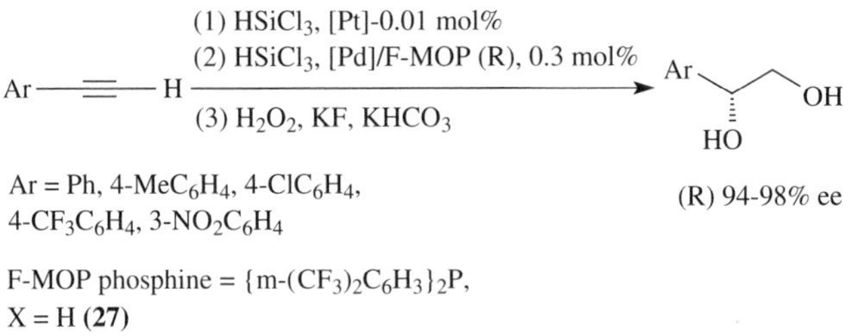

R = n-Bu, n-hex, n-dec, CH_2CH_2Ph, cyclo-C_6H_{11}

$$\text{Ar-C}{\equiv}\text{C-H} \xrightarrow[\text{(3) H}_2\text{O}_2\text{, KF, KHCO}_3]{\text{(1) HSiCl}_3\text{, [Pt]-0.01 mol\%; (2) HSiCl}_3\text{, [Pd]/F-MOP (R), 0.3 mol\%}} \text{Ar-CH(OH)-CH}_2\text{OH}\ \text{(R) 94-98\% ee} \quad (24)$$

Ar = Ph, 4-MeC_6H_4, 4-ClC_6H_4, 4-$CF_3C_6H_4$, 3-$NO_2C_6H_4$

F-MOP phosphine = $\{m\text{-}(CF_3)_2C_6H_3\}_2P$, X = H **(27)**

Great latitude in tuning Pd complexes toward achieving phenomenal stereoselectivity has been demonstrated *via* use of a ferrocene-anchored *P,N*-ligand **28**. For the hydrosilation of norbornene, increasing the bulk of the pyrazole substituents increases yield and ee significantly, and electron-withdrawing 3,5-bis(trifluoromethyl)phenyl groups on the phosphine lead to >99.5% enantiomeric excess of the *exo*-norborneol derived *via* oxidation of the silyl-adduct.[126] Substrate electronic effect on stereoselectivity, using 4-substituted styrenes, showed *para*-substituent influences on both ee as well as chiral sense of the product that follow the Hammett linear free-energy relationship. Palladium complexes containing a bidentate ligand **29** with either a P or As donor and a sulfonamide group (O-donor postulated) exhibited moderate stereoselectivity control for the hydrosilation of conjugated cyclic dienes with $HSiMeCl_2$.[127]

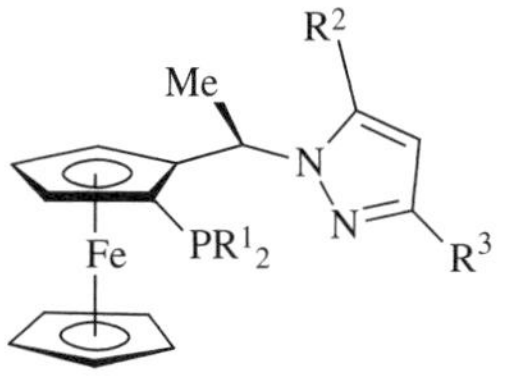

28

R^1 = Ph, 3,5-$(CF_3)_2$-C_6H_3
R^2 = H, Me
R^3 = Me, Ph, 9-Anthryl,
2,4,6-Me_3-C_6H_2, 2,4,6-$(OMe)_3$-C_6H_2

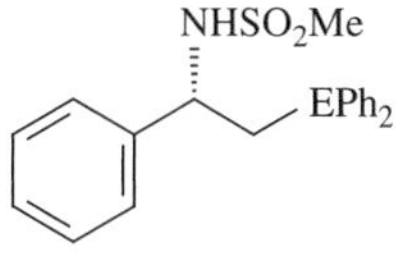

29

E = P,As

Wiedenhofer and coworkers have employed optically active bisoxazoline **30**–**32** and pyridine–oxazoline **33** complexes $NN'PdCl_2$ for the first asymmetric cyclization–hydrosilation of dienes, primarily diallylmalonates, with tertiary silanes.[128] Whereas the bisoxazoline complexes achieve moderate yields of silylated and non-silylated cyclic mixtures with high diastereoselectivity and moderate enantioselectivity, the pyridine–oxazoline complexes afford the silylmethyl-carbocycle in high yield, excellent diastereoselectivity and high enatiomeric execess [Eq. (25)]. Using a Pd catalyst crafted from phosphoramidite-MOP ligands **34** and **35** Johannsen *et al.* demonstrated moderate to exceptionally high stereoselective hydrosilation of aromatic alkenes with $HSiCl_3$ [Eq. (26)].[129]

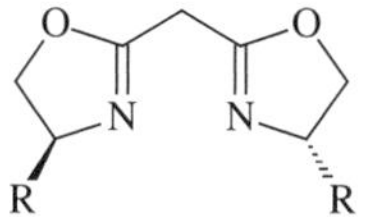

30 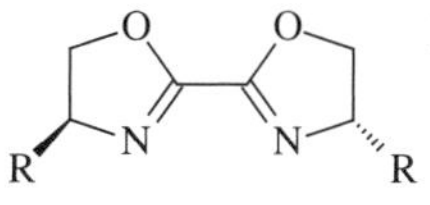 **31** 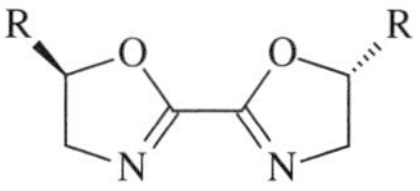 **32**

R = alkyl, aryl (*R, R* and *S,S* forms used)

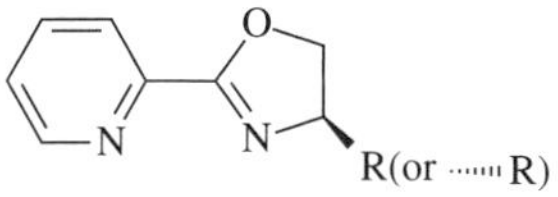

33

R = alkyl, aryl (*R* and *S* forms used)

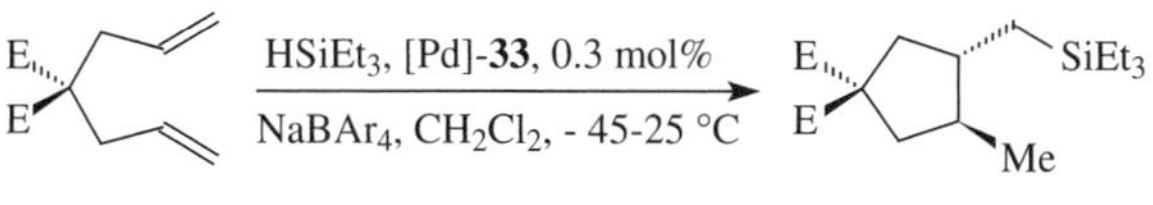

$HSiEt_3$, [Pd]-**33**, 0.3 mol%; $NaBAr_4$, CH_2Cl_2, - 45-25 °C

Exclusive product (25)

E = CO_2R, etc. [Pd] = Pd(COD)MeCl, or preformed complex with **33** used
Ar = 3,5-bis(trifluoromethyl)phenyl

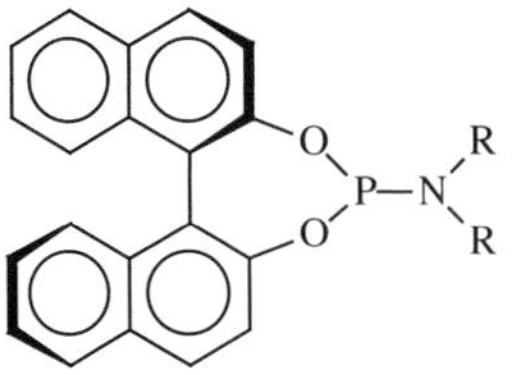

34
R = Me,^{i}Pr

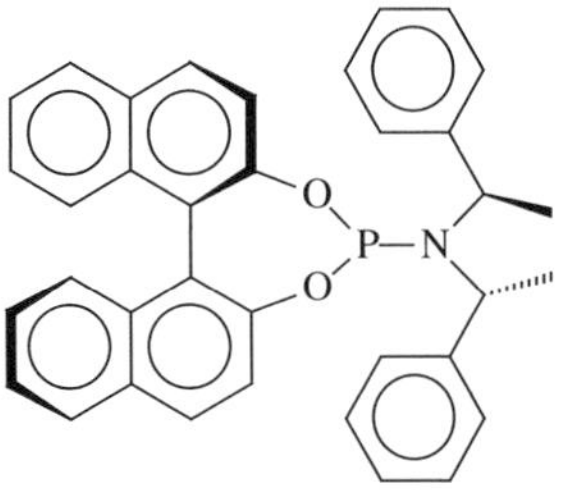

35

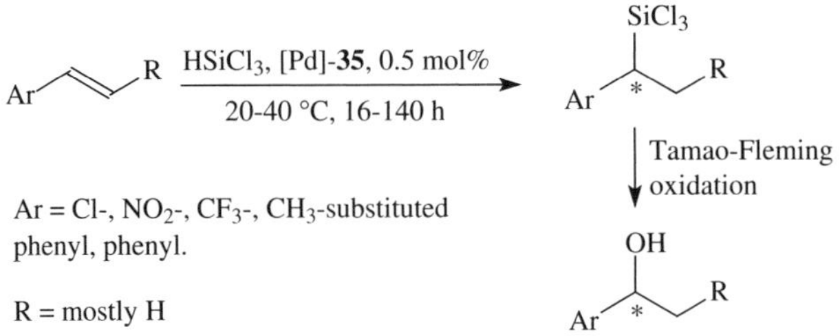

$$Ar\text{-}CH{=}CH\text{-}R \xrightarrow[20\text{-}40\ ^\circ C,\ 16\text{-}140\ h]{HSiCl_3,\ [Pd]\text{-}\mathbf{35},\ 0.5\ mol\%} Ar\text{-}C^*H(SiCl_3)\text{-}CH_2R \xrightarrow{\text{Tamao-Fleming oxidation}} Ar\text{-}C^*H(OH)\text{-}CH_2R \quad (26)$$

Ar = Cl-, NO_2-, CF_3-, CH_3-substituted phenyl, phenyl.

R = mostly H

74-95% isolated yield of silane, 86-99% ee of alcohol

Employing a number of Pd–MOP complexes, with 3,5-disubstituted aryls on the phosphorus and H, MeO or CN at the 2′ position, Pregosin and coworkers achieved quantitative yields and up to 92% ee in the hydrosilation of styrene with $HSiCl_3$.[130] Pregosin's group also examined metal–ligand bonding in a number of Pd–MOP complexes, and the significant variation in bonding modes and the resulting diversity in spatial orientation of these MOP complexes observed led them to conclude that selection and design of MOP ligands for predictable improvements in enantioselective hydrosilation is not a trivial or even simple task.[131] Indeed, many papers published on novel chiral ligands and their metal complexes that only achieve low to moderate enantiomeric excesses support this assertion.

Unlike the asymmetric hydrosilation of alkenes and acetylenes where Pd catalysts seem to dominate activity, for enantioselective hydrosilation of ketones a preponderance of Rh-based catalysts have been used, although a few other metals including Pd have been examined and some ligand templates such as binaphthyl are in common.

Using the (*R*)-[$TiCl_2(\eta^5:\eta^5\text{-}C_5Me_4SiMe_2C_5H_3$-menthyl)]/2-*n*BuLi catalyst system, White and coworkers accomplished hydrosilation of acetophenone and substituted acetophenones with $PhSiH_3$ that afford chiral alcohols (following hydrolysis of adduct) in moderate enantiomeric excesses.[132] The (*S*)-neomenthyl catalyst gave very low enantioselectivity. A number of heteroaromatic ketones, containing nitrogen/oxygen/sulfur heterocycles, have been asymmetrically hydrosilated with poly(methylhydrosiloxane) using a Cu(I)-hydride complex of the ligands **36** and **37**.[133] Several such substrates (acetyl pyridines, acetyl furan, acetyl thiazoles and others) provide high yields and very good to excellent stereoselectivity of the corresponding chiral alcohols, in toluene or toluene–THF solvent at −78 to −35 °C and at 1 mol% catalyst.

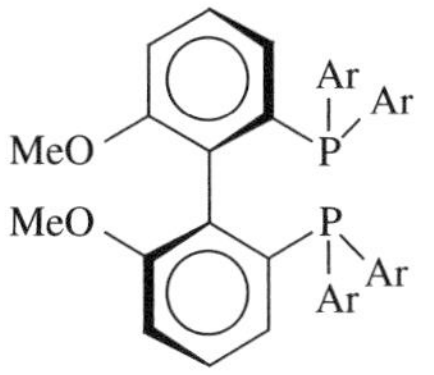

MeO MeO Ar Ar P P Ar Ar **36**

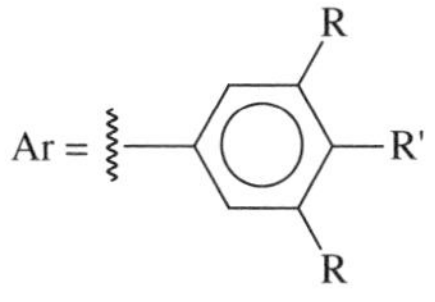

Ar = R R R'

R = t-Bu, R' = OMe;
R = Me, R' = H

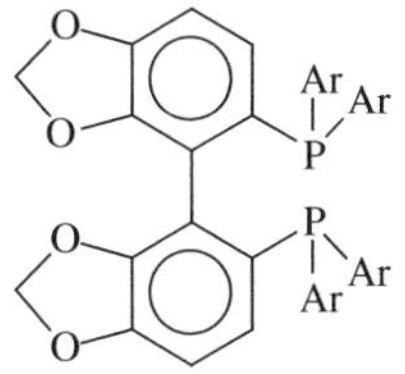

37

A Rh(III) complex **38** assembled from a bis-NHC-functionalized binaphthyl ligand was found to provide excellent asymmetric induction for the hydrosilation of aryl methyl ketones and dialkyl ketones with Ph_2SiH_2, affording yields and ee's of 85–96% and 67–98%, respectively [Eq. (27)].[134] Aryl substitutions include F, Br, Me, OMe and CF_3. Crabtree and coworkers also reported Rh- and Ir-mono-NHC-binaphthyl complexes for the hydrosilation of acetophenone with Ph_2SiH_2, but with these latter complexes ee is moderate (Ir) or low (Rh), although yields are high.[135] Recently, the use of Rh complexes **39** with a novel oxazolyl-NHC ligand for the enantioselective hydrosilation of both aryl/alkyl ketones and dialkyl ketones with Ph_2SiH_2 was reported.[136] Amongst four variants, one (**39c**) proved an excellent catalyst for the hydrosilation [Eq. (28)].

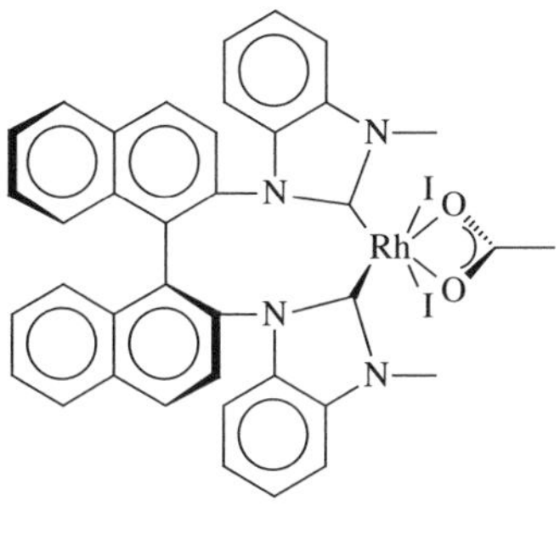

38

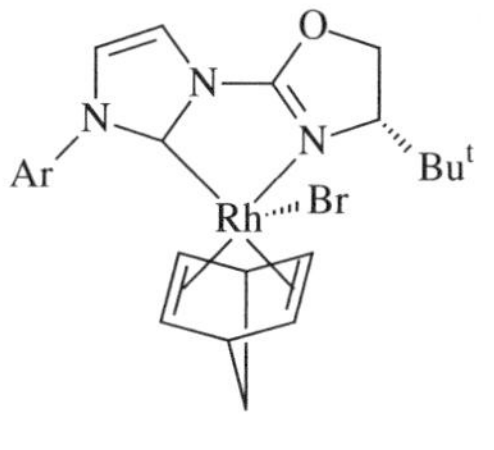

39

Ar = Ph(**a**); *o*-Tol(**b**); Mes(**c**); 2,6-$^{i}Pr_2$-C_6H_3(**d**)

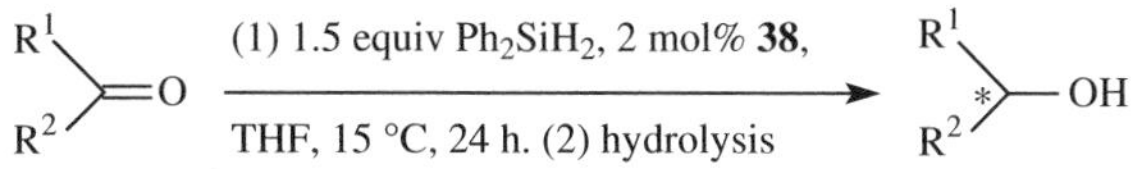

$$R^1R^2C{=}O \xrightarrow[\text{THF, 15 °C, 24 h. (2) hydrolysis}]{\text{(1) 1.5 equiv Ph}_2\text{SiH}_2\text{, 2 mol\% \textbf{38},}} R^1R^2C^*H{-}OH \quad (27)$$

$$R^1R^2C{=}O \xrightarrow[\text{CH}_2\text{Cl}_2\text{, - 60 °C, 10 h. (2) hydrolysis}]{\text{(1) Ph}_2\text{SiH}_2\text{, 1 mol\% \textbf{39c}, 1.2 mol\% AgBF}_4\text{,}} R^1R^2C^*H{-}OH \quad (28)$$

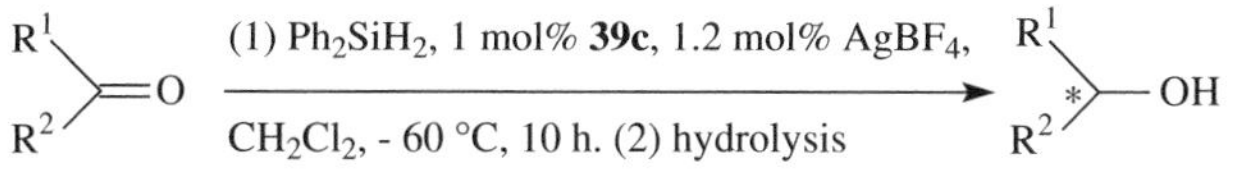

Rhodium complexes derived from chelating bisphosphines anchored to an asymmetric, ancillary Cp–Re complex exhibit moderate enantioselectivity in the hydrosilation of phenyl/alkyl ketones with Ph_2SiH_2.[137] A *trans*-chelating, planar–chiral bisphosphine **40**, based on a biferrocenyl motif, forms Rh complexes that are highly effective for the enantioselective hydrosilation of phenyl/alkyl or dialkyl ketones with Ar_2SiH_2 (Ar = Ph or 3-F-C_6H_4). Excellent yields (73–99%) and high ee (60–92%) are obtained, especially with the bis-Et_2P ligand.[138] P,N-chelating ligands **41** built on the TADDOL scaffold, form cationic Rh(I)(COD) complexes that catalyze hydrosilation of acetophenone and derivatives with good to excellent enantiomeric excess (up to 93%).[139]

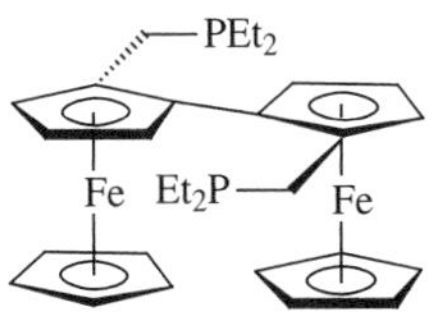

40

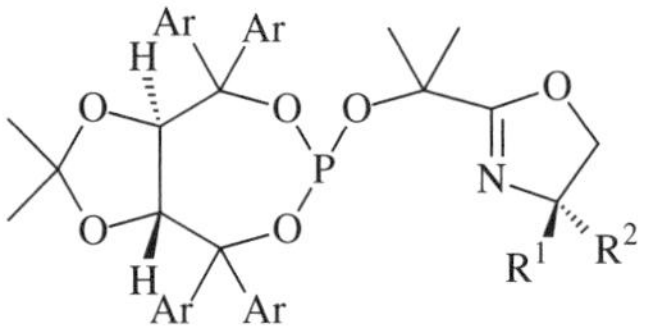

41

Ar = Ph, 2-naphthyl
R^1/R^2 = H/^{i}Pr

A number of Rh complexes derived from bis(pyridyl)- silane or methanol were found to provide low enatioselectivity in the hydrosilation of acetophenone.[140] Chiral (*R*,*R*) 1,2-diaminocyclohexane-based N,P and P,P ligated Rh complexes also achieved low enantioselectivity for the hydrosilation of acetophenone.[141] *Cis*-"A-frame" phosphine Rh complexes based on bi-naphthol and resorcinol anchors gave products from the hydrosilation of acetophenone in moderate yields and low enantioselectivity.[142] Rh(I) phosphine-(sulfinyl)phosphonium ylide complexes that bind through the carbon center of the ylide have been examined in acetophenone hydrosilation, but enantiomeric excesses are rather low even though yields are moderate to good.[143] Some monothio- and dithioureas, with high ligand-to-metal ratio (up to 10:1) have been investigated in Ir-catalyzed hydrosilation of acetophenone, where yields and enantioselectivities vary from low to moderate.[144] Korostylev *et al.* have reported up to 50% ee for hydrosilation of acetophenone with Ph_2SiH_2 using Rh–P,N complexes based on ligands assembled from phosphites and amino-alcohols.[145]

A second novel Re(V) complex **42** was employed by Toste and coworkers for the reduction of a wide range of aromatic, heteroaromatic and α,β-unsaturated *N*-phosphinyl imines with either $HSiMe_2Ph$ or $HSiMePh_2$.[146] Good to high yields and high to often exceptional (95 to >99%) enantioselectivity were achieved, at room temperature in CH_2Cl_2. Again, the Re(V) complex stands out as a high oxidation state metal catalyst that is insensitive to air and moisture exposure. In another example of highly enatioselective hydrosilation of aromatic *N*-phosphinyl imines, Ar–C=N–P(O)(xylyl)$_2$, Lipshutz and coworkers demonstrated the use of a Cu catalyst, derived from the ligand **43** and CuCl–NaOMe–3.3^{t}BuOH, to achieve excellent yields and 94–99.3% ee, for the reductions with *sym*-tetramethyldisiloxane.[147] This group also reported excellent asymmetric induction (91–99% ee) in the 1,4-hydrosilation of α,β-unsaturated esters with poly(methylhydrosiloxane), using Cu catalysts derived from **43** or chiral bisphosphinoferrocenes.[148] In a rare example of Ru-catalyzed asymmetric hydrosilation, Ru(PPh_3)(oxazolinylferrocenylphosphine) complexes have been shown to convert ketoximes to chiral amines in good yields and high enantioselectivities (up to 89%).[149]

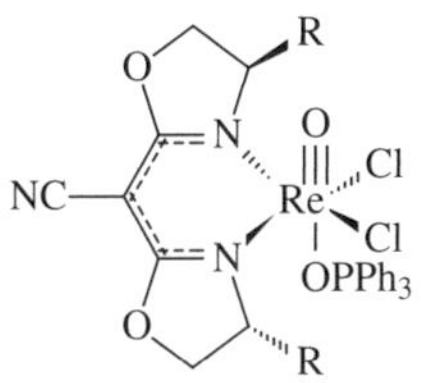

42
R = 4-^{t}Bu-C_6H_4

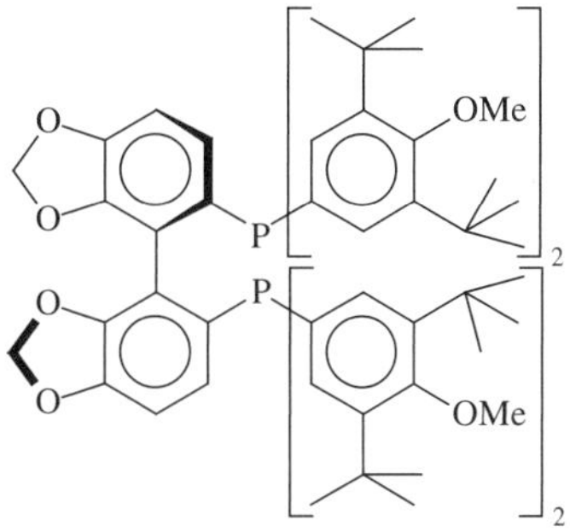

43

It is apparent from the above brief excerpts from enantioselective hydrosilations of acetophenone alone, that chiral ligand selection for predictable high asymmetric induction is non-trivial. One of the promising and increasingly successful approaches to new catalyst development is the adoption of combinatorial methods. Using this tool, and the analytical technique of "mass spectrometry enantiomeric excess determination" (MSEED) developed in their group, Finn and coworkers assembled an initial array of 21 P,N ligands based on various amino alcohols, phosphite tether group and scaffolds related to BINOL and TADDOL.[150] Rh complexes prepared from these ligands in 2:1 and 4:1 ligand:Rh ratio were then tested for enantiomeric excess in the hydrosilation of 1-naphthyl methyl ketone, comparing results against (*R*)-*p*-tolyl-BINAP and (*S*)-PyBOX. Following two further iterations with refined ligand sets for other ketone hydrosilations, ligand **44** emerged as the most effective, affording 70–94% ee for five of the seven ketone substrates tested. The authors conclude that the diol base, amino-alcohol, and the latter's *N*-alkyl bulk and chiral sense are all structural features of the ligand that are important for overall performance. The coordination of the nitrogen to the Rh could not be determined with certainty, and either a chloro-complex [RhCl(**44**)(COD)] with no *N*-bonding or a cationic complex [Rh(**44**)(COD)]Cl, with chelating P,N was presumed to be the active precatalyst.

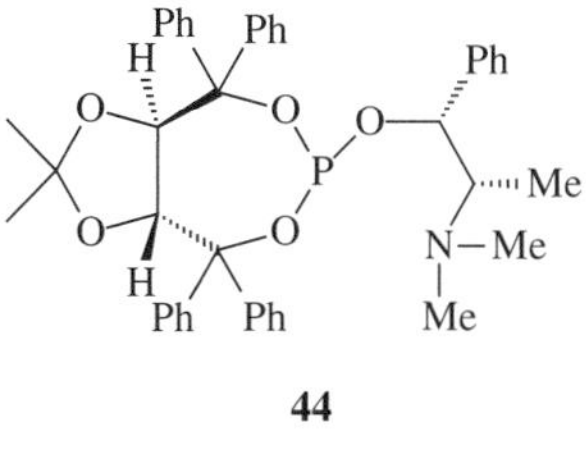

44

IX

APPLIED ASPECTS – SYNTHESIS OF POLYMERS AND MATERIALS

One of the very first polymer applications to incorporate and benefit from hydrosilation is of course silicone rubber. However, this section will not cover the traditional aspects of using the reaction, as practiced in silicone commerce. Instead, glimpses are provided of the myriad ways hydrosilation has been used creatively to develop or advance "simple" linear polymers and more structured or ordered materials. Because of the nearly explosive growth in this area, only limited and representative structures are provided, as can be accommodated within the scope of this review, and surface treatments *via* hydrosilation are omitted in general, except for special or unusual cases. The types of polymers and materials synthesized using hydrosilation can be roughly divided into three or four broad classes: (A) linear polymers containing primarily silicon and carbon (and sometimes other atoms) in the backbone – many are σ–π "conjugated" systems, (B) linear polymers with pendent groups that confer particular properties, (C) dendritic/hyperbranched systems or materials based on scaffolds and (D) siloxane or organic polymers that have been functionalized or otherwise modified through hydrosilation, and studies on the hydrosilation aspect of reactions with polymeric reactants.

(A) The coupling of π-conjugation in polyacetylene (and related polyphenylenevinylenes) and σ-conjugation in polysilanes to produce hybrid materials with potentially new or superior optical or optoelectronic properties, and/or to generate efficient ceramic precursors, has been an important area of research in silicon physics, chemistry and materials science. Interests have been driven by the advancing electronics industry and by new structural material needs, with a very broad range of potential applications. A variety of alkyne and hydrosilane difunctional monomers of the AA/BB and AB types, differing in substitution, positional isomerism and initial π-conjugation length have been subjected to hydrosilation–polymerization using a number of different catalysts.

Barton and coworkers carried out the first hydrosilation polymerization of an ethynylhydrosilane that led to soluble, well-characterized, low-medium molecular weight poly(methylphenylsilylenevinylene) as a glassy polymer in 95% yield, using H_2PCl_6 in THF as catalyst.[151] Poly(dimethylsilylenevinylene) was also prepared similarly [Eq. (29)]. Interestingly, only terminal (anti-Markovnikov) addition of SiH was observed for the polymers **45**, as supported by NMR data, and it also appears that the hydrosilation was highly stereoselective towards one isomeric backbone structure. Flexible fibers drawn from the Ph/Me polymer melt could be UV-crosslinked and pyrolyzed to yield ceramic fibers.

$$\text{H–Si(Me)(R)–C}\equiv\text{CH} \xrightarrow[\text{Neat or THF (preferred)}]{\text{cat. } H_2PtCl_6} \left[\text{Si(Me)(R)–CH=CH}\right]_n \quad (29)$$

R = Me, Ph; **45**

Kim and Shim reported polymerization of *o*-, *m*- and *p*-(dimethylsilyl)phenylacetylene, using H_2PtCl_6 catalysis, where polymer yields of 69%, 89% and 67%, respectively, were obtained.[152] All three polymers **46a–c** exhibited similar UV–visible absorption profiles with λ_{max} 260–270 nm. Fluorescence emissions were also observed, with an unusual red-shifted emission for **46a** which the authors attribute to possible charge-transfer interaction between Si and the neighboring alkene group in each unit. Polymer **46a** decomposed almost completely on heating to 800 °C (possibly *via* silaindene type cyclization), while **46b** and **46c** afforded 40% and 65% char yields. The same authors also polymerized 4,4′-diethynylbiphenyl with Ph_2SiH_2 or $PhMeSiH_2$.[153] λ_{max} values and UV absorption profiles of the polymers vs. model monomers were nearly identical, indicating that extension of conjugation through Si–CH=CH_2 σ–π interaction is not significant. Ceramic fibers could be processed from the Ph_2Si-based polymers.

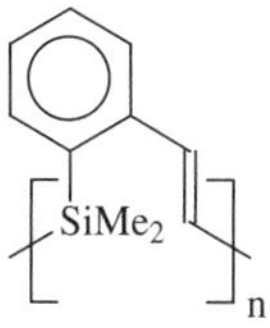

SiMe2 n

46a

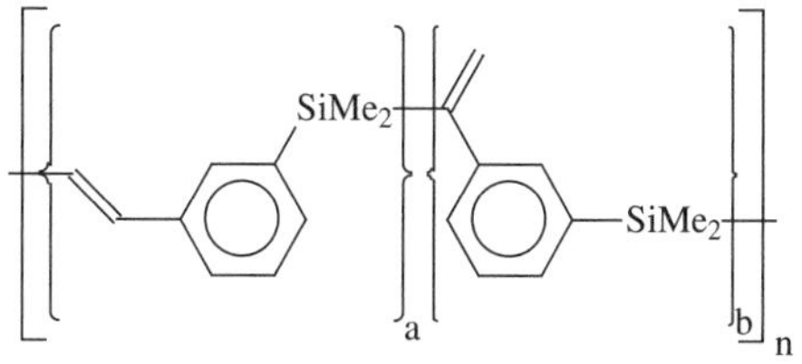

SiMe2 SiMe2 a b n

46b

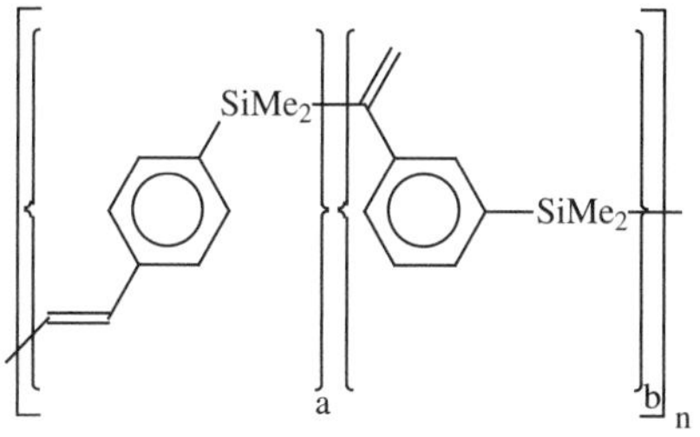

SiMe2 SiMe2 a b n

46c

Excellent stereo- and regio-control in hydrosilation–polymerization with *m*- or *p*-diethynylbenzene and *m*- or *p*-bis(dimethylsilyl)benzene have been exercised, using $[RhI(PPh_3)_3]$ or $[RhI(COD)]_2$ (COD = cycloocta-1,5-diene) as catalyst.[154] Heating reactants and catalyst together yields *E*-polymers, while pre-reacting catalyst and bis-silane followed by dialkyne addition affords *Z*-polymers, in both cases with mostly >95% stereoselectivity and very good to excellent yields. No α-addition is detectable. As with the similar polymers of Barton and of Shim, UV–Vis spectroscopic studies showed no evidence of extended σ–π conjugation, and absorption, emission and λ_{max} characteristics were correlated with the various stereo- and positional isomeric combinations present within the polymer backbones. Synthesis of a number of poly(silylene–divinylene)s, containing Me, Ph or vinyl substituents on silicon, *via* hydrosilation of *m*- or *p*-diethynylbenzene with secondary silanes, employing $Pd_2(dba)_3/PCy_3$ (dba = dibenzylideneacetone) as catalyst has been reported.[155] Because Pd catalysis promotes C≡C hydrosilation selectively in the presence of C=C, polymers containing Si-vinyl groups could be made. Char yields in the range 56–75% (highest, for Si-vinyl) at 980 °C were obtained for the polymers which also exhibited expected UV–vis absorption and emission behavior.

Neckers and coworkers have prepared linear and star polymers containing silylene–vinylene–phenylene units from *p*-dimethylsilylphenylacetylene and 1,4-diethynylbenzene or 1,3,5-triethynylbenzene, respectively, *via* $Pt(acac)_2$-catalyzed photohydrosilation.[156] Pi-conjugation chain lengths were varied *via* incorporation of consecutive ethynylbenzene units. For these photoluminescent polymers, it is mainly the vinylenephenylene segments that serve as the fluorescent chromophore. Kang, Ko and their coworkers have synthesized a number of linear and star-like (**47**) chromophores, based on polyaromatic cores, that are separated by silicon units.[157] The syntheses are based on hydrosilation of di- or triethynylbenzene with $HSiMe_2Cl$ using Karstedt's complex as the preferred catalyst, followed by reaction with various aryl lithium compounds. High $\beta(E)$-selectivity for the hydrosilation is reported. Preliminary UV–visible spectroscopic studies of some of the bis-silyl and the tris-silyl fluorophores are also presented.

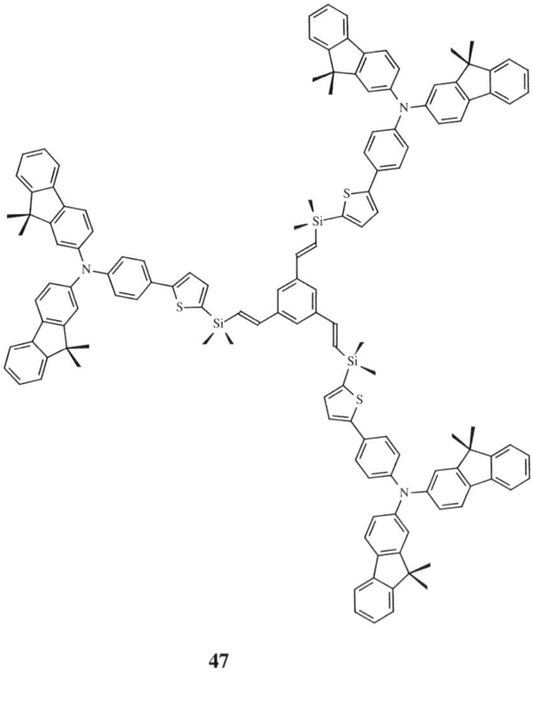

47

A number of silylene–vinylene–phenylene type polymers containing 1–3 ferrocene moieties per repeat unit have been synthesized by Sheridan's group *via* hydrosilation of various diynes with 1,1′-bis(dimethylsilyl)ferrocene.[158] Either Karstedt's complex or $RhI(PPh_3)_3$ was used as catalyst. Varying levels of regio- and stereochemical distribution of the repeat units were obtained (Rh provided mostly *β*-(*Z*) product), and the polymers were generally of relatively low molecular weight. Cyclic voltammetry studies showed single redox waves that are suggestive of no through-chain interaction between neighboring ferrocene units. Liquid crystalline di- and trisiloxanes **48** exhibiting smectic C phases and ferroelectric electro-optic switching have been prepared *via* hydrosilation of a ferrocene-containing substrate with a 1-undecylenoxy substituent, using Karstedt's catalyst.[159]

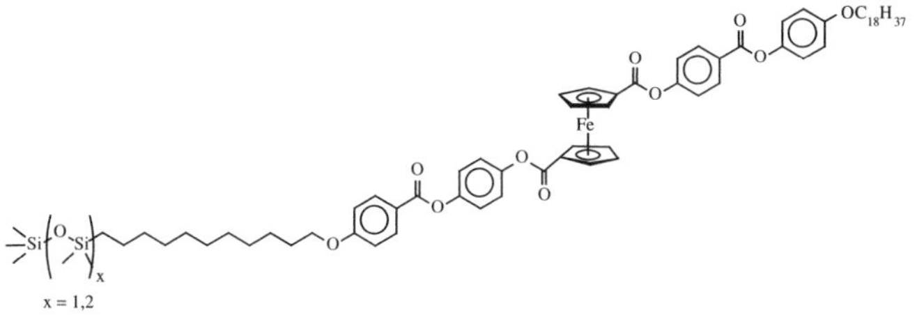

48

(B) The hydrosilation reaction has been very effectively used to functionalize both siloxanes and organic molecules to confer properties of the one species on the other, or simply to modify or tune certain characteristics. Side-chain and chain-end modification of siloxanes is a powerful synthetic tool for this purpose. Hydrosilation of methacrylate-based hindered amine light stabilizers (HALS) with *sym*-tetramethyldisiloxane or tris(dimethylsiloxy)methylsilane, leads to products with two or three light-stabilizer groups in a single molecule.[160] Benzophenone-based UV absorbers have been attached to polysiloxanes *via* hydrosilation with poly(methylhydro-*co*-dimethylsiloxane), using Cp_2PtCl_2 as catalyst.[161] The polymeric actives **49** showed essentially unchanged absorbance wavelengths and extinction coefficients when compared with the active small molecules. Sulfonation of silyl ketene acetal modified siloxanes, prepared *via* hydrosilation,[162] yields water-soluble siloxanes **50**.[163] Carbohydrate-functional siloxanes have been prepared *via* two different approaches using hydrosilation. In one, acrolein diethyl acetal was hydrosilated with SiH-terminal siloxanes, followed by transacetalation with glucose.[164] SiH rake polymers initially yielded soluble materials which were not isolable in pure form. The second approach employed hydrosilation of *p*-allyloxy-benzaldehyde glucose acetal, *N*-allylgluconamide or 1-allylglucose with terminal SiH-functional siloxanes.[165] The OH groups of glucose were protected prior to hydrosilation and then deprotected *via* methanolysis after hydrosilation.

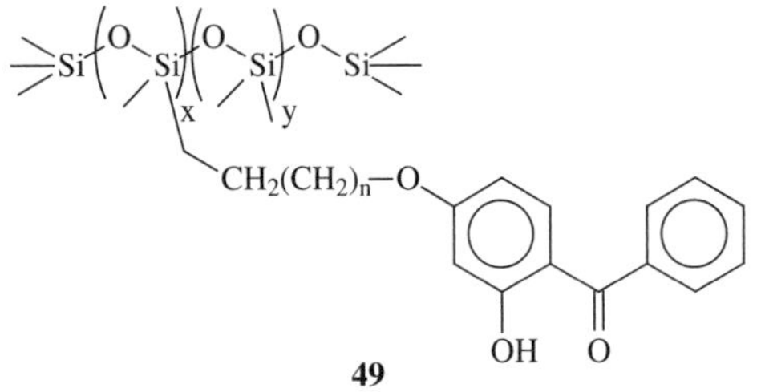

Liquid crystals are of course of immense importance to the electronics industry. Side-chain liquid-crystalline siloxanes prepared *via* hydrosilation of mesogens with poly(methylhydrosiloxane) [Eq. (30), e.g.], and their phase behavior studies using DSC, optical microscopy and X-ray diffraction have been reported.[166,167]

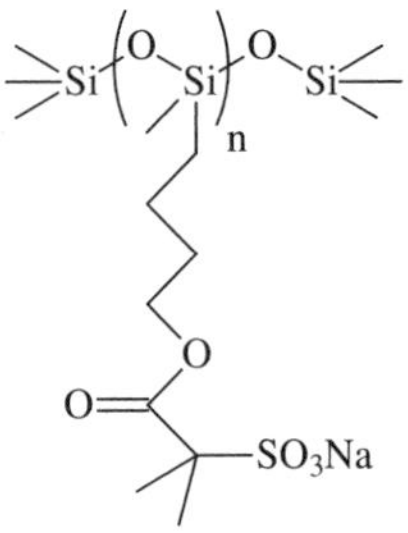

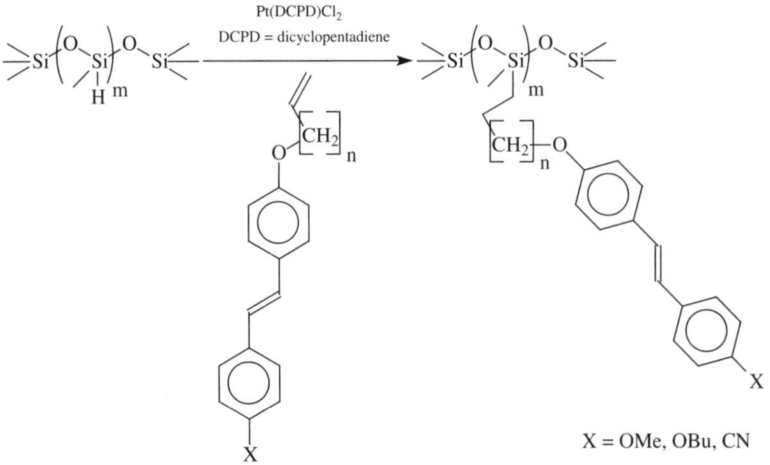

(30)

Fullerene C_{60} undergoes hydrosilation with several chloro- and alkoxyhydrosilanes.[168] The silyl-C_{60} can then be cohydrolyzed with TEOS to generate a "homogeneous" hybrid silica-gel. A variety of isocyanate-functional silanes, cyclosiloxanes and polysiloxanes have been synthesized *via* reaction of *m*-TMI [*m*-($CH_2{=}C(Me){-}C_6H_4{-}C(Me)_2N{=}C{=}O$] with the corresponding SiH components, using Karstedt's catalyst and without any detrimental reaction at $N{=}C{=}O$.[169] The "cyclotetrasiloxane–tetraisocyanate" was successfully converted to polyurethanes *via* reaction with MPEG. Silicone-urethanes/ureas are extremely interesting materials with potential for a number of commercial applications, and this alternative to allyl-PEG hydrosilation offers a second approach to synthesizing this class of silicone-organic hybrids. A unique ladder-like, soluble polysiloxane with

4,4′-oxybis(benzamidopropyl) rungs and a naphthyl substituent on each silicon has been synthesized *via* hydrosilation followed by H-bonding-templated hydrolysis-condensation of the monomeric bis(diethoxysilane).[170]

Silicone polyethers (SPE) **51**, also known as silicone-copolyols (SCP), which are readily prepared *via* hydrosilation [Eq. (31)], form a very important class of co-polymers in the area of commercial surfactants. Their utility, however, goes beyond the surface properties as they are also highly suited for ion conduction in polymeric liquid, gel or solid battery applications. A good number of publications have appeared on this utility, and ionic conductivities approaching 1.0×10^{-2} S/cm appear within reach in the near future.[171–174] Even oligoethylenoxysilanes and short-chain siloxanes show promise for good liquid electrolyte conductance.[175] Polysiloxanes with pendent cyclic carbonate[176] and sulfonate groups[177] were also reported earlier for this application.

$$Me_3Si(OSi(Me)(H))_x(OSiMe_2)_yOSiMe_3 \xrightarrow[\text{[Pt]}]{\text{Excess } CH_2{=}CHCH_2O(EO)_a(PO)_b{-}H} Me_3Si(OSi(Me)(CH_2CH_2CH_2(EO)_a(PO)_b{-}H))_x(OSiMe_2)_yOSiMe_3 \quad (31)$$

$EO = CH_2CH_2O$, $PO = CH_2CH(Me)O$

Dimethicone copolyol (DMC)
or Silicone polyether (SPE)

51

(C) Amongst macromolecular architectures, dendrimers, hyperbranched polymers and high molecular weight species "grown" on scaffolds are materials from which the scientific and technological communities have great expectations. A number of simple and efficient chemistries of silicon lend themselves well toward the synthesis of the above type polymeric systems. Among these, hydrosilation as a catalyzed addition reaction is a very well-suited transformational tool at the chemist's disposal. A review on heteroatom-containing dendrimers has covered hydrosilation-based materials.[178] For example, carbosilane dendrimers of the type **52** have been prepared *via* alternating Grignard and hydrosilation reactions, starting with $SiCl_4$. This method allows great versatility as successive chlorosilanes can be $HSiCl_3$ or $HSiRCl_2$ and the alkenyl Grignard reagent can be many, containing various methylene spacer lengths, $CH_2{=}CH{-}(CH_2)_nMgX$. Not only carbosilane but also silane and siloxane dendrimers with up to 7th generation layer have been prepared using hydrosilation with judicious control over reaction conditions. Despite the immense potential, however, dendrimer synthesis is tedious and wide-spectrum commercial manufacture will likely be possible only with developments of peptide-synthesis-like automated or semi-automated processes.

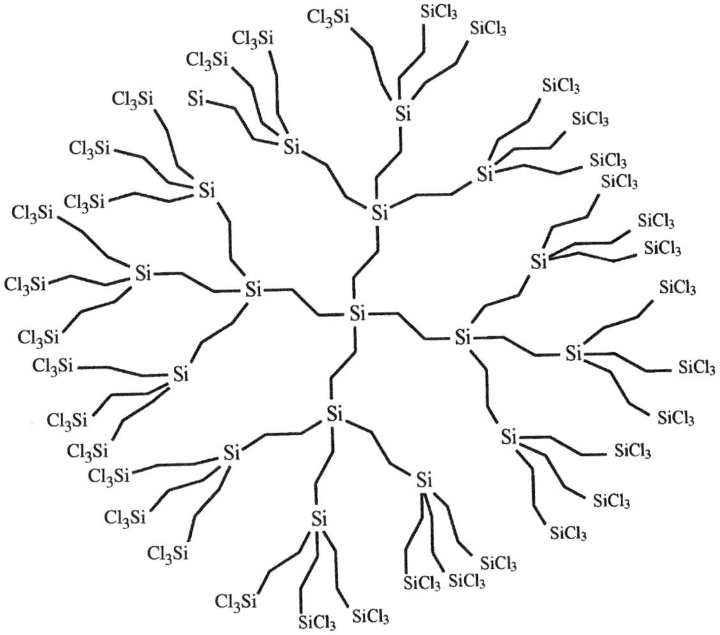

52
Shown to G_4 level

Hyperbranched polymers, which are close cousins of dendrimers, are much more amenable to synthesis, since they only require an AB_2 monomer for "chain" growth. In this area, synthesis of a σ–π conjugated polymer from the AB_2 monomer *m*-$(HMe_2Si)_2$–C_6H_3–C≡CH has been reported, using $RhCl(PPh_3)_3/NaI$ as the stereo- and regioselective catalyst.[179] The SiH-ended polymer was capped with excess phenylacetylene to generate a stable material. Without capping, air oxidation/hydrolysis and subsequent crosslinking occurs to produce gels and insoluble films which emit an intense blue light on excitation at the charge-transfer absorption wavelength.

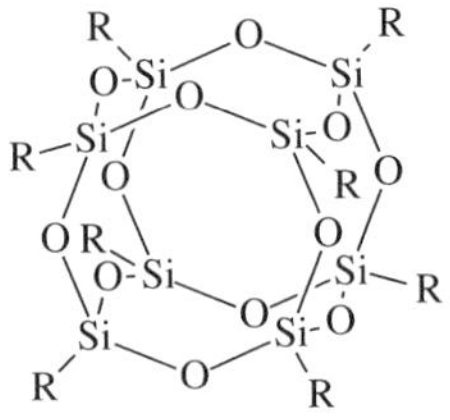

54

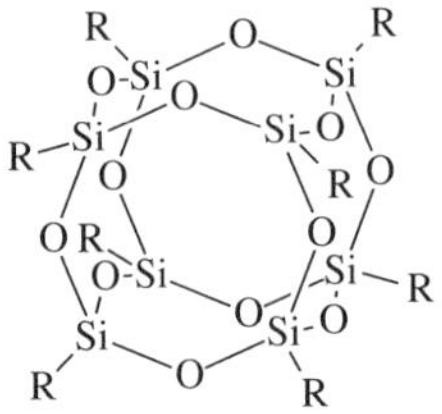

53

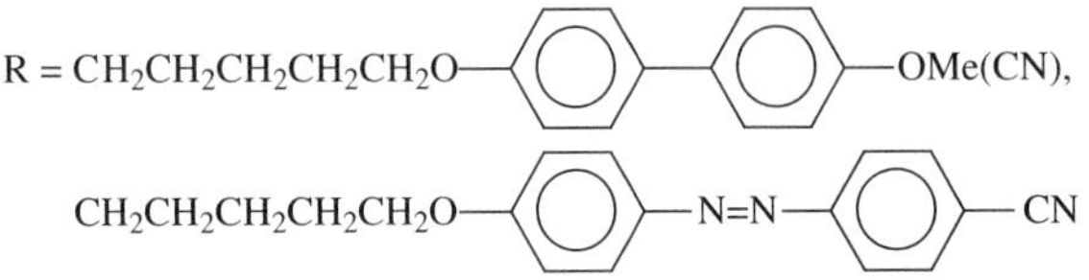

R = $CH_2CH_2CH_2CH_2CH_2O$–C_6H_4–C_6H_4–OMe(CN), $CH_2CH_2CH_2CH_2CH_2O$–C_6H_4–N=N–C_6H_4–CN

The cubic silsesquioxane unit $[RSiO_{1.5}]_8$ **53** (R = organic, inorganic or organometallic substituent) presents a three-dimensional inorganic scaffold that is amenable to a wide variety of materials synthesis. Laine and coworkers have investigated this arena extensively. Using hydrosilation to form a network structure [e.g., Eq. (32)] from [vinyl-$(Me_2SiO)_{0\ or\ 1}$–$SiO_{1.5}]_8$ and [H–$(Me_2SiO)_{0\ or\ 1}$–$SiO_{1.5}]_8$, where the vinyl or H is attached directly or through a siloxane spacer unit to the cage Si, variation of the pore structure of the material (0.3 nm diameter – cube interior; 1–50 nm diameter – between cubes) has been studied.[180] Silsesquioxanes, bearing liquid crystalline mesogens[181–183] (e.g., **54**) and photopolymerizable glycidyl groups[181,182] have been made. Completely nano-homogeneous epoxy nanocomposites are derived from either glycidyl- or cyclohexeneoxide-substituted T_8 and aromatic diamines.[184] Thermomechanical properties of these materials have been studied in detail to understand structure–property relationships. Synthesis of alcohol-soluble silsequioxanes with amine-terminal groups has been reported.[185] Here, a chlorosilane precursor to the silsesquioxane was first prepared *via* hydrosilation using Cp_2PtCl_2 as catalyst. In general, for hydrosilation directly performed on silsequioxanes, Karstedt's complex has been the most commonly employed catalyst, although Speier's catalyst and Pd complexes have also been used. The promise of materials based on silsesquioxanes remains high, but ready availability of core T_8 raw materials such as T_8-H in large quantities and at economical cost will likely determine true commercial application of this unique, yet versatile building block.

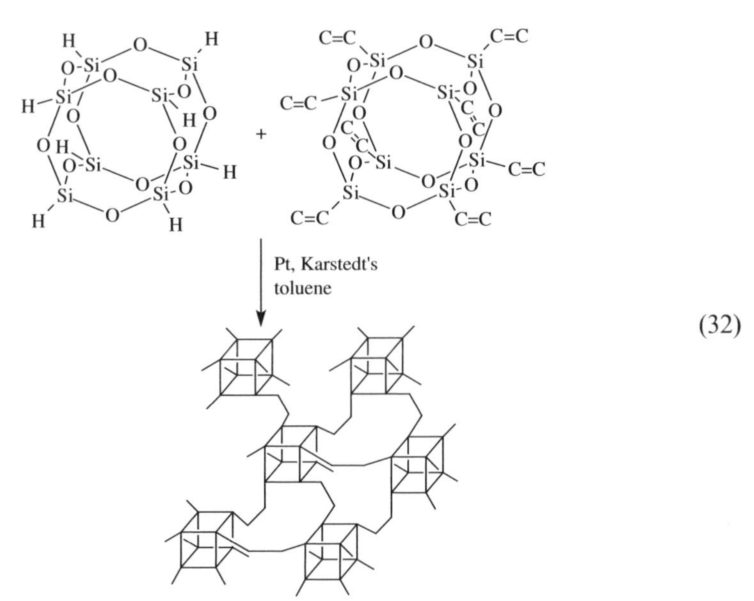

(32)

(D) The polymer literature is replete with examples of silicone-organic hybrid macromolecules. Just about every major organic polymer class, from polyolefins to polyesters to polyimides, has been covalently linked to polysiloxanes. More often than not, hydrosilation is involved in either monomer synthesis or in catalyzing the copolymerization reaction itself. Only glimpses of the utility of hydrosilation in polymer synthesis are provided here.

Direct hydrosilation of vinyl-containing organic monomers or macromers with hydrosilane or hydrosiloxane can lead to block or graft copolymers. Mathias and Lewis,[186] and later Boileau and coworkers,[187] reported hydrosilation-polymerization with 2,2′-diallyl bisphenol A to prepare phenol-functional carbosilane- or siloxane–bisphenol A multiblock copolymers [Eq. (33)]. The main side reaction, isomerization of the allyl group (which is addressed in Sections X and XI), thwarts achievement of high molecular weight not only in this system but for similar allylic substrates as well. Silicone-grafted polybutadiene and polyethylene were prepared *via* hydrosilation of commercial polybutadiene with $HSiMe_2(SiMe_2O)_nSiMe_3$ in toluene solution, using $Pt(SEt_2)_2Cl_2$ as catalyst, followed by hydrogenation of the unsaturated *graft*-copolymer.[188] Extensive NMR analysis was performed to determine extent of silicone incorporation. Weber and Sargent report that end-group analysis of polymers obtained *via* reaction between $(HSiMe_2)_2O$ and α,ω-dienes, catalyzed by Karstedt's complex, indicates isomerization to internal C=C in the diene as the reason behind the arrest of chain growth during hydrosilation-polymerization.[189] Bis(vinyldimethylsilyl)dichloromethane undergoes hydrosilation-polymerization with α,ω-dihydrosiloxanes to afford halogenated silmethylene–silethylene–siloxane terpolymers.[190]

H—(Si-X-)$_m$Si—H + [2,2′-diallyl bisphenol A, RO / OR]

X = O, CH_2CH_2
m = 1, 2, 8, 17, 52 for O,
and 1 for CH_2CH_2

R = H,$SiMe_3$

Pt(0) | CH_2Cl_2 (33)

[—(Si-X-)$_m$—$SiCH_2CH_2CH_2$—(RO)bisphenol A(OR)—$CH_2CH_2CH_2$—]$_n$

Omega-allyl polycarbonates have been hydrosilated with either tertiary silanes or *sym*-tetramethyldisiloxane to yield silylated polycarbonates or polycarbonate–disiloxane–polycarbonate triblock copolymers.[191] The siloxane-containing polymers exhibit relatively lower T_g and higher thermooxidative stability compared with bisphenol A polycarbonate. Hydrosilation of allyl-terminal poly(alkyleneoxide-*co*-sulfone) in 1:1 or 2:1 ratio with hydride-terminal polysiloxane leads to ABC and $(AB)_2C$ type block-terpolymers, respectively.[192] DSC studies indicate microphase separation, while TGA data point to higher thermal stability for the siloxane terpolymers. In a series of papers Crivello and Bi have reported synthesis of a host of epoxy-functional siloxanes (and silanes), where controlled reaction using Wilkinson's catalyst leads to mono- or diepoxy-substituted di-, tri-, and tetrasiloxanes [Eq. (34)].[193] These epoxy-siloxanes have been polymerized photochemically, thermally, or using Pt catalyst and the resulting polymers further derivatized *via* a second hydrosilation [Eq. (35)]. Depending on the derivatization, the epoxy polymers can also be cured as a final step.

$H-Si(OSi)_nOSi-H$ + (vinylcyclohexene oxide) $\xrightarrow{RhCl(PPh_3)_3}$ $H-Si(OSi)_nOSi-$(epoxycyclohexylethyl); n = 0-2 (34)

$H-Si(OSi)_mOSi-$(epoxycyclohexylethyl) $\xrightarrow{h\nu,\ toluene,\ \text{diaryliodonium } SbF_6^{-}\ (OC_{10}H_{21})}$ poly(epoxide) with pendant $-SiO(SiO)_mSi-H$ $\xrightarrow{RhCl(PPh_3)_3,\ \text{vinylcyclohexene oxide}}$ poly(epoxide) with pendant $-SiO(SiO)_mSi-$(epoxycyclohexylethyl) $\xrightarrow{h\nu,\ toluene,\ \text{diaryliodonium } SbF_6^{-}\ (OC_{10}H_{21})}$ Insoluble, intractable solid (35)

The physical union of a carbosilane polymer and a ladder silsesquioxane forms the basis for an interpenetrating polymer network.[194] Careful selection of precursors and judicious choice of catalysts for silanol condensation and hydrosilation allowed hydrosilation-polymerization and hydrolysis-condensation to occur at compatible rates so that transparent gels were obtained in several cases. The thermal and mechanical properties of the IPNs are reported. Polysilanes with sila-alkyl substituents, introduced *via* hydrosilation of allylmethyldichlorosilane using Karstedt's catalyst, have been reported.[195] Poly(silethylene-*co*-phenylenealkylene)s have been prepared from the AB monomers *p*-$CH_2{=}CH{-}C_6H_4{-}(CH_2)_mSiMe_2H$ using Speier's catalyst.[196] An optically active 1-hydro-3-vinyl-disiloxane was synthesized and polymerized *via* hydrosilation to afford an optically active isotactic poly(silethylenesiloxane) [Eqs. (36)].[197] A poly(silethylene) with a chiral silicon center has also been prepared.[198] In this case, an optically active hydrosilane was used to hydrosilate acetylene for the synthesis of the vinyl monomer. Paulasaari and Weber reported the preparation of a highly regular poly(1-hydrido-1,3,3,5,5-pentamethyltrisiloxane) and its substitution *via* hydrosilation using either a Ru-carbonyl- or Karstedt's complex.[199]

$$CH_2{=}CH{-}Si(Ph)(Np){-}OMen \xrightarrow[60\,^{\circ}C,\ 5\,h,\ toluene]{excess\ KOH} CH_2{=}CH{-}Si(Ph)(Np){-}OK$$

$$\xrightarrow[0\,^{\circ}C,\ xylene]{ClMe_2SiH} CH_2{=}CH{-}Si(Ph)(Np){-}O{-}Si{-}H \xrightarrow[80\,^{\circ}C]{Karstedt's\ Pt} \left[{-}CH_2{-}CH_2{-}Si^{\star}(Ph)(Np){-}O{-}Si{-}\right]_n \quad (36)$$

X
MECHANISTIC STUDIES

For a reaction that has been widely practiced for over a half century, development of a clear understanding of the catalytic cycle that is the engine behind its scientific and commercial utility has been difficult and slow. This has not been due to lack of insight or effort, but simply because characterization and isolation of the extremely reactive and fleeting catalytic intermediates, with the highly efficient catalytic metals, have evaded numerous attempts and foiled strategies. In order to convey a historical perspective that provides connectivity between earlier work and more recent developments in this area, salient research prior to 1990 is discussed in brief, with greater coverage of the significant and rapid progress that occurred during the 1990s and is continuing into the new century. Excellent, more in-depth coverage of mechanistic work up to ca. 1998 is provided in the Ph.D. thesis of Chan.[200]

Not long after transition-metal catalyzed hydrosilation took root in the 1950s as a new and powerful tool for Si–C bond formation, based on parallels drawn from TM-catalyzed hydrogenation and on the intrinsic redox properties of these metals, Chalk and Harrod proposed a simple and yet very logical catalytic cycle for TM-catalyzed hydrosilation (Scheme 1).[201] The key steps of this cycle are: (1) oxidative addition of Si–H to the metal center, (2) coordination of the olefin or other unsaturated species to the metal, (3) insertion of this usually-η^2-coordinated molecule into the M–H bond and finally, (4) reductive elimination of the Si–C pair to regenerate the catalytically active metal center.

This fundamentally sound model of the catalytic cycle continues to be widely accepted, but its simplistic picture cannot explain some common observations and the formation of certain products in hydrosilations catalyzed by several transition metals other than platinum. The most difficult to explain and accommodate *via* the Chalk–Harrod mechanism are formation of vinylsilanes, highly-colored solutions often associated with gray–black metal precipitation, and frequently an induction period before the catalytic cycle kicks off in earnest. The deeper question of whether the catalysis occurs homogeneously or on the surface of colloidal metal has been an additional mystery that has shrouded the catalysis, especially in light of the fact that

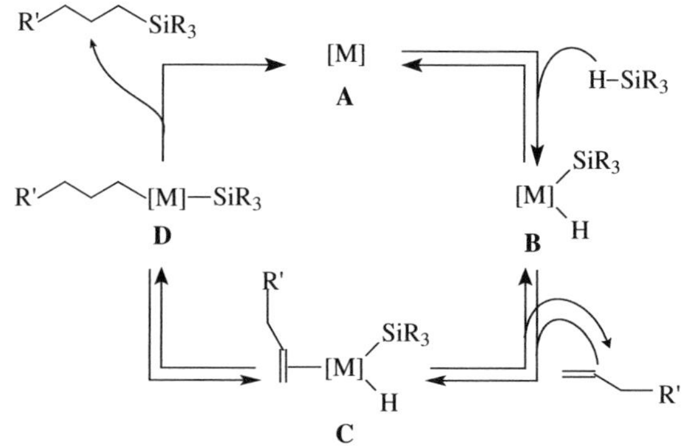

SCHEME 1. The Chalk–Harrod mechanism.

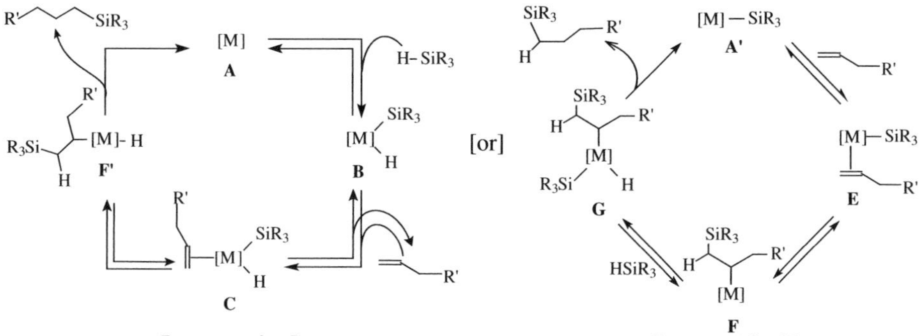

SCHEME 2. The modified Chalk–Harrod mechanism.

many transition metals in finely divided form (on a support) are excellent or at least effective catalysts for the reaction.

Since the advent of the Chalk–Harrod model, many excellent mechanistic studies have been put forth that examined detailed kinetics and/or intermediate characterization, most often using the less active transition metals to analyze and identify intermediate species. Perhaps the most salient of the earlier probing studies are those of Wrighton, Seitz–Wrighton, Perutz, Brookhart and their coworkers. These brought to light an important mechanistic feature of the catalysis – that the sequence of the basic steps of the cycle might partially determine what products and byproducts are formed. Based on the formation of alkylsilane, vinylsilane and alkane in investigations with $Fe(CO)_5$ as the precatalyst, Wright and coworkers first proposed the concept that a modification of the Chalk–Harrod cycle to allow C=C insertion into the M–Si bond to form a M–C–C–Si species, followed by Si–H oxidative addition and reductive elimination of alkane from H–M–C–C–Si could explain vinylsilanes formation.[202] This original notion forms the basis for the Modified Chalk–Harrod Mechanism (Scheme 2) where intermediates C or E could lead to olefin isomerization and intermediates *F* or *F′*could lead to vinylsilane and alkane formation *via* β-hydride transfer mechanisms outside the catalytic cycle. For vinylsilane formation, this author is of the opinion that a β-proton transfer or a concerted process rather than a β-hydride shift is more likely, since silicon is better capable of stabilizing an anion or transient negative charge at an adjacent carbon.

Wrighton's mechanistic scheme is well-supported by later research: *via* Seitz and Wrighton's further work using $[Co(SiEt_3)(CO)_4]$ as precatalyst,[203] where direct insertion of ethylene into Co–Si was evidenced, *via* Perutz and Duckett's proposed "Two Silicon Cycle" with $[CpRh(H)(SiR_3)(C_2H_4)]$ as the precatalyst,[204] and from Brookhart and Grant's detailed kinetic and spectroscopic studies using $[Cp^*Co(CH_2CH_2\text{-}\mu\text{-}H)\{P(OMe)_3\}]$ as precatalyst.[205] Detection in the Brookhart study of a catalytic resting state and direct insertion product of olefin into the Co–Si bond *via* low-temperature NMR spectroscopy, together with deuterium-labeling and kinetic experiments unequivocally established the modified Chalk–Harrod pathway to be operative in the systems studied. Further corroborating evidence was produced by Brookhart's group in 1997, *via* use of two cationic Pd precatalysts, $[Pd(Me)(phen)L]^+BAr_4$ (phen = 1,10-phenanthroline; Ar = 3,5-$(CF_3)_2$–C_6H_3; L = Et_2O or $Me_3SiC{\equiv}CSiMe_3$).[206] Low-temperature NMR and deuterium-labeling studies showed alkene–Pd–Si and H–Pd–Si_2 species as reactive intermediates with the former leading to alkene insertion into Pd–Si, which was the rate determining step in the catalytic cycle. Murai *et al.* have also produced evidence from Co_2CO_8-catalyzed hydrosilation of oxygen-containing olefins that supports olefin insertion into M–Si.[207] Thus, the total body of evidence for a modified Chalk–Harrod mechanism followed by a number of Group VIII metals (except Pt) in hydrosilation catalysis is overwhelming.

In 1998 Ozawa *et al.* examined the hydrosilation of 1-(trimethylsilyl)-1-butene-3-yne using $[RuHCl(CO)(PPh_3)_3]$ as precatalyst and concluded from their results that both the Chalk–Harrod pathway and its modified version were tenable with this precatalyst.[208] Further studies indicated that the steric environment of the Ru center dictates the course of the preferred catalytic pathway.[209] Over the last 10 years Ozawa and coworkers have investigated the catalytic aspects of hydrosilation in great and fine detail. The group's work, particularly with respect to the last step of the catalytic cycle, namely, the reductive elimination, from a number of angles of structure and kinetics, has been of enormous value to a fuller understanding of the composite of steps in the catalytic cycle.[210] Exclusive C–Si reductive elimination from Pt was observed by Milstein's group and attributed to probable *trans* influence of ancillary ligands.[211]

The large body of mechanistic work with the less active catalytic transition metals has been instrumental to broadening understanding of not only hydrosilation but also the closely related and important reaction dehydrogenative silylation. However, little progress was made in unraveling the catalytic pathway traversed by the most commonly employed platinum catalysts until 1986 when Lewis and Lewis proposed that colloidal Pt formed *via* reduction by the hydrosilane reactant was the active catalytic species in hydrosilation.[212] This was a stirring conclusion, backed also by further studies with colloidal Rh, Pt and other metals,[213] that hurled enormous implications at homogeneous catalysis. Lewis and coworkers' original conclusions were based on lightscattering, TEM and ESCA for characterization and the inhibitory effect of metallic mercury on hydrosilation catalyzed by colloidal metal. However, later more refined studies by Lewis's group, using Karstedt's catalyst as well as H_2PtCl_6 led to their conclusion that most likely mononuclear and soluble complexes are the catalytic species, and that Pt–Pt and Pt–Si bonded insoluble species form at the end of hydrosilation when Pt-stabilizing olefins are absent.[214] In this author's opinion, formed during the last few years from reviewing the many excellent publications that have appeared on mechanistic studies, Lewis's latter studies most likely represent the truest general picture of how the catalysis proceeds and ends. As will be evident from the brief reference below to emerging nanoparticles as catalysts, the line between heterogeneous and homogeneous catalysis has begun to blur, and that there are elements of truth in Lewis's conclusions about colloidal catalysts that may well be applicable beyond hydrosilation catalysis. Some very recent work in other areas has shown convincingly the ability of one precatalyst to facilitate reactions homogeneously in one case and heterogeneously in another.[215,216]

One of the first studies that actually corroborated Lewis's revised conclusions was conducted by Osborn *et al.* who reported that Pt catalysts containing an e-withdrawing olefin such as methylnaphthoquinone **13**, were not only much more active than Karstedt's catalyst, but were more stable and catalyzed hydrosilation in a homogeneous manner.[68] Since then, others have made similar observations with these types of Pt catalysts.[69,71] Thus, contrary to the long-held belief that electron-withdrawing olefins retard catalysis (and, indeed they do in "inhibited" commercial one-part-cure formulations at lower temperatures, with *lower* being the operative word here), a picture was emerging that stabilizing the metal to aggregation with ancillary ligands of just the right coordination strength is conducive to higher activity and homogeneous reaction.

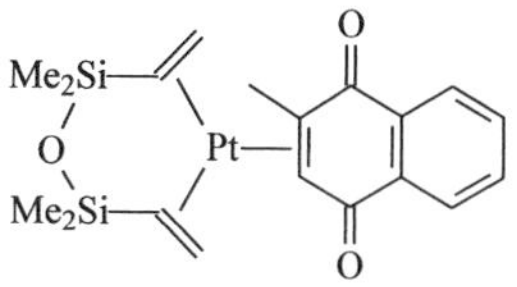

13

As mentioned earlier in Section IV, in 2002 Roy and Taylor reported the first identification and characterization of a true Pt catalytic species.[76] With COD (1,5-cyclooctadiene) as a simple dienic stabilizing ligand, a series of long-lived precatalytic intermediates **55** and stable catalytic species **19** representing the Pt-catalyst resting-state were identified and characterized in solution or *via* isolation at room temperature. The simple bis(silyl)PtCOD complexes were found to be highly active catalysts for both COD isomerization (in the absence of hydrosilatable olefins) and hydrosilation of 1-hexene and styrene. Not only could the same charge of catalyst be reused, but the one based on $SiCl_3$ ligands (which was normally an insoluble solid, but dissolved in a hydrosilation medium), precipitated out at the end of the reaction, redissolved to catalyze a fresh charge of olefin and silane and reprecipitated at the end of the second run. Since all hydrosilation with these catalysts were found to remain fully homogeneous (light yellow solutions) throughout the reaction, the "insoluble" member of this series points toward prospects of homogeneous Pt hydrosilation catalysts that can be separated from products by filtration and recycled.

55

R based on Cl, Me, Et, OSi, Ph

19

R_3 = Me/Cl_2 (**19a**), Cl_3, Me_2/Cl, Me_2/Ph

With catalysts **19** three direct observations bore powerful implications for the catalytic cycle with Pt. First, the catalysts themselves (with different silyl groups on Pt) were "living" proof that olefin insertion into the Pt–Si bond is not facile. The second observation was that a number of hydrosilatable olefins by themselves (no silane present) showed no interaction with the catalysts at room temperature; therefore, COD is not uniquely resistant to Pt–Si insertion. Third, during the preparative steps to **19**, which involved both COD and silane, only isomerization and hydrogenation of COD (implicating intermediate cyclooct-4-enyl-Pt species, i.e., alkyl–Pt intermediates) were observed, but no hydrosilation. That meant olefin insertion into Pt–H was more facile than into Pt–Si. These observations when taken together clearly supported the proposed chemistry of the insertive and reductive elimination steps of the Chalk–Harrod catalytic cycle. Accordingly, a clearer picture of the Chalk–Harrod mechanism with at least Pt(II) complexes emerged. The elaborated Chalk–Harrod cycle proposed by Roy and Taylor (abbreviated as CHaRT) is shown in Scheme 3. It is of interest to note, and probably not

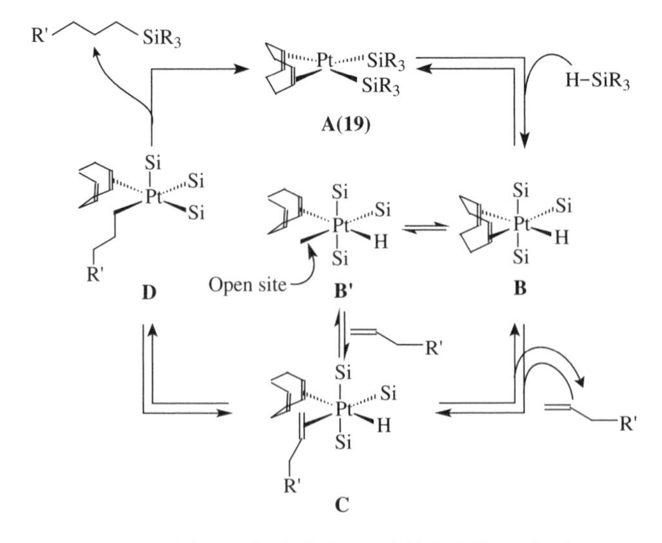

SCHEME 3. Elaborated Chalk–Harrod (CHaRT) mechanism.

coincidental, that the resting states proposed by Perutz and by Brookhart for the modified Chalk–Harrod mechanism and some reactive intermediates are Rh–Si and Co–Si/Pd–Si and bis(silyl)-metal species, respectively, and these are represented by Roy and Taylor's bis(silyl)-Pt catalyst for the Chalk–Harrod cycle.

Thus, it appears that for Pt, two redox paths, 0$\leftrightarrow$II and II$\leftrightarrow$IV, are equally capable of sustaining a catalytic cycle, since numerous Pt^0 precatalysts (including the supported forms) mediate the reaction well and for these a simple change in oxidation state between zero and two suffices to accomplish the addition of silane to an unsaturation.

For over a decade now, Tilley's group has methodically synthesized and studied the reactivity of a large number of metal–silylene complexes. Naturally, they examined the possibility of M$=$Si intermediacy in hydrosilation catalysis. In 1998, Mitchell and Tilley reported a reversible 1,2-migration of hydrogen between Pt and ligated Si that indicated the possibility of a Pt–silylene (Pt$=$Si) intermediate.[217] Later, Tilley, Bergman and Klei showed that the silylene complex $[Ir(H)(Me_5C_5)(SiPh_2)(PMe_3)]^+B(C_6F_5)_4^-$ acts as a precatalyst for the hydrosilation of acetophenone, but its bulkier relative based on Pt$=$$Si(Mes)_2$ was inert likely because of the severe congestion around the metal.[54] When, a metal–silylene based on the more catalytically active metal Ru was employed as precatalyst, Glaser and Tilley observed an unusually selective and fairly general hydrosilation of a number of olefins with primary silanes and primary silanes only.[218] The sole (anti-Markovnikov) product was a secondary silane; no side product or further hydrosilation by the secondary silane was observed. This work represents the first cases of facile olefin hydrosilation using a metal–silylene complex. The mechanism would be expected to be novel as well, and the authors have proposed a catalytic cycle that incorporates earlier observations of the M$=$Si 1,2 hydrogen shift and a concerted reaction between the olefin and the silylene-derived Si–H (Scheme 4).

Unlike Group VIII metals, Group IIIB and lanthanide/actinide metals cannot expand or contract their coordination sphere *via* redox reactions in any general way that is useful to catalysis. It is widely agreed that catalysts based on these metals mediate hydrosilation *via* σ-bond metathesis-type reactions. Detailed early mechanistic studies by Marks's and by Molander's groups and various observations by many research groups studying hydrosilation with these metals, strongly support the mechanism depicted in Scheme 5. The catalytic cycle is analogous to that of hydrogenation with these metal centers, although there are small differences. There has been no direct evidence for olefin or alkyne insertion into a M–Si bond; hence, the resting state is most likely a M–H species for which both monomeric and H-bridged dimeric species have been proposed. Olefin or alkyne insertion into M–H is fast and irreversible, while the M–C/H–Si metathetic reaction is slow and rate determining, with regioselectivity consistent with the C–C multiple bond polarity and steric interactions in the transition state that also regenerates the M–H catalytic intermediate.

Almost in parallel with experimental work during the 1990s and now in the new century, theoretical investigations of the mechanism of hydrosilation have contributed very significantly to advancing understanding of not only reactive intermediates that might be involved but also the energy changes associated with

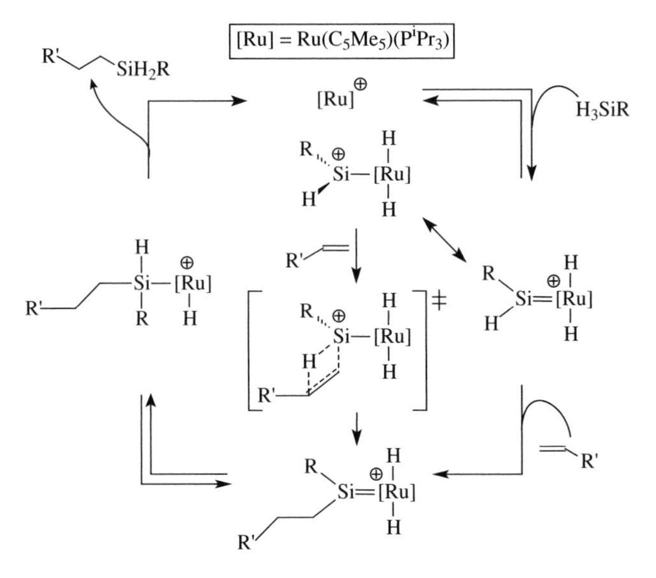

SCHEME 4. Proposed Ru–silylene-based catalytic cycle.

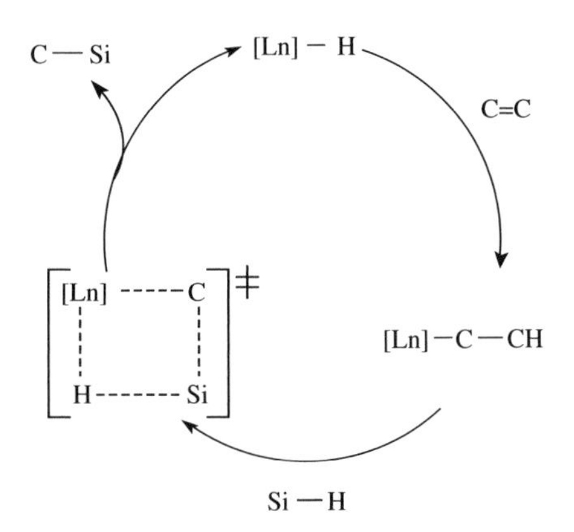

SCHEME 5. d^0-metal center catalytic cycle.

activations and transition states. The most extensive computational effort, and perhaps the most valuable, has been mounted by Sakaki and coworkers but others have also contributed to unraveling aspects of the catalytic cycle in specific areas.

The results of the extensive *ab initio* calculations performed by Sakaki's group[219] using MO/MP2–MP4(SDQ) and CCD methods, for Pt-catalyzed Si–H addition to ethylene, can be summarized as follows: (1) Oxidative addition of Si–H to Pt is facile as is ethylene coordination. (2) The rate determining step for a Chalk–Harrod pathway is the coordinated-ethylene-assisted *cis–trans* isomerization of the Si–Pt–H complex, with a barrier of ca. 22–26 kcal/mol. The isomerization is proposed to occur through a distorted square-pyramidal intermediate *via* the Berry pseudo-rotation process. (3) For a modified Chalk–Harrod mechanism, the rate-determining step is the insertion of ethylene into the Pt–Si bond, with a barrier of ca. 44–60 kcal/mol depending on the nature of the substituent on silicon. However, this barrier is lower if a *cis–trans* isomerization also occurs here prior to ethylene insertion into Pt–Si. (4) Taking into consideration both isomerization as well as reductive elimination activation energies (where Si–C elimination is lower in energy than C–H), the Chalk–Harrod mechanism is predicted to be the favored pathway for Pt-catalyzed hydrosilation.

Sgamelloti, Giorgi and coworkers recently employed DFT methods to investigate Pt-catalyzed hydrosilation of ethylene.[220] Their results, with computed barriers of 25.1 and 38.2 kcal/mol for the ethylene insertion–isomerization step for the Chalk–Harrod catalytic cycle and its modified version, respectively, also support the former being the preferred pathway for catalysis by Pt.

Interestingly, while present experimental evidence for Pd-catalyzed hydrosilation favors the modified Chalk–Harrod mechanism, first-principles and hybrid QM/MM molecular dynamics simulations for an enantioselective Pd-catalyzed hydrosilation of styrene suggest the Chalk–Harrod path as the operative mechanism.[221] For Rh-catalyzed hydrosilation, however, calculations by Sakaki *et al.* show a clear preference for the modified Chalk–Harrod pathway.[222] The difference arises from much lower activation energy for ethylene insertion into Rh–Si and a higher Si–C elimination barrier compared to Pt–Si. Trost, Wu and their coworkers have used DFT methods to investigate alkyne hydrosilation using cationic CpRu complexes.[223] From this effort, they deduce that neither the Chalk–Harrod nor the modified mechanism directly explains the "abnormal" regio-and stereochemical aspects of adduct formation. The authors find that a hydride transfer is favored overall. But the oxidative addition of silane and hydride transfer are concerted, and a stereospecific rotation of the C(1)–C(2) bond occurs following a Markovnikov hydride transfer that together is able to explain the experimental *anti*-addition and internal adduct outcome.

TiH_2-catalyzed ethylene hydrosilation was modeled *via* several *ab initio* methods by Gordon and coworkers.[224] The reaction was found to be a barrierless transformation (because of the highly exothermic formation of titanacyclopropane initially, which readily reacts with silane). The net reaction of SiH addition across ethylene is found exothermic by 28 kcal/mol. Very recently, Sakaki *et al.* examined hydrosilation of ethylene using Cp_2Zr as catalyst and DFT and MP2–MP4(SDQ) computational methods.[225] They concluded that the most favorable pathway to the adduct is a coupling reaction of $CpZr(C_2H_4)$ with SiH_4 followed by an ethylene-assisted C–H reductive elimination.

Mechanisms involved in asymmetric induction during hydrosilation have been reviewed by Pavlov[226] and ligand effects in asymmetric hydrosilation of acetophenone using $[Rh(Cl)(COD)]_2$-chiral diphosphine were examined recently by Giering, Prock and Reyes.[227] The mechanism of cyclization of *o*-allylstyrene *via* hydrosilation was explored by Hayashi and coworkers.[228] In a study that examined the effect of adjacent SiH units in poly(methylhydrosiloxane-*co*-dimethylsiloxane) on catalysis by Pt, Sauvet and coworkers showed that an SiH that has another SiH unit as immediate neighbor is much more reactive than an isolated SiH unit on the backbone.[229] The first instance of a strong regioselectivity effect of oxygen, in addition to rate acceleration effect, was observed by Frejd and Gustafsson for the hydrosilation of 2-substituted 1,3-dienes using Rh complex catalysts.[230] The 1-adduct was favored in the presence of oxygen and the 4-adduct favored when oxygen was excluded during catalysis.

It was alluded to above that the operational differences between heterogeneous and homogeneous catalysis may be narrowing. Several reports have appeared in the last few years describing the preparation of metal nanoparticles in a variety of liquid media, including polymers, ionic liquids and even water. For instance, Chauhan and Latif have prepared spherical Ag and Pd nanoparticles ca. 5 nm in diameter in copolymers of polyols and cylic- or cage-siloxanes.[231] Rhodium nanoparticles of average 3 nm diameter were synthesized within a copolymer of *N*-vinyl-2-pyrrolidone and 1-vinyl-3-alkylimidazolium salt.[232] Gold, palladium, Fe_2O_3 and CdSe/ZnS particles roughly 5 nm in diameter were transferred to water (without change in particle characteristics) *via* use of a phosphine oxide gel (based on polyethylene glycol).[233] Water-soluble Pt nanoclusters have been prepared using PEG-substituted T_8 cage siloxanes.[234] Pd nanoclusters with as few as seven Pd atoms were found to be excellent catalysts for important organic reactions.[235] Pt–Pd bimetallic "core-shell" colloids with nanometer Pt shell efficiently catalyzed the hydrosilation of 1-octene.[236] Platinum nanoclusters, 1–3 nm in diameter have also been synthesized *via* reduction of Pt(II) with poly(methylhydrosiloxane).[237] The Pt particles were shown to be excellent hydrosilation catalysts for a variety of olefin types, exhibiting high selectivity for β-addition for aliphatic substrates and good tolerance for oxygen containing substrates such as epoxyolefins and unsaturated ketones. High recyclability was also demonstrated. Although a "homogeneity/heterogeneity test" was not reported for the Pt clusters, catalysis by the above-mentioned Rh particles was inhibited by mercury, thus pointing toward a heterogeneous mode for the latter.

It is apparent that this developing preparative and efficient catalytic chemistry of metal nanoparticles that are only a few nm in diameter is likely to lead to even smaller species that approach metal-cluster complexes with clearly identifiable ligands, many of which are soluble compounds. Thus, the mechanistic question of homogeneous or heterogeneous at that point will become moot or nearly irrelevant, except for the specific advantageous properties of each type of catalyst that are of practical utility and for studying metal–metal species near atomic dimensions.

XI

CHALLENGES AND THE FUTURE OF HYDROSILATION

Any area of catalysis, no matter how old in practice or how far advanced has its "wish list" for improvements, either to overcome existing shortcomings or to meet developing and future requirements. For hydrosilation, the two primary areas that would benefit from significant advances, in this author's opinion, are: (1) development of commercially viable non-precious metal catalysts in the longer term, but more efficient use of existing catalysts based on Pt and Rh in particular in the near future, including homogeneous catalyst recyclability. (2) Synthesis of catalysts with higher selectivity for the desired product, especially for many commercial hydrosilations that generate wasteful byproducts and/or require higher levels of catalyst loading to offset activity loss *via* one or more of several inactivation pathways. Precious metal catalysts are extremely expensive with diminishing reserves, and the cost for waste disposal can be a significant fraction of production cost.

The research covered in this review has revealed remarkable creativity of the silicon-smitten chemist to design and develop many new ligands, at least some of which may well be suitable for practical, large-scale preparations of better catalysts. Asymmetric hydrosilation in particular leads the field in this area. It is also apparent that more catalyst development effort has gone into metals other than Pt, and this is certainly a welcome sign. The use of $AlCl_3$ as catalyst for high-selectivity hydrosilation, despite some current drawbacks, shows significant promise for the further development of Lewis acidic catalysts in general. It is possibly now established beyond doubt that for metals that catalyze hydrosilation *via* a redox cycle, stabilization of the metal center to reduction to metal^0 and subsequent aggregation is essential to maintaining activity. The classes of olefins that promote good back-donation from the metal (especially Pt) appear to have the right balance of π-bonding strength and π-acceptor ability to extend catalyst life. These should help guide the development of other ligands that, in addition to this feature, also add better solubility and resistance to hydrosilation themselves as useful properties. On the other hand, metals in higher oxidation state, such as Re(V), that are not susceptible to detrimental reduction, or those such as the lanthanides that do not rely on a redox process for catalysis, are also candidates for further development to produce general, highly active catalysts.

One of the present challenges, of particular relevance to commercial hydrosilation, is side reactions that lead to olefin isomerization and the formation of byproducts derived from the olefin. The cost associated with these nuisance reactions is huge, both in terms of raw material waste as well as waste-disposal expenses, not to mention production capacity loss and sometimes detrimental effects on the catalysis itself. All of these in turn drive up the price of the products affected, in an industry (chemical) that has become highly competitive.

Some of the specific unsaturated substrates suffering from the above shortcomings, which also leverage many commercial applications as well as various synthetic transformations useful to research, are allyl chloride, allyl alcohol derivatives, and a number of compounds with a electron-withdrawing group at the allylic position. Also handicapped are many 1,3-dienes, and other dienes where the desirable terminal olefinic product either cannot be obtained (butadiene, for instance), or suffers partial isomerization to the internal olefin which then is not useful for the targeted application.

To illustrate the serious problem, a case in point is the hydrosilation of allyl chloride which produces equimolar quantities of propene and a chlorosilane as the primary byproducts [Eq. (37)]. The propene can undergo hydrosilation itself leading to a secondary byproduct. Some of the best yields of 3-chloropropylsilane lie in the range 70–80% with a few combinations of hydrosilane and catalyst, but for many useful hydrosilanes, the primary byproducts can be 60% or higher.

$$CH_2{=}CH{-}CH_2Cl + R_3SiH \xrightarrow[M\,=\,Pt,\ Rh,\ etc.]{[M]\ catalyst} \underbrace{R_3SiCH_2CH_2CH_2Cl}_{product} + \underbrace{CH_2{=}CHCH_3 + R_3SiCl}_{byproducts} \quad (37)$$

$$CH_2{=}CHCH_3 \xrightarrow{R_3SiH} R_3SiCH_2CH_2CH_3 \ (byproduct)$$

R = alkyl, aryl, halide, alkoxy, acyloxy, siloxy

In an effort to understand rogue allylic substrate behavior in hydrosilation, Roy, Taylor and Naasz undertook a designed study that included hydrosilation of several allyl-chloride-like allylic-chloro-substituted olefin substrates and a deuterium labeling experiment on a specifically chosen olefin, crotyl chloride.[238] The results strongly indicated that electronic effects drive byproduct formation in olefinic substrates with a electron-withdrawing group at the allylic position. Unlike many ordinary olefins, such as 1-hexene, which exhibit extensive deuterium scrambling along the chain, crotyl chloride showed clean deuterium incorporation only at specific positions (Scheme 6). This, together with another later study (unpublished) that confirmed the allylic rearrangement (from clear chloride transfer to Pt), is highly indicative of a "Pt-assisted" and concerted allylic rearrangement reaction during hydrosilation.

This facile pathway is likely responsible for not only the byproduct(s) formation in allyl chloride hydrosilation but also for all related isomerizations, and probably for olefin isomerization during hydrosilation in general (given that deuterium scrambles readily along the chain). This finding has similarity of principle to the mechanistic suggestion forwarded by Boudjouk for exclusive β-selectivity in Cu-catalyzed hydrosilation of methyl acrylate described earlier.[105] Since allylic rearrangement is a facile and intrinsic process, its suppression certainly poses a formidable challenge in hydrosilation. A catalysis solution would be expected to bring enormous benefits to the science and technology of hydrosilation, not just for

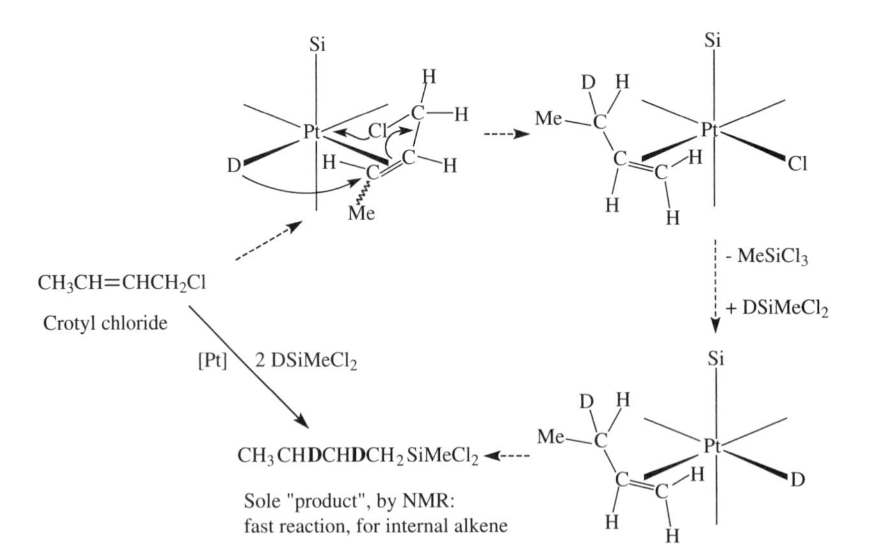

SCHEME 6. Proposed S_N2' concerted path for byproduct formation.

today, but also for the future with the ability to hydrosilate a number of substrates that could provide much more economical access to several newer intermediates and polymers. Perhaps solutions to some of these problems lie in modified versions of some catalysts that have shown high regio-and stereoselectivity in studies covered in this review.

A handful of new catalysts covered in this review have shown promise toward varied functional-group tolerance in hydrosilation. This is an area that is becoming increasingly important in synthetic chemistry, as multistep reactions leading to complex molecules (often of biological importance) are sought to be performed *via* several catalyzed steps. Hydrosilation catalysts that are tolerant of highly basic oxygen- and nitrogen-containing functionalities during addition to C–C multiple bonds would provide distinct advantages.

The last 15 years have seen great strides made in multiple areas of hydrosilation chemistry, as is evident from the research covered in this review and others. Because of the increasing importance of both small molecule and polymeric silicon compounds in commerce as well as in scientific areas such as in biological fields and emerging materials science research, accelerated scientific and technological growth of catalyzed hydrosilation in the future is clearly foreseeable.

Acknowledgments

There is no doubt of some published research that I missed, and I apologize for that to the authors. I am deeply indebted to Ms. Karen Oye of Public Relations with Case Western Reserve University Library, Cleveland, Ohio, for her permission to access articles electronically – without her generosity and understanding this review simply could not have been written by me as an industrial scientist. My son, Ishan Roy, searched for and copied a large number of papers from the CWRU library, often at a few hours' notice. I am glad that he enjoys chemistry. Librarians Rhonda Kidner and Lynne Stamm at Lubrizol Corporation and Joseph Hecht at Noveon were extremely helpful with document and interlibrary book loan requests and an initial literature search. Last but not least, Professor West's enormous patience with me on multiple delays cannot be acknowledged enough. It has been a special privilege working with him to report on this endearing area of science and technology for me.

References

(1) Ojima, I.; Li, Z.; Zhu, J. (Z. Rappoport, Y. Apeloig Eds.), *The Chemistry of Organic Silicon Compounds* Volume 2, Part 2 **1998**, Wiley, New York, pp. 1687–1792.
(2) Ojima, I. (S. Patai, Z. Rappoport, Eds.), *The Chemistry of Organic Silicon Compounds* Volume 1, Part 2 **1989**, Wiley, New York, pp. 1479–1526.
(3) Marciniec, B., Ed., *Comprehensive Handbook on Hydrosilylation*, Pergamon, Oxford, 1992.
(4) Speier, J. L. (F. G. A. Stone, R. West, Eds.), *Advances in Organometallic Chemistry* Volume 17, **1979**, Academic Press, New York, pp. 407–447.
(5) Lukevics, E.; Belyakova, Z. V.; Pomerantseva, M. G.; Voronkov, M. G. (D. Seyferth, A. G. Davies, E. O. Fisher, J. F. Normant, O. A. Reutov, Eds.), *Organometallic Chemistry Review*, Journal of Organometallic Chemistry Library 5, Elsevier, Amsterdam, **1977**, pp. 1–179.
(6) Harrod, J. F.; Chalk, A. J. (I. Wender, P. Pino, Eds.), *Organic Synthesis via Metal Carbonyls*, Wiley, New York, **1977**, pp. 673–704.
(7) Brook, M. A. *Silicon in Organic, Organometallic and Polymer Chemistry*, Wiley-Interscience, Toronto, **2000**, p. 401.
(8) Marciniec, B. *Silicon Chem.* **2002**, *1*, 155.
(9) Bart, S. C.; Lobkovsky, E.; Chirik, P. J. *J. Am. Chem. Soc.* **2004**, *126*, 13794.
(10) Na, Y.; Chang, S. *Org. Lett.* **2000**, *2*, 1887.

(11) Katayama, H.; Taniguchi, K.; Kobayashi, M.; Sagawa, T.; Minami, T.; Ozawa, F. *J. Organomet. Chem.* **2002**, *645*, 192.
(12) (a) Martin, M.; Sola, E.; Lahoz, F. J.; Oro, L. A. *Organometallics* **2002**, *21*, 4027. (b) Esteruelas, M. A.; Herrero, J.; Oro, L. A. *Organometallics* **1993**, *12*, 2377.
(13) Yardy, N. M.; Lemke, F. R. *Main Group Chem.* **2000**, *3*, 143.
(14) (a) Trost, B. M.; Machacek, M. R.; Ball, Z. T. *Org. Lett.* **2003**, *5*, 1895. (b) Trost, B. M.; Ball, Z. T. *J. Am. Chem. Soc.* **2001**, *123*, 12726.
(15) Aricó, C. S.; Cox, L. R. *Org. Biomol. Chem.* **2004**, *2*, 2558.
(16) Tanaka, M.; Hayashi, T.; Mi, Z.-Y. *J. Mol. Catal.* **1993**, *81*, 207.
(17) Hashimoto, H.; Aratani, I.; Kabuto, C.; Kira, M. *Organometallics* **2003**, *22*, 2199.
(18) Igarashi, M.; Mizuno, R.; Fuchikami, T. *Tet. Lett.* **2001**, *42*, 2149.
(19) Esteruelas, M. A.; Oro, l. A.; Valero, C. *Organometallics* **1991**, *10*, 462.
(20) Esteruelas, M. A.; López, A. M.; Oro, L. A.; Tolosa, J. I. *J. Mol. Catal.* **1995**, *96*, 21.
(21) Hilt, G.; Lüers, S.; Schmidt, F. *Synthesis* **2003**, 634.
(22) (a) Ojima, I.; Clos, N.; Donovan, R. J.; Ingallina, P. *Organometallics* **1990**, *9*, 3127. (b) Ojima, I.; Donovan, R. J.; Clos, N. *Organometallics* **1991**, *10*, 2606.
(23) Mori, A.; Takahasi, E.; Nishihara, Y.; Hiyama, T. *Can. J. Chem.* **2001**, *79*, 1522.
(24) Comte, V.; Le Gendre, P.; Richard, P.; Moïse, C. *Organometallics* **2005**, *24*, 1439.
(25) (a) Takeuchi, R.; Ebata, I. *Organometallics* **1997**, *16*, 3707. (b) Takeuchi, R.; Nitta, S.; Watanabe, D. *J. Org. Chem.* **1995**, *60*, 3045. (c) Takeuchi, R.; Tanouchi, N. *J. Chem. Soc. Perkin Trans.* **1994**, 2909. (d) Takeuchi, R.; Tanouchi, N. *J. Chem. Soc. Chem. Commun.* **1993**, 1319.
(26) Basato, M.; Biffis, A.; Martinati, G.; Zecca, M.; Ganis, P.; Benetollo, F.; Aronica, L. A.; Caporusso, A. M. *Organometallics* **2004**, *23*, 1947.
(27) Faller, J. W.; D'Alliessi, D. G. *Organometallics* **2002**, *21*, 1743.
(28) Baruah, J. B.; Osakada, K.; Yamamoto, T. *J. Mol. Catal. A* **1995**, *101*, 17.
(29) Betley, T. A.; Peters, J. C. *Angew. Chem. Int. Ed.* **2003**, *42*, 2385.
(30) Shibata, T.; Kadowaki, S.; Takagi, K. *Organometallics* **2004**, *23*, 4116.
(31) Frería, M.; Whitehead, A. J.; Tocher, D. A.; Motherwell, W. B. *Tetrahedron* **2004**, *60*, 2673.
(32) Park, K. H.; Kim, S. Y.; Son, S. U.; Chung, Y. K. *Eur. J. Org. Chem.* **2003**, 4341.
(33) Liu, C.; Widenhoefer, R. A. *Organometallics* **2002**, *21*, 5666.
(34) (a) Muraoka, T.; Matsuda, I.; Itoh, K. *Organometallics* **2002**, *21*, 3650. (b) Muraoka, T.; Matsuda, I.; Itoh, K. *Tet. Lett.* **1998**, *39*, 7325.
(35) (a) Ojima, I.; Vu, A. T.; Lee, S.-Y.; McCullagh, J. V.; Moralee, A. C.; Fujiwara, M.; Hoang, T. H. *J. Am. Chem. Soc.* **2002**, *124*, 9164. (b) Ojima, I.; Lee, S.-Y. *J. Am. Chem. Soc.* **2000**, *122*, 2385. (c) Ojima, I.; Vu, A. T.; McCullagh, J. V.; Kinoshita, A. *J. Am. Chem. Soc.* **1999**, *121*, 3230. (d) Ojima, I.; Zhu, J.; Vidal, E. S.; Kass, D. F. *J. Am. Chem. Soc.* **1998**, *120*, 6690. (e) Ojima, I.; Donovan, R. J.; Shay, W. R. *J. Am. Chem. Soc.* **1992**, *114*, 6580.
(36) Tamao, K.; Kobayashi, K.; Ito, Y. *Synlett.* **1992**, 539.
(37) Tius, M.; Pal, S. K. *Tet. Lett.* **2001**, *42*, 2605.
(38) Mori, A.; Kato, T. *Synlett.* **2002**, 1167.
(39) Rivera, G.; Crabtree, R. H. *J. Mol. Catal. A* **2004**, 59.
(40) Hewitt, G. W.; Somers, J. J.; Sieburth, S. M. c. N. *Tet. Lett.* **2000**, *41*, 10175.
(41) Niyomura, O.; Tokunaga, M.; Obora, Y.; Iwasawa, T.; Tsuji, Y. *Angew Chem. Int. Ed.* **2003**, *42*, 1287.
(42) (a) Marciniec, B.; Błażejewska-Chadyniak, P.; Kubicki, M. *Can. J. Chem.* **2003**, *81*, 1292. (b) Marciniec, B.; Krzyżanowski, P.; Walczuk-Guściora, E.; Duczmal, W. *J. Mol. Catal. A* **1999**, *144*, 263.
(43) Ganicz, T.; Mizerska, U.; Moszner, M.; O'Brien, M.; Perry, R.; Stańczyk, W. A. *Appl. Catal.* **2004**, *259*, 49.
(44) Dinh, L. V.; Gladysz, J. A. *Tet. Lett.* **1999**, *40*, 8995.
(45) de Wolf, E.; Speets, E. A.; Deelman, B.-J.; van Koten, G. *Organometallics* **2001**, *20*, 3686.
(46) van den Broeke, J.; Winter, F.; Deelman, B.-J.; van Koten, G. *Org. Lett.* **2002**, *4*, 3851.
(47) Dioumaev, V.; Bullock, R. M. *Nature* **2000**, *424*, 530.
(48) Behr, A.; Toslu, N. *Chem. Eng. Technol.* **2000**, *23*, 122.

(49) Roy, A. K.; Taylor, R. B. Partly published results from another study which is discussed in the section on Mechanism.
(50) Tanke, R. S.; Crabtree, R. H. *J. Am. Chem. Soc.* **1990**, *112*, 7984.
(51) Doherty, S.; Knight, J. G.; Scanlan, T. H.; Elsegood, M. R. J.; Clegg, W. *Organomet. Chem.* **2002**, *650*, 231.
(52) Esteruelas, M.; Oliván, M.; Oro, L. A.; Tolosa, J. I. *J. Organomet. Chem.* **1995**, *487*, 143.
(53) Martín, M.; Sola, E.; Torres, O.; Pablo, P.; Oro, L. A. *Organometallics* **2003**, *22*, 5406.
(54) Klei, S. R.; Tilley, T. D.; Bergman, R. G. *Organometallics* **2002**, *21*, 4648.
(55) Field, L. D.; Messerle, B. A.; Wren, S. L. *Organometallics* **2003**, *22*, 4393.
(56) (a) Groux, L. F.; Zargarian, D. *Organometallics* **2003**, *22*, 4759. (b) Fontaine, F.-G.; Nguyen, R.-V.; Zargarian, D. *Can. J. Chem.* **2003**, *81*, 1299. (c) Chen, Y.; Sui-Seng, C.; Boucher, S.; Zargarian, D. *Organometallics* **2005**, *24*, 149.
(57) Bartik, T.; Nagy, G.; Kvintovics, P.; Happ, B. *J. Organomet. Chem.* **1993**, *453*, 29.
(58) Tillack, A.; Pulst, S.; Baumann, W.; Baudisch, H.; Kortus, K.; Rosenthal, U. *J. Organomet. Chem.* **1997**, *532*, 117.
(59) Maciejewski, H.; Marciniec, B.; Kownacki, I. *J. Organomet. Chem.* **2000**, *597*, 175.
(60) Guliński, J.; James, B. R. *J. Mol. Catal.* **1992**, *72*, 167.
(61) Motoda, D.; Shinokubo, H.; Oshima, K. *Synlett.* **2002**, 1529.
(62) (a) Widenhoefer, R. A. *Acc. Chem. Res.* **2002**, *35*, 905. (b) Widenhoefer, R. A.; Vadehra, A. *Tet. Lett.* **1999**, *40*, 8499.
(63) Sabourault, N.; Mignani, G.; Wagner, A.; Mioskowski, C. *Org. Lett.* **2002**, *4*, 2117.
(64) Lewis, L. N.; Sumpter, C. A. *J. Mol. Catal. A* **1996**, *104*, 293.
(65) Kishi, K.; Ishimaru, T.; Ozono, M.; Tomita, I.; Endo, T. *Macromolecules* **1998**, *31*, 9392.
(66) Denmark, S. E.; Wang, Z. *Org. Lett.* **2001**, *3*, 1073.
(67) Perales, J. B.; Vranken, D. L. *J. Org. Chem.* **2001**, *66*, 7270.
(68) Steffanut, P.; Osborn, J. A.; DeCian, A.; Fisher, J. *Chem. Eur. J.* **1998**, *4*, 2008.
(69) Hopf, A.; Dötz, K. H. *J. Mol. Catal. A* **2000**, *164*, 191.
(70) Yamamoto, Y.; Ohno, T.; Itoh, K. *Organometallics* **2003**, *22*, 2267.
(71) Sprengers, J. W.; Agerbeek, M. J.; Elsevier, C. J. *Organometallics* **2004**, *23*, 3117.
(72) Markó, I. E.; Stérin, S.; Buisine, O.; Mignani, G.; Branlard, P.; Tinant, B.; Declercq, J.-P. *Science* **2002**, *298*, 204.
(73) Sprengers, J. W.; Mars, M. J.; Duin, M. A.; Cavell, K. J.; Elsevier, C. J. *J. Organomet. Chem.* **2003**, *679*, 149.
(74) (a) Nishihara, Y.; Itazaki, M.; Osakada, K. *Tet. Lett.* **2002**, *43*, 2059. (b) Itazaki, M.; Nishihara, Y.; Osakada, K. *J. Org. Chem.* **2002**, *67*, 6889.
(75) Maruyama, Y.; Yoshiuchi, K.; Ozawa, F. *J. Organomet. Chem.* **2000**, *609*, 130.
(76) Roy, A. K.; Taylor, R. B. *J. Am. Chem. Soc.* **2002**, *124*, 9510.
(77) Tsukada, N.; Hartwig, J. F. *J. Am. Chem. Soc.* **2005**, *127*, 5022.
(78) Corey, J. Y.; Zhu, X.-H. *Organometallics* **1992**, *11*, 672.
(79) Kesti, M. R.; Waymouth, R. M. *Organometallics* **1992**, *11*, 1095.
(80) Ura, Y.; Gao, G.; Bao, F.; Ogasawara, M.; Takahashi, T. *Organometallics* **2004**, *23*, 4804.
(81) Takahashi, T.; Bao, F.; Gao, G.; Ogasawara, M. *Org. Lett.* **2003**, *5*, 3479.
(82) Jia, L.; Zhao, J.; Ding, E.; Brennessel, W. W. *J. Chem. Soc. Dalton Trans.* **2002**, 2608.
(83) Hao, L.; Harrod, J. F.; Lebuis, A.-M.; Mu, Y.; Shu, R.; Samuel, E.; Woo, H.-G. *Angew. Chem. Int. Ed.* **1998**, *37*, 3126.
(84) Shu, R.; Harrod, J. F.; Lebuis, A.-M. *Can. J. Chem.* **2002**, *80*, 489.
(85) Beletskaya, I. P.; Voskoboynikov, A. Z.; Parshina, I. N.; Magomedov, G. K.-I. *Bull. Acad. Sci. USSR Div. Chem. Sci.* **1990**, *39*, 613.
(86) Sakakura, T.; Lautenschlager, H.-J.; Tanaka, M. *J. Chem. Soc. Chem. Commun.* **1991**, 40.
(87) Molander, G. A.; Romero, J. A. C. *Chem. Rev.* **2002**, *102*, 2161.
(88) Molander, G. A.; Julius, M. *J. Org. Chem.* **1992**, *57*, 6347.
(89) Molander, G. A.; Romero, J. A. C.; Corrette, C. P. *J. Organomet. Chem.* **2002**, *647*, 225.
(90) Molander, G. A.; Retsch, W. H. *Organometallics* **1995**, *14*, 4570.
(91) Fu, P.-F.; Brard, L.; Li, Y.; Marks, T. J. *J. Am. Chem. Soc.* **1995**, *117*, 7157.

(92) Koo, K.; Fu, P.-F.; Marks, T. J. *Macromolecules* **1999**, *32*, 981.
(93) Onozawa, S.-Y.; Sakakura, T.; Tanaka, M. *Tet. Lett.* **1994**, *35*, 8177.
(94) Voskoboynikov, A. Z.; Shestakova, A. K.; Beletskaya, I. P. *Organometallics* **2001**, *20*, 2794.
(95) Trifonov, A.; Spaniol, T. P.; Okuda, J. *Organometallics* **2001**, *20*, 4869.
(96) Tardif, O.; Nishiura, M.; Hou, Z. *Tetrahedron* **2003**, *59*, 10525.
(97) Horino, Y.; Livinghouse, T. *Organometallics* **2004**, *23*, 12.
(98) (a) Takaki, K.; Komeyama, K.; Takehira, K. *Tetrahedron* **2003**, *59*, 10381. (b) Takaki, K.; Sonoda, K.; Kousaka, T.; Shishido, T.; Takehira, K. *Tet. Lett.* **2001**, *42*, 9211.
(99) (a) Dash, A. K.; Gurevizt, Y.; Wang, J. Q.; Kapon, M.; Eisen, M. S. *J. Alloy Compounds* **2002**, *344*, 65. (b) Dash, A. K.; Gourevich, I.; Wang, J. Q.; Wang, J.; Kapon, M.; Eisen, M. S. *Organometallics* **2001**, *20*, 5084.
(100) Dash, A. K.; Wang, J. Q.; Eisen, M. S. *Organometallics* **1999**, *18*, 4724.
(101) Mao, Z.; Gregg, B. T.; Cutler, A. R. *J. Am Chem. Soc.* **1995**, *117*, 10139.
(102) Cavanaugh, M. D.; Gregg, B. T.; Cutler, A. R. *Organometallics* **1996**, *15*, 2764 and references therein on earlier work by this group in the 1990s.
(103) Kennedy-Smith, J. J.; Nolin, K. A.; Gunterman, H. P.; Toste, F. D. *J. Am. Chem. Soc.* **2003**, *125*, 4056.
(104) Thiel, W. R. *Angew. Chem. Int. Ed.* **2003**, *42*, 5390.
(105) Boudjouk, P.; Kloos, S.; Rajkumar, A. B. *J. Organomet. Chem.* **1993**, *443*, C41.
(106) Díez-González, S.; Kaur, H.; Zinn, F. K.; Stevens, E. D.; Nolan, S. P. *J. Org. Chem.* **2005**, *70*, 4784.
(107) Fujita, M.; Fukuzumi, S.; Otera, J. *J. Mol. Catal.* **1993**, *85*, 143.
(108) (a) Asao, N.; Sudo, T.; Yamamoto, Y. *J. Org. Chem.* **1996**, *61*, 7654. (b) Sudo, T.; Asao, N.; Gevorgyan, V.; Yamamoto, Y. *J. Org. Chem.* **1999**, *64*, 2494.
(109) Song, Y.-S.; Yoo, B. R.; Lee, G.-H.; Jung, I. N. *Organometallics* **1999**, *18*, 3109.
(110) Rubin, M.; Schwier, T.; Gevorgyan, V. *J. Org. Chem.* **2002**, *67*, 1936.
(111) Harrison, D. J.; McDonald, R.; Rosenberg, L. *Organometallics* **2005**, *24*, 1398.
(112) Liu, Y.; Yamazaki, S.; Yamabe, S. *J. Org. Chem.* **2005**, *70*, 556.
(113) Kawanami, Y.; Yuasa, H.; Toriyama, F.; Yoshida, S.-I.; Baba, T. *Catal. Comm.* **2003**, *4*, 455.
(114) Boardman, L. D. *Organometallics* **1992**, *11*, 4194.
(115) Lewis, F. D.; Salvi, G. D. *Inorg. Chem.* **1995**, *34*, 3182.
(116) Fry, B. E.; Neckers, D. C. *Macromolecules* **1996**, *29*, 5306.
(117) Stefanac, T. M.; Brook, M. A.; Stan, R. *Macromolecules* **1996**, *29*, 4549.
(118) Mayer, T.; Burget, D.; Mignani, G.; Fouassier, J. P. *J. Polym. Sci. A* **1996**, *34*, 3141.
(119) Felföldi, K.; Szőri, K.; Török, B.; Bartók, M. *Ultrasonics Sonochem* **2000**, *7*, 15.
(120) (a) Nishiyama, H.; Itoh, K. (I. Ojima, Ed.), *Catalytic Asymmetric Synthesis* Chap. 2, **2000**, Wiley-VCH, New York, pp. 111–143. (b) Hayashi, T. (E. N. Jacobsen; A. Pfaltz; H. Yamamoto, Eds.), *Comprehensive Asymmetric Catalysis*, **1999**, Volume 1, Springer, Berlin, Chap. 7. (c) Yamamoto, K.; Hayashi, T. (M. Beller; C. Bolm, Eds.), *Transition Metals for Organic Synthesis: Building Blocks and Fine Chemicals*, **1998**, Wiley-VCH, Weinheim. (d) Hayashi, T. *Acc. Chem. Res.* **2000**, *33*, 354.
(121) Hayashi, T. *Catal. Today* **2000**, *62*, 3.
(122) (a) Uozumi, Y.; Kitayama, K.; Hayashi, T.; Yanagi, K.; Fukuyo, E. *Bull. Chem. Soc. Jpn.* **1995**, *68*, 713. (b) Uozumi, Y.; Hayashi, T. *J. Am. Chem. Soc.* **1991**, *113*, 9887.
(123) Uozumi, Y.; Lee, S.-Y.; Hayashi, T. *Tet. Lett.* **1992**, *33*, 7185.
(124) (a) Han, J. W.; Hayashi, T. *Tet. Asymm.* **2002**, *13*, 325. (b) Han, J. W.; Hayashi, T. *Chem. Lett.* **2001**, 977.
(125) Shimada, T.; Mukaide, K.; Shinohara, A.; Han, J. W.; Hayashi, T. *J. Am Chem. Soc.* **2002**, *124*, 1584.
(126) Pioda, G.; Togni, A. *Tet. Asymm.* **1998**, *9*, 3903.
(127) Gustafsson, M.; Bergqvist, K.-E.; Frejd, T. *J. Chem. Soc. Perkin Trans.* **2001**, *1*, 1452.
(128) (a) Perch, N. S.; Widenhoefer, R. A. *J. Am. Chem. Soc.* **1999**, *121*, 6960. (b) Perch, N. S.; Pei, T.; Widenhoefer, R. A. *J. Org. Chem.* **2000**, *65*, 3836.
(129) Jensen, J. F.; Svendsen, B. Y.; la Cour, T. V.; Pedersen, H. L.; Johanssen, M. *J. Am. Chem. Soc.* **2002**, *124*, 4558.

(130) Dotta, P.; Kumar, P. G. A.; Pregosin, P. S.; Albinati, A.; Rizzato, S. *Organometallics* **2004**, *23*, 2295.
(131) Dotta, P.; Kumar, P. G. A.; Pregosin, P. S. *Organometallics* **2004**, *23*, 4247.
(132) Beagley, P.; Davies, P. J.; Blacker, A. J.; White, C. *Organometallics* **2002**, *21*, 5852.
(133) Lipshutz, B. H.; Lower, A.; Noson, K. *Org. Lett.* **2002**, *4*, 4045.
(134) Duan, W.-L.; Shi, M.; Rong. G.-B. *Chem. Commun.* **2003**, 2916.
(135) Chianese, A. R.; Crabtree, R. H. *Organometallics* **2005**, *24*, 4432.
(136) Gade, L. H.; César, V.; Bellemin-Laponnaz, S. *Angew. Chem. Int. Ed.* **2004**, *43*, 1014.
(137) Kromm, K.; Osburn, P. L.; Gladysz, J. A. *Organometallics* **2002**, *21*, 4275.
(138) Kuwano, R.; Uemura, T.; Saitoh, M.; Ito, Y. *Tet. Asymm.* **2004**, *15*, 2263.
(139) Heldmann, D. K.; Seebach, D. *Helv. Chim. Acta* **1999**, *82*, 1096.
(140) Wright, M. E.; Svejda, S.; Jin, M.-J.; Peterson, M. *Organometallics* **1990**, *9*, 136.
(141) Arena, C. G.; Pattacini, R. *J. Mol. Catal. A* **2004**, *222*, 47.
(142) Ini, S.; Oliver, A. G.; Tilley, T. D.; Bergman, R. G. *Organometallics* **2001**, *20*, 3839.
(143) Zurawinski, R.; Donnadieu, B.; Mikolajczyk, M.; Chauvin, R. *Organometallics* **2003**, *22*, 4810.
(144) Karamé, I.; Tommasino, L.; Lemaire, M. *J. Mol. Catal. A* **2003**, *196*, 137.
(145) Gavrilov, K. N.; Bondarev, O. G.; Lbedev, R. V.; Shiryaev, A. A.; Lyubimov, S. E.; Polosikhin, A. I.; Grintselev-Knyazev, G. V.; Lyssenko, K. A.; Moiseev, S. K.; Ikonnokov, N. S.; Kalinin, V. N.; Davankov, V. A.; Korostylev, A. V.; Gais, H.-J. *Eur. J. Inorg. Chem.* **2002**, 1367.
(146) Nolin, K. A.; Ahn, R. W.; Toste, F. D. *J. Am. Chem. Soc.* **2004**, *127*, 12462.
(147) Lipshutz, B. H.; Shimizu, H. *Angew. Chem. Int. Ed.* **2004**, *43*, 2228.
(148) Lipshutz, B. H.; Servesko, J. M.; Taft, B. R. *J. Am. Chem. Soc.* **2004**, *126*, 8352.
(149) Takei, I.; Nishibayashi, Y.; Mizobe, Y.; Uemura, S.; Hidai, M. *Chem. Commun.* **2001**, 2360.
(150) Yao, S.; Meng, J.-C.; Siuzdak, G.; Finn, M. G. *J. Org. Chem.* **2003**, *68*, 2540.
(151) Pang, Y.; Ijadi-Maghsoodi, S.; Barton, T. J. *Macromolecules* **1993**, *26*, 5671.
(152) Kim, D. S.; Shim, S. C. *J. Polym. Sci. A. Polym. Chem.* **1999**, *37*, 2263.
(153) Kim, D. S.; Shim, S. C. *J. Polym. Sci. A. Polym. Chem.* **1999**, *37*, 2933.
(154) (a) Mori, A.; Takahisa, E.; Yamamura, Y.; Kato, T.; Mudalige, A. P.; Kajiro, H.; Hirabayashi, K.; Nishihara, Y.; Hiyama, T. *Organometallics* **2004**, *23*, 1755. (b) Mori, A.; Takahisa, E.; Kajiro, H.; Nishihara, Y. *Macromolecules* **2000**, *33*, 1115.
(155) Yamashita, H.; de Leon, M. S.; Channasanon, S.; Suzuki, Y.; Uchimaru, Y.; Takeuchi, K. *Polymer* **2003**, *44*, 7089.
(156) Wang, F.; Kaafarani, B. R.; Neckers, D. C. *Macromolecules* **2003**, *36*, 8225.
(157) Lee, T.; Jung, I.; Song, K. H.; Baik, C.; Kim, S.; Kim, D.; Kang, S. O.; Ko, J. *Organometallics* **2004**, *23*, 4184.
(158) (a) Jain, R.; Lalancette, R. A.; Sheridan, J. B. *Organometallics* **2005**, *24*, 1458. (b) Jain, R.; Choi, H.; Lalancette, R. A.; Sheridan, J. B. *Organometallics* **2005**, *24*, 1468.
(159) Coles, H. J.; Meyer, S.; Lehmann, P.; Deshenaux, R.; Jauslin, I. *J. Mater. Chem.* **1999**, *9*, 1085.
(160) Pan, J.; Lau, W. W. Y.; Lee, C. S. *J. Polym. Sci. Part A: Polym Chem.* **1994**, *32*, 997.
(161) Riedel, J.-H.; Höcker, H. *J. Appl. Polym. Sci.* **1994**, *51*, 573.
(162) (a) Yang, C. Y.; Wnek, G. E. *Polymer* **1992**, *22*, 4191. (b) Zhu, Z.; Rider, J.; Yang, C. Y.; Gilmartin, M. E.; Wnek, G. E. *Macromolecules* **1992**, *25*, 7330.
(163) Liu, J.-K.; Wnek, G. E. *Macromolecules* **1994**, *27*, 4080.
(164) Ogawa, T. *J. Polym. Sci. Part A: Polym. Chem.* **2003**, *41*, 3336.
(165) Henkensmeier, D.; Abele, B. C.; Candussio, A.; Thiem, J. *Macromol. Chem. Phys.* **2004**, *205*, 1851.
(166) Castelvetro, V.; Galli, G.; Ciardelli, F.; Chiellini, E. *J. M. S.- Pure Appl. Chem.* **1997**, A34, 2205.
(167) Chien, G.-P. C.; Kuo, J.-F.; Chen, C.-Y. *J. Polym. Sci. A Polym. Chem.* **1993**, *31*, 2423.
(168) Gunji, T.; Ozawa, M.; Abe, Y. *J. Sol–Gel Sci. Tech.* **2001**, *22*, 219.
(169) Zhou, G.; Smid, J. *J. Polym. Sci. A: Polym. Chem.* **1991**, *29*, 1097.
(170) Sun, J.; Tang, H.; Jiang, J.; Zhou, X.; Xie, P.; Zhang, R.; Fu, P.-F. *J. Polym. Sci. A Polym. Chem.* **2003**, *41*, 636.
(171) Rossi, N. A. A.; Zhang, Z.; Wang, Q.; Amine, K.; West, R. *Polymer Preprints* **2005**, *46*, 723 and references therein.
(172) (a) Lee, J.; Kang, Y.; Suh, D. H.; Lee, C. *Electrochim. Acta* **2004**, *50*, 351. (b) Kang, Y.; Lee, W.; Suh, D. H.; Lee, C. *J. Power Sources* **2004**, *119–121*, 448.

(173) Noda, K.; Yasuda, T.; Nishi, Y. *Electrochim. Acta* **2004**, *50*, 243.
(174) (a) Kuo, P.-L.; Hou, S.-S.; Lin, C.-Y.; Chen, C.-C.; Wen, T.-C. *J. Polym. Sci. A: Polym Chem.* **2004**, *42*, 2051. (b) Liang, W.-J.; Wu, C.-P.; Kuo, P.-L. *J. Polym. Sci. B: Polym Phys.* **2004**, *42*, 1928. (c) Liang, W.-J.; Kuo, P.-L. *Polymer* **2004**, *45*, 1617. (d) Liang, W.-J.; Kuo, P.-L. *Macromolecules* **2004**, *37*, 840.
(175) (a) Rossi, N. A. A.; Zhang, Z.; Schneider, Y.; Morkom, K.; Lyons, L. J.; Wang, Q.; Amine, K.; West, R. *Chem. Mater.* **2006**, *18*, 1289. (b) Zhang, Z.; Rossi, N. A. A.; Olson, C.; Wang, Q.; Amine, K.; West, R. *Polymer Preprints* **2005**, *46*, 662.
(176) Zhu, Z.; Einset, A. G.; Yang, C.-Y.; Chen, W.-X.; Wnek, G. E. *Macromolecules* **1994**, *27*, 4076.
(177) Zhou, G.; Khan, I. M.; Smid, J. *Macromolecules* **1993**, *26*, 2202.
(178) Majoral, J.-P.; Caminade, A.-M. *Chem. Rev.* **1999**, *99*, 845.
(179) Kwak, G.; Takagi, A.; Fujiki, M.; Masuda, T. *Chem. Mater.* **2004**, *16*, 781.
(180) Zhang, C.; Babonneau, F.; Bonhomme, C.; Laine, R. M.; Soles, C. L.; Hristov, H. A.; Yee, A. F. *J. Am. Chem. Soc.* **1998**, *120*, 8380.
(181) Sellinger, A.; Laine, R. M.; Chu, V.; Viney, C. *J. Polym. Sci. A Polym. Chem.* **1994**, *32*, 3069.
(182) Laine, R. M.; Zhang, C.; Sellinger, A.; Viculis, L. *Appl. Organomet. Chem.* **1998**, *12*, 715.
(183) Kim, K.-M.; Chujo, Y. *Polym. Bull.* **2001**, *46*, 15.
(184) (a) Choi, J.; Yee, A. F.; Laine, R. M. *Macromolecules* **2003**, *36*, 5666 and references therein. (b) Laine, R. M.; Choi, J.; Lee, I. *Adv. Mater.* **2001**, *13*, 800.
(185) Liu, C.; Liu, Y.; Xie, P.; Zhang, R.; He, C.; Chung, N. T. *React. Func. Polym.* **2000**, *46*, 175.
(186) Mathias, L. J.; Lewis, C. M. *Macromolecules* **1993**, *26*, 4070.
(187) Tronc, F.; Lestel, L.; Boileau, S. *Polymer* **2000**, *41*, 5039.
(188) Ciolino, A. E.; Pieroni, O. I.; Vuano, B. M.; Villar, M. A.; Vallés, E. M. *J. Polym. Sci. A: Polym. Chem.* **2004**, *42*, 2920.
(189) Weber, W. P.; Sargent, J. R. *Macromolecules* **1999**, *32*, 2826.
(190) Son, D. Y.; Hu, J. *Macromolecules* **1998**, *31*, 4645.
(191) Kim, S. H.; Woo, H.-G.; Kim, S.-H.; Kim, J.-S.; Kang, H.-G.; Kim, W.-G. *Macromolecules* **1999**, *32*, 6363.
(192) Giurgiu, D.; Hamciuc, V.; Butuc, E.; Cozan, V.; Stoleriu, A.; Marcu, M.; Ionescu, C. *J. Appl. Polym. Sci.* **1996**, *59*, 1507.
(193) (a) Chung, P. H.; Crivello, J. V.; Fan, M. *J. Polym. Sci. A: Polym. Chem.* **1993**, *31*, 1741. (b) Crivello, J. V.; Bi, D. *J. Polym. Sci. A: Polym. Chem.* **1993**, *31*, 2563. (c) Crivello, J. V.; Bi, D. *J. Polym. Sci. A: Polym. Chem.* **1993**, *31*, 2729. (d) Crivello, J. V.; Bi, D. *J. Polym. Sci. A: Polym. Chem.* **1993**, *31*, 3109. (e) Crivello, J. V.; Bi, D. *J. Polym. Sci. A: Polym. Chem.* **1993**, *31*, 3121. (f) Crivello, J. V.; Bi, D. *J. Polym. Sci. A: Polym. Chem.* **1994**, *32*, 683.
(194) Tsumura, M.; Ando, K.; Kotani, J.; Hiraishi, M.; Iwahara, T. *Macromolecules* **1998**, *31*, 2716.
(195) Shankar, R.; Joshi, A. *Macromolecules* **2005**, *38*, 4176.
(196) Itsuno, S.; Chao, D.; Ito, K. *J. Polym. Sci. A Polym. Chem.* **1993**, *31*, 287.
(197) Li, Y.; Kawakami, Y. *Macromolecules* **1998**, *31*, 5592.
(198) Tang, B. Z.; Wan, X.; Kwok, H. S. *Eur. Polym. J.* **1998**, *34*, 341.
(199) Paulasaari, J. K.; Weber, W. P. *Macromolecules* **1999**, *32*, 6574.
(200) Chan, D. Ph. D. Dissertation, University of York, York, U. K., **1999**.
(201) Chalk, A. J.; Harrod, J. F. *J. Am. Chem. Soc.* **1965**, *87*, 16.
(202) (a) Mitchener, J. C.; Wrighton, M. S. *J. Am. Chem. Soc.* **1981**, *103*, 975. (b) Schroeder, M. A.; Wrighton, M. S. *J. Organomet. Chem.* **1977**, *128*, 345.
(203) Seitz, F.; Wrighton, M. S. *Angew Chem. Int. Ed.* **1988**, *27*, 289.
(204) Duckett, S. B.; Perutz, R. N. *Organometallics* **1992**, *11*, 90.
(205) Brookhart, M.; Grant, B. E. *J. Am. Chem. Soc.* **1993**, *115*, 2151.
(206) LaPointe, A. M.; Rix, F. C.; Brookhart, M. *J. Am. Chem. Soc.* **1997**, *119*, 906.
(207) Chatani, N.; Kodama, T.; Kajikawa, Y.; Murakami, H.; Kakiuchi, F.; Ikeda, S.-I.; Murai, S. *Chem. Lett.* **2000**, 14.
(208) Maruyama, Y.; Yamamura, K.; Nakayama, I.; Yoshiuchi, K.; Ozawa, F. *J. Am. Chem. Soc.* **1998**, *120*, 1421.
(209) Maruyama, Y.; Yamamura, K.; Sagawa, T.; Katayama, H.; Ozawa, F. *Organometallics* **2000**, *19*, 1308.

(210) (a) Ozawa, F.; Hikida, T.; Hayashi, T. *J. Am. Chem. Soc.* **1994**, *116*, 2844. (b) Ozawa, F.; Kamite, J. *Organometallics* **1998**, *17*, 5630. (c) Ozawa, F. *J. Organomet. Chem.* **2000**, *611*, 332. (d) Ozawa, F.; Mori, T. *Organometallics* **2003**, *22*, 3593. (e) Ozawa, F.; Tani, T.; Katayama, H. *Organometallics* **2005**, *24*, 2511.
(211) van der Boom, M. E.; Ott, J.; Milstein, D. *Organometallics* **1998**, *17*, 4263.
(212) Lewis, L. N.; Lewis, N. *J. Am. Chem. Soc.* **1986**, *108*, 7228.
(213) (a) Lewis, L. N.; Uriarte, R. J.; Lewis, N. *J. Mol. Catal.* **1991**, *66*, 105. (b) Lewis, L. N.; Uriarte, R. J. *Organometallics* **1990**, *9*, 621. (c) Lewis, L. N. *J. Am. Chem. Soc.* **1990**, *112*, 5998.
(214) (a) Stein, J.; Lewis, L. N.; Gao, Y.; Scott, R. A. *J. Am. Chem. Soc.* **1999**, *121*, 3693. (b) Lewis, L. N.; Stein, J.; Gao, Y.; Colborn, R. E. *Platinum Metals Rev.* **1997**, *41*, 66.
(215) Manners, I.; Jaska, C. A. *J. Am. Chem. Soc.* **2004**, *126*, 1334.
(216) Hagen, C. M.; Widegren, J. A.; Maitlis, P. M.; Finke, R. G. *J. Am. Chem. Soc.* **2005**, *127*, 4423.
(217) Mitchell, G. P.; Tilley, T. D. *J. Am. Chem. Soc.* **1998**, *120*, 7635.
(218) Glaser, P. B.; Tilley, T. D. *J. Am. Chem. Soc.* **2003**, *125*, 13640.
(219) (a) Sakaki, S.; Mizoe, N.; Sugimoto, M.; Musashi, Y. *Coord. Chem. Rev.* **1999**, *190–192*, 933. (b) Sakaki, S.; Mizoe, N.; Musashi, Y.; Sugimoto, M. *J. Mol. Struct. (Theochem.)* **1999**, *461–462*, 533. (c) Sakaki, S.; Mizoe, N.; Sugimoto, M. *Organometallics* **1998**, *17*, 2510. (d) Sakaki, S.; Ogawa, M.; Musashi, Y.; Arai, T. *J. Am. Chem. Soc.* **1994**, *116*, 7258.
(220) (a) Giorgi, G.; De Angelis, F.; Re, N.; Sgamellotti, A. *Future Generation Computer Systems* **2004**, *20*, 781. (b) Giorgi, G.; De Angelis, F.; Re, N.; Sgamellotti, A. *J. Mol. Struct. (Theochem.)* **2003**, *623*, 277. (c) Giorgi, G.; De Angelis, F.; Re, N.; Sgamellotti, A. *Chem. Phys. Lett.* **2002**, *364*, 87.
(221) Magistrato, A.; Woo, T. K.; Togni, A.; Rothlisberger, U. *Organometallics* **2004**, *23*, 3218.
(222) Sakai, S.; Sumimoto, M.; Fukuhara, M.; Sugimoto, M.; Fujimoto, H.; Matsuzaki, S. *Organometallics* **2002**, *21*, 3788.
(223) Chung, L. W.; Wu, Y.-D.; Trost, B. M.; Ball, Z. T. *J. Am. Chem. Soc.* **2003**, *125*, 11578.
(224) Bode, B. M.; Day, P. N.; Gordon, M. S. *J. Am. Chem. Soc.* **1998**, *120*, 1552.
(225) Sakaki, S.; Takayama, T.; Sumimoto, M.; Sugimoto, M. *J. Am. Chem. Soc.* **2004**, *126*, 3332.
(226) (a) Pavlov, V. A. *Russ. Chem. Rev.* **2002**, *71*, 33. (b) Pavlov, V. A. *Russ. Chem. Rev.* **2001**, *70*, 1037.
(227) Reyes, C.; Prock, A.; Giering, W. P. *J. Organomet. Chem.* **2003**, *671*, 13.
(228) Uozumi, Y.; Tsuji, H.; Hayashi, T. *J. Org. Chem.* **1998**, *63*, 6137.
(229) Cancouët, P.; Pernin, S.; Hélary, G.; Sauvet, G. *J. Polym. Sci. A Polym. Chem.* **2000**, *38*, 837.
(230) Gustafsson, M.; Frejd, T. *J. Chem. Soc. Perkin Trans.* **2002**, *1*, 102.
(231) Chauhan, B. P. S.; Latif, U. *Macromolecules* **2005**, *38*, 6231.
(232) Mu, X.-D.; Meng, J.-Q.; Li, Z.-C.; Kou, Y. *J. Am. Chem. Soc.* **2005**, *127*, 9694.
(233) Kim, S.-W.; Kim, S.; Tracy, J. B.; Jasanoff, A.; Bawendi, M. G. *J. Am. Chem. Soc.* **2005**, *127*, 4556.
(234) Huang, J.; He, C.; Liu, X.; Xiao, Y.; Mya, K. Y.; Chai, J. *Langmuir* **2004**, *20*, 5145.
(235) Okamoto, K.; Akiyama, R.; Yoshida, H.; Yoshida, T.; Kobayashi, S. *J. Am. Chem. Soc.* **2005**, *127*, 2125.
(236) Schmid, G.; West, H.; Mehles, H.; Lehnert, A. *Inorg. Chem.* **1997**, *36*, 891.
(237) Chauhan, B. P. S.; Rathore, J. S. *J. Am. Chem. Soc.* **2005**, *127*, 5790.
(238) Roy, A. K.; Taylor, R. B.; Naasz, B. M. Presented in part at the 29th Silicon Symposium, Evanston, IL, March **1996**.

Organotransition Metal Complexes for Nonlinear Optics

JOSEPH P. MORRALL,[a,b] GULLIVER T. DALTON,[a,b] MARK G. HUMPHREY[a,*] and MAREK SAMOC[b]

[a]*Department of Chemistry, Australian National University, Canberra, ACT 0200, Australia*
[b]*Laser Physics Centre, Research School of Physical Sciences and Engineering, Australian National University, Canberra, ACT 0200, Australia*

I
INTRODUCTION

Nonlinear optical (NLO) behavior arises from interaction of electromagnetic fields with matter. The consequent generation of new field components (differing in amplitude, phase, frequency, path, polarization, etc.) is of enormous technological importance for optical devices, with potential applications in data storage, communication, switching, image processing, and computing. These applications have generated a need for materials with exceptional NLO properties and satisfactory materials properties (processing, stability, etc.). In the last two decades, extensive investigations have been carried out into the NLO properties of organotransition metal complexes. Since our earlier reviews,[1,2] the field has expanded enormously. As a consequence, this review of the field to the end of 2005 is somewhat more selective, the focus being on the most popular area, namely molecular properties measured from solution studies. This survey excludes computational studies, measurements of bulk material susceptibilities, and reports of optical limiting investigations, an overview of which is available in our earlier reviews.

An understanding of the theory and experimental procedures that are involved is necessary to understand the outcomes from the NLO studies thus far. Only an abbreviated overview of the theory and procedures will be presented here due to the multitude of reviews on both of these topics that exist,[3–24] several reviews being of particular relevance in the field of organometallics for nonlinear optics.[1,2,8,12,21] Qualitatively, NLO effects are optical phenomena occurring when strong electrical

*Corresponding author. Tel.: +61261252927
E-mail: mark.humphrey@anu.edu.au (M.G. Humphrey).

ADVANCES IN ORGANOMETALLIC CHEMISTRY
VOLUME 55 ISSN 0065-3055/DOI 10.1016/S0065-3055(07)55002-1

fields, such as those present in high-intensity laser beams, interact with matter. The NLO effects are of extreme importance for photonic and biophotonic technologies. The characteristic feature of these effects is that at high electric field strengths, the response of matter is no longer linearly dependent on the magnitude of the field. In the case of a field containing several components, the response does not follow the superposition principle, that is, it is no longer a simple sum of responses due to the individual field components. When a light beam interacts with an NLO-active material (i.e. a material possessing certain NLO properties), the electromagnetic field components existing in the beam may be modified and new ones, e.g. at new frequencies or phase-shifted ones, may be created as the result. Intensity-dependent light refraction and absorption are simple manifestations of NLO properties. These effects may lead to a variety of phenomena, including formation of solitons, power limiting of light transmission, and multiphoton fluorescence. When more than one light beam is present, NLO effects lead to interactions between the beams that are usually referred to as wave mixing phenomena. All of these effects are of interest in areas of optics and photonics such as laser technologies, where the generation of coherent beams with new frequencies of light is needed, in optical signal processing, e.g. for the rapid modulation and demodulation of optical signals for fiber-optic communication, and for multiphoton information storage and retrieval, etc. Biological applications are also of considerable interest, e.g. NLO microscopy or multiphoton photodynamic therapy.

The technological requirements for devices and processes that involve NLO effects necessitate optimized NLO materials (with maximized appropriate nonlinear susceptibilities), in order to provide good performance at reduced light intensities. At the same time, such materials usually need other properties such as a high damage threshold and good photochemical and thermal stability (for materials that need to work inside a device in a photonic system) or a high fluorescence efficiency (for a dye intended for multiphoton microscopy). At present, crystals of inorganic salts (e.g. borates of barium and lithium, tantalates, KH_2PO_4, $LiNbO_3$) dominate in such NLO applications as electro-optic modulation and frequency mixing (which are examples of so-called second-order NLO effects). Glasses (e.g. silica, chalcogenides, fluorides) are most often considered as materials for third-order NLO applications involving refractive nonlinearities, and organic dyes or semiconductor nanoparticles are common candidates for applications involving two-photon absorption, formally also a third-order NLO process. All of these materials have their advantages and disadvantages. For example, the electronic NLO effects in ionic crystals of inorganic salts are accompanied by effects derived from the lattice distortion within the crystals. The dielectric properties of such materials involve the presence of large lattice contributions, which leads to large values of the dielectric constant. This influences the capability of such a material to be used for applications such as electro-optic modulation occurring at frequencies of ca. 100 GHz. On the other hand, inorganic ionic crystals possess significant advantages, e.g. a wide optical transparency range, and low optical losses; in addition, they are robust, and they can be grown as large single crystals and made into wafers for integrated optics fabrication. All of these are important considerations in device development.

The feasibility of obtaining enhanced optical nonlinearities from organic molecules and organometallic complexes, and thereby overcoming some of the limitations of inorganic materials, has been studied. The main advantage of organic molecules appears to be the possibility of obtaining large NLO responses of a primarily electronic character.[3,4,17–19] Since organic molecules form solids with van der Waals interactions, the materials containing them possess low dielectric constants, helping to alleviate the electro-optic modulation speed limit. An important benefit of organic molecules is the design flexibility that is available, and the relative ease with which they can be synthesized. Materials based on organic molecules are easy to process. In particular, it is easy to fabricate thin films of polymeric or hybrid glass materials incorporating organic chromophores, an important consideration for device applications. Organic molecules have several major drawbacks that do not allow them to be the "perfect" NLO material. One of these drawbacks is the trade-off in optical transparency vs. NLO efficiency: an increase of nonlinearity is usually accompanied by a decrease in the transmission window. Another shortcoming is the thermal and photochemical stability of the materials.

In proceeding to organometallics, one hopes to combine the advantages of organic molecules (design flexibility, fast response) with those of inorganic salts (robustness, thermal stability). The flexibility available for organometallics is significant: metal variation, ligand variation, coordination geometry, oxidation state, electron donating/withdrawing capabilities of the metal or ligand, and stabilization of unstable organic fragments can all, in principle, be modified to optimize targeted responses.[1,2,8,12,22,23]

II

THEORY AND TECHNIQUES

A. *Theory of Nonlinear Optics*

As indicated above, NLO properties of matter are manifested when strong electrical fields act upon the charge distribution inside a material. For materials such as organic or organometallic molecules, it is convenient to consider such an interaction at the molecular level and then derive the macroscopic NLO properties of a material through a simple summation of molecular effects, taking into account the need to correct the electrical field for effects due to the polarization of the molecules (local field correction) and considering various orientations of the molecules with respect to the direction of the external field. Such a procedure, called the oriented gas approach, is usually justified for liquids and solids containing organometallic molecules. On a molecular level, the optical properties can be considered in a dipole approximation. When a local electric field $\mathbf{E}_{loc}$ acts upon a molecule it distorts the molecule's electron density distribution $\rho(\mathbf{r})$. In the dipole approximation one limits the considerations to the changes in the dipole moment $\boldsymbol{\mu} = \int \mathbf{r}\rho(\mathbf{r})\,d\mathbf{r}$, although higher electron distribution moments (quadrupole, etc.) as well as magnetic moments may also be of significance in special cases (for example, when considering chiroptic properties of optically active molecules). When $|\mathbf{E}_{loc}|$ is weak (in comparison to internal electric fields inside a molecule), the dipole varies in a linear relationship to the external field. However, if the field is strong then the linear relationship does not suffice. It is usual to represent the dependence of the dipole moment on the electric field as a power series:

$$\boldsymbol{\mu} = \boldsymbol{\mu}_0 + \alpha \mathbf{E}_{\text{loc}} + \beta \mathbf{E}_{\text{loc}} \mathbf{E}_{\text{loc}} + \gamma \mathbf{E}_{\text{loc}} \mathbf{E}_{\text{loc}} \mathbf{E}_{\text{loc}} + \cdots \tag{1}$$

where $\boldsymbol{\mu}_0$ is the static dipole moment and α, β, and γ are the polarizabilities specific to the linear, second-order, and third-order interactions of the electric field with the molecule, respectively. The second-order polarizability is often referred to as the first hyperpolarizability or quadratic hyperpolarizability and, similarly, the third-order polarizability is the second hyperpolarizability or cubic hyperpolarizability. Since $\boldsymbol{\mu}$ and $\mathbf{E}_{\text{loc}}$ are vectors, the polarizabilities are tensors of appropriate ranks: α is a second-rank tensor (represented with a 3×3 matrix), β is a third-rank tensor ($3 \times 3 \times 3$ matrix), and γ is a fourth-rank tensor ($3 \times 3 \times 3 \times 3$ matrix). The large number of molecular tensor components can be reduced through the use of permutations,[24] and the requirement that polarizabilities be invariant with respect to all point group symmetry operations of a molecule. This is especially important for β, as it is found that all its components must vanish for centrosymmetric point groups. As will be discussed below, a knowledge of all the components of the molecular polarizabilities may not be needed in practice, because of orientational averaging.

Eq. (1) is an oversimplification, being only strictly correct for a static field. Optical fields are time-dependent, in the simplest case containing an oscillatory term at a frequency ω and the envelope function (amplitude) $\mathbf{E}_0$ which may be time-dependent: :

$$\mathbf{E}(t) = \mathbf{E}_0 \cos(\omega t) = \frac{\mathbf{E}_0}{2}[\exp(\mathrm{i}\omega t) + \exp(-\mathrm{i}\omega t)] \tag{2}$$

Substituting this into Eq. (1), one obtains:

$$\begin{aligned} \boldsymbol{\mu}(t) &= \boldsymbol{\mu}_0 + \alpha \mathbf{E}_0 \cos(\omega t) + \beta \mathbf{E}_0^2 \cos^2(\omega t) + \gamma \mathbf{E}_0^3 \cos^3(\omega t) + \cdots \\ &= \boldsymbol{\mu}_0 + \tfrac{1}{2}\alpha \mathbf{E}_0 \exp(\mathrm{i}\omega t) + \tfrac{1}{2}\beta \mathbf{E}_0^2 + \tfrac{1}{4}\beta \mathbf{E}_0^2 \exp(2\mathrm{i}\omega t) + \tfrac{3}{8}\gamma \mathbf{E}_0^3 \exp(\mathrm{i}\omega t) \\ &\quad + \tfrac{1}{8}\gamma \mathbf{E}_0^3 \exp(3\mathrm{i}\omega t) + \text{c.c.} + \cdots \end{aligned} \tag{3}$$

with c.c. denoting complex conjugate terms. It is evident from the above that the presence of higher-order terms leads to generation of new frequencies of molecular dipole oscillation. The "β term" causes the appearance of frequency doubling (2ω) (second-harmonic generation) while the "γ term" yields frequency tripling (3ω). In addition, the second-order term generates a time-independent contribution to the dipole (called "optical rectification") and there is also a cubic term, due to γ, oscillating at the frequency ω.

A plethora of NLO effects appears when one considers that electromagnetic fields can contain several components with different frequencies and also propagating in different directions in space. In general, a field acting on an NLO material may be expressed as:

$$\mathbf{E}(\mathbf{r}, t) = \sum_j \mathbf{E}_j(t) \exp[i(\omega_j t - \mathbf{k}_j \mathbf{r})] + \text{c.c.} \tag{4}$$

By substituting Eq. (4) into Eq. (1), it is easy to prove that the result of several field components acting on a nonlinear material is "wave mixing", i.e. generation of oscillating dipole contributions containing combinations of frequencies ω_j and wave vectors $\mathbf{k}_j$. Among the many combinations that are possible, that of an optical field oscillating at ω_1 with a static field (for which $\omega_2 = 0$) is of special importance. This combination leads to the electro-optic (Pockels) effect, which is a second-order (deriving from β) NLO effect where a static (or oscillating at a suboptical frequency) field can provide a phase-shift to an optical field and thus modulate it.

In general, Eq. (3) is only a good approximation of the behavior of the oscillating dipole if the frequency of the optical field is far from any material resonances. If this is not the case, one needs to take into account the damping of the dipole oscillation resulting in its change in amplitude and phase lag in respect to the field. This is accomplished by considering Fourier components of the field and the dipole oscillation at particular frequencies. To account for the damping, one needs to treat the linear and higher-order frequency-dependent polarizabilities as being complex (that is, composed of real and imaginary parts, both being frequency dependent). The linear and nonlinear Fourier components of the induced dipole are then written as:

$$\Delta\mu^{(1)}(\omega) = \alpha(\omega)E(\omega) \tag{5}$$

$$\Delta\mu^{(2)}(\omega_3) = \beta(-\omega_3; \omega_1, \omega_2)E(\omega_1)E(\omega_2) \tag{6}$$

$$\Delta\mu^{(3)}(\omega_4) = \gamma(-\omega_4; \omega_1, \omega_2, \omega_3)E(\omega_1)E(\omega_2)E(\omega_3) \tag{7}$$

(the vector and tensor nature of the quantities in these equations are not indicated explicitly, but it needs to be understood that frequency-dependent quantities preserve their tensor properties). For the linear component of the oscillation, $\Delta\mu^{(1)}(\omega)$, one considers the linear polarizability $\alpha(\omega)$, which is a complex quantity in which the real part is responsible for refractive properties of matter and the imaginary part is related to the absorption coefficient. The nonlinear terms require more complicated notation for the frequency dependence of the polarizabilities. The hyperpolarizabilities referring to different combinations of input frequencies are different, and one therefore needs to specify all the input frequencies explicitly, together with a sign denoting how a given frequency enters the combination giving the output frequency. For example, the hyperpolarizability responsible for the electro-optic effect is specified as $\beta(-\omega;\omega,0)$ while that for the second-harmonic generation process is $\beta(-2\omega;\omega,\omega)$. The important case of a degenerate cubic hyperpolarizability responsible for nonlinear refraction at a frequency ω is denoted as $\gamma(-\omega;\omega,-\omega,\omega)$.

Strong dispersion of the real parts of the polarizabilities and large values of the imaginary parts are expected when input frequencies or the output frequency (or, in some cases, certain combinations of the frequencies) approach resonance frequencies of the molecule. In particular, for the degenerate case of the cubic nonlinearity described by $\gamma(-\omega;\omega,-\omega,\omega)$, the imaginary part of γ is related to nonlinear absorption, often defined as two-photon absorption and expected to be significant for 2ω in the vicinity of molecular two-photon allowed transitions. It should be noted that two-photon absorption of molecules is often quantified using the so-called two-photon absorption cross-section σ_2, which is thus related to γ_{imag}.

Eqs. (1)–(7) deal with microscopic NLO effects. On a macroscopic scale, one needs to consider the macroscopic polarization **P** and its relation to the input field **E**. A macroscopic analogue of Eq. (1) uses the susceptibility tensors, χ, of corresponding orders, affording (in the cgs system of units):

$$\mathbf{P} = \chi^{(1)}\mathbf{E} + \chi^{(2)}\mathbf{EE} + \chi^{(3)}\mathbf{EEE} + \cdots \tag{8}$$

As is the case with molecular quantities, Fourier components of **E** and **P** are accompanied by frequency-dependent, complex susceptibilities χ. The macroscopic susceptibilities are used in the physical description of NLO effects, such effects typically being analyzed using wave equations in which the nonlinear polarization produced by a given type of interaction constitutes a source term. Quantities other than the susceptibilities are often used for describing specific NLO interactions. The most useful of these are the electro-optic coefficient r related to $\chi^{(2)}(-\omega;\omega,0)$, the nonlinear refractive index n_2, related to the real component of the degenerate third-order susceptibility $\text{Re}(\chi^{(3)}(-\omega;\omega,-\omega,\omega))$, and the two-photon absorption coefficient β_2, related to the corresponding imaginary component $\text{Im}(\chi^{(3)}(-\omega;\omega,-\omega,\omega))$.

The nonlinear susceptibilities in molecular solids and liquids can be calculated from the "oriented gas" approximation. For example, in the oriented gas approximation, $\chi^{(2)}$ for an organic crystal built of molecules possessing a molecular second-order hyperpolarizability β is given by:[24]

$$\chi^{(2)}_{IJK}(-\omega_3;\omega_1,\omega_2) = L_I(\omega_3)L_J(\omega_1)L_K(\omega_2)\sum_{t=1}^{p} N_t b^t_{IJK}(-\omega_3;\omega_1,\omega_2) \tag{9}$$

where L_I, L_J, and L_K are local field factors, approximated with the Lorenz–Lorentz expression:

$$L = \frac{n^2+2}{3} \tag{10}$$

where n is the refractive index. The summation in Eq. (9) is over the number of inequivalent positions of a molecule in a unit cell, p. The hyperpolarizability transformed to the unit cell coordinates, b_{IJK}, is given by:

$$b^t_{IJK}(-\omega_3;\omega_1,\omega_2) = \frac{1}{N_g}\sum_{ijk}\sum_{s=1}^{N_g} \cos\theta^{(s)}_{Ii_t}\,\cos\theta^{(s)}_{Jj_t}\,\cos\theta^{(s)}_{Kk_t}\beta_{ijk}(-\omega_3;\omega_1,\omega_2) \tag{11}$$

with *ijk* being the molecular coordinates, and *IJK* those of the unit cell for the crystal. N_t represents the number of molecules occupying inequivalent sites within a unit cell, and N_g is the number of equivalent sites.

The situation is simpler for random collections of molecules as in, for example, liquids or glasses. As mentioned above, because isotropic media possess a statistical center of symmetry, the second-order susceptibility $\chi^{(2)}$ vanishes. For the third-order susceptibility, only two tensor components, $\chi^{(3)}_{1111}$ and $\chi^{(3)}_{1122}$, are obtained as a result of orientational averaging. $\chi^{(3)}_{1111}$ can be related to the molecular hyperpolarizability as follows:

$$\chi^{(3)}_{1111}(-\omega_4;\omega_1,\omega_2,\omega_3) = L_{\omega_1}L_{\omega_2}L_{\omega_3}L_{\omega_4}\mathrm{N}\langle\gamma(-\omega_4;\omega_1,\omega_2,\omega_3)\rangle \qquad (12)$$

with L_ω being the Lorenz–Lorentz expression [Eq. (10)] and the orientationally averaged third-order polarizability being:

$$\langle\gamma\rangle = \tfrac{1}{5}(\gamma_{1111} + \gamma_{2222} + \gamma_{3333} + 2\gamma_{1122} + 2\gamma_{1133} + 2\gamma_{2233}) \qquad (13)$$

In the simplest case, there may be only one dominant component of γ, e.g. γ_{1111} in linear π-conjugated systems with the dominant hyperpolarizability along the molecular axis.

An important case is also that of a random system of molecules that has been "poled", that is, subjected to an external field resulting in the symmetry being broken. The macroscopic nonlinear properties of such a system can be calculated from a knowledge of the degree of orientation (the order parameter) and the properties of the NLO-active chromophores.

B. *Unit Systems*

Two unit systems are commonly employed in describing NLO properties: the SI (MKS) and the Gaussian (cgs) systems. One should note that many equations have a different form when written in these two systems of units, and that conversion between these two systems is frequently cumbersome.[1,2] One should also be aware that definitions of hyperpolarizabilities and susceptibilities may differ between different authors because of the lack of general agreement whether the complex field amplitude in Eq. (2) should include a factor of 1/2 and whether the multiplying degeneracy factors [such as those in Eq. (3)] should be included in the hyperpolarizabilities.

C. *Experimental Techniques for the Determination of NLO Properties of Organometallic Molecules*

A number of techniques have been used for the measurement of second- and third-order NLO properties of organometallic molecules, the most popular of which are described below. A more complete description of these and other techniques that are used for crystals, polymers, and other forms of material can be found in Ref. 28.

1. Electric Field-Induced Second Harmonic Generation

One of the methods of determining the second-order polarizability β of noncentrosymmetric molecules is by measuring the second harmonic of laser radiation generated by a system that has been made macroscopically noncentrosymmetric by application of an external electric field. The second harmonic can be obtained on solid samples (e.g. poled polymers), but the most common implementation of the technique called electric field-induced second-harmonic generation (EFISH) is that in which molecules are simply dissolved in common solvents. EFISH has been used to determine β for a variety of organometallic compounds (see later), and is widely used in the study of organic molecules. However, much of the work in the domain of organometallic complexes has shifted to the more widely applicable hyper-Rayleigh scattering (HRS) technique (see below).

Application of an electric field to a liquid solution results in two distinct effects that contribute to the generation of the second harmonic. Firstly, all materials possess a third-order nonlinearity. The particular component responsible for generation of the second-harmonic under the action of a dc field is described by $\chi^{(3)}(-2\omega;\omega,\omega,0)$. Even in the absence of asymmetry, as in, for example, pure CCl_4, there is a small EFISH effect due entirely to the cubic hyperpolarizability $\gamma(-2\omega;\omega,\omega,0)$. However, as a rule, a stronger contribution to second-harmonic generation is obtained if the molecules in the solution have a non-zero dipole moment and a non-zero second-order polarizability β. The field attempts to orient dipolar molecules along its direction while thermal motions act against the orientation, resulting in a partial orientation. It can be shown that the effective nonlinearity obtained under such conditions is then dependent upon the $\boldsymbol{\mu} \cdot \boldsymbol{\beta}_{\text{vec}}$ product, where $\boldsymbol{\mu}$ is the permanent dipole moment of the molecule and $\boldsymbol{\beta}_{\text{vec}}$ is the vectorial component of the second-order hyperpolarizability (the hyperpolarizability β is a symmetric third-rank tensor that can be treated as being composed of a vector and a septor part).[25] Since the directions of $\boldsymbol{\beta}_{\text{vec}}$ and of $\boldsymbol{\mu}$ are in general not coincident, the effective hyperpolarizability determined from the technique, β_{EFISH}, is given by $\boldsymbol{\mu} \cdot \boldsymbol{\beta}_{\text{vec}} = \mu\beta_{\text{EFISH}}$. For dipolar molecules containing strong electron donor and acceptor groups, β_{CT} (the hyperpolarizability along the charge-transfer axis) usually accounts for most of β_{EFISH}.

In the practical implementation of the EFISH experiment, the static electric field is usually applied in the form of a voltage pulse that precedes the laser pulse sufficiently to allow the dipoles to orient. The use of pulses instead of a constant electric field helps to avoid electric breakdown and electrolysis of the solution in an EFISH cell. Even so, highly conducting solutions such as those containing ionic species cannot be measured by this technique.

One of the complications in analyzing experiments in which a NLO interaction generates a new light beam is the fact that the newly generated wave (at a frequency 2ω for second-harmonic generation) usually travels at a speed different to that of the fundamental wave because of the wavelength dispersion of the refractive index. One defines a coherence length l_c as the length of the interaction at which the newly generated wave reaches a maximum amplitude. For sample lengths longer than l_c, the amplitude exhibits an oscillatory behavior as a function of the interaction length (the period of the oscillation being twice the coherence length of the interaction). For the case of EFISH, the coherence length is $l_c = \lambda/(n_{2\omega}-n_\omega)$, typically of the order of $10\,\mu m$. To avoid uncertainty in the relative phase of this oscillatory behavior of the second harmonic, the EFISH cell is often made wedge-shaped and the wedge is translated in a direction perpendicular to the incident laser beam. The second-harmonic output as a function of the position of the cell creates so-called Maker fringes whose periodicity is related to the wedge design and to the coherence length; the latter can therefore be determined and the maximum amplitude of the second-harmonic signal can be established. Measurement of a pure solvent (often chloroform) or of a solution of a well-known chromophore (e.g. *p*-nitroaniline) may be used to calibrate the system. The EFISH susceptibility $\Gamma = 3\chi^{(3)}(-2\omega;\omega,\omega,0)$ is then related to the effective molecular second hyperpolarizability γ' by local field factors and the molecule number density, and β can then be obtained from $\gamma' = \gamma + \mu\beta_{EFISH}/(5k_bT)$, where γ' is the effective second hyperpolarizability, γ is the intrinsic second hyperpolarizability (consisting of electronic and vibrational parts), k_b is Boltzmann's constant, and T is the temperature. One needs to perform the measurements as a function of the concentration of solute in a well-characterized solvent, to properly account for the contribution of the solvent to the EFISH signal. The concentration dependence study is necessary to establish whether the $\mu\beta_{EFISH}$ products of the solvent and the solute are of the same or opposite signs, i.e. if addition of a small amount of solute to the solvent leads to an increase or decrease in Γ. The second-harmonic signal obtained at a single concentration of the solution does not provide this information because the SHG intensity is proportional to the square of Γ, and therefore gives no information about its sign.

The calculation of β_{EFISH} requires knowledge of various physical constants for the molecule and the solutions, such as the dielectric constant, the permanent dipole moment of the solute, and the intrinsic second hyperpolarizability of the solute (found from a separate experiment or sometimes ignored). Often results of EFISH are given as $\mu\beta_{EFISH}$ rather than β itself, due to the difficulties in measuring dipole moments.

EFISH has usually been used to evaluate nonlinearities of neutral organometallic complexes, the presence of ionic species rendering it difficult to apply high electric fields to a solution. It is also not possible to utilize EFISH when the complex has no net dipole, and so the method is not useful for molecules of trigonal symmetry, for example, although they can have a strong second-order nonlinearity of octupolar type (i.e. leading to a non-zero septor part of β) with a zero dipolar (vector) part of β.

2. Hyper-Rayleigh Scattering Technique

HRS is a technique that is widely utilized to measure molecular quadratic nonlinearities of organometallic complexes. The HRS technique involves detecting the incoherently scattered second-harmonic light generated from an isotropic solution. An isotropic solution is normally thought to be a random collection of molecules with no preferred orientation, and therefore having all tensor components of $\chi^{(2)}$ equal to zero (an exception is a liquid containing chiral molecules); it is thus unable to generate a coherent second-harmonic beam. However, orientational fluctuations

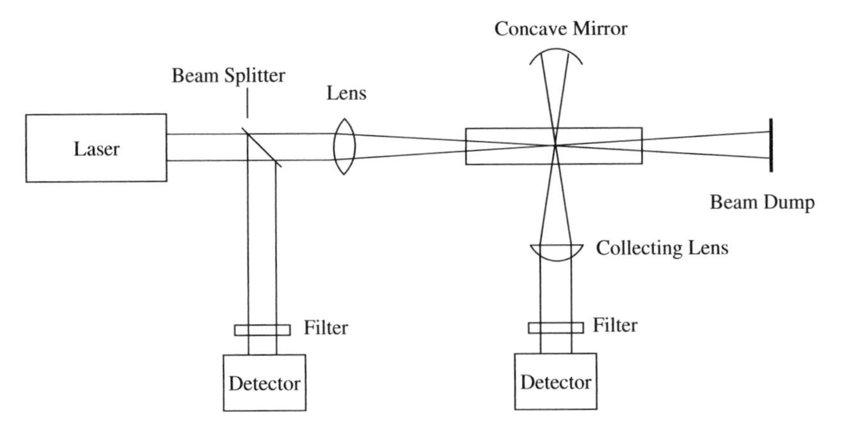

FIG. 1. Schematic diagram of the HRS experiment. (Adapted with permission from Ref. 26, Copyright 1992. American Institute of Physics, Copyright 2002.)

of unsymmetrical molecules in solution do result in local asymmetry and give rise to scattering of light at the second-harmonic frequency.[26] The intensity of the second-harmonic component of the scattered light depends on certain tensor components of the first hyperpolarizability of the solute molecules, and varies quadratically with the incident light intensity. A schematic of a typical HRS experiment is shown in Fig. 1.

A beam from a pulsed laser (nanosecond, picosecond, or femtosecond lasers have been used) is tightly focused into the HRS cell, the incident power being monitored by a photodiode or energy meter. The relatively weak second harmonic is scattered in all directions, and therefore it is important to use an optical system with good collection efficiency. Fig. 1 shows collection effected using a concave mirror and a lens. The second-harmonic light is then isolated with interference filters or a monochromator and is detected by a photomultiplier tube. Gated or phase-sensitive detection is often used to increase the sensitivity and to provide a way to separate the second harmonic (which is generated instantaneously) from upconversion fluorescence which can often be much stronger than the second harmonic, but which has different temporal characteristics than the true HRS signal.

HRS has a number of advantages over EFISH: it is simpler (a dc electric field is not required, nor is knowledge of μ or γ), and it is possible to perform measurements on both octupolar (no dipole moment) and ionic species, the latter being of particular importance to organometallic complexes, many of which possess multiple accessible oxidation states. The disadvantage of HRS is the need to use high intensities of the fundamental, which often results in stimulated Raman or Brillouin scattering, self-focusing, or dielectric breakdown.

3. *Z*-Scan

The *Z*-scan technique[27] is a very simple and convenient way of investigating self-focusing or self-defocusing phenomena as well as nonlinear absorption in a nonlinear material. *Z*-scan has been used for the measurement of the cubic NLO

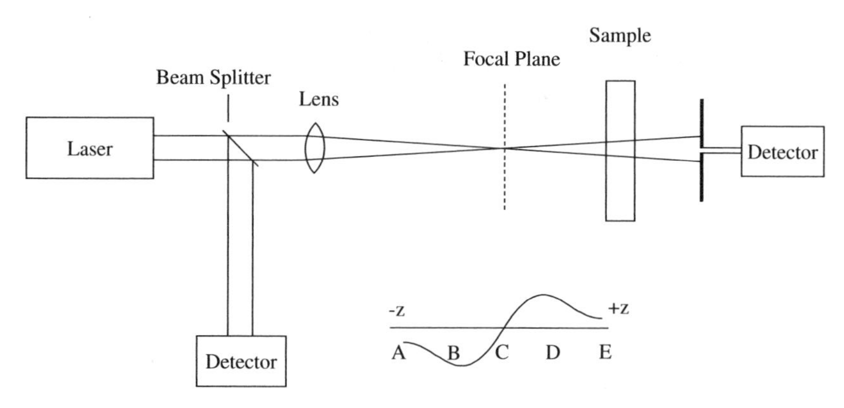

FIG. 2. Schematic diagram of the Z-scan experiment.

properties of the majority of organometallic complexes to date. Fig. 2 provides a schematic diagram of a Z-scan experiment. A laser beam with a well-defined geometry (usually Gaussian, but modifications of the experiment using a top-hat beam or a truncated Airy pattern beam are also in use) is focused with a lens, such that the focal plane is at $z = 0$, and forming a spot with the $1/e^2$ intensity radius w_0. A thin sample (i.e. thinner than the Rayleigh length of the Gaussian beam defined as $z_R = \pi w_0^2/\lambda$) is moved along the z-axis from $-z$ to $+z$, usually scanning several Rayleigh lengths in both directions. As the sample travels, the light intensity that it experiences changes as $I(z) = I_{max}/[1+(z/z_R)^2]$; thus, the scan is started at a low intensity, the highest intensity is reached at $z = 0$ and then it decreases again. Due to the intensity changes, the beam passing through the sample is subject to a varying degree of focusing or defocusing by the virtual lens created by the sample nonlinearity (for positive values of the nonlinear refractive index n_2, the sample acts as a positive lens, and for $n_2 < 0$ as a negative lens), and may also be modified in intensity if the sample acts as a nonlinear absorber. The intensity of the central part of the beam can be monitored as a function of the position of the sample by an aperture that is placed in the far field. Therefore, both distortion of the beam and the changes of its intensity due to changes in the absorption can be detected.

The sample is far from the focal plane at the start (A) (and the finish (E)) of the scan, and so the intensity of the beam is low, and lensing and modification of absorption are not observed. However, for higher intensities the sample acts as a thin lens due to the fact that the refractive index in the sample varies as $n = n_0 + n_2 I(r,z)$, where n_2 is the nonlinear refractive index (related to $\chi^{(3)}$) and $I(r,z)$ is the intensity. For a positive value of n_2, as the sample approaches the focal plane (B), lensing results in additional focusing of the beam and its increased size in the far field, and so the transmittance through the aperture is reduced. The focal plane (C) corresponds to the maximum intensity, but the presence of a lens at this plane does not influence the focused beam wavefront and there is no change in transmittance. At point D lensing has an effect that is the reverse to that of (B): the size of the beam in the far field is decreased and the transmission through the aperture is increased. A characteristic S-shaped Z-scan curve (a valley followed by a maximum of transmittance) is obtained. This shape is reversed in the case of negative n_2.

The shape of the Z-scan curve is additionally modified if nonlinear transmission (absorption bleaching) or nonlinear absorption occur. Due to the presence of the intensity-dependent absorption coefficient, which can often be expressed as $\alpha = \alpha_0 + \beta_2 I(z)$, the curves become unsymmetrical because of increased transmission or absorption close to the focal plane. The absorption changes can also be monitored by measuring the total intensity of the transmitted beam without the use of an aperture in an "open-aperture Z-scan". Materials with potential optical limiting properties are often studied using this technique.

Z-scan experiments for determining the NLO properties of organometallic complexes are usually carried out in cells containing solutions of the complexes in common solvents. The analysis of data must take into account the fact that the effect observed for such samples incorporates contributions from the cell walls and from the solvent, in addition to those from the dissolved complex. The closed-aperture and open-aperture Z-scan curves are usually analyzed using equations derived by Sheikh-Bahae *et al.*,[27] and the sample nonlinearity as a function of concentration is determined. The concentration dependence is then used to derive the extrapolated nonlinear parameters of the pure solute: they can be expressed as the real and imaginary part of $\chi^{(3)}(-\omega;\omega,-\omega,\omega)$, the components of the complex hyperpolarizability $\gamma(-\omega;\omega,-\omega,\omega)$, or the macroscopic parameters, n_2 and β_2. In the nonlinear absorption case, an alternative way of expressing the molecular NLO parameters is by quoting the appropriate molecular absorption cross-sections, and referring to the absorption process responsible for the observed effects, for example, two-photon absorption cross-section or excited-state absorption cross-section.

There are numerous advantages in the use of the Z-scan technique for the determination of the third-order NLO properties: the sign and the magnitude of the nonlinear refractive index can be determined, both the real and the imaginary parts of $\chi^{(3)}$ are obtained, and the single-beam configuration results in simplicity (compared with, e.g., DFWM: see below). The main difficulty is that Z-scan provides data that often lead to uncertainties about the origin of nonlinear effects. Thermal effects (changes in the refractive index and the absorption coefficient due to local heating of the sample by the laser beam), photochemical changes, and other cumulative effects are difficult to distinguish from the (practically) instantaneous changes due to the electronic third-order nonlinearity. It is of crucial importance to use a short pulse laser system at a sufficiently low repetition rate that any long-lived changes in the sample can completely relax in the time interval between the consecutive laser pulses. Additional experiments are often needed to establish the microscopic mechanism of the NLO effects.

4. Degenerate Four-Wave Mixing

Degenerate four-wave mixing (DFWM) is an elegant technique for investigating cubic nonlinearities in various materials that has been used to measure molecular cubic nonlinearities of organometallic complexes. The principle of DFWM is the interaction of three laser beams at the same frequency ω to generate a fourth beam at the same frequency. One of the ways to understand the principle of DFWM is to consider it as a "transient grating" experiment. When two coherent laser beams are crossed at a certain angle within a material, they create an interference pattern of light intensity and, through the dependence of the refractive index of a third-order NLO material on the light intensity, a periodic modulation (a 3D grating) of the (complex) refractive index:

$$\Delta n(\mathbf{r}) = n_2 I(\mathbf{r}) = n_2 I_0 \sin(\Delta \mathbf{k} \mathbf{r}) \tag{14}$$

where $\Delta\mathbf{k}$ is the difference between the wave vectors of the two beams creating the grating. A third beam of light at the same frequency can then be diffracted at the grating, generating a fourth beam. To maximize the intensity of the fourth beam, it is necessary to fulfill the Bragg diffraction condition, which is equivalent to the phase matching condition, i.e. $\sum_i \mathbf{k}_i = 0$, where $\mathbf{k}_i$ are the wave vectors of the four interacting waves. Under phase-matching conditions, the intensity of the diffracted wave is proportional to the product of all the input intensities and to the square of the absolute value of the complex third-order susceptibility and to the square of the interaction length:

$$I_4 \propto |\chi^{(3)}|^2 I_1 I_2 I_3 L^2 \tag{15}$$

In practice, one laser beam is split to provide the three interacting beams. Short laser pulses (picosecond or femtosecond scale) are typically used, so it is very important that their propagation paths are carefully adjusted so that they arrive at the sample at exactly the same time. Probing the temporal behavior of the transient grating is accomplished by directing one of the beams through a computer-controlled variable delay line and recording the DFWM signal as a function of the delay. The DFWM measurements can be performed on cells containing solutions of the compounds being investigated. It is very important to carry out concentration dependence studies, so that contributions to the signal from the real and imaginary parts of the third-order susceptibility can be distinguished. The concentration dependence of the DFWM signal is given by:

$$I_{\mathrm{DFWM}} \propto |\chi^{(3)}|^2 \propto [N_{\mathrm{solvent}}\gamma_{\mathrm{solvent}} + N_{\mathrm{solute}}\mathrm{Re}(\gamma_{\mathrm{solute}})]^2 + [N_{\mathrm{solute}}\mathrm{Im}(\gamma_{\mathrm{solute}})]^2 \tag{16}$$

where it is assumed that the solvent contributes only to the real part of the solution susceptibility, whereas the solute can contribute to both the real (refractive) and the imaginary (absorptive) components. The concentration dependence is thus, in general, nonlinear, complicating the derivation of the molecular hyperpolarizabilities from the experimental data.

Despite the experimental complexity, DFWM is very useful as a technique complementary to *Z*-scan, in that it can be used to verify that the origin of the observed nonlinearity is electronic.

5. Optical Kerr Gate and Pump–Probe Measurements

A variety of two-beam experiments have been devised to measure NLO properties, the principle being that a high-intensity beam, the "pump", causes a change in the optical properties of a material and the change is then (usually after a variable delay) detected by the second beam, the "probe". The pump and probe beams can be at different frequencies (for nondegenerate nonlinear effect studies) or at the same frequency, similar to the DFWM experiment. The simplest modification of a pump–probe experiment can be used to determine nonlinear absorption, which can be related to the imaginary part of $\chi^{(3)}$. Another modification is called the optical Kerr gate (OKG). In the OKG experiment, the pump is linearly polarized and its action on the material induces optical birefringence, that is, a change in the refractive index for directions parallel to the polarization and perpendicular to it.[28] The probe beam is also linearly polarized (usually at 45° to that of the pump). As the probe passes through the material, its polarization becomes elliptic and thus a part of its intensity will be transmitted through a crossed polarizer. The Kerr gate transmittance is proportional to the square of the induced birefringence, and this allows one to determine the appropriate effective $\chi^{(3)}$. However, the OKG experiment measures a specific combination of the tensor components ($\chi^{(3)}_{xxyy} + \chi^{(3)}_{xyyx}$), and both the real and the imaginary parts of $\chi^{(3)}$ contribute to the signal.

Pump–probe experiments are of similar complexity to DFWM. Various modifications of the pump–probe principle allow one to resolve both the real and the imaginary parts of $\chi^{(3)}$ and the temporal dependence of the response.

6. Third-Harmonic Generation

Third-harmonic generation (THG) is an NLO process occurring in all materials and depending on the $\chi^{(3)}(-3\omega;\omega,\omega,\omega)$ susceptibility that is always present. THG measurements can be performed on solid samples (e.g. thin films) and on solutions, the experiment being similar to that of second-harmonic generation (SHG) with a similar consideration of the complications arising from the nonlinear interaction being non-phase matched.[28] The values of the electronic molecular second hyperpolarizability $\gamma(-3\omega;\omega,\omega,\omega)$ obtained from THG experiments can be quite different from those obtained from the determination of n_2, due to different frequency dispersion of the third harmonic γ and the degenerate γ, and also due to the fact that the nonlinear refractive index often contains sizable contributions from nonelectronic mechanisms giving rise to intensity-dependent refractive index changes. The dispersion of the THG nonlinearity often leads to the hyperpolarizability being complex, with a sizable imaginary part, particularly if the frequency of the third harmonic is close to the frequencies corresponding to electronic transitions of the molecule. THG has been infrequently used in the determination of the third-order susceptibility of organometallics, one of the difficulties being the need for a long wavelength high-power laser source, the third harmonic of which should not be in the range of one-photon absorption of the substance being investigated.

III

MOLECULAR SECOND-ORDER MEASUREMENTS

In the majority of reports, researchers have employed the EFISH or HRS techniques to measure molecular second-order nonlinearities of organometallics. The earliest studies in this field were of metal carbonyl-derived complexes, but since these reports, the majority of the measurements have been of metallocenyl, metal alkynyl or related complexes, which can be synthesized in high yield by well-established methodologies, and for which early studies afforded large molecular nonlinearities. A further important consideration is that ferrocenyl and alkynyl complexes are (comparatively) thermally stable, and therefore have potential for device applications.

Table I contains molecular second-order data for ferrocenyl, ruthenocenyl, and related complexes, and some efficient complexes are depicted in Fig. 3. Metallocenyl compounds are amongst the most efficient organometallic complexes for bulk material SHG, and large molecular nonlinearities are anticipated.[10] However, the largest reported molecular data for ferrocenyl complexes contain contributions from fluorescence that could not be deconvoluted from those from HRS, and so the data represent an upper bound.[29] Almost all examples in Table I involve coupling the metallocenyl unit to a π-chromophore; several silyl-functionalized ferrocenes have also been studied, but they have uniformly low nonlinearities.[30] Overall, the ferrocenyl group has been found to act as a moderate donor in these donor–bridge–acceptor complexes, observed nonlinearities being comparable to those of analogous methoxyphenyl organic compounds.[31]

The low binding energy of the metal electrons results in a strong donor character of the ferrocenyl group,[32] oxidation from ferrocene to ferrocenium being facile. However, the MLCT axis and organic chromophore are orthogonal, possibly resulting in poor coupling between the metal center and the substituent.[35]

As is the case for many of the organometallic complexes in subsequent Tables, the UV-visible spectra of many of the metallocenyl complexes in Table I reveal absorptions near the second-harmonic of the laser fundamental frequency, resulting in resonance enhancement of the β coefficient. A two-level correction can be employed to calculate frequency-independent nonlinearities, but Marder *et al.* suggest that both the π–π^* and the MLCT transitions are significant for metallocenyl complexes,[35] so the validity of a two-level approximation is questionable; in instances where structural changes affect both the MLCT and the π–π^* transitions significantly,[31] the two-level model is not justified. Note, though, that it has also been suggested that the MLCT is dominant for monometallic ferrocenyl derivatives;[49] indeed, two-level-derived β_0 for some bimetallic ferrocenyl derivatives are similar, consistent with a dominant MLCT and a less significant π–π^* transition.

The possibility of tuning quadratic nonlinearity by metal replacement has been probed; replacing iron by ruthenium in $[M(\eta^5\text{-}C_5H_5)\{\eta^5\text{-}C_5H_4\text{-}(E)\text{-}CH{=}CHC_6H_4\text{-}4\text{-}NO_2\}]$ (M = Fe, Ru) results in reduced β, which can be rationalized by the higher ionization potential of ruthenium. $[FcCH{=}C(SCH_2Ph)C{=}NCMe_2CH_2O]$ has been coupled to all of the group 10 metals, the second metal usually enhancing β, with an efficiency series $PtCl_2 \approx PdCl_2NiCl_2NiBr_2$.[49] This series confirms that the ferrocenyl group is a comparable donor to the best organic groups, coordination to a ligated metal resulting in a larger β_0 value than coordination of dimethylanilino to the same metal. Alkylating the cyclopentadienyl rings increases the donor strength of the iron, resulting in an increase in nonlinearity over that of the related ferrocenyl complexes; a similar result is seen with $[Ru(\eta^5\text{-}C_5R_5)\{\eta^5\text{-}C_5H_4\text{-}(E)\text{-}CH{=}CHC_6H_4\text{-}4\text{-}NO_2\}]$ (R = H, Me), where permethylating only the ring remote from the chromophore results in a significant increase in nonlinearity. Consistent with this observation, oxidizing the iron significantly reduces the ferrocenyl unit

TABLE I

MOLECULAR SECOND-ORDER NLO RESULTS FOR FERROCENYL, RUTHENOCENYL AND RELATED COMPLEXES

Complex	Solvent	λ_{max} (nm)	β_{exp} (10^{-30} esu)	β_0 (10^{-30} esu)	Technique	Fundamental (μm)	Ref.
[FcCOCH$_3$]	Dioxane	—	0.3	—	EFISH	1.91	21,31,32
[FcSiMe$_3$]	[a]	448	—	0.5	EFISH	[a]	30
[FcSiMe$_2$Ph]	[a]	456	—	0.5	EFISH	[a]	30
[Fc(SiMe$_2$)$_2$Ph]	[a]	455	—	−0.6	EFISH	[a]	30
[Fc(SiMe$_2$)$_3$Ph]	[a]	457	—	0.4	EFISH	[a]	30
[Fc(SiMe$_2$)$_4$Ph]	[a]	456	—	6.4	EFISH	[a]	30
[Fc(SiMe$_2$)$_5$Ph]	[a]	456	—	1.8	EFISH	[a]	30
[Fc(SiMe$_2$)$_6$Ph]	[a]	456	—	1.0	EFISH	[a]	30
[Fc(SiMe$_2$)$_2$-4-C$_6$H$_4$Cl]	[a]	456	—	1.6	EFISH	[a]	30
[Fc(SiMe$_2$)$_2$-4-C$_6$H$_4$OMe]	[a]	456	—	−0.3	EFISH	[a]	30
[Fc(SiMe$_2$)$_2$-4-C$_6$H$_4$NMe$_2$]	[a]	455	—	3.5	EFISH	[a]	30
[Fc(SiMe$_2$)$_2$-3-C$_6$H$_4$CF$_3$]	[a]	457	—	2.2	EFISH	[a]	30
[Fc(SiMe$_2$)$_2$-4-C$_6$H$_4$CH=C(CN)$_2$]	[a]	342	—	7	EFISH	[a]	30
[FcC(=CH$_2$)Si(OCH$_2$CH$_2$)$_3$N]	CH$_2$Cl$_2$	446	4.8[b]	—	HRS	1.30	33
[Fc{(*E*)-CH=CHSi(OCH$_2$CH$_2$)$_3$N}]	CH$_2$Cl$_2$	446	8.4[b]	—	HRS	1.30	33
[Fc{(*E*)-CH=NPh}]	MeCN	448	13.8	3.3	HRS	1.06	34
[Fc{(*E*)-CH=N-4-C$_6$H$_4$OMe}]	MeCN	457	11.9	2.5	HRS	1.06	34
[Fc{(*E*)-CH=N-4-C$_6$H$_4$Cl}]	MeCN	463	20.9	4.1	HRS	1.06	34
[Fc{(*E*)-CH=N-4-C$_6$H$_4$NO$_2$}]	MeCN	490	51.2	6.1	HRS	1.06	34
[Fc{(*E*)-CH=CH-4-C$_6$H$_4$CHO}]	Dioxane	474	12–34	—	EFISH	1.91	21,31,35
[Fc{(*Z*)-CH=CH-4-C$_6$H$_4$NO$_2$}]	Dioxane	480	13	—	EFISH	1.91	21,31,32,36,37
[Fc{(*E*)-CH=CH-4-C$_6$H$_4$NO$_2$}]	Dioxane	536	31	—	EFISH	1.91	21,31,32,36,37
	[a]	500	34	—	EFISH	1.91	38
[Fc{(*E*)-CH=CC(O)NEtC(S)NEtC(O)}]	CHCl$_3$	576[c]	26	—	EFISH	1.91	163
[Fc{(*E,E*)-CH=CHCH=CC(O)NEtC(S)NEtC(O)}]	CHCl$_3$	617[c]	177	—	EFISH	1.91	163
[Fc{(*E,E,E*)-CH=CHCH=CHCH=CC(O)NEtC(S)NEtC(O)}]	CHCl$_3$	632[c]	380	—	EFISH	1.91	163
[Fc{(*E*)-CH=CC[=C(CN)$_2$]C$_6$H$_4$-2-SO$_2$}]	CHCl$_3$	667[c]	29	—	EFISH	1.91	163
[Fc{(*E,E*)-CH=CHCH=CC[=C(CN)$_2$]C$_6$H$_4$-2-SO$_2$}]	CHCl$_3$	721[c]	192	—	EFISH	1.91	163
[Fc{(*E,E,E*)-CH=CHCH=CHCH=CC[=C(CN)$_2$]C$_6$H$_4$-2-SO$_2$}]	CHCl$_3$	746[c]	405	—	EFISH	1.91	163
[Fc{C(OH)H(4-C$_6$H$_4$-(*E*)-CH=CH-4-C$_6$H$_4$NMe$_2$)}]	CHCl$_3$	[a]	18 ± 1	—	HRS	1.06	39
	CF$_2$HCOOH	[a]	200 ± 10	—	HRS	1.06	39
[Fc{C(OH)(Ph)(4-C$_6$H$_4$-(*E*)-CH=CH-4-C$_6$H$_4$NMe$_2$)}]	CHCl$_3$	[a]	10 ± 1	—	HRS	1.06	39
	CF$_2$HCOOH	[a]	320 ± 15	—	HRS	1.06	39

[Fc{C(OH)(4-C_6H_4-(*E*)-CH=CH-4-$C_6H_4NMe_2$)$_2$}]	$CHCl_3$	a	10±1	—	HRS	1.06	39
	CF_2HCOOH	a	260±20	—	HRS	1.06	39
[Fc{(*E*)-CH=CH-4-C_6H_4C(OH)(Ph)(1-naphthyl)}]	$CHCl_3$	a	55±1	—	HRS	1.06	39
	CF_2HCOOH	a	400±200	—	HRS	1.06	39
[Fe(η^5-C_5H_4Me){η^5-C_5H_3Me-(*E*)-CH=CH-4-$C_6H_4NO_2$}]	Dioxane	533	40	—	EFISH	1.91	21,31,32,36,37
[Fc{(*E*)-CH=CH-(*E*)-CH=CH-4-$C_6H_4NO_2$}]	Dioxane	500	66	—	EFISH	1.91	21,31,32,35
[Fc{(*E*)-CH=CH-2,4-$C_6H_3(NO_2)_2$}]	a	545	20	—	EFISH	1.91	38
	Dioxane	536	23		EFISH	1.91	32
[Fc{(*Z*)-CH=CH-4-C_6H_4CN}]	—	460	4	—	EFISH	1.91	31
[Fc{(*E*)-CH=CH-4-C_6H_4CN}]	$CHCl_3$	473	203	34	HRS	1.06	40
	Dioxane	466	10	—	EFISH	1.91	21,31
[Fc-(*Z*)-CH=C(CN)-4-$C_6H_4NO_2$]	Dioxane	560	35	—	EFISH	1.91	32
[Fc-(*E*)-CH=C(CN)-4-$C_6H_4NO_2$]	Dioxane	526	21	—	EFISH	1.91	32
[Fc-(*E*)-CH=CH-4-C_6H_4-(*E*)-CH=CH-4-$C_6H_4NO_2$]	$CHCl_3$	387	403	—	HRS	1.06	41
[Fc-(*E*)-CH=CH-4-C_6H_4-(*Z*)-CH=CH-4-$C_6H_4NO_2$]	$CHCl_3$	315	122	—	HRS	1.06	41
[Fc-(*E*)-CH=CH-4-C_6H_4-(*E*)-CH=CH-4-C_6H_4-(*Z*)-CH=CH-4-$C_6H_4NO_2$]	$CHCl_3$	364	209	—	HRS	1.06	41
[Fc{(*E*)-CH=CH-4-C_6H_4-(*E*)-CH=CH-4-C_5H_4N}]	$CHCl_3$	459	146	—	HRS	1.06	42
[Fc{(*E*)-CH=CH-4-C_5H_4NMe}](PF_6)	Acetone	553	40	—	HRS	1.06	42
[Fc{(*E*)-CH=CH-4-C_6H_4-(*E*)-CH=CH-4-C_5H_4NMe}](PF_6)	Acetone	503	197	—	HRS	1.06	42
[Fc{(*E*)-(CH=CH-4-C_6H_4)}$_2$-(*E*)-CH=CH-4-C_5H_4NMe}](PF_6)	Acetone	462	458	—	HRS	1.06	42
[Fc{(*E*)-CH=CH(bipy)-4′-C_9H_{19}}]	a	469	9	—	EFISH	1.34	43
[FcCH=$NC_{14}H_7O_2$]	$CHCl_3$	495	939	90	a	a	44
[FcCH=NC(CN)=C(CN)NH_2]	$CHCl_3$	508	476	40	a	a	44
[Fc{(*E*)-CH=CH-4-C_5H_4N}]	$MeNO_2$	468	185	—	HRS	1.06	45
	CH_2Cl_2	466	16	3	HRS	1.06	46
	$CHCl_3$	468	21	—	HRS	1.06	40,42
[Fc{(*E*)-CH=CH-2-C_4H_2S-5-CHO}]	CH_2Cl_2	397	148	13	HRS	1.06	46
[Fc{(*E*)-CH=CH-2-C_4H_2S-5-NO_2}](PF_6)	CH_2Cl_2	563	25	10	HRS	1.06	46
[Fc{(*E*)-CH=CH-2-C_4H_2S-5-NO_2}]	CH_2Cl_2	646	316	95	HRS	1.06	46
[Fc-(azulene)](BF_4)	CH_2Cl_2	716	326	145	HRS	1.06	29
[Fc-(*E*,*E*)-CH=CHCH=CH(azulene)](BF_4)	CH_2Cl_2	866	810[d]	451	HRS	1.06	29
[Fc-(1,4-dimethyl-7-isopropylazulene)](PF_6)	CH_2Cl_2	724	220	101	HRS	1.06	29
[Fc-(*E*)-CH=CH(1,4-dimethyl-7-isopropylazulene)](BF_4)	CH_2Cl_2	782	1539[d]	821	HRS	1.06	29
[Fc-(*E*,*E*)-CH=CHCH=CH-(1,4-dimethyl-7-isopropylazulene)](BF_4)	CH_2Cl_2	827	1373[d]	770	HRS	1.06	29
			360[e]	132	HRS	1.30	29
[Fc{(*E*)-CH=CHB(mes)$_2$}]	$CHCl_3$	336[f]	−24	—	EFISH	1.91	47
[FcC≡CB(mes)$_2$]	$CHCl_3$	336[f]	4.4	—	EFISH	1.91	47
[FcCH=C(SCH_2Ph)C=NCMe$_2$$CH_2$O]	$CHCl_3$	472	4.5	2.0	EFISH	1.34	48–50
	$CHCl_3$	472	2.9	2.07	EFISH	1.91	49,50
[FcCH=C(S(O)CH_2Ph)C=NCMe$_2$$CH_2$O]	$CHCl_3$	470	8.4	3.7	EFISH	1.34	48–50
[FcCH=C(S(O)$_2$$CH_2$Ph)C=NCMe$_2$$CH_2$O]	$CHCl_3$	477	6.3	1.6	EFISH	1.34	49,50

TABLE I
(CONTINUED)

Complex	Solvent	λ_{max} (nm)	β_{exp} (10^{-30} esu)	β_0 (10^{-30} esu)	Technique	Fundamental (μm)	Ref.
$[Fc\{(E)\text{-}CH=CHCH=C(SCH_2Ph)\overline{C=NCMe_2CH_2O}\}]$	$CHCl_3$	482	13.3	5.6	EFISH	1.34	49,50
	$CHCl_3$	482	8.8	6.14	EFISH	1.91	49,50
$[Fc\{(E)\text{-}CH=CHCH=C(S(O)CH_2Ph)\overline{C=NCMe_2CH_2O}\}]$	$CHCl_3$	494	11.0	4.6	EFISH	1.34	49,50
$[Fc\{C\equiv CCH=C(SCH_2Ph)\overline{C=NCMe_2CH_2O}\}]$	$CHCl_3$	460	2.6	1.2	EFISH	1.34	49,50
$[Fe(\eta^5\text{-}C_5H_5)\{\eta^5\text{-}C_5H_3\text{-}1\text{-}CH_2NMe_2\text{-}2\text{-}Si(OCH_2CH_2)_3N\}]$	$CHCl_3$	400	8.5^b	—	HRS	1.3	33
$[Fe(\eta^5\text{-}C_5H_5)(\eta^5\text{-}C_5H_3\text{-}1\text{-}Me\text{-}2\text{-}\{(E)\text{-}CH=CH\text{-}4\text{-}C_6H_4NO_2\})]$	Dioxane	359	24	—	EFISH	1.9	36
$[Fe(\eta^5\text{-}C_5H_5)(\eta^5\text{-}C_5H_3\text{-}1\text{-}CH_2OH\text{-}2\text{-}\{(E)\text{-}CH=CH\text{-}4\text{-}C_6H_4NO_2\})]$	Dioxane	362	43	—	EFISH	1.9	36
$[Fe(\eta^5\text{-}C_5H_5)(\eta^5\text{-}C_5H_3\text{-}1\text{-}SiMe_3\text{-}2\text{-}\{(E)\text{-}CH=CH\text{-}4\text{-}C_6H_4NO_2\})]$	Dioxane	357	36	—	EFISH	1.9	36
$[Fe(\eta^5\text{-}C_5HMe_4)\{\eta^5\text{-}C_5Me_4\text{-}(E)\text{-}CH=CH\text{-}2\text{-}C_4H_2S\text{-}5\text{-}NO_2\}](PF_6)$	CH_2Cl_2	851	25	10	HRS	1.06	53
$[Fe(\eta^5\text{-}C_5HMe_4)\{\eta^5\text{-}C_5Me_4\text{-}(E)\text{-}CH=CH\text{-}2\text{-}C_4H_2S\text{-}5\text{-}NO_2\}]$	CH_2Cl_2	644	316	95	HRS	1.06	53
$[Fe(\eta^5\text{-}C_5HMe_4)(\eta^5\text{-}C_5Me_4CH=N\text{-}4\text{-}C_5H_4N)]$	CH_2Cl_2	503	67	5	HRS	1.06	46
$[Fe(\eta^5\text{-}C_5HMe_4)\{\eta^5\text{-}C_5Me_4\text{-}(E)\text{-}CH=CH\text{-}4\text{-}C_5H_4N\}]$	CH_2Cl_2	503	76	6	HRS	1.06	46
$[Fe(\eta^5\text{-}C_5HMe_4)\{\eta^5\text{-}C_5Me_4\text{-}(E)\text{-}CH=CH\text{-}2\text{-}C_4H_2S\text{-}5\text{-}(E)\text{-}CH=CH\text{-}4\text{-}C_5H_4N\}]$	CH_2Cl_2	538	410	8	HRS	1.06	46
$[Fe(\eta^5\text{-}C_5Me_5)\{\eta^5\text{-}C_5H_4\text{-}(E)\text{-}CH=CH\text{-}4\text{-}C_5H_4N\}]$	CH_2Cl_2	489	73	9	HRS	1.06	46
$[Ru(\eta^5\text{-}C_5Me_5)\{\eta^5\text{-}C_5H_4NO_2\}]$	CH_2Cl_2	—	0.6	—	EFISH	1.91	21,31,32
$[Ru(\eta^5\text{-}C_5H_5)\{\eta^5\text{-}C_5H_4\text{-}(E)\text{-}CH=CH\text{-}4\text{-}C_6H_4NO_2\}]$	Dioxane	390	12–16	—	EFISH	1.91	21,31,32,35
$[Ru(\eta^5\text{-}C_5Me_5)\{\eta^5\text{-}C_5H_4\text{-}(E)\text{-}CH=CH\text{-}4\text{-}C_6H_4NO_2\}]$	Dioxane	424	24	—	EFISH	1.91	21,31,32,35
$[Ru(\eta^5\text{-}C_5Me_5)\{\eta^5\text{-}C_5H_4\text{-}(E)\text{-}CH=C(CN)\text{-}4\text{-}C_6H_4NO_2\}]$	Dioxane	443	24	—	EFISH	1.91	32
$[Ru\{\eta^5\text{-}2\text{-}C_4H_3S\text{-}(E)\text{-}CH=CH\text{-}4\text{-}C_6H_4OMe\}(\eta^6\text{-}C_6Me_6)](OTf)_2$	$MeNO_2$	447	68	16	HRS	1.06	54
$[Ru\{\eta^5\text{-}2\text{-}C_4H_3S\text{-}(E)\text{-}CH=CH\text{-}4\text{-}C_6H_4Me\}(\eta^6\text{-}C_6Me_6)](OTf)_2$	$MeNO_2$	416	61	20	HRS	1.06	54
$[Ru\{\eta^5\text{-}2\text{-}C_4H_3S\text{-}(E)\text{-}CH=CHPh\}(\eta^6\text{-}C_6Me_6)](OTf)_2$	$MeNO_2$	386	66	27	HRS	1.06	54
$[Ru\{\eta^5\text{-}2\text{-}C_4H_3S\text{-}(E)\text{-}CH=CH\text{-}4\text{-}C_6H_4Br\}(\eta^6\text{-}C_6Me_6)](OTf)_2$	$MeNO_2$	412	98	33	HRS	1.06	54
$[Ru\{\eta^5\text{-}2\text{-}C_4H_3S\text{-}(E)\text{-}CH=CH\text{-}4\text{-}C_6H_4NO_2\}(\eta^6\text{-}C_6Me_6)](OTf)_2$	$MeNO_2$	410	226	78	HRS	1.06	54
$[Ru\{\eta^5\text{-}2\text{-}C_4H_3S\text{-}(E)\text{-}CH=CH\text{-}4\text{-}C_6H_4NO_2\}(\eta^6\text{-}C_6H_4\text{-}1\text{-}Me\text{-}4\text{-}iPr)](OTf)_2$	$MeNO_2$	407	226	80	HRS	1.06	54
$[Ru\{\eta^5\text{-}2\text{-}C_4H_3S\text{-}(E)\text{-}CH=CH\text{-}4\text{-}C_6H_4NO_2\}(\eta^5\text{-}C_5Me_5)](OTf)$	$MeNO_2$	408	389	137	HRS	1.06	54
$[Ru(\eta^5\text{-}C_5H_5)(\eta^5\text{-}C_5H_4\text{-}C_7H_6)](PF_6)$	$MeNO_2$	536	378	4	HRS	1.06	55,56
$[Ru(\eta^5\text{-}C_5H_5)\{\eta^5\text{-}C_5H_4\text{-}(E)\text{-}CH=CH\text{-}C_7H_6\}](PF_6)$	$MeNO_2$	600	649	120	HRS	1.06	55,56
	CH_2Cl_2	640	362	105	HRS	1.06	55,56
$[Ru(\eta^5\text{-}C_5H_5)\{\eta^5\text{-}C_5H_4\text{-}(E)\text{-}CH=CH(bipy)\text{-}4'\text{-}Me\}]$	a	345	7	—	EFISH	1.34	43
$[FcC\equiv C(\eta^7\text{-}C_7H_6)Cr(CO)_3]$	CH_2Cl_2	600	570	105	HRS	1.06	51
$[Fc\{(E)\text{-}CH=CH(\eta^7\text{-}C_7H_6)Cr(CO)_3\}]$	CH_2Cl_2	670	320	113	HRS	1.06	51
$[Fc(\eta^6\text{-}BC_5H_5)Co(\eta^5\text{-}C_5H_5)](PF_6)$	CH_2Cl_2	650		90	HRS	1.06	52
$[Fc(\eta^6\text{-}BC_6H_6)Co(\eta^5\text{-}C_5H_5)]$	CH_2Cl_2	650	—	90	HRS	1.06	52
$[Fc\{CH=C(SCH_2Ph)\overline{C=NCMe_2CH_2O}(N,S\text{-}NiCl_2)\}]$	$CHCl_3$	500	10.3	3.92	EFISH	1.34	49,50

[Fc{CH =C(SCH_2Ph)$\overline{\mathrm{C=NCMe_2CH_2O}}$(*N*, *S*-$NiBr_2$)}]	$CHCl_3$	511	7.14	2.55	EFISH	1.34	49,50
[Fc{CH =C(S(O)CH_2Ph)$\overline{\mathrm{C=NCMe_2CH_2O}}$(*N*, *S*-$NiCl_2$)}]	$CHCl_3$	498	15.8	6.0	EFISH	1.34	49,50
[Fc{CH =C(S(O)CH_2Ph)$\overline{\mathrm{C=NCMe_2CH_2O}}$(*N*, *S*-$NiBr_2$)}]	$CHCl_3$	494	5.85	2.3	EFISH	1.34	49,50
[Fc{CH =C(SCH_2Ph)$\overline{\mathrm{C=NCMe_2CH_2O}}$(*N*, *S*-$PdCl_2$)}]	$CHCl_3$	530	12.2	4.13	EFISH	1.34	48–50
	$CHCl_3$	530	9	5.8	EFISH	1.91	49,50
[Fc{CH =C(S(O)CH_2Ph)$\overline{\mathrm{C=NCMe_2CH_2O}}$(*N*, *S*-$PdCl_2$)}]	$CHCl_3$	540	30.2	7.6	EFISH	1.34	48–50
[Fc{(*E*)-CH =CHCH =C(SCH_2Ph)$\overline{\mathrm{C=NCMe_2CH_2O}}$(*N*, *S*-$PdCl_2$)}]	$CHCl_3$	567	58	13.5	EFISH	1.34	49,50
	$CHCl_3$	567	18.2	10.8	EFISH	1.91	49,50
[Fc{(*E*)-CH =CHCH =C(S(O)CH_2Ph)$\overline{\mathrm{C=NCMe_2CH_2O}}$(*N*, *S*-$PdCl_2$)}]	$CHCl_3$	602	123.6	19	EFISH	1.34	49,50
[Fc{C ≡CCH =C(SCH_2Ph)$\overline{\mathrm{C=NCMe_2CH_2O}}$(*N*, *S*-$PdCl_2$)}]	$CHCl_3$	558	33.6	8.5	EFISH	1.34	49,50
[Fc{CH =C(SCH_2Ph)$\overline{\mathrm{C=NCMe_2CH_2O}}$(*N*, *S*-$PtCl_2$)}]	$CHCl_3$	523	16	5.2	EFISH	1.34	49,50

[a] Not stated.
[b] β_{333} values calculated by applying the two-level model on the π–π* absorption band.
[c] CH_2Cl_2 solvent.
[d] Two-photon absorption fluorescence enhanced.
[e] Obtained from high-frequency demodulation.
[f] Cyclohexane solvent.

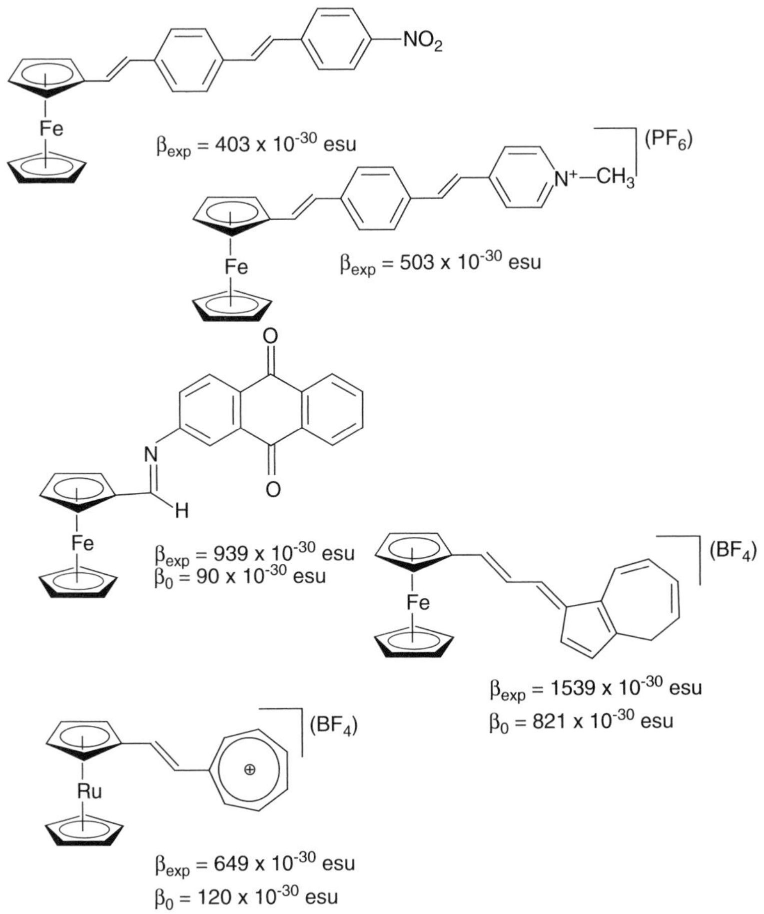

NO_2

Fe

$\beta_{exp} = 403 \times 10^{-30}$ esu

(PF_6)

N^{+}–CH_3

Fe

$\beta_{exp} = 503 \times 10^{-30}$ esu

O

N

O

H

Fe

$\beta_{exp} = 939 \times 10^{-30}$ esu
$\beta_0 = 90 \times 10^{-30}$ esu

(BF_4)

Fe

$\beta_{exp} = 1539 \times 10^{-30}$ esu
$\beta_0 = 821 \times 10^{-30}$ esu

(BF_4)

⊕

Ru

$\beta_{exp} = 649 \times 10^{-30}$ esu
$\beta_0 = 120 \times 10^{-30}$ esu

FIG. 3. Selected ferrocenyl and ruthenocenyl complexes with large β.

donor strength, resulting in a dramatic decrease in β value in proceeding from $n = 0$ to 1 in $[Fc\{(E)\text{-}CH{=}CH\text{-}2\text{-}C_4H_2S\text{-}5\text{-}NO_2\}]^{n+}$.[46]

Because the metallocenyl group is the donor in a donor–bridge–acceptor construction, increasing the chromophore acceptor strength generally results in an increase in nonlinearity (e.g. $[Fc\{(E)\text{-}CH{=}N\text{-}4\text{-}C_6H_4X\}]$ [β values: X = OMe < H < Cl < NO_2][34] and $[Fc\{(E)\text{-}CH{=}CH\text{-}2\text{-}C_4H_2S\text{-}5\text{-}X\}]$ [β values: X = CHO < NO_2][46]); the same observation can be noted upon alkylating the 4-pyridyl N atom in $[Fc\{(E)\text{-}CH{=}CH(4\text{-}C_6H_4\text{-}(E)\text{-}CH{=}CH)_n\text{-}4\text{-}C_5H_4N\}]$ ($n = 0$, 1). The 4-nitro complex $[Fc\text{-}(E)\text{-}CH{=}CHC_6H_4\text{-}4\text{-}NO_2]$ has a larger β than the 2,4-dinitro complex $[Fc\text{-}(E)\text{-}CH{=}CHC_6H_3\text{-}2,4\text{-}(NO_2)_2]$. It has been suggested that the second nitro group may saturate the polarizability, with the nitro groups competing for oscillator strength in the excited state.[38] It has also been noted that EFISH measures the β component in the dipolar direction; the two complexes have different dipole moments, and the dipole of the dinitro complex is less aligned with β_{CT} than is the dipole of the 4-nitro complex.

The relative efficiencies of the bridging groups in the metallocenyl–bridge–acceptor complexes have been assayed: The *E*-ene isomer [Fc-(*E*)-CH=CHC$_6$H$_4$-4-NO$_2$] has a higher nonlinearity than the *Z* isomer, and the ene-linked complex [Fc-(*E*)-CH=CHB(mes)$_2$] has a greater β value than the yne-linked complex [FcC≡CB(mes)$_2$], structure–property observations that have been established previously with organic compounds. Similarly, lengthening the chromophore π-system (e.g. proceeding from [Fc-(*E*)-CH=CH-4-C$_6$H$_4$-(*Z*)-CH=CH-4-C$_6$H$_4$NO$_2$] to [Fc-(*E*)-CH=CH-4-C$_6$H$_4$-(*E*)-CH=CH-4-C$_6$H$_4$-(*Z*)-CH=CH-4-C$_6$H$_4$NO$_2$][41]) generally results in increased nonlinearity, as is also seen with organic compounds.

Table II contains $\boldsymbol{\mu} \cdot \boldsymbol{\beta}$ data for ferrocenyl complexes for which dipole moments are not available, so only an internal comparison is possible. Some efficient examples are shown in Fig. 4.

Not surprisingly, $\boldsymbol{\mu} \cdot \boldsymbol{\beta}$ and $\boldsymbol{\mu} \cdot \boldsymbol{\beta}_0$ values increase upon increasing acceptor strength (replacing formylmethylene by dicyanomethylene) and upon π-system lengthening the ferrocenyl oligo-ene complexes; in the optical spectra of the latter, there are two absorption bands assigned to MLCT and π–π* transitions that become closer with increasing chain length, and, for sufficiently long chain length, the bands overlap, suggesting increased coupling between the metal center and the acceptor.

Table III contains results from metal carbonyl complexes, some efficient examples being depicted in Fig. 5. In many cases, the most important contribution to the nonlinearity derives from other ligands (N-donor ligands, carbocycles, and carbenes). Many pyridine and arene complexes have EFISH-derived negative β values, consistent with these complexes undergoing a decrease in dipole moment or change in direction of dipole moment between the ground and excited states.

Ligated metal variation has been examined. Little difference in β value is seen in N-ligand complexes across the series [M(CO)$_4$(1,10-phen)] (M = Cr, Mo, W),[61] in contrast to C-ligand compounds; varying the metal in the carbene complexes [M{C(OMe)-(*E*)-CH=CHNMe$_2$}(CO)$_5$] (M = Cr, W) results in a substantial change in β. The carbonyl ligand is a π-acceptor, and electron density at the metal center can be increased by replacing carbonyl by phosphine or phosphite; ligand variation in the complexes [Cr(CO)$_2$(L)(η^6-PhC≡C-4-C$_6$H$_4$NO$_2$)] (L = CO, PPh$_3$) and [Mn(CO)$_2$(L){η^5-C$_5$H$_4$(C≡C)$_n$C$_7$H$_6$}](BF$_4$) [L = CO, P(OMe)$_3$, PPh$_3$; $n = 0$, 1] generally (but not always) results in an increase in nonlinearity on replacement of carbonyl, consistent with increasing the donor strength of the ligated metal center. Neither metal nor halogen replacement in [MX(CO)$_2${NC$_5$H$_4$-4-(*E*)-CH=CHC$_6$H$_4$-4-OCH$_2$CHMeEt}] results in significant change in β. π-System lengthening [e.g. in the series of complexes [Cr(CO)$_3${η^6-Ph-(*E*)$_n$-(CH=CH)$_n$-4-C$_6$H$_4$NO$_2$}] ($n = 1$, 2, 3) and [Cr(CO)$_3${η^6-Ph-(*E*)$_n$-(CH=CH)$_n$-4-C$_6$H$_4$NMe$_2$}] ($n = 1$, 2)] results in an increase in nonlinearity, as observed in other organometallic and organic systems.

A considerable number of chromium arene complexes have been studied, none of which have particularly large second-order responses, and the most efficient of which are matched by cumulenyl carbene complexes in which the pentacarbonylchromium unit functions as an acceptor. Nonlinearities for the (arene)chromium

TABLE II

MOLECULAR SECOND-ORDER NLO RESULTS AS $\mu \cdot \beta$ PRODUCT FOR FERROCENYL AND RUTHENOCENYL COMPLEXES

Complex	Solvent	λ_{max} (nm)	$\mu\beta_{exp}$ (10^{-48} esu)	$\mu\beta_0$ (10^{-48} esu)	Technique	Fundamental (μm)	Ref.
[Fc{(*E*)-CH=CHCHO}]	Acetone	480	60	42	EFISH	1.91	57,163
[Fc{(*E*)-CH=CH-(*E*)-CH=CHCHO}]	Acetone	494	215	147	EFISH	1.91	57,163
[Fc{(*E*)-CH=CH-(*E*)-CMe=CH-(*E*)-CH=CH-(*E*)-CH=CMeCHO}]	Acetone	398[a]	560	440	EFISH	1.91	57
[Fc{(*E*)-CH=CH-(*E*)-CH=CH-(*E*)-CMe=CH-(*E*)-CH=CH-(*E*)-CH=CMe-(*E*)-CH=CHCHO}]	Acetone	430[a]	1150	870	EFISH	1.91	57
[FcCH=C(CN)$_2$]	Acetone	526	92	60	EFISH	1.91	57,163
[Fc{(*E*)-CH=CHCH=C(CN)$_2$}]	Acetone	556	420	250	EFISH	1.91	57,163
[Fc{(*E*)-CH=CH-(*E*)-CH=CHCH=C(CN)$_2$}]	Acetone	568	1120	660	EFISH	1.91	57,163
[Fc{(*E*)-CH=CH-(*E*)-CMe=CH-(*E*)-CH=CH-(*E*)-CH=CMeCH=C(CN)$_2$}]	Acetone	458[a]	4600	3300	EFISH	1.91	57
[Fc{(*E*)-CH=CH-4-C_6H_4CH=C(CN)$_2$}]	CH_2Cl_2	543	450	200	EFISH	1.54	58
[Fc{(*E*)-CH=CH-4-C_6H_4-(*E*)-CH=CH-4-C_6H_4CH=C(CN)$_2$}]	CH_2Cl_2	500	570	370	EFISH	1.54	58
[Fc{(*Z*)-CH=CH-4-C_6H_4CH(2,5,7-trinitrofluorene-4-carboxylic acid triethyleneglycol monomethylether ester)}]	CH_2Cl_2	570	1300	830	EFISH	1.54	58
[Fc{(*E*)-CH=CH-4-C_6H_4-(*E*)-CH=CH-4-C_6H_4CH(2,5,7-trinitrofluorene-4-carboxylic acid triethyleneglycol monomethylether ester)}]	CH_2Cl_2	570	410	240	EFISH	1.54	58
[Fc{(*E*,*E*)-(CH=CH)$_2$CH(2,5,7-trinitrofluorene-4-carboxylic acid methylether ester)}]	CH_2Cl_2	663	5000	900	EFISH	1.54	58
[Fc{(*E*,*E*)-(CH=CH)$_2$CH(2,5,7-trinitrofluorene-4-carboxylic acid triethyleneglycol monomethylether ester)}]	CH_2Cl_2	660	2700	470	EFISH	1.54	58
[Fc{C≡CCH(2,5,7-trinitrofluorene-4-carboxylic acid triethyleneglycol monomethylether ester}]	CH_2Cl_2	616	700	170	EFISH	1.54	58
[Fc{(*E*,*E*,*E*,*E*)-CH=CHCH=CHCH=CHCH=CC[=C(CN)$_2$]C_6H_4-2-SO_2]	$CHCl_3$	743[b]	11200	—	EFISH	1.91	163
[Fc(3-dicyanomethyleneindan-2-ylidene-1-one)]	$CHCl_3$	632	141	70	EFISH	1.91	59
[Fe(η^5-3,4-$PC_4H_2Me_2$){η^5-3,4-PC_4HMe_2-2-CH=C(CN)$_2$}]	$CHCl_3$	498	67	45	EFISH	1.91	59
[Fe(η^5-3,4-$PC_4H_2Me_2$){η^5-3,4-PC_4HMe_2-2-(*N*,*N*′-diethylbarbiturate)}]	$CHCl_3$	550	102	68	EFISH	1.91	59
[Fe(η^5-3,4-$PC_4H_2Me_2$){η^5-2,3-PC_4HMe_2-2-(3-dicyanomethyleneindan-2-ylidene-1-one)}]	$CHCl_3$	606	104	56	EFISH	1.91	59
[Ru(η^5-C_5H_5){η^5-C_5H_4-(*E*,*E*)-CH=CHCH=CHO}]	$CHCl_3$	381[b]	140	—	EFISH	1.91	163
[Ru(η^5-C_5H_5){η^5-C_5H_4CH=C(CN)$_2$}]	$CHCl_3$	411[b]	65	—	EFISH	1.91	163
[Ru(η^5-C_5H_5){η^5-C_5H_4-(*E*)-CH=CHCH=C(CN)$_2$}]	$CHCl_3$	449[b]	410	—	EFISH	1.91	163
[Ru(η^5-C_5H_5){η^5-C_5H_4-(*E*,*E*)-CH=CHCH=CHCH=C(CN)$_2$}]	$CHCl_3$	472[b]	750	—	EFISH	1.91	163
[Ru(η^5-C_5H_5){η^5-C_5H_4CH=CC(O)NEtC(S)NEtC(O)}]	$CHCl_3$	458[b]	90	—	EFISH	1.91	163
[Ru(η^5-C_5H_5){η^5-C_5H_4-(*E*)-CH=CHCH=CC(O)NEtC(S)NEtC(O)}]	$CHCl_3$	508[b]	390	—	EFISH	1.91	163
[Ru(η^5-C_5H_5){η^5-C_5H_4-(*E*,*E*)-CH=CHCH=CHCH=CC(O)NEtC(S)NEtC(O)}]	$CHCl_3$	530[b]	1000	—	EFISH	1.91	163
[Ru(η^5-C_5H_5){η^5-C_5H_4CH=C[=C(CN)$_2$]C_6H_4-2-SO_2}]	$CHCl_3$	541[b]	105	—	EFISH	1.91	163
[Ru(η^5-C_5H_5){η^5-C_5H_4-(*E*)-CH=CHCH=C[=C(CN)$_2$]C_6H_4-2-SO_2}]	$CHCl_3$	595[b]	630	—	EFISH	1.91	163
[Ru(η^5-C_5H_5){η^5-C_5H_4-(*E*,*E*)-CH=CHCH=CHCH=C[=C(CN)$_2$]C_6H_4-2-SO_2}]	$CHCl_3$	623[b]	1900	—	EFISH	1.91	163

[a]Lower energy band obscured by a band to higher energy.
[b]CH_2Cl_2 solvent.

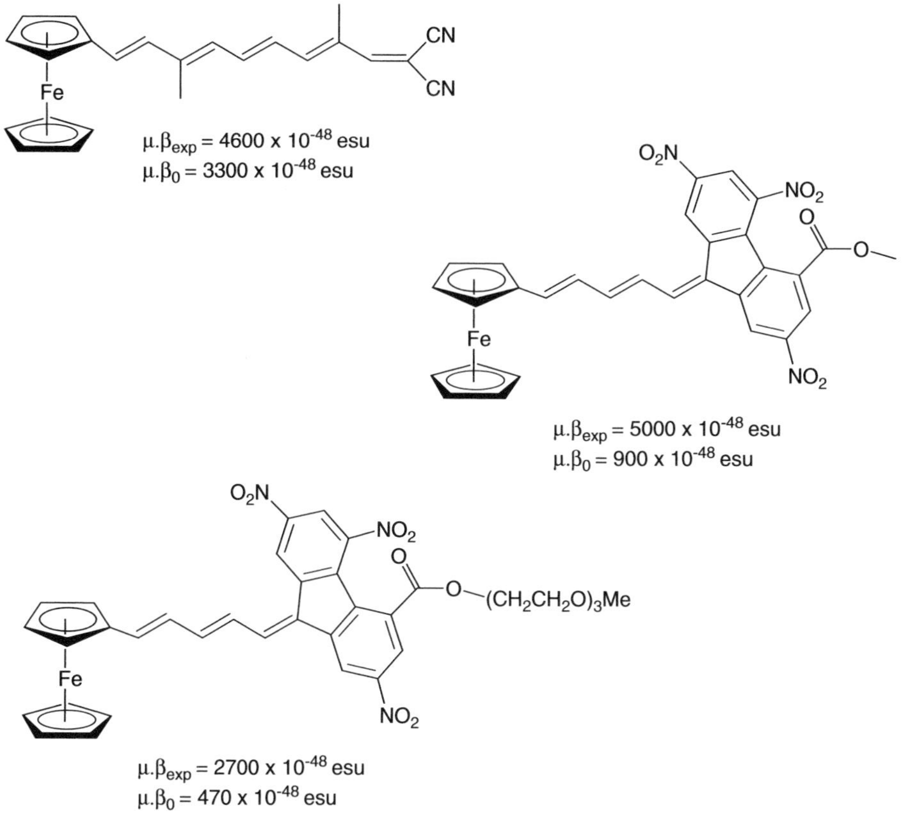

$\mu.\beta_{exp} = 4600 \times 10^{-48}$ esu
$\mu.\beta_0 = 3300 \times 10^{-48}$ esu

$\mu.\beta_{exp} = 5000 \times 10^{-48}$ esu
$\mu.\beta_0 = 900 \times 10^{-48}$ esu

$\mu.\beta_{exp} = 2700 \times 10^{-48}$ esu
$\mu.\beta_0 = 470 \times 10^{-48}$ esu

FIG. 4. Selected ferrocenyl complexes with large $\boldsymbol{\mu} \cdot \boldsymbol{\beta}$.

complexes increase upon introduction of strongly polarizing substituents (NO_2, NMe_2) and π-system lengthening. In contrast to the carbocycle complexes, the carbene complexes are consistent with a proposed efficient NLO design, namely that the metal should be located collinear with the chromophore, preferably with M–C multiple bonding.[31]

Theoretical studies predicted that sesquifulvalenes should have large nonlinearities, but these compounds are very reactive, and so experimental confirmation was not possible. However, coordination to transition metal(s) stabilizes the sesquifulvalenes; experimental studies on the resultant sesquifulvalene complexes revealed large nonlinearities.

Because the $[(\text{thiophene})Mn(CO)_3]^+$ center is an effective excited state acceptor, the (thiophene)manganese complex cations possess substantial β values,[78] although these are significantly resonance enhanced. The $\beta_{1.06}$ and β_0 data for the (thiophene)manganese complexes increase as the styryl substituent increases in acceptor strength (proceeding from OMe to Me, H, Br, and NO_2) and increasing the electron density at the manganese (replacing co-ligand CO with PPh_3). The (indenyl)ruthenium complex data are resonance enhanced; the two-level-corrected β_0 values of the

TABLE III
MOLECULAR SECOND-ORDER NLO RESULTS FOR METAL CARBONYL COMPLEXES

Complex	Solvent	λ_{max} (nm)	β (10^{-30} esu)	β_0 (10^{-30} esu)	Technique	Fundamental (μm)	Ref.
$[Cr(CO)_3(\eta^6\text{-}C_6H_6)]$	Toluene	310	−0.8	—	EFISH	1.91	32
$[Cr(CO)_3(\eta^6\text{-}PhOMe)]$	Toluene	310	−0.9	—	EFISH	1.91	32,60,61
$[Cr(CO)_3(\eta^6\text{-}PhNH_2)]$	Dioxane	313	−0.6	—	EFISH	1.91	32,60,61
$[Cr(CO)_3(\eta^6\text{-}PhNMe_2)]$	Toluene	318	−0.4	—	EFISH	1.91	32,60,61
$[Cr(CO)_3(\eta^6\text{-}PhCO_2Me)]$	Toluene	318	−0.7	—	EFISH	1.91	32,60,61
$[Cr(CO)_3\{\eta^6\text{-}Ph\text{-}(E)\text{-}CH{=}CHPh\}]$	Dioxane	410	−2.2	—	EFISH	1.91	32,60,61
$[Cr(CO)_3\{\eta^6\text{-}Ph\text{-}(E)\text{-}CH{=}CH\text{-}4\text{-}C_6H_4NO_2\}]$	$CHCl_3$	455	15[a]	12[a]	HRS	1.50	62
$[Cr(CO)_3\{\eta^6\text{-}Ph\text{-}(E,E)\text{-}(CH{=}CH)_2\text{-}4\text{-}C_6H_4NO_2\}]$	$CHCl_3$	457	33[a]	25[a]	HRS	1.50	62
$[Cr(CO)_3\{\eta^6\text{-}Ph\text{-}(E,E,E)\text{-}(CH{=}CH)_3\text{-}4\text{-}C_6H_4NO_2\}]$	$CHCl_3$	453	44[a]	31[a]	HRS	1.50	62
$[Cr(CO)_3\{\eta^6\text{-}Ph\text{-}(E)\text{-}CH{=}CH\text{-}4\text{-}C_6H_4NMe_2\}]$	$CHCl_3$	418	20[a]	14[a]	HRS	1.50	62
$[Cr(CO)_3\{\eta^6\text{-}Ph\text{-}(E,E)\text{-}(CH{=}CH)_2\text{-}4\text{-}C_6H_4NMe_2\}]$	$CHCl_3$	438	38[a]	27[a]	HRS	1.50	62
$[Cr(CO)_3\{\eta^6\text{-}Ph\text{-}(E)\text{-}CH{=}CH\text{-}2\text{-}C_4H_2S\text{-}5\text{-}NO_2\}]$	$CHCl_3$	488	30[a]	22[a]	HRS	1.50	62
$[Cr(CO)_2(PPh_3)(\eta^6\text{-}PhC{\equiv}C\text{-}4\text{-}C_6H_4NO_2)]$	$CHCl_3$	481	52[a]	41[a]	HRS	1.50	62
$[Cr(CO)_3(\eta^6\text{-}C_6H_4\text{-}2\text{-}CHO\text{-}1\text{-}C{\equiv}CC_6H_4\text{-}4\text{-}NMe_2)]$	$CHCl_3$	378	52[a]	39[a]	HRS	1.50	62
$[Cr(CO)_3(\eta^6\text{-}PhC{\equiv}CC{\equiv}C\text{-}4\text{-}C_6H_4NO_2)]$	$CHCl_3$	437	18[a]	14[a]	HRS	1.50	62
$[Cr(CO)_3(\eta^6\text{-}PhC{\equiv}CC{\equiv}C\text{-}4\text{-}C_6H_4NMe_2)]$	$CHCl_3$	414	34[a]	26[a]	HRS	1.50	62
$[Cr(CO)_3(\eta^6\text{-}PhC{\equiv}C\text{-}4\text{-}C_6H_4NO_2)]$	$CHCl_3$	435	10[a]	8[a]	HRS	1.50	62
$[Cr(CO)_3(\eta^6\text{-}PhC{\equiv}C\text{-}4\text{-}C_6H_4NMe_2)]$	$CHCl_3$	405	11[a]	9[a]	HRS	1.50	62
$[Cr(CO)_3(\eta^6\text{-}PhC{\equiv}C\text{-}4\text{-}C_6H_4CN)]$	$CHCl_3$	422	28	—	HRS	1.06	63
$[Cr(CO)_3(\eta^6\text{-}PhC{\equiv}C\text{-}2\text{-}C_4H_2S\text{-}5\text{-}CN)]$	$CHCl_3$	430	59	—	HRS	1.06	63
$[Cr(CO)_3(\eta^6\text{-}PhC{\equiv}C\text{-}2,5\text{-}C_4H_2S\text{-}2\text{-}C_4H_2S\text{-}5\text{-}CN)]$	$CHCl_3$	424	88	—	HRS	1.06	63
$[Cr\{C(OMe)\text{-}(E)\text{-}CH{=}CHNMe_2\}(CO)_5]$	$CHCl_3$	420	16	5	HRS	1.06	64
$[Cr\{C{=}C{=}C(NMe_2)_2\}(CO)_5]$	DMF	372	21	9.5	HRS	1.06	65
$[Cr\{C{=}C{=}C(NMe_2)CH{=}C(NMe_2)_2\}(CO)_5]$	DMF	388	22	9	HRS	1.06	65
$[Cr\{C{=}C{=}C{=}C{=}C(NMe_2)_2\}(CO)_5]$	DMF	424	100	31	HRS	1.06	65
$[Cr\{C{=}C{=}C{=}C{=}C(NEt_2)CMe{=}C(NMe_2)_2\}(CO)_5]$	DMF	418	125	40	HRS	1.06	65
$[Cr\{C\overline{C(Me)C(NEt_2)C(CH_2)_4}CH_2\}(CO)_5]$	DMF	394	22	85	HRS	1.06	66
$[Cr(CO)_4(1,10\text{-}phen)]$	CH_2Cl_2	500	−13	—	EFISH	1.91	61
$[Cr(CO)_5\{C(N\text{-}pyrrole)\text{-}5\text{-}(2,2'\text{-}bithiophene)\}]$	$CHCl_3$	335	18	10	HRS	1.06	64
$[Mn(CO)_3(\eta^5\text{-}C_5H_4C_7H_6)](BF_4)$	CH_2Cl_2	536	45	—	HRS	1.06	70
	MeCN	506	38				
$[Mn(CO)_2\{P(OMe)_3\}(\eta^5\text{-}C_5H_4C_7H_6)](BF_4)$	CH_2Cl_2	612	54	—	HRS	1.06	70
	MeCN	581	53				
$[Mn(CO)_2(PPh_3)(\eta^5\text{-}C_5H_4C_7H_6)](BF_4)$	CH_2Cl_2	643	29	—	HRS	1.06	70
	MeCN	630	39				
$[Mn(CO)_3(\eta^5\text{-}C_5H_4C{\equiv}CC_7H_6)](BF_4)$	CH_2Cl_2	537	40	—	HRS	1.06	70
	MeCN	491	94				

[Mn(CO)$_2${P(OMe)$_3$}(η^5-C$_5$H$_4$C≡CC$_7$H$_6$)](BF$_4$)	CH$_2$Cl$_2$	649	50	—	HRS	1.06	70
	MeCN	583	151				
[Mn(CO)$_2$(PPh$_3$)(η^5-C$_5$H$_4$C≡CC$_7$H$_6$)](BF$_4$)	CH$_2$Cl$_2$	740	44	—	HRS	1.06	70
	MeCN	650	73				
[Mn(CO)$_3$(η^5-C$_5$H$_4$C$_7$H$_6$)](BF$_4$)	MeNO$_2$	510	240	15	HRS	1.06	71
[Mn(CO)$_2$(PPh$_3$)(η^5-C$_5$H$_4$C≡CC$_7$H$_6$)](BF$_4$)	CH$_2$Cl$_2$	751	226	113	HRS	1.06	71
[Mn(CO)$_3$(η^5-C$_4$H$_3$SCH=CHC$_6$H$_4$-4-OMe)](BF$_4$)	MeNO$_2$	415	42	14	HRS	1.06	72,78
[Mn(CO)$_3$(η^5-C$_4$H$_3$SCH=CHC$_6$H$_4$-4-Me)](BF$_4$)	MeNO$_2$	405	59	21	HRS	1.06	72,78
[Mn(CO)$_3$(η^5-C$_4$H$_3$SCH=CHPh)](BF$_4$)	MeNO$_2$	390	68	27	HRS	1.06	72,78
[Mn(CO)$_3$(η^5-C$_4$H$_3$SCH=CHC$_6$H$_4$-4-Br)](BF$_4$)	MeNO$_2$	393	88	35	HRS	1.06	72,78
[Mn(CO)$_3$(η^5-C$_4$H$_3$SCH=CHC$_6$H$_4$-4-NO$_2$)](BF$_4$)	MeNO$_2$	400	101	38	HRS	1.06	72,78
[Mn(CO)$_2$(PPh$_3$)(η^5-C$_4$H$_3$SCH=CHC$_6$H$_4$-4-NO$_2$)](BF$_4$)	MeNO$_2$	384	115	48	HRS	1.06	72,78
[Mn(CO)$_3$(η^6-C$_6$H$_4$-4-NMe$_2$-1-CH=CH-2-C$_4$H$_3$S)](BF$_4$)	MeNO$_2$	470	62	11	HRS	1.06	72,78
[Mo(CO)$_4$(1,10-phen)]	CH$_2$Cl$_2$	496	−13	—	EFISH	1.91	61
[Ru{(*Z*)-HC=CHPh}(CO)$_2$(η^5-C$_5$Ph$_5$)]	THF	289	22	14	HRS	1.06	73
[Ru{(*Z*)-HC=CH-4-C$_6$H$_4$NO$_2$}(CO)$_2$(η^5-C$_5$Ph$_5$)]	THF	364	33	15	HRS	1.06	73
[Ru{(*Z*)-HC=CH-4-C$_6$H$_4$NO$_2$}(CO)(PMe$_2$Ph)(η^5-C$_5$Ph$_5$)]	THF	434	526	135	HRS	1.06	73
[RhCl(CO)$_2${NC$_5$H$_4$-4-(*E*)-CH=CH-4-C$_6$H$_4$OCH$_2$CH(Me)Et}]	CHCl$_3$	355	20.1	—	EFISH	1.91	76
[RhBr(CO)$_2${NC$_5$H$_4$-4-(*E*)-CH=CH-4-C$_6$H$_4$OCH$_2$CH(Me)Et}]	CHCl$_3$	357	23.9	—	EFISH	1.91	76
[W(CO)$_5${C(OMe)-(*E*)-CH=CHNMe$_2$}]	CHCl$_3$	411	34	12	HRS	1.06	64
[W{C=C=C(NMe$_2$)$_2$}(CO)$_5$]	DMF	368	25	11	HRS	1.06	65
[W{C=C=C=C=C(NMe$_2$)$_2$}(CO)$_5$]	DMF	424	102	31	HRS	1.06	65
[W(CO)$_5$(NC$_5$H$_5$)]	Toluene	332	−4.4	—	EFISH	1.91	32,60,61
[W(CO)$_5$(NC$_5$H$_4$-4-NH$_2$)]	DMSO	290	−2.1	—	EFISH	1.91	32,60,61
[W(CO)$_5$(NC$_5$H$_4$-4-Bu^n)]	Dioxane	328	−3.4	—	EFISH	1.91	32,60,61
[W(CO)$_5$(NC$_5$H$_4$-4-Ph)]	CHCl$_3$	330–340	−4.5	—	EFISH	1.91	32,60,61
[W(CO)$_5$(NC$_5$H$_4$-4-COMe)]	CHCl$_3$	420–440	−9.3	—	EFISH	1.91	32,60,61
[W(CO)$_5$(NC$_5$H$_4$-4-CHO)]	CHCl$_3$	420–440	−12	—	EFISH	1.91	32,60,61
[W(CO)$_5${NC$_5$H$_4$-4-(*E*)-CH=CHPh}]	CHCl$_3$	440	−5.7	—	EFISH	1.91	61
[W(CO)$_5${NC$_5$H$_4$-4-(*E*)-CH=CH-4-C$_6$H$_4$CHO}]	CHCl$_3$	420	−17	—	EFISH	1.91	61
[W(CO)$_5${NC$_5$H$_4$-4-(*E*)-CH=CH-4-C$_6$H$_4$NO$_2$}]	CHCl$_3$	425	−20	—	EFISH	1.91	37,61
[W(CO)$_5${NC$_5$H$_4$-4-(*E*)-CH=CH-4-C$_5$H$_4$N}]	—	—	−7	−5.2	EFISH	1.907	77
[W(CO)$_5$(pyrazine)]	CHCl$_3$	—	6.0	—	EFISH	1.91	61
[W(CO)$_4$(1,10-phen)]	CH$_2$Cl$_2$	492	−13	—	EFISH	1.91	61
[W(CO)$_4$(5-NO$_2$-1,10-phen)]	CH$_2$Cl$_2$	—	−18	—	EFISH	1.91	61
[ReBr(CO)$_3${bipy-4-(*E*)-CH=CH-4-C$_6$H$_4$NBu$_2$-4′-Me}]	CH$_2$Cl$_2$	475	71	31	EFISH	1.34	43
[ReBr(CO)$_3${bipy-4-(*E*)-CH=CH-4-C$_6$H$_4$OOct-4′-Me}]	CH$_2$Cl$_2$	367	11	7	EFISH	1.34	43
[ReBr(CO)$_3${NC$_6$H$_4$-4-(*E,E*)-CH=CHCH=CHC$_6$H$_4$-4-NMe$_2$}$_2$]	CHCl$_3$	438[b]	208	—	HRS	1.50	164
[IrCl(CO)$_2${NC$_5$H$_4$-4-(*E*)-CH=CH-4-C$_6$H$_4$OCH$_2$CH(Me)Et}]	CHCl$_3$	364	24.4	—	EFISH	1.91	76
[FcC≡C(η^7-C$_7$H$_6$)Cr(CO)$_3$]	CH$_2$Cl$_2$	600	570	105	HRS	1.06	51
[Fc{(*E*)-CH=CH(η^7-C$_7$H$_6$)Cr(CO)$_3$]	CH$_2$Cl$_2$	670	320	113	HRS	1.06	51
[Ru{C≡NCr(CO)$_5$}(PPh$_3$)$_2$(η^5-indenyl)]	CH$_2$Cl$_2$	392	25	10	HRS	1.06	74,75
[Ru{C≡C-(*E*)-CH=CH-4-C$_5$H$_4$N-1-Cr(CO)$_5$}(PPh$_3$)$_2$(η^5-indenyl)]	CH$_2$Cl$_2$	451	260	60	HRS	1.06	74,75
[Ru{C≡C-(*E*)-CH=CH-4-C$_6$H$_4$C≡NCr(CO)$_5$}(PPh$_3$)$_2$(η^5-indenyl)]	CH$_2$Cl$_2$	442	465	119	HRS	1.06	74,75

TABLE III
(CONTINUED)

Complex	Solvent	λ_{max} (nm)	β (10^{-30} esu)	β_0 (10^{-30} esu)	Technique	Fundamental (μm)	Ref.
$[\{Fe(CO)(\eta^5\text{-}C_5H_5)\}_2(\mu\text{-}CO)(\mu\text{-}C{=}CHCHC_{10}H_7)](BF_4)$	CH_2Cl_2	732	[c]	[c]	HRS	1.50	67
$[\{Fe(CO)(\eta^5\text{-}C_5H_5)\}_2(\mu\text{-}CO)(\mu\text{-}C{=}CHCHC_{10}H_4Me_2Pr^i)](BF_4)$	CH_2Cl_2	771	274	79	HRS	1.50	67
$[\{Fe(CO)(\eta^5\text{-}C_5H_5)\}_2(\mu\text{-}CO)\{\mu\text{-}C{=}CHCH{=}CH\text{-}4\text{-}C_6H_4CH{=}C(CN)_2\}]$	CH_2Cl_2	523	156	—	HRS	1.50	68
$[\{Fe(CO)(\eta^5\text{-}C_5H_5)\}_2(\mu\text{-}CO)\{\mu\text{-}C{=}CHCH{=}CH\text{-}2,5\text{-}C_4H_2SCH{=}C(CN)_2\}]$	CH_2Cl_2	565	227	—	HRS	1.50	68
$[\{Fe(CO)(\eta^5\text{-}C_5H_5)\}_2\{\mu\text{-}CO)(\mu\text{-}CCH{=}CH\text{-}4\text{-}C_6H_4\text{-}(E)\text{-}CH{=}CH\text{-}4\text{-}C_6H_4NO_2\}](BF_4)$	CH_2Cl_2	463	346[d]	68	HRS	1.06	69
$[\{Fe(CO)(\eta^5\text{-}C_5H_5)\}_2(\mu\text{-}CO)\{\mu\text{-}CCH{=}CH\text{-}4\text{-}C_6H_4\text{-}(E)\text{-}CH{=}CHPh\}](BF_4)$	CH_2Cl_2	513	792	43	HRS	1.06	69
$[\{Fe(CO)(\eta^5\text{-}C_5H_5)\}_2(\mu\text{-}CO)\{\mu\text{-}CCH{=}CH\text{-}2,5\text{-}C_4H_2S\text{-}(E)\text{-}CH{=}CH\text{-}4\text{-}C_6H_4NMe_2\}](BF_4)$	CH_2Cl_2	765	2443[d]	1260	HRS	1.06	69
	MeCN	696	1623[d]	661			
$[Ru\{C{\equiv}NW(CO)_5\}(PPh_3)_2(\eta^5\text{-indenyl})]$	CH_2Cl_2	392	40	15	HRS	1.06	74,75
$[Ru\{C{\equiv}C\text{-}(E)\text{-}CH{=}CH\text{-}4\text{-}C_5H_4N\text{-}1\text{-}W(CO)_5\}(PPh_3)_2(\eta^5\text{-indenyl})]$	CH_2Cl_2	462	535	71	HRS	1.06	74,75
$[Ru\{C{\equiv}C\text{-}(E)\text{-}CH{=}CH\text{-}4\text{-}C_6H_4C{\equiv}NW(CO)_5\}(PPh_3)_2(\eta^5\text{-indenyl})]$	CH_2Cl_2	456	700	150	HRS	1.06	74,75

[a] β_{333} values calculated by applying the two-level model on the π–π^* absorption band.
[b] CH_2Cl_2 solvent.
[c] Not detected.
[d] Fluorescence contribution.

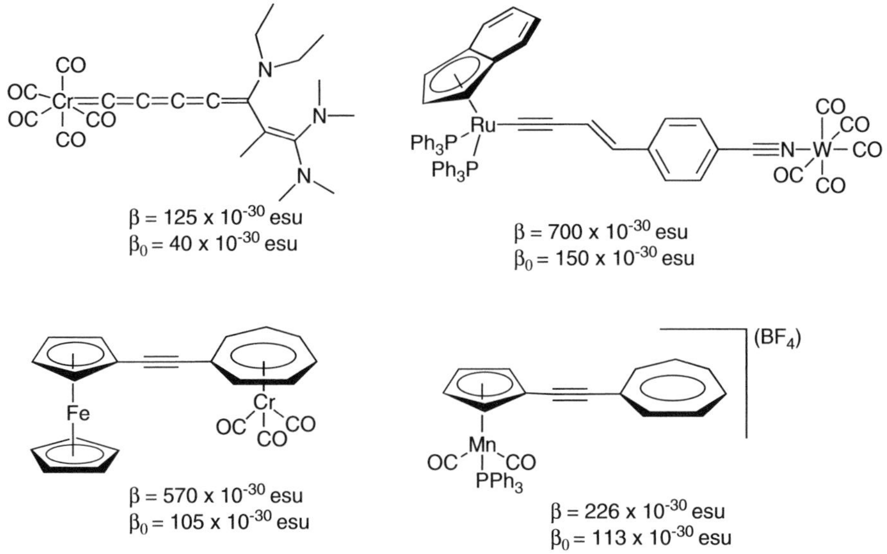

$\beta = 125 \times 10^{-30}$ esu
$\beta_0 = 40 \times 10^{-30}$ esu

$\beta = 700 \times 10^{-30}$ esu
$\beta_0 = 150 \times 10^{-30}$ esu

$\beta = 570 \times 10^{-30}$ esu
$\beta_0 = 105 \times 10^{-30}$ esu

$\beta = 226 \times 10^{-30}$ esu
$\beta_0 = 113 \times 10^{-30}$ esu

FIG. 5. Selected examples of metal carbonyl complexes with large β.

chromium- and tungsten-containing complexes do not differ significantly and are amongst the largest of the carbonyl-containing complexes.

While results for almost all metal carbonyl complexes have been reported as β values, data for several complexes are available as $\boldsymbol{\mu} \cdot \boldsymbol{\beta}$ products, and these are collected in Table IV, with efficient examples depicted in Fig. 6. Nonlinearities for these diiron complexes with oligo-ene barbiturate units increase on increasing the π-system length, as expected.

Table V contains results from selected and related carbonyl complexes at three specific wavelengths, together with data at one or two wavelengths for related complexes. The data for $[MCl(CO)_2(4\text{-}NC_5H_4NMe_2)]$ (M = Rh, Ir) show minimal dispersion of the quadratic nonlinearity; in contrast, β values for $[RhCl(CO)_2(4\text{-}NC_5H_4CH{=}CH\text{-}4\text{-}C_6H_4NMe_2)]$ at 1.06 and 1.34 μm vary significantly, due to substantial resonance enhancement at the former wavelength.

Table VI contains data from vinylidene, alkynyl, and nitrile complexes, with Fig. 7 showing representative examples with large nonlinearities. Although ruthenium complexes have been the most intensively investigated, iron, cobalt, nickel, osmium, and gold complexes have also been of interest. In several cases, identical alkynyl group and co-ligands permit assessment of the effect of metal variation on quadratic NLO response. Experimentally obtained quadratic nonlinearities for (cyclopentadienyl)bis(phosphine)ruthenium complexes at 1.064 μm are dispersion-enhanced, but both these data and two-level-corrected values increase on π-bridge lengthening, proceeding from one-phenyl-ring to biphenyl, yne-linked and then ene-linked two-phenyl-ring alkynyl ligand, with the azo-linked analogue the most efficient. The resonance-enhanced experimentally observed nonlinearities for ruthenium complexes are significantly larger than those for analogous gold

TABLE IV

MOLECULAR SECOND-ORDER NLO RESULTS AS $\mu \cdot \beta$ PRODUCTS FOR METAL CARBONYL COMPLEXES

Complex	Solvent	λ_{max} (nm)	$\mu \cdot \beta$ (10^{-48} esu)	$\mu \cdot \beta_0$ (10^{-48} esu)	Technique	Fundamental (μm)	Ref.
[{Fe(CO)$_2$(η^5-C_5H_5)}$_2$(μ-CO){μ-(*E*)-CH=CH(*N,N'*-diethylbarbiturate}]	a	474	170	120	EFISH	a	79
[{Fe(CO)$_2$(η^5-C_5H_5)}$_2$(μ-CO){μ-(*E,E*)-(CH=CH)$_2$(*N,N'*-diethylbarbiturate)}]	a	564	1100	660	EFISH	a	79
[{Fe(CO)$_2$(η^5-C_5H_5)}$_2$(μ-CO){μ-(*E,E,E*)-(CH=CH)$_3$(*N,N'*-diethylbarbiturate)}]	a	646	3100	1500	EFISH	a	79

[a]Not stated.

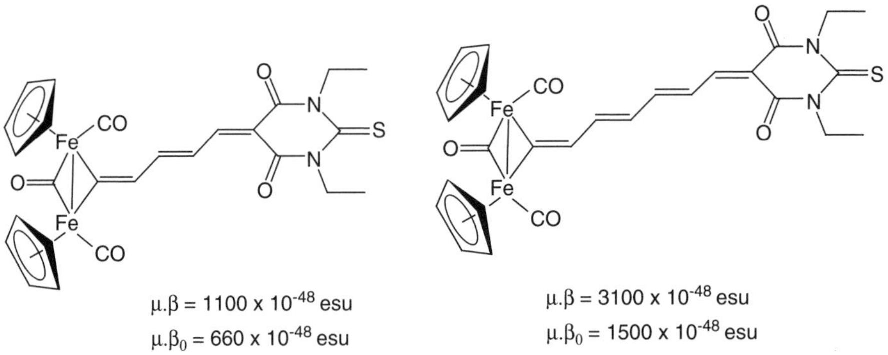

μ.β = 1100 x 10^{-48} esu
μ.β$_0$ = 660 x 10^{-48} esu

μ.β = 3100 x 10^{-48} esu
μ.β$_0$ = 1500 x 10^{-48} esu

FIG. 6. Selected examples of metal carbonyls with large $\mu \cdot \beta$.

complexes, but the same trend is seen with two-level-corrected values, a not-unexpected result because the 18 valence electron ruthenium(II) should be a better donor than the 14 valence electron gold(I). The (triphenylphosphine)gold-containing alkynyl complexes have nonlinearities similar to analogous compounds with the strongest organic donors. This suggests that the more NLO-efficient 18 valence electron organometallic complexes have stronger ligated metal donors than are possible in purely organic systems. A third series of related complexes incorporating (cyclopentadienyl)(triphenylphosphine)nickel units possesses an 18 electron metal center that is less easily oxidizable than ruthenium; their data are also substantially resonance enhanced (although the relative orderings were maintained with two-level-corrected values), and their nonlinearities are greater than those of related gold complexes, but smaller than those of the analogous ruthenium complexes. The Ru(PPh_3)$_2$(η^5-C_5H_5) unit is on average 10 and 3 times as efficient as the Au(PPh_3) and Ni(PPh_3)(η^5-C_5H_5) moieties, respectively, which has permitted "figures of merit" for these donor groups to be defined.[89] The only alkynyl complexes for which 3d, 4d, and 5d metal examples from the same group have been explored, namely *trans*-[M(C≡C-4-$C_6H_4NO_2$)Cl{(*R,R*)-diph}$_2$] and [M(C≡C-4-$C_6H_4NO_2$)(dppe)(η^5-C_5H_5)] [M = Fe, Ru, Os; diph = 1,2-bis(methylphenylphosphino)

TABLE V

MOLECULAR SECOND-ORDER NLO RESULTS FOR RELATED METAL CARBONYL COMPLEXES AT UP TO THREE WAVELENGTHS

Complex	Solvent	λ_{max} (nm)	$\beta_{1.06}$ (10^{-30} esu)	$\beta_{1.34}$ (10^{-30} esu)	$\beta_{1.91}$ (10^{-30} esu)	Technique	Ref.
$[RhCl(CO)_2(NC_5H_5)]$	$CHCl_3$	—	2.6	—	—	EFISH	80
$[RhCl(CO)_2(4\text{-}NC_5H_4Bu^t)]$	$CHCl_3$	—	~0	—	—	EFISH	80
$[RhCl(CO)_2(4\text{-}NC_5H_4NMe_2)]$	$CHCl_3$	289	8.7	12	6.4	EFISH	80
$[RhCl(CO)_2(4\text{-}NC_5H_4CN)]$	$CHCl_3$	—	−4.3	—	—	EFISH	80
$[RhCl(CO)_2(4\text{-}NC_5H_4CH{=}CH\text{-}4\text{-}C_6H_4NMe_2)]$	$CHCl_3$	421	177	111	—	EFISH	80
$[RhCl(CO)_2\{4\text{-}NC_5H_4(CH{=}CH)_2\text{-}4\text{-}C_6H_4NMe_2\}]$	$CHCl_3$	442	—	131	—	EFISH	80
$[OsCl_2(CO)_3(NC_5H_5)]$	$CHCl_3$	—	~0	—	—	EFISH	80
$[OsCl_2(CO)_3(4\text{-}NC_5H_4NMe_2)]$	$CHCl_3$	290	6	—	4.4	EFISH	80
$[OsCl_2(CO)_3\{4\text{-}NC_5H_4C(O)Me\}]$	$CHCl_3$	—	~2	—	—	EFISH	80
$[OsCl_2(CO)_3(4\text{-}NC_5H_4CN)]$	$CHCl_3$	—	~2	—	—	EFISH	80
$[OsCl_2(CO)_3(4\text{-}NC_5H_4CH{=}CH\text{-}4\text{-}C_6H_4NMe_2)]$	$CHCl_3$	435	160	—	—	EFISH	80
$[OsCl_2(CO)_3\{4\text{-}NC_5H_4(CH{=}CH)_2\text{-}4\text{-}C_6H_4NMe_2\}]$	$CHCl_3$	456	—	116	—	EFISH	80
$[IrCl(CO)_2(4\text{-}NC_5H_4Bu^t)]$	$CHCl_3$	—	0.1	—	—	EFISH	80
$[IrCl(CO)_2(4\text{-}NC_5H_4NMe_2)]$	$CHCl_3$	291	9	8	6.5	EFISH	80
$[IrCl(CO)_2(4\text{-}NC_5H_4CN)]$	$CHCl_3$	—	−9	—	—	EFISH	80
$[IrCl(CO)_2(4\text{-}NC_5H_4CH{=}CH\text{-}4\text{-}C_6H_4NMe_2)]$	$CHCl_3$	431	—	128	—	EFISH	80
$[IrCl(CO)_2\{4\text{-}NC_5H_4(CH{=}CH)_2\text{-}4\text{-}C_6H_4NMe_2\}]$	$CHCl_3$	449	—	135	—	EFISH	80

TABLE VI
MOLECULAR SECOND-ORDER NLO RESULTS FOR VINYLIDENE, ALKYNYL AND NITRILE COMPLEXES

Complex	Solvent	λ_{max} (nm)	β (10^{-30} esu)	β_0 (10^{-30} esu)	Technique	Fundamental (μm)	Ref.
$[Fe(C{\equiv}CPh)(dppe)(\eta^5\text{-}C_5Me_5)](PF_6)$	CH_2Cl_2	663	80	—	HRS	1.06	82
$[Fe(C{\equiv}CPh)(dppe)(\eta^5\text{-}C_5Me_5)]$	CH_2Cl_2	348	52	24	HRS	1.06	82
$[Fe(C{\equiv}C\text{-}4\text{-}C_6H_4NO_2)(CO)_2(\eta^5\text{-}C_5H_5)]$	THF	370	49	22	HRS	1.06	81
$[Fe(C{\equiv}C\text{-}4\text{-}C_6H_4NO_2)(dppe)(\eta^5\text{-}C_5H_5)]$	THF	498	665	64	HRS	1.06	81
	$CHCl_3$	504	1160	—	HRS	1.06	83
$[Fe(C{\equiv}C\text{-}4\text{-}C_6H_4\text{-}4\text{-}C_6H_4NO_2)(dppe)(\eta^5\text{-}C_5H_5)]$	$CHCl_3$	479	1150	—	HRS	1.06	83
$[Fe\{C{\equiv}CC_6H_4\text{-}4\text{-}(E)\text{-}CH{=}CHC_6H_4\text{-}4\text{-}NO_2\}(dppe)(\eta^5\text{-}C_5H_5)]$	$CHCl_3$	506	2315	—	HRS	1.06	83
$(-)_{436}$-*trans*-$[Fe(C{\equiv}C\text{-}4\text{-}C_6H_4NO_2)Cl\{(R,R)\text{-}diph\}_2]$	THF	543	440	−14	HRS	1.06	84
$[1,3\text{-}C_6H_4\{(C{\equiv}C)Fe(dppe)(\eta^5\text{-}C_5Me_5)\}_2](PF_6)$	CH_2Cl_2	650	150	—	HRS	1.06	82
$[1,3\text{-}C_6H_4\{(C{\equiv}C)Fe(dppe)(\eta^5\text{-}C_5Me_5)\}_2](PF_6)_2$	CH_2Cl_2	662	200	—	HRS	1.06	82
$[1,3\text{-}C_6H_4\{(C{\equiv}C)Fe(dppe)(\eta^5\text{-}C_5Me_5)\}_2]$	CH_2Cl_2	349	210	98	HRS	1.06	82
$[1,3,5\text{-}C_6H_3\{(C{\equiv}C)Fe(dppe)(\eta^5\text{-}C_5Me_5)\}_3](PF_6)$	CH_2Cl_2	710	190	—	HRS	1.06	82
$[1,3,5\text{-}C_6H_3\{(C{\equiv}C)Fe(dppe)(\eta^5\text{-}C_5Me_5)\}_3](PF_6)_2$	CH_2Cl_2	688	170	—	HRS	1.06	82
$[1,3,5\text{-}C_6H_3\{(C{\equiv}C)Fe(dppe)(\eta^5\text{-}C_5Me_5)\}_3](PF_6)_3$	CH_2Cl_2	662	53	—	HRS	1.06	82
$[1,3,5\text{-}C_6H_3\{(C{\equiv}C)Fe(dppe)(\eta^5\text{-}C_5Me_5)\}_3]$	CH_2Cl_2	351	175	87	HRS	1.06	82
$[1,4\text{-}C_6H_4\{(C{\equiv}C)Fe(dppe)(\eta^5\text{-}C_5Me_5)\}_2](PF_6)$	CH_2Cl_2	702	400	—	HRS	1.06	82
$[1,4\text{-}C_6H_4\{(C{\equiv}C)Fe(dppe)(\eta^5\text{-}C_5Me_5)\}_2](PF_6)_2$	CH_2Cl_2	702	200	—	HRS	1.06	82
$[1,4\text{-}C_6H_4\{(C{\equiv}C)Fe(dppe)(\eta^5\text{-}C_5Me_5)\}_2]$	CH_2Cl_2	413	180	60	HRS	1.06	82
$[Fe(N{\equiv}C\text{-}4\text{-}C_6H_4NO_2)(dppe)(\eta^5\text{-}C_5H_5)](PF_6)$	CH_2Cl_2	460	395	115	HRS	1.06	85
	MeOH	468	410	—	HRS	1.06	86
	$CHCl_3$	460	375	—	HRS	1.06	86
$[Fe(N{\equiv}C\text{-}4\text{-}C_6H_4\text{-}4\text{-}C_6H_4NO_2)(dppe)(\eta^5\text{-}C_5H_5)](PF_6)$	MeOH	375	276	—	HRS	1.06	86
	$CHCl_3$	372	240	—	HRS	1.06	86
$[Fe\{N{\equiv}C\text{-}(E)\text{-}CH{=}CH\text{-}4\text{-}C_6H_4NO_2\}(dppe)(\eta^5\text{-}C_5H_5)](PF_6)$	CH_2Cl_2	484	570	112	HRS	1.06	85
$[Fe(N{\equiv}C\text{-}4\text{-}C_6H_4NO_2)\{(+)\text{-}diop\}(\eta^5\text{-}C_5H_5)](PF_6)$	CH_2Cl_2	454	380	86	HRS	1.06	85
$[Fe(N{\equiv}C\text{-}4\text{-}C_6H_4\text{-}4\text{-}C_6H_4NO_2)\{(+)\text{-}diop\}(\eta^5\text{-}C_5H_5)](PF_6)$	CH_2Cl_2	403	190	78	HRS	1.06	85
$[Fe\{N{\equiv}C\text{-}(E)\text{-}CH{=}CH\text{-}4\text{-}C_6H_4NO_2\}\{(+)\text{-}diop\}(\eta^5\text{-}C_5H_5)](PF_6)$	CH_2Cl_2	466	520	105	HRS	1.06	85
$[Fe(N{\equiv}C\text{-}4\text{-}C_6H_4NO_2)\{(R)\text{-}prophos\}(\eta^5\text{-}C_5H_5)](PF_6)$	$CHCl_3$	475	545	—	HRS	1.06	87
$[Co(N{\equiv}C\text{-}4\text{-}C_6H_4NMe_2)(dppe)(\eta^5\text{-}C_5H_5)](PF_6)_2$	MeOH	332	<40	—	HRS	1.06	86
$[Co(N{\equiv}C\text{-}4\text{-}C_6H_4NO_2)(dppe)(\eta^5\text{-}C_5H_5)](PF_6)_2$	MeOH	419	≈ 45	—	HRS	1.06	86
$[Co(N{\equiv}C\text{-}4\text{-}C_6H_4Ph)(dppe)(\eta^5\text{-}C_5H_5)](PF_6)_2$	MeOH	298	<25	—	HRS	1.06	86
$[Co(N{\equiv}C\text{-}4\text{-}C_6H_4\text{-}4\text{-}C_6H_4NO_2)(dppe)(\eta^5\text{-}C_5H_5)](PF_6)_2$	[a]	420	≈ 35	—	HRS	1.06	86
$[Ni(C{\equiv}CPh)(PPh_3)(\eta^5\text{-}C_5H_5)]$	THF	307	24	15	HRS	1.06	88,89
$[Ni(C{\equiv}C\text{-}2\text{-}C_5H_4N)(PPh_3)(\eta^5\text{-}C_5H_5)]$	THF	415	25	8	HRS	1.06	90
$[Ni(C{\equiv}C\text{-}2\text{-}C_5H_4N\text{-}5\text{-}NO_2)(PPh_3)(\eta^5\text{-}C_5H_5)]$	THF	456	186	41	HRS	1.06	90
$[Ni(C{\equiv}C\text{-}4\text{-}C_6H_4NO_2)(PPh_3)(\eta^5\text{-}C_5H_5)]$	THF	439	221	59	HRS	1.06	89
$[Ni(C{\equiv}C\text{-}4\text{-}C_6H_4\text{-}4\text{-}C_6H_4NO_2)(PPh_3)(\eta^5\text{-}C_5H_5)]$	THF	413	193	65	HRS	1.06	89

[Ni{C≡C-4-C_6H_4-(*E*)-CH=CH-4-$C_6H_4NO_2$}(PPh_3)(η^5-C_5H_5)]	THF	437	445	120	HRS	1.06	89
[Ni{C≡C-4-C_6H_4-(*Z*)-CH=CH-4-$C_6H_4NO_2$}(PPh_3)(η^5-C_5H_5)]	THF	417	145	47	HRS	1.06	89
[Ni(C≡C-4-C_6H_4C≡C-4-$C_6H_4NO_2$)(PPh_3)(η^5-C_5H_5)]	THF	417	326	106	HRS	1.06	89
[Ni(C≡C-4-C_6H_4N=CH-4-$C_6H_4NO_2$)(PPh_3)(η^5-C_5H_5)]	THF	448	387	93	HRS	1.06	89
[Ni(N≡C-4-$C_6H_4NMe_2$)(PPh_3)(η^5-C_5H_5)](PF_6)	MeOH	291	18.4	—	HRS	1.06	86,91
	$CHCl_3$	298	14	—	HRS	1.06	86,91
[Ni(N≡C-4-$C_6H_4NO_2$)(PPh_3)(η^5-C_5H_5)](PF_6)	$CHCl_3$	419	93	—	HRS	1.06	86
[Ni(N≡C-4-C_6H_4Ph)(PPh_3)(η^5-C_5H_5)](PF_6)	MeOH	267	17	—	HRS	1.06	86,91
	$CHCl_3$	282	18	—	HRS	1.06	86,91
[Ni(N≡C-4-C_6H_4-4-$C_6H_4NO_2$)(PPh_3)(η^5-C_5H_5)](PF_6)	$CHCl_3$	299	45	—	HRS	1.06	86
[1-(HC≡C)-3,5-C_6H_3{(C≡C)Ni(PPh_3)(η^5-C_5H_5)}$_2$]	THF	316	94	55	HRS	1.06	88
[Ru(C≡CPh)(PPh_3)$_2$(η^5-C_5H_5)]	THF	310	89	45	HRS	1.06	75,92
[Ru(C≡C-4-C_6H_4CHO)(PPh_3)$_2$(η^5-C_5H_5)]	THF	400	120	45	HRS	1.06	93
[Ru(C≡C-4-C_6H_4CH{OC(O)Me}$_2$)(PPh_3)$_2$(η^5-C_5H_5)]	THF	326	68	38	HRS	1.06	93
[Ru(C≡C-4-$C_6H_4NO_2$)(CO)$_2$(η^5-C_5H_5)]	THF	364	58	27	HRS	1.06	81
[Ru(C≡C-4-$C_6H_4NO_2$)(PMe_3)$_2$(η^5-C_5H_5)]	THF	477	248	39	HRS	1.06	92,94
[Ru(C≡C-4-$C_6H_4NO_2$)(PPh_3)$_2$(η^5-C_5H_5)]	THF	460	468	96	HRS	1.06	81,92,94
[Ru(C≡C-4-$C_6H_4NO_2$)(dppe)(η^5-C_5H_5)]	THF	447	664	161	HRS	1.06	81
[Ru(C≡C-4-C_6H_4-4-$C_6H_4NO_2$)(PPh_3)$_2$(η^5-C_5H_5)]	THF	448	560	134	HRS	1.06	92
[Ru(C≡C-4-C_6H_4C≡C-4-$C_6H_4NO_2$)(PPh_3)$_2$(η^5-C_5H_5)]	THF	446	865	212	HRS	1.06	92
[Ru{C≡C-4-C_6H_4-(*E*)-CH=CH-4-$C_6H_4NO_2$}(PPh_3)$_2$(η^5-C_5H_5)]	THF	476	1455	232	HRS	1.06	92
	$CHCl_3$	484	2270	—	HRS	1.06	83
	THF	476	1464	234	EFISH	1.06	92,94
[Ru(C≡C-4-C_6H_4N=CH-4-$C_6H_4NO_2$)(PPh_3)$_2$(η^5-C_5H_5)]	THF	496	840	86	HRS	1.06	92
	THF	496	760	78	EFISH	1.06	92,94
[Ru(C≡C-4-C_6H_4C≡C-4-C_5H_4NMe)(PPh_3)$_2$(η^5-C_5H_5)](PF_6)	CH_2Cl_2	558	1400	102	HRS	1.06	95
[Ru{C≡C-4-C_6H_4-(*E*)-CH=CH-4-C_5H_4NMe}(PPh_3)$_2$(η^5-C_5H_5)](PF_6)	CH_2Cl_2	582	1600	154	HRS	1.06	95
[Ru{C≡C-4-C_6H_4-(*E*)-CH=CH-4-$C_6H_4NO_2$}(PPh_3)$_2$(η^5-C_5H_5)]	THF	476	1455	232	HRS	1.06	94
			186	105	HRS	1.56	96
[Ru(C≡C-4-C_6H_4N=CH-4-$C_6H_4NO_2$)(PPh_3)$_2$(η^5-C_5H_5)]	THF	496	840	86	HRS	1.06	94
[Ru(C≡C-4-C_6H_4C≡C-2-C_4H_2S-5-NO_2)(PPh_3)$_2$(η^5-C_5H_5)]	CH_2Cl_2	505	210	109	HRS	1.56	96
[Ru{C≡C-4-C_6H_4-(*E*)-CH=CH-2-C_4H_2S-5-NO_2}(PPh_3)$_2$(η^5-C_5H_5)]	CH_2Cl_2	533	294	138	HRS	1.56	96
[Ru{C≡C-2-C_4H_2S-5-(*E*)-CH=CH-4-$C_6H_4NO_2$}(PPh_3)$_2$(η^5-C_5H_5)]	CH_2Cl_2	522	333	163	HRS	1.56	96
[Ru{C≡C-2-C_4H_2S-5-(*E*)-CH=CH-2-C_4H_2S-5-(*E*)-CH=CH-2-C_4H_2S-5-NO_2}(PPh_3)$_2$(η^5-C_5H_5)]	CH_2Cl_2	536	419	195	HRS	1.56	96
[Ru(C≡C-2-C_5H_4N)(PPh_3)$_2$(η^5-C_5H_5)]	THF	331	18	10	HRS	1.06	90
[Ru(C≡C-4-C_5H_4NMe)(PPh_3)$_2$(η^5-C_5H_5)](PF_6)	CH_2Cl_2	460	80	16	HRS	1.06	95
[Ru(C≡C-2-C_5H_3N-5-NO_2)(PPh_3)$_2$(η^5-C_5H_5)]	THF	468	622	113	HRS	1.06	90
[Ru(C≡C-4-C_6H_4N=CH-2-C_4H_2S-5-NO_2)(PPh_3)$_2$(η^5-C_5H_5)]	CH_2Cl_2	562	308	129	HRS	1.56	96
[Ru{C≡C-4-C_6H_4N=$\overline{\text{CCH}=\text{CBu}^t\text{C(O)CBu}^t=\text{CH}}$}($PPh_3$)$_2$($\cdot^5$-$C_5H_5$)]	THF	622	658	159	HRS	1.06	97
[Ru(C=CPhN=NPh)(PPh_3)$_2$(η^5-C_5H_5)](BF_4)	Acetone	363	14	6.6	HRS	1.06	98
[Ru(C=CPhN=N-2-C_6H_4OMe)(PPh_3)$_2$(η^5-C_5H_5)]Cl	Acetone	373	22	10	HRS	1.06	98
[Ru(C=CPhN=N-3-C_6H_4OMe)(PPh_3)$_2$(η^5-C_5H_5)](BF_4)	Acetone	382	23	10	HRS	1.06	98
[Ru(C=CPhN=N-4-C_6H_4OMe)(PPh_3)$_2$(η^5-C_5H_5)]Cl	Acetone	370	26	12	HRS	1.06	98

TABLE VI
(CONTINUED)

Complex	Solvent	λ_{max} (nm)	β (10^{-30} esu)	β_0 (10^{-30} esu)	Technique	Fundamental (μm)	Ref.
$[Ru(C{=}CPhN{=}N\text{-}4\text{-}C_6H_4NO_2)(PPh_3)_2(\eta^5\text{-}C_5H_5)](BF_4)$	CH_2Cl_2	413	184	62	HRS	1.06	98
$[Ru(C{=}CPhN{=}N\text{-}4\text{-}C_6H_4NO_2)(PPh_3)_2(\eta^5\text{-}C_5H_5)]Cl$	CH_2Cl_2	413	137	46	HRS	1.06	98
$[Ru(C{=}CPhN{=}N\text{-}4\text{-}C_6H_4NO_2)(PPh_3)_2(\eta^5\text{-}C_5H_5)]Br$	CH_2Cl_2	413	136	45	HRS	1.06	98
$[Ru(C{=}CPhN{=}N\text{-}4\text{-}C_6H_4NO_2)(PPh_3)_2(\eta^5\text{-}C_5H_5)]I$	CH_2Cl_2	413	134	45	HRS	1.06	98
	Acetone	417	150	48			
	THF	415	101	33			
$[Ru(C{=}CPhN{=}N\text{-}4\text{-}C_6H_4NO_2)(PPh_3)_2(\eta^5\text{-}C_5H_5)](4\text{-}MeC_6H_4SO_3)$	CH_2Cl_2	413	164	55	HRS	1.06	98
$[Ru(C{=}CPhN{=}N\text{-}4\text{-}C_6H_4NO_2)(PPh_3)_2(\eta^5\text{-}C_5H_5)](NO_3)$	CH_2Cl_2	413	181	61	HRS	1.06	98
$[Ru\{C{=}CPhN{=}N\text{-}3,5\text{-}C_6H_3(NO_2)_2\}(PPh_3)_2(\eta^5\text{-}C_5H_5)]Cl$	Acetone	395	33	13	HRS	1.06	98
$[Ru\{C{\equiv}C\text{-}4\text{-}C_6H_4\text{-}(E)\text{-}N{=}N\text{-}4\text{-}C_6H_4NO_2\}(PPh_3)_2(\eta^5\text{-}C_5H_5)]$	THF	565	1627	149	HRS	1.06	99
$[Ru(C{\equiv}C\text{-}4\text{-}C_6H_4NO_2)(dppm)(\eta^5\text{-indenyl})]$	CH_2Cl_2	456	540	117	HRS	1.06	100
$[Ru(C{\equiv}C\text{-}4\text{-}C_6H_4NO_2)(dppe)(\eta^5\text{-indenyl})]$	CH_2Cl_2	459	516	107	HRS	1.06	100
$[Ru(C{=}CH\text{-}4\text{-}C_6H_4NO_2)(PPh_3)_2(\eta^5\text{-indenyl})](PF_6)$	CH_2Cl_2	379	116	50	HRS	1.06	74,100
$[Ru(C{\equiv}C\text{-}4\text{-}C_6H_4NO_2)(PPh_3)_2(\eta^5\text{-indenyl})]$	CH_2Cl_2	476	746	119	HRS	1.06	74,100
$[Ru(C{\equiv}C\text{-}4\text{-}C_6H_4C{\equiv}C\text{-}4\text{-}C_6H_4NO_2)(PPh_3)_2(\eta^5\text{-indenyl})]$	CH_2Cl_2	463	1027	202	HRS	1.06	100
$[Ru\{C{\equiv}CCH{=}C(3\text{-}C_6H_4NO_2)_2\}(PPh_3)_2(\eta^5\text{-indenyl})]$	CH_2Cl_2	398	177	67	HRS	1.06	74
$[Ru\{C{\equiv}CCH{=}C(4\text{-}C_6H_4NO_2)_2\}(PPh_3)_2(\eta^5\text{-indenyl})]$	CH_2Cl_2	345	48	25	HRS	1.06	100
$[Ru\{C{\equiv}C\text{-}(E)\text{-}CH{=}CH\text{-}4\text{-}C_5H_4N\}(PPh_3)_2(\eta^5\text{-indenyl})]$	CH_2Cl_2	399	100	37	HRS	1.06	74,100
$[Ru\{C{\equiv}C\text{-}(E)\text{-}CH{=}CH\text{-}4\text{-}C_6H_4NO_2\}(PPh_3)_2(\eta^5\text{-indenyl})]$	CH_2Cl_2	507	1257	89	HRS	1.06	74,100
$[Ru\{C{\equiv}C\text{-}(E)\text{-}CH{=}CH\text{-}4\text{-}C_6H_4CN\}(PPh_3)_2(\eta^5\text{-indenyl})]$	CH_2Cl_2	427	238	71	HRS	1.06	74,100
$[Ru\{C{\equiv}C\text{-}(E,E)\text{-}(CH{=}CH)_2\text{-}4\text{-}C_6H_4NO_2\}(PPh_3)_2(\eta^5\text{-indenyl})]$	CH_2Cl_2	523	1320	34	HRS	1.06	74,100
$[Ru(C{\equiv}C\text{-}4\text{-}C_6H_4N{=}CH\text{-}4\text{-}C_6H_4NO_2)(PPh_3)_2(\eta^5\text{-indenyl})]$	CH_2Cl_2	509	1295	85	HRS	1.06	100
$[Ru(C{\equiv}CCH{=}CH\text{-}2\text{-}C_4H_2O\text{-}5\text{-}NO_2)(PPh_3)_2(\eta^5\text{-indenyl})]$	CH_2Cl_2	550	908	43	HRS	1.06	100
$[Ru\{C{\equiv}C\text{-}(E)\text{-}CH{=}CH\text{-}2\text{-}C_4H_2S\text{-}5\text{-}NO_2\}(PPh_3)_2(\eta^5\text{-indenyl})]$	CH_2Cl_2	598	487	88	HRS	1.06	100
$[Ru\{C{\equiv}C\text{-}(E)\text{-}CH{=}CH\text{-}4\text{-}C_5H_4NCr(CO)_5\}(PPh_3)_2(\eta^5\text{-indenyl})]$	CH_2Cl_2	451	260	60	HRS	1.06	74,75
$[Ru\{C{\equiv}C\text{-}(E)\text{-}CH{=}CH\text{-}4\text{-}C_5H_4NW(CO)_5\}(PPh_3)_2(\eta^5\text{-indenyl})]$	CH_2Cl_2	462	535	71	HRS	1.06	74,75
$[Ru\{C{\equiv}C\text{-}(E)\text{-}CH{=}CH\text{-}4\text{-}C_6H_4C{\equiv}NCr(CO)_5\}(PPh_3)_2(\eta^5\text{-indenyl})]$	CH_2Cl_2	442	465	119	HRS	1.06	74,75
$[Ru\{C{\equiv}C\text{-}(E)\text{-}CH{=}CH\text{-}4\text{-}C_6H_4C{\equiv}NW(CO)_5\}(PPh_3)_2(\eta^5\text{-indenyl})]$	CH_2Cl_2	456	700	150	HRS	1.06	74,75
$[Ru\{C{\equiv}C\text{-}(E)\text{-}CH{=}CH\text{-}4\text{-}C_6H_4C{\equiv}NRu(NH_3)_5\}(PPh_3)_2(\eta^5\text{-indenyl})]$	Acetone	442	315	80	HRS	1.06	74,75
trans-$[Ru(C{=}CHPh)Cl(dppm)_2](PF_6)$	THF	320	24	16	HRS	1.06	101
trans-$[Ru(C{\equiv}CPh)Cl(dppm)_2]$	THF	308	20	12	HRS	1.06	88,102
trans-$[Ru\{C{=}CH\text{-}2\text{-}C_6H_4CHO\}Cl(dppm)_2](PF_6)$	THF	555	27	2	HRS	1.06	93
trans-$[Ru\{C{=}CH\text{-}3\text{-}C_6H_4CHO\}Cl(dppm)_2](PF_6)$	THF	320	45	26	HRS	1.06	93
trans-$[Ru\{C{\equiv}C\text{-}3\text{-}C_6H_4CHO\}Cl(dppm)_2]$	THF	321	58	34	HRS	1.06	93
trans-$[Ru(C{=}CH\text{-}4\text{-}C_6H_4CHO)Cl(dppm)_2](PF_6)$	THF	403	108	39	HRS	1.06	101
trans-$[Ru(C{\equiv}C\text{-}4\text{-}C_6H_4CHO)Cl(dppm)_2]$	THF	405	106	38	HRS	1.06	101

trans-[Ru{C=CH-4-$C_6H_4\overline{CHO(CH_2)_3O}$}Cl$(dppm)_2$]$(PF_6)$	THF	317	64	38	HRS	1.06	93
trans-[Ru{C≡C-4-$C_6H_4\overline{CHO(CH_2)_3O}$}Cl$(dppm)_2$]	THF	320	61	35	HRS	1.06	93
trans-[Ru(C=CH-4-C_6F_4OMe)Cl$(dppm)_2$](PF_6)	THF	334	32	17	HRS	1.06	103
trans-[Ru(C≡C-4-C_6F_4OMe)Cl$(dppm)_2$]	THF	337	26	14	HRS	1.06	103
trans-[Ru(C=CH-4-$C_6H_4NO_2$)Cl$(dppm)_2$](PF_6)	THF	470	721	127	HRS	1.06	101
trans-[Ru(C≡C-4-$C_6H_4NO_2$)Cl$(dppm)_2$]	THF	473	767	129	HRS	1.06	102
trans-[Ru(C≡C-2-C_5H_4N)Cl$(dppm)_2$]	THF	351	35	19	HRS	1.06	102
trans-[Ru(C≡C-2-C_5H_4N-5-NO_2)Cl$(dppm)_2$]	THF	490	468	56	HRS	1.06	102
trans-[Ru(C≡C-4-C_6H_4-4-$C_6H_4NO_2$)Cl$(dppm)_2$]	THF	465	933	178	HRS	1.06	102
trans-[Ru(C=CH-4-C_6H_4C≡CPh)Cl$(dppm)_2$](PF_6)	THF	380	64	31	HRS	1.06	101
trans-[Ru(C≡C-4-C_6H_4C≡CPh)Cl$(dppm)_2$]	THF	381	101	43	HRS	1.06	101
trans-[Ru(C=CH-4-C_6H_4C≡C-4-$C_6H_4NO_2$)Cl$(dppm)_2$](PF_6)	THF	326	424	122	HRS	1.06	101
trans-[Ru(C≡C-4-C_6H_4C≡C-4-$C_6H_4NO_2$)Cl$(dppm)_2$]	THF	464	833	161	HRS	1.06	101
trans-[Ru{C=CH-4-C_6H_4-(*E*)-CH=CH-4-$C_6H_4NO_2$}Cl$(dppm)_2$](PF_6)	THF	369	1899	314	HRS	1.06	101
trans-[Ru{C≡C-4-C_6H_4-(*E*)-CH=CH-4-$C_6H_4NO_2$}Cl$(dppm)_2$]	THF	490	1964	235	HRS	1.06	101,102
trans-[Ru{C≡C-4-C_6H_4N=N-4-$C_6H_4NO_2$}Cl$(dppm)_2$]	THF	583	1649	232	HRS	1.06	99
trans-[Ru{C≡C-4-$C_6H_4N\overline{CCH=CBu^tC(O)CBu^t=CH}$}Cl$(dppm)_2$]	THF	645	417	124	HRS	1.06	97
trans-[Ru(C≡C-4-C_6H_4C≡C-4-C_6H_4C≡C-4-$C_6H_4NO_2$)Cl$(dppm)_2$]	THF	439	1379	365	HRS	1.06	101
trans-[Ru(C=CHPh)Cl$(dppe)_2$](PF_6)	THF	317	[b]	—	HRS	1.06	101
trans-[Ru(C≡CPh)Cl$(dppe)_2$]	THF	319	6	3	HRS	1.06	101,104
trans-[Ru(C=CH-4-C_6H_4CHO)Cl$(dppe)_2$](PF_6)	THF	412	181	61	HRS	1.06	101
trans-[Ru(C≡C-4-C_6H_4CHO)Cl$(dppe)_2$]	THF	413	120	40	HRS	1.06	101
trans-[Ru(C=CH-4-$C_6H_4NO_2$)Cl$(dppe)_2$](PF_6)	THF	476	1130	180	HRS	1.06	101
trans-[Ru(C≡C-4-$C_6H_4NO_2$)Cl$(dppe)_2$]	THF	477	351	55	HRS	1.06	101
trans-[Ru{C=CH-4-C_6H_4-(*E*)-CH=CH-4-$C_6H_4NO_2$}Cl$(dppe)_2$](PF_6)	THF	473	441	74	HRS	1.06	101
trans-[Ru{C≡C-4-C_6H_4-(*E*)-CH=CH-4-$C_6H_4NO_2$}Cl$(dppe)_2$]	THF	489	2676	342	HRS	1.06	101
trans-[Ru(C≡CPh)(C≡C-4-C_6H_4C≡CPh)$(dppe)_2$]	THF	383	34	—	HRS	1.06	104
$(-)_{578}$-*trans*-[Ru(C≡CPh)Cl{(*R*,*R*)-diph}$_2$]	THF	292	~0	~0	HRS	1.06	84
$(-)_{589}$-*trans*-[Ru(C≡C-4-$C_6H_4NO_2$)Cl{(*R*,*R*)-diph}$_2$]	THF	467	530	97	HRS	1.06	84
$(-)_{589}$-*trans*-[Ru{C≡C-4-C_6H_4-(*E*)-CH=CH-4-$C_6H_4NO_2$}Cl{(*R*,*R*)-diph}$_2$]	THF	481	2795	406	HRS	1.06	84
[1-(HC≡C)-3,5-C_6H_3{*trans*-(C≡C)RuCl$(dppm)_2$}$_2$]	THF	323	42[c]	24	HRS	1.06	88
[1,3,5-{*trans*-[RuCl$(dppe)_2$(C≡C-4-C_6H_4C≡C)]}$_3C_6H_3$]	THF	414	94	—	HRS	1.06	104
[1,3,5-{*trans*-[Ru(C≡CPh)$(dppe)_2$(C≡C-4-C_6H_4C≡C)]}$_3C_6H_3$]	THF	411	93	—	HRS	1.06	104
[Ru(N≡C-4-C_6H_4-4-$C_6H_4NO_2$)(dppe)(η^5-C_5H_5)](PF_6)	MeOH	288	96	—	HRS	1.06	86
	$CHCl_3$	293	85				
[Ru(N≡C-4-$C_6H_4NO_2$)(dppe)(η^5-C_5H_5)](PF_6)	MeOH	358	138	—	HRS	1.06	86
	$CHCl_3$	358	126				
[Os(C≡C-4-$C_6H_4NO_2$)$(PPh_3)_2$(η^5-C_5H_5)]	THF	474	1051	174	HRS	1.06	81
[Os(C≡C-4-$C_6H_4NO_2$)(dppe)(η^5-C_5H_5)]	THF	461	929	188	HRS	1.06	81
$(-)_{365}$-*trans*-[Os(C≡C-4-$C_6H_4NO_2$)Cl{(*R*,*R*)-diph}$_2$]	THF	490	620	74	HRS	1.06	84
[Au(C≡CPh)(PPh_3)]	THF	296	6	4	HRS	1.06	88,105
[Au(C≡C-2-C_5H_4N-5-NO_2)(PMe_3)]	THF	340	12	6	HRS	1.06	90
[Au(C≡C-2-C_5H_4N-5-NO_2)(PPh_3)]	THF	339	38	20	HRS	1.06	90
[Au(C≡C-4-C_6H_4CHO)(PPh_3)]	THF	322	14	8	HRS	1.06	93

TABLE VI
(CONTINUED)

Complex	Solvent	λ_{max} (nm)	β (10^{-30} esu)	β_0 (10^{-30} esu)	Technique	Fundamental (μm)	Ref.
$[Au(C\equiv C\text{-}4\text{-}C_6F_4OMe)(PMe_3)]$	THF	324	44	25	HRS	1.06	103
$[Au(C\equiv C\text{-}4\text{-}C_6F_4OMe)(PPh_3)]$	THF	292	20	13	HRS	1.06	103
$[Au\{C\equiv C\text{-}4\text{-}C_6H_4\overline{CHO(CH_2)_3O}\}(PMe_3)]$	THF	292	48	13	HRS	1.06	93
$[Au\{C\equiv C\text{-}4\text{-}C_6H_4\overline{CHO(CH_2)_3O}\}(PPh_3)]$	THF	296	15	4	HRS	1.06	93
$[Au(C\equiv C\text{-}4\text{-}C_6H_4NO_2)(PMe_3)]$	THF	339	50	27	HRS	1.06	106
$[Au(C\equiv C\text{-}4\text{-}C_6H_4NO_2)(PPh_3)]$	THF	338	22	12	HRS	1.06	105
$[Au(C\equiv C\text{-}4\text{-}C_6H_4NO_2)(PCy_3)]$	THF	342	31	16	HRS	1.06	106
$[Au(C\equiv C\text{-}4\text{-}C_6H_4\text{-}4\text{-}C_6H_4NO_2)(PPh_3)]$	THF	350	39	20	HRS	1.06	105
$[Au\{C\equiv C\text{-}4\text{-}C_6H_4\text{-}(Z)\text{-}CH=CH\text{-}4\text{-}C_6H_4NO_2\}(PPh_3)]$	THF	362	58	28	HRS	1.06	75,105
$[Au\{C\equiv C\text{-}4\text{-}C_6H_4\text{-}(E)\text{-}CH=CH\text{-}4\text{-}C_6H_4NO_2\}(PPh_3)]$	THF	386	120	49	HRS	1.06	105
$[Au\{C\equiv C\text{-}4\text{-}C_6H_4N=N\text{-}4\text{-}C_6H_4NO_2\}(PPh_3)]$	THF	398	180	68	HRS	1.06	99
$[Au\{C\equiv C\text{-}4\text{-}C_6H_4\text{-}(Z)\text{-}CH=CH\text{-}4\text{-}C_6H_4NO_2\}(PPh_3)]$	THF	362	58	28	HRS	1.06	105
$[Au(C\equiv C\text{-}4\text{-}C_6H_4\text{-}C\equiv C\text{-}4\text{-}C_6H_4NO_2)(PPh_3)]$	THF	362	59	28	HRS	1.06	105
$[Au(C\equiv C\text{-}4\text{-}C_6H_4N=CH\text{-}4\text{-}C_6H_4NO_2)(PPh_3)]$	THF	392	85	34	HRS	1.06	105
$[Au(C\equiv C\text{-}2\text{-}C_5H_4N)(PPh_3)]$	THF	300	7	4	HRS	1.06	90
$[1,3,5\text{-}C_6H_3\{(C\equiv C)Au(PPh_3)\}_3]$	THF	298	6[c]	4	HRS	1.06	88

[a]Not stated.

[b]Not determined (solution showed excessive light scattering).

[c]Upper bound only: deconvolution of SHG and fluorescence contribution could not be effected.

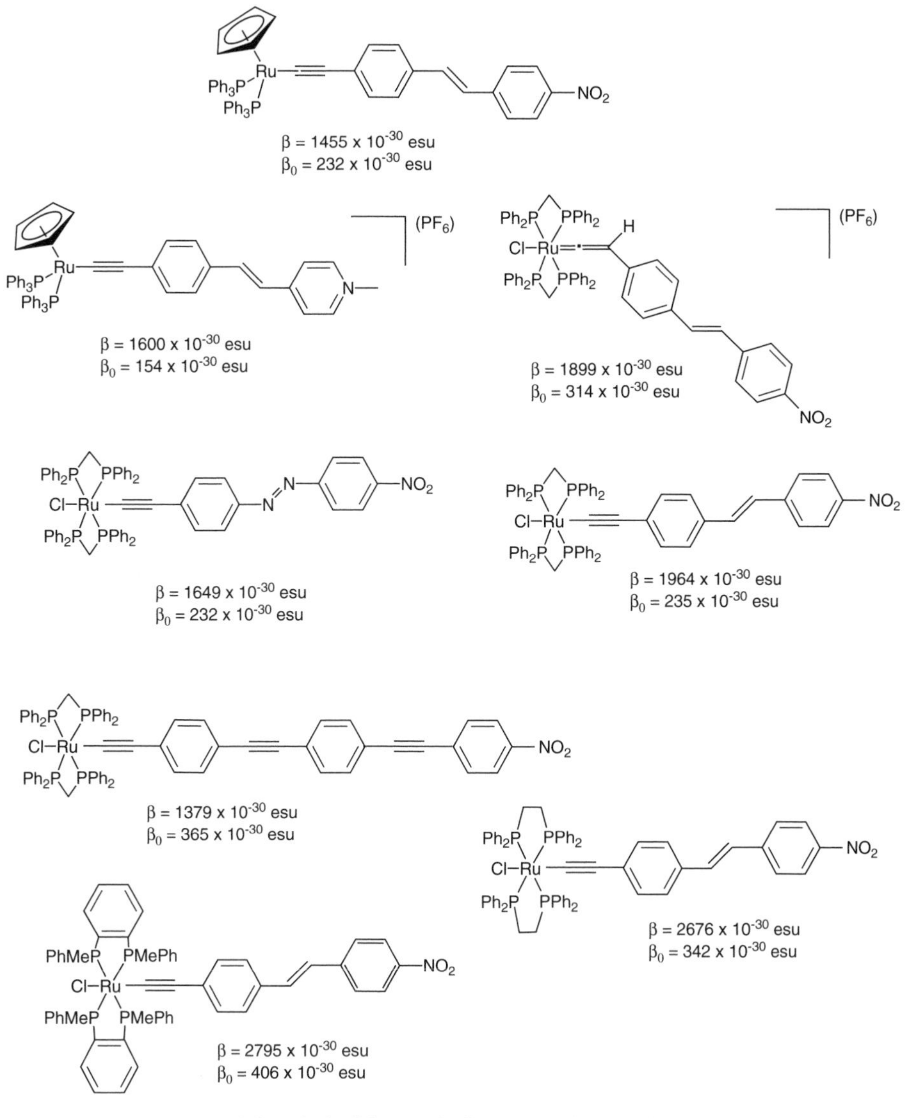

$\beta = 1455 \times 10^{-30}$ esu
$\beta_0 = 232 \times 10^{-30}$ esu

(PF$_6$)

$\beta = 1600 \times 10^{-30}$ esu
$\beta_0 = 154 \times 10^{-30}$ esu

(PF$_6$)

$\beta = 1899 \times 10^{-30}$ esu
$\beta_0 = 314 \times 10^{-30}$ esu

$\beta = 1649 \times 10^{-30}$ esu
$\beta_0 = 232 \times 10^{-30}$ esu

$\beta = 1964 \times 10^{-30}$ esu
$\beta_0 = 235 \times 10^{-30}$ esu

$\beta = 1379 \times 10^{-30}$ esu
$\beta_0 = 365 \times 10^{-30}$ esu

$\beta = 2676 \times 10^{-30}$ esu
$\beta_0 = 342 \times 10^{-30}$ esu

$\beta = 2795 \times 10^{-30}$ esu
$\beta_0 = 406 \times 10^{-30}$ esu

FIG. 7. Selected vinylidene and alkynyl complexes with large β.

benzene], show an increase in nonlinearity as Fe $\leqslant$ Ru $\leqslant$ Os, mirroring the increase in π-backbonding in complexes of this type suggested by DFT calculations.[81]

Substituting cyclopentadienyl by the more electron-rich indenyl group in [Ru(C$\equiv$CC$_6$H$_4$-4-NO$_2$)(PPh$_3$)$_2$(η^5-C$_5$H$_4$R)] (R = H, C$_4$H$_3$) results in the expected increase in the molecular hyperpolarizability, further modification to include four phosphorus donors (as in *trans*-[Ru(C$\equiv$CC$_6$H$_4$-4-NO$_2$)Cl(dppm)$_2$]) also leading to an increase. In contrast, replacing triphenylphosphine by the stronger base trimethylphosphine for the complexes [Ru(C$\equiv$CC$_6$H$_4$-4-NO$_2$)L$_2$(η^5-C$_5$H$_5$)] (L = PPh$_3$, PMe$_3$) results in a decrease in hyperpolarizability, suggesting that greater π-delocalization through the phenyl groups of the former is more important for quadratic NLO merit. Substituting phosphines or diphosphines with carbonyl ligands results in a significant decrease in electron density at the metal donor unit in $[M(C{\equiv}C\text{-}4\text{-}C_6H_4NO_2)(L)_2(\eta^5\text{-}C_5H_5)]$ [M = Fe, Ru, Os; L = CO, PPh_3, (1/2)dppe], and a corresponding decrease in molecular hyperpolarizability.

A wide variety of bridging units has been used to connect ligated ruthenium centers to a nitro acceptor group. Bridge lengthening (proceeding from $[Ru(C{\equiv}CC_6H_4\text{-}4\text{-}NO_2)(PPh_3)_2(\eta^5\text{-indenyl})]$ to $[Ru\{C{\equiv}C\text{-}(E)\text{-}CH{=}CHC_6H_4\text{-}4\text{-}NO_2\}(PPh_3)_2(\eta^5\text{-indenyl})]$ and then $[Ru\{C{\equiv}C\text{-}(E)\text{-}CH{=}CH\text{-}(E)\text{-}CH{=}CHC_6H_4\text{-}4\text{-}NO_2\}(PPh_3)_2(\eta^5\text{-indenyl})]$) results in a large increase in nonlinearity upon introduction of the first ene-linkage, but no significant difference upon introduction of the second alkene group. In contrast, proceeding from *trans*-$[Ru(C{\equiv}C\text{-}4\text{-}C_6H_4NO_2)Cl(dppm)_2]$ to *trans*-$[Ru(C{\equiv}C\text{-}4\text{-}C_6H_4C{\equiv}C\text{-}4\text{-}C_6H_4NO_2)Cl(dppm)_2]$ and then to *trans*-$[Ru(C{\equiv}C\text{-}4\text{-}C_6H_4C{\equiv}C\text{-}4\text{-}C_6H_4C{\equiv}C\text{-}4\text{-}C_6H_4NO_2)Cl(dppm)_2]$ results in no significant increase for the first bridge lengthening, but a sizable increase for the second bridge lengthening. Structural modifications to the bridging group have a similar impact to that observed in purely organic systems: replacing phenylene with thienylene groups in proceeding from $[Ru\{C{\equiv}C\text{-}4\text{-}C_6H_4\text{-}(E)\text{-}CH{=}CH\text{-}4\text{-}C_6H_4NO_2\}(PPh_3)_2(\eta^5\text{-}C_5H_5)]$ to $[Ru\{C{\equiv}C\text{-}4\text{-}C_6H_4\text{-}(E)\text{-}CH{=}CH\text{-}2\text{-}C_4H_2S\text{-}5\text{-}NO_2\}(PPh_3)_2(\eta^5\text{-}C_5H_5)]$ or $[Ru\{C{\equiv}C\text{-}2\text{-}C_4H_2S\text{-}5\text{-}(E)\text{-}CH{=}CH\text{-}4\text{-}C_6H_4NO_2\}(PPh_3)_2(\eta^5\text{-}C_5H_5)]$, or from $[Ru(C{\equiv}C\text{-}4\text{-}C_6H_4N{=}CH\text{-}4\text{-}C_6H_4NO_2)(PPh_3)_2(\eta^5\text{-}C_5H_5)]$ to $[Ru(C{\equiv}C\text{-}4\text{-}C_6H_4N{=}CH\text{-}2\text{-}C_4H_2S\text{-}5\text{-}NO_2)(PPh_3)_2(\eta^5\text{-}C_5H_5)]$ affords an increase in β value. One difference to that seen in the organic system is that the azo linkage is as efficient as the *E*-ene linkage in complexes of the type $[L_nM(C{\equiv}C\text{-}4\text{-}C_6H_4X{=}X\text{-}4\text{-}C_6H_4NO_2)]$ (ML_n = ligated metal; X=CH, N).

Many acceptor groups have been explored, particularly with ruthenium complexes, but the nitro group has been by far the most common. Nonlinearities of pyridylalkynyl complexes are not large, but in many instances can be enhanced on alkylation of the pyridyl N atom, consistent with an increase in acceptor strength on proceeding to an alkylpyridinium unit. Similarly, coordination of the free pyridine in $[Ru\{C{\equiv}C\text{-}(E)\text{-}CH{=}CH\text{-}4\text{-}C_5H_4N\}(PPh_3)_2(\eta^5\text{-indenyl})]$ to a $M(CO)_5$ unit (M = Cr, W) leads to a substantial increase in nonlinearity. Coordination of the same pentacarbonylmetal groups to $[Ru\{C{\equiv}C\text{-}(E)\text{-}CH{=}CHC_6H_4\text{-}4\text{-}C{\equiv}N\}(PPh_3)_2(\eta^5\text{-indenyl})]$ has also been investigated, resulting in β(tungsten-containing complex) > β(chromium-containing complex). In addition to the aforementioned nitro, pyridyl, and nitrile, other acceptor groups have been explored including trifluoromethyl and formyl, both being weaker acceptors than nitro and both affording complexes with lower nonlinearities than the related nitro-containing complexes.

Cationic nitrile complexes investigated for their quadratic NLO performance have frequently possessed the same bridge and acceptor groups as the isoelectronic neutral alkynyl complex analogues; where analogues exist, data are consistent with the trend β(alkynyl complex) > β(nitrile complex), not unexpected because the latter are cationic and the efficacy of the ligated metal donor is significantly reduced. Data for the nitrile complexes listed in Table VI reveal some trends upon metal variation; for example, β increases as Co < Ni < Ru < Fe. Reaction of alkynyl complexes with a range of electrophiles affords vinylidene complexes. A number of aryldiazovinylidene complexes have been studied, with those possessing nitrophenyldiazo units having the largest nonlinearities.

Table VI also summarizes results from recent studies of "switching" molecular nonlinearities. Such switching has merit if it is facile and reversible, and switching by both oxidation/reduction and protonation/deprotonation sequences has been explored. The quadratic nonlinearities of several iron complexes have been examined in more than one oxidation state, conversion of the iron(II) resting state to the iron(III) state "switching on" an optical transition at long wavelengths that also results in a modification of nonlinearity, with differences of up to a factor of 4 being seen. Switching can also be achieved chemically, protonation of an alkynyl ligand affording a vinylidene ligand; a considerable number of alkynyl/vinylidene complex pairs have been examined, with differences of up to a factor of 6 in β value being observed.

Table VII contains data for alkynyl/vinylidene complex pairs at two wavelengths. Nonlinearities are strongly wavelength-dependent, as is clear from the data; in both cases, proceeding from alkynyl to vinylidene complex at both wavelengths leads to a reduction in nonlinearity. Table VIII contains data for iron and ruthenium alkynyl complexes as $\boldsymbol{\mu} \cdot \boldsymbol{\beta}$ values, examples of which are shown in Fig. 8. Table VIII also includes several examples of differing oxidation state pairs. Several compounds have been examined as both M^{II} neutral and M^{III} cationic complexes. In all cases except $[Fe(C{\equiv}C\text{-}4\text{-}C_6H_4NO_2)(dppe)(\eta^5\text{-}C_5Me_5)]^{n+}$ ($n = 0, 1$), there is a change in sign of $\boldsymbol{\mu} \cdot \boldsymbol{\beta}$ and $\boldsymbol{\mu} \cdot \boldsymbol{\beta}_0$ products upon oxidation, indicating a decrease or change in sign of dipole moment between ground and excited states.

Table IX collects data as the square root of the orientationally averaged square of the nonlinearity, permitting a comparison between linear and trigonal ("octupolar") complexes. Within the error margins of the experiment ($\pm$10–15%), proceeding from linear to octupolar complexes (Fig. 9) results in a tripling of NLO response, consistent with the increase in number of ligated metal centers, but extending the π-system by replacing the *trans*-disposed chloro by a phenylalkynyl ligand results in no further increase.

TABLE VII

MOLECULAR SECOND-ORDER NLO RESULTS FOR ALKYNYL AND VINYLIDENE COMPLEXES AT TWO WAVELENGTHS

Complex	Solvent	λ_{max} (nm)	$\beta_{1.06}$ (10^{-30} esu)	$\beta_{0.8}$ (10^{-30} esu)	Technique	Ref.
trans-$[Ru(C{=}CH\text{-}4\text{-}C_6H_4\text{-}(E)\text{-}CH{=}CHPh)Cl(dppm)_2](PF_6)$	THF	360	70	33	HRS	107
trans-$[Ru(C{\equiv}C\text{-}4\text{-}C_6H_4\text{-}(E)\text{-}CH{=}CHPh)Cl(dppm)_2]$	THF	397	200	⩽920	HRS	107
[1,3,5-(*trans*-$[(dppm)_2ClRu\{C{=}CH\text{-}4\text{-}C_6H_4\text{-}(E)\text{-}CH{=}CH\}])_3C_6H_3](PF_6)_3$	THF	396	165	483	HRS	107
[1,3,5-(*trans*-$[(dppm)_2ClRu\{C{\equiv}C\text{-}4\text{-}C_6H_4\text{-}(E)\text{-}CH{=}CH\}])_3C_6H_3]$	THF	415	244	935	HRS	107

TABLE VIII

MOLECULAR SECOND-ORDER NLO RESULTS AS $\mu \cdot \beta$ PRODUCTS FOR ALKYNYL COMPLEXES

Complex	Solvent	λ_{max} (nm)	$\mu \cdot \beta$ (10^{-48} esu)	$\mu \cdot \beta_0$ (10^{-48} esu)	Technique	Fundamental (μm)	Ref.
$[Fe(C{\equiv}CPh)(dppe)(\eta^5\text{-}C_5Me_5)](PF_6)$	$CHCl_3$	662	−12	−5	EFISH	1.907	108
$[Fe(C{\equiv}CPh)(dppe)(\eta^5\text{-}C_5Me_5)]$	CH_2Cl_2	350	269	225	EFISH	1.907	108
$[Fe(C{\equiv}C\text{-}4\text{-}C_6H_4OMe)(dppe)(\eta^5\text{-}C_5Me_5)](PF_6)$	$CHCl_3$	718	−145	−53	EFISH	1.907	108
$[Fe(C{\equiv}C\text{-}4\text{-}C_6H_4OMe)(dppe)(\eta^5\text{-}C_5Me_5)]$	CH_2Cl_2	333	44	37	EFISH	1.907	108
$[Fe(C{\equiv}C\text{-}4\text{-}C_6H_4NH_2)(dppe)(\eta^5\text{-}C_5Me_5)](PF_6)$	$CHCl_3$	789	−334	−87	EFISH	1.907	108
$[Fe(C{\equiv}C\text{-}4\text{-}C_6H_4NH_2)(dppe)(\eta^5\text{-}C_5Me_5)]$	CH_2Cl_2	322	54	47	EFISH	1.907	108
$[Fe(C{\equiv}C\text{-}2\text{-}C_5H_4N)(dppe)(\eta^5\text{-}C_5Me_5)](PF_6)$	$CHCl_3$	521	−9	−6	EFISH	1.907	108
$[Fe(C{\equiv}C\text{-}2\text{-}C_5H_4N)(dppe)(\eta^5\text{-}C_5Me_5)]$	CH_2Cl_2	403	−139	−110	EFISH	1.907	108
$[Fe(C{\equiv}C\text{-}3\text{-}C_5H_4N)(dppe)(\eta^5\text{-}C_5Me_5)]$	CH_2Cl_2	401	−169	−133	EFISH	1.907	108
$[Fe(C{\equiv}C\text{-}4\text{-}C_5H_4N)(dppe)(\eta^5\text{-}C_5Me_5)]$	CH_2Cl_2	391	−137	−109	EFISH	1.907	108
$[Fe(C{\equiv}C\text{-}4\text{-}C_5H_4N\text{-}1\text{-}Me)(dppe)(\eta^5\text{-}C_5Me_5)](BF_4)$	$CHCl_3$	580	948	552	EFISH	1.907	108
$[Fe(C{\equiv}C\text{-}2\text{-}C_5H_4N\text{-}1\text{-}Me)(dppe)(\eta^5\text{-}C_5Me_5)](PF_6)$	CH_2Cl_2	538	662	415	EFISH	1.907	108
$[Fe(C{\equiv}C\text{-}4\text{-}C_6H_4CF_3)(dppe)(\eta^5\text{-}C_5Me_5)]$	CH_2Cl_2	387	265	212	EFISH	1.907	108
$[Fe(C{\equiv}C\text{-}4\text{-}C_6H_4CN)(dppe)(\eta^5\text{-}C_5Me_5)](PF_6)$	$CHCl_3$	652	−56	−26	EFISH	1.907	108
$[Fe(C{\equiv}C\text{-}4\text{-}C_6H_4CN)(dppe)(\eta^5\text{-}C_5Me_5)]$	CH_2Cl_2	445	789	583	EFISH	1.907	108
$[Fe(C{\equiv}C\text{-}4\text{-}C_6H_4NO_2)(dppe)(\eta^5\text{-}C_5Me_5)](PF_6)$	$CHCl_3$	650	1	1	EFISH	1.907	108
$[Fe(C{\equiv}C\text{-}4\text{-}C_6H_4NO_2)(dppe)(\eta^5\text{-}C_5Me_5)]$	CH_2Cl_2	595	2100	1155	EFISH	1.907	108
$[Ru(C{\equiv}C\text{-}4\text{-}C_6H_4CN)(dppe)(\eta^5\text{-}C_5Me_5)]$	CH_2Cl_2	384	155	124	EFISH	1.907	108
$[Ru(C{\equiv}C\text{-}4\text{-}C_6H_4NO_2)(dppe)(\eta^5\text{-}C_5Me_5)]$	CH_2Cl_2	500	690	465	EFISH	1.907	108

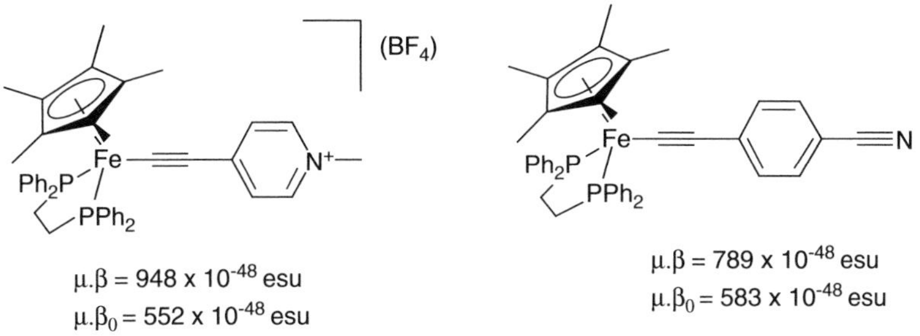

(BF$_4$)

$\mu.\beta$ = 948 x 10^{-48} esu
$\mu.\beta_0$ = 552 x 10^{-48} esu

$\mu.\beta$ = 789 x 10^{-48} esu
$\mu.\beta_0$ = 583 x 10^{-48} esu

FIG. 8. Selected examples of alkynyl complexes with large $\mu \cdot \beta$.

TABLE IX
MOLECULAR SECOND-ORDER NLO RESULTS FOR ALKYNYL COMPLEXES AS ORIENTATIONALLY AVERAGED DATA

Complex	Solvent	λ_{max} (nm)	$\sqrt{\langle \beta^2 \rangle}$ (10^{-30} esu)	Technique	Fundamental (μm)	Ref.
trans-[Ru(C≡CPh)Cl(dppe)$_2$]	THF	319	6	HRS	1.06	104
trans-[Ru(C≡C-4-C$_6$H$_4$C≡CPh)(C≡CPh)(dppe)$_2$]	THF	383	34	HRS	1.06	104
[1,3,5-(*trans*-[(dppe)$_2$ClRu(C≡C-4-C$_6$H$_4$C≡C)])$_3$C$_6$H$_3$]	THF	414	94	HRS	1.06	104
[1,3,5-(*trans*-[(dppe)$_2$(PhC≡C)Ru(C≡C-4-C$_6$H$_4$C≡C)])$_3$C$_6$H$_3$]	THF	411	93	HRS	1.06	104

Incorporating one metal atom into organic chromophores often results in increased nonlinearities, so introduction of two metal atoms is a logical step, and this has been an area of tremendous growth in the past few years. A number of bimetallics have been listed in earlier tables for which the emphasis was metal–ligand groupings, but Table X contains collected data for all bimetallic and trimetallic complexes for which β values have been reported. Fig. 10 displays some of the complexes with large nonlinearities.

The vast majority of bi- and trimetallic complexes for which quadratic optical nonlinearities have been reported contain at least one metallocenyl unit, paralleling the considerable interest in monometallic metallocenyl complexes. Large nonlinearities are found in complexes coupling ferrocenyl to diiron-stabilized carbocation groups by a variety of π-delocalized linkages, significantly smaller β values being seen in complexes coupling metallocenyl to N-ligand ligated metal centers. Several complexes coupling ferrocenyl to pentacarbonylmetal carbene units have been examined in a range of solvents, with a doubling of response observed in proceeding from hexane to acetonitrile. A significant increase in β value is seen for diferrocenyl complexes with π-delocalizable bridges when charge-transfer salts with DDQ, TCNE, and other reagents are formed.

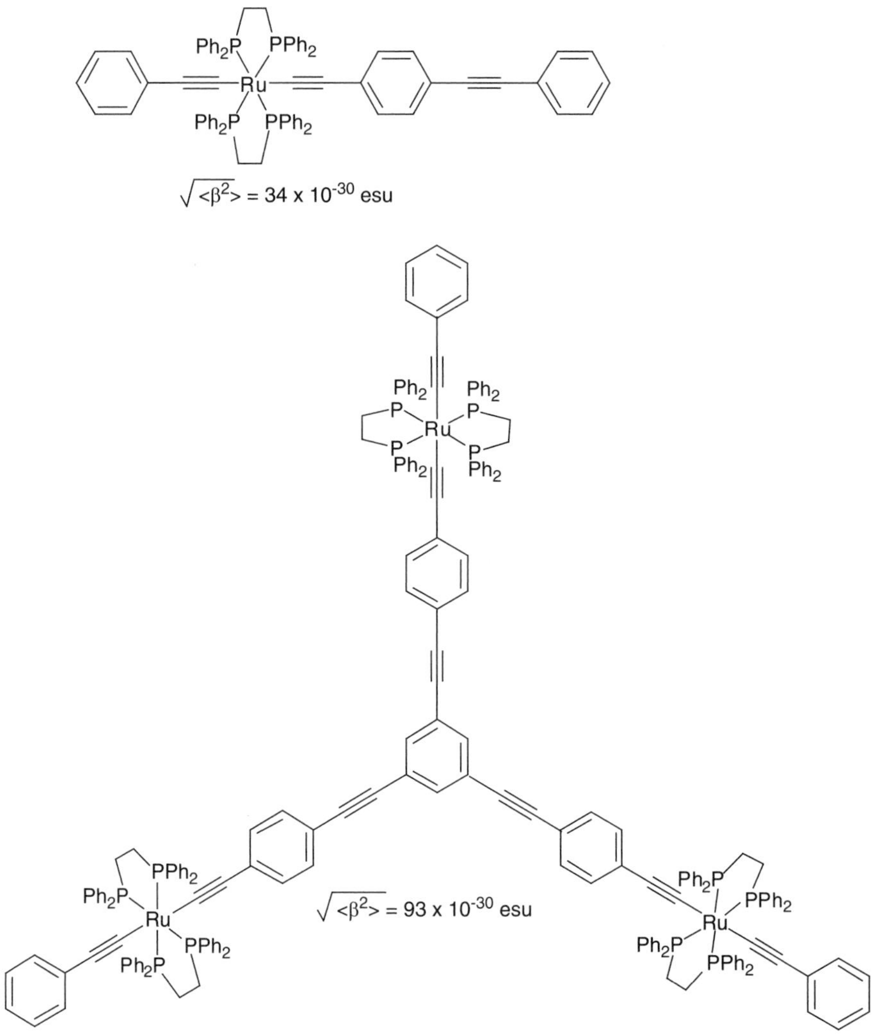

$\sqrt{\langle\beta^2\rangle} = 34 \times 10^{-30}$ esu

$\sqrt{\langle\beta^2\rangle} = 93 \times 10^{-30}$ esu

FIG. 9. Selected alkynyl complexes with large orientationally averaged NLO response.

The mixed-valence Ru^{II}/Ru^{III} complex $[Ru\{C{\equiv}NRu(NH_3)_5\}(PPh_3)_2(\eta^5\text{-}C_5H_5)](CF_3SO_3)_3$ was originally reported to have a very large β value,[109] but it was subsequently found to have a significant fluorescence contribution, drawing attention to the problems associated with this phenomenon.[110]

Table XI contains data for other organometallic complexes. The chloro complexes are of interest because they were reacted with terminal alkynes to afford alkynyl complexes, and the greater nonlinearity of the alkynyl complex compared to the sum of the precursor alkyne and chloro complex suggests the importance of electronic communication between ligated metal and alkynyl ligand.[89,92,105] The aryl complexes have a donor–bridge–acceptor composition, but nonlinearities are

Table X

Molecular Second-Order NLO Results for Bimetallic and Trimetallic Complexes

Complex	Solvent	λ_{max} (nm)	β (10^{-30} esu)	β_0 (10^{-30} esu)	Technique	Fundamental (μm)	Ref.
CrMn							
$[Mn(CO)_3(\eta^5\text{-}C_5H_4\text{-}\eta^7\text{-}C_7H_6)Cr(CO)_3](BF_4)$	$MeNO_2$	445	77	19	HRS	1.06	71
$[Mn(CO)_3(\eta^5\text{-indenyl-}\eta^7\text{-}C_7H_6)Cr(CO)_3](BF_4)$	$MeNO_2$	435	112	31	HRS	1.06	71
$[Mn(CO)_3(\eta^5\text{-}C_5H_4C{\equiv}C\text{-}\eta^7\text{-}C_7H_6)Cr(CO)_3](BF_4)$	CH_2Cl_2	449	244	58	HRS	1.06	71
CrFe							
$[Cr(CO)_3\{\eta^6\text{-}PhC{\equiv}C\text{-}4\text{-}C_6H_4CNFe(dppe)(\eta^5\text{-}C_5H_5)\}]$	$CHCl_3$	429	94	—	HRS	1.06	63
$[Cr(CO)_3\{\eta^6\text{-}PhC{\equiv}C\text{-}2\text{-}C_4H_2S\text{-}5\text{-}CNFe(dppe)(\eta^5\text{-}C_5H_5)\}]$	$CHCl_3$	444	170	—	HRS	1.06	63
$[Fc\{(E)\text{-}CH{=}CHC(OMe)Cr(CO)_5\}]$	Hexane	543	47	—	HRS	1.06	111
	CH_2Cl_2	—	69	—	HRS	1.06	
	Acetone	—	89	—	HRS	1.06	
	MeCN	—	110	—	HRS	1.06	
$[Fc\{(E,E)\text{-}(CH{=}CH)_2C(OMe)Cr(CO)_5\}]$	Hexane	546	161	—	HRS	1.06	111
	CH_2Cl_2	—	240	—	HRS	1.06	
	Acetone	—	306	—	HRS	1.06	
	MeCN	—	410	—	HRS	1.06	
$[Fc\{(E,E,E)\text{-}(CH{=}CH)_3C(OMe)Cr(CO)_5\}]$	Hexane	550	342	—	HRS	1.06	111
	CH_2Cl_2	—	440	—	HRS	1.06	
	Acetone	—	760	—	HRS	1.06	
	MeCN	—	1030	—	HRS	1.06	
$[Fc\{C{\equiv}C(\eta^7\text{-}C_7H_6)Cr(CO)_3\}]$	CH_2Cl_2	600	570	105	HRS	1.06	51
$[Fc\{(E)\text{-}CH{=}CH(\eta^7\text{-}C_7H_6)Cr(CO)_3\}]$	CH_2Cl_2	670	320	113	HRS	1.06	51
$[Fc\{(E)\text{-}CH{=}CH\text{-}(\eta^6\text{-}Ph)Cr(CO)_3\}]$	$CHCl_3$	304	193	119	HRS	1.06	40
$[Fc\{(E,E)\text{-}(CH{=}CH)_2\text{-}(\eta^6\text{-}Ph)Cr(CO)_3\}]$	$CHCl_3$	334	300	164	HRS	1.06	40
$[Cr(CO)_3(\eta^6\text{-}PhC{\equiv}CFc)]$	$CHCl_3$	387	9[a]	7[a]	HRS	1.5	62
$[Cr(CO)_3(\eta^6\text{-}PhC{\equiv}CC{\equiv}CFc)]$	$CHCl_3$	419	11[a]	9[a]	HRS	1.5	62
$[Fc\{(E)\text{-}CH{=}CH\text{-}4\text{-}C_5H_4NCr(CO)_5\}]$	$CHCl_3$	401	63	23	HRS	1.06	40
	$MeNO_2$	480	242	—	HRS	1.06	45
$[Fc\{(E)\text{-}CH{=}CH\text{-}4\text{-}C_6H_4CNCr(CO)_5\}]$	$CHCl_3$	481	271	39	HRS	1.06	40
$[Fc\{(E)\text{-}CH{=}CH\text{-}4\text{-}C_6H_4\text{-}(E)\text{-}CH{=}CH\text{-}4\text{-}C_5H_4NCr(CO)_5\}]$	$CHCl_3$	462	369	—	HRS	1.06	42
$[Cr(CO)_3\{\eta^6\text{-}PhC{\equiv}C\text{-}2,5\text{-}C_4H_2S\text{-}2\text{-}C_4H_2S\text{-}5\text{-}CNFe(dppe)(\eta^5\text{-}C_5H_5)\}]$	$CHCl_3$	398	197	—	HRS	1.06	63
$[\{Fe(CO)(\eta^5\text{-}C_5H_5)\}_2(\mu\text{-}CO)(\mu\text{-}C{=}CHCH{=}CH\text{-}\eta^7\text{-}C_7H_6)Cr(CO)_3](BF_4)$	CH_2Cl_2	635	451	117	HRS	1.5	67
CrRu							
$[Cr(CO)_3\{\eta^6\text{-}PhC{\equiv}C\text{-}4\text{-}C_6H_4CNRu(dppe)(\eta^5\text{-}C_5H_5)\}]$	$CHCl_3$	317	42	—	HRS	1.06	63
$[Cr(CO)_3(\eta^6\text{-}PhC{\equiv}C\text{-}2\text{-}C_4H_2S\text{-}5\text{-}CNRu(dppe)(\eta^5\text{-}C_5H_5)\}]$	$CHCl_3$	414	83	—	HRS	1.06	63
$[Cr(CO)_3\{\eta^6\text{-}PhC{\equiv}C\text{-}2,5\text{-}C_4H_2S\text{-}2\text{-}C_4H_2S\text{-}5\text{-}CNRu(dppe)(\eta^5\text{-}C_5H_5)\}]$	$CHCl_3$	426	143	—	HRS	1.06	63

TABLE X
(CONTINUED)

Complex	Solvent	λ_{max} (nm)	β (10^{-30} esu)	β_0 (10^{-30} esu)	Technique	Fundamental (μm)	Ref.
$[Ru\{CNCr(CO)_5\}(PPh_3)_2(\eta^5\text{-indenyl})]$	CH_2Cl_2	392	25	10	HRS	1.06	74,100
$[Ru\{C{\equiv}CCH{=}CH\text{-}4\text{-}C_6H_4C{\equiv}NCr(CO)_5\}(PPh_3)_2(\eta^5\text{-indenyl})]$	CH_2Cl_2	442	465	119	HRS	1.06	74,100
$[Ru\{C{\equiv}C\text{-}(E)\text{-}CH{=}CH\text{-}4\text{-}C_5H_4NCr(CO)_5\}(PPh_3)_2(\eta^5\text{-indenyl})]$	CH_2Cl_2	451	260	60	HRS	1.06	100
$[Ru(\eta^5\text{-}C_5H_5)(\eta^5\text{-}C_5H_4CH{=}CH\text{-}\eta^6\text{-}C_7H_7)Cr(CO)_3](PF_6)$	$MeNO_2$	490	497	59	HRS	1.06	112
$[Ru\{C{\equiv}C\text{-}(E)\text{-}CH{=}CH\text{-}4\text{-}C_5H_4NCr(CO)_5\}(PPh_3)_2(\eta^5\text{-indenyl})]$	CH_2Cl_2	451	260	60	HRS	1.06	74
MnFe							
$[Fc\{(E)\text{-}CH{=}CH\text{-}4\text{-}C_5H_4NMn(CO)_5\}](BF_4)$	$MeNO_2$	502	602	—	HRS	1.06	45
$[Fc\text{-}2\text{-}\eta^5\text{-}SC_4H_3Mn(CO)_3](BF_4)$	$MeNO_2$	514	260	13	HRS	1.06	72,78
$[(FcCH{=}CH\text{-}2\text{-}\eta^5\text{-}SC_4H_3)Mn(CO)_3](BF_4)$	$MeNO_2$	536	670	−8	HRS	1.06	72,78
$[\{Fc(CH{=}CH)_2\text{-}2\text{-}\eta^5\text{-}SC_4H_3\}Mn(CO)_3](BF_4)$	$MeNO_2$	548	771	−34	HRS	1.06	72,78
$[(FcCH{=}CH\text{-}3\text{-}\eta^5\text{-}SC_4H_3)Mn(CO)_3](BF_4)$	$MeNO_2$	470	377	67	HRS	1.06	72,78
$[(FcCH_2\text{-}2\text{-}\eta^5\text{-}SC_4H_3)Mn(CO)_3](BF_4)$	$MeNO_2$	—	220	—	HRS	1.06	72,78
Fe_2							
$[(Fc\{(E)\text{-}CH{=}CH\text{-}4\text{-}C_6H_4\})_2C(OH)Ph]$	CF_3COOH	[b]	900±150	—	HRS	1.06	39
$[\{Fe(CO)(\eta^5\text{-}C_5H_5)\}_2(\mu\text{-}CO)_2\{\mu\text{-}C\text{-}(E)\text{-}CH{=}CH\text{-}2\text{-}C_4H_2S\text{-}5\text{-}N(CH_2)_4CH_2\}]$	[b]	658[c]	826	—	HRS	1.06	156
$[\{Fe(CO)(\eta^5\text{-}C_5H_5)\}_2(\mu\text{-}CO)_2\{\mu\text{-}C\text{-}(E)\text{-}CH{=}CH\text{-}2\text{-}C_4H_2S\text{-}5\text{-}NCH(CH_2OMe)(CH_2)_2CH_2\}]$	[b]	655[c]	964	—	HRS	1.06	156
$[Fc\{(E)\text{-}CH{=}CH\text{-}4\text{-}C_6H_4C{\equiv}NFe(CO)_2(\eta^5\text{-}C_5H_5)\}](PF_6)$	$CHCl_3$	474	171	28	HRS	1.06	115
$[Fc\{C{\equiv}C\text{-}(\eta^6\text{-}Ph)Fe(CO)_3\}](BF_4)$	[b]	500	100	—	HRS	1.06	116
$[Fc\{C{\equiv}C\text{-}\eta^6\text{-}4\text{-}C_6H_4OMe)Fe(CO)_3\}](BF_4)$	[b]	540	⩾140	—	HRS	1.06	116
$[FcCH{=}NC_6H_3\text{-}3\text{-}OMe\text{-}4\text{-}N{=}CHFc]$	MeCN	453	18	4	HRS	1.06	34
$[FcCH{=}NC_6H_3\text{-}3\text{-}OMe\text{-}4\text{-}N{=}CHFc]\cdot I_5$	MeCN	514	127	7	HRS	1.06	34
$[FcCH{=}NC_6H_3\text{-}3\text{-}OMe\text{-}4\text{-}N{=}CHFc]\cdot(DDQ)_2$	MeCN	546	174	7	HRS	1.06	34
$[FcCH{=}NC_6H_3\text{-}3\text{-}OMe\text{-}4\text{-}N{=}CHFc]\cdot(TCNE)_2$	MeCN	498	259	25	HRS	1.06	34
$[FcCH{=}NC_6H_3\text{-}3\text{-}OMe\text{-}4\text{-}N{=}CHFc]\cdot(TCNQ)_2$	MeCN	482	166	24	HRS	1.06	34
$[FcCH{=}NC_6H_3\text{-}3\text{-}OMe\text{-}4\text{-}N{=}CHFc]\cdot(CA)_2$	MeCN	634	135	37	HRS	1.06	34
$[FcCH{=}NC_6H_4\text{-}4\text{-}N{=}CHFc]$	MeCN	460	23	5	HRS	1.06	34
$[FcCH{=}NC_6H_4\text{-}4\text{-}N{=}CHFc]\cdot I_{4.5}$	MeCN	510	137	9	HRS	1.06	34
$[FcCH{=}NC_6H_4\text{-}4\text{-}N{=}CHFc]\cdot(DDQ)_2$	MeCN	582	150	21	HRS	1.06	34
$[FcCH{=}NC_6H_4\text{-}4\text{-}N{=}CHFc]\cdot(TCNE)_2$	MeCN	479	181	27	HRS	1.06	34
$[FcCH{=}NC_6H_4\text{-}4\text{-}N{=}CHFc]\cdot(TCNQ)_2$	MeCN	476	144	23	HRS	1.06	34
$[FcCH{=}NC_6H_4\text{-}4\text{-}N{=}CHFc]\cdot(CA)_2$	MeCN	570	263	28	HRS	1.06	34
$[FcCH{=}NC_6H_3\text{-}3\text{-}Cl\text{-}4\text{-}N{=}CHFc]$	MeCN	474	33	6	HRS	1.06	34
$[FcCH{=}NC_6H_3\text{-}3\text{-}Cl\text{-}4\text{-}N{=}CHFc]\cdot I_{4.5}$	MeCN	521	160	5	HRS	1.06	34
$[FcCH{=}NC_6H_3\text{-}3\text{-}Cl\text{-}4\text{-}N{=}CHFc]\cdot(DDQ)_4$	MeCN	588	175	27	HRS	1.06	34

[FcCH=NC$_6$H$_3$-3-Cl-4-N=CHFc]·(TCNE)$_2$	MeCN	490	126	34	HRS	1.06	34
[FcCH=NC$_6$H$_3$-3-Cl-4-N=CHFc]·(TCNQ)$_2$	MeCN	476	204	33	HRS	1.06	34
[FcCH=NC$_6$H$_3$-3-Cl-4-N=CHFc]·(CA)$_2$	MeCN	601	279	53	HRS	1.06	34
[FcCH=NC$_6$Me$_4$-4-N=CHFc]	MeCN	455	23	5	HRS	1.06	34
[FcCH=NC$_6$Me$_4$-4-N=CHFc]·I$_3$	MeCN	502	121	10	HRS	1.06	34
[FcCH=NC$_6$Me$_4$-4-N=CHFc]·(DDQ)$_2$	MeCN	494	153	17	HRS	1.06	34
[FcCH=NC$_6$Me$_4$-4-N=CHFc]·(TCNE)$_2$	MeCN	478	161	25	HRS	1.06	34
[FcCH=NC$_6$Me$_4$-4-N=CHFc]·(TCNQ)$_2$	MeCN	475	171	28	HRS	1.06	34
[FcCH=NC$_6$Me$_4$-4-N=CHFc]·(CA)$_3$	MeCN	438	170	46	HRS	1.06	34
[FcCH=NC$_6$H$_4$-4-C$_6$H$_4$N=CHFc]	MeCN	465	36	7	HRS	1.06	34
[FcCH=NC$_6$H$_4$-4-C$_6$H$_4$N=CHFc]·I$_3$	MeCN	524	149	3	HRS	1.06	34
[FcCH=NC$_6$H$_4$-4-C$_6$H$_4$N=CHFc]·(DDQ)$_2$	MeCN	590	132	21	HRS	1.06	34
[FcCH=NC$_6$H$_4$-4-C$_6$H$_4$N=CHFc]·(TCNE)$_2$	MeCN	505	288	22	HRS	1.06	34
[FcCH=NC$_6$H$_4$-4-C$_6$H$_4$N=CHFc]·(TCNQ)$_2$	MeCN	478	173	27	HRS	1.06	34
[FcCH=NC$_6$H$_4$-4-C$_6$H$_4$N=CHFc]·(CA)$_3$	MeCN	448	197	47	HRS	1.06	34
[FcCH=NC(CN)=C(CN)N=CHFc]	CHCl$_3$	587	576	52	—	—	44
[FcCH=NC$_{14}$H$_6$O$_2$N=CHFc]	CHCl$_3$	498	889	45	—	—	44
FeCo							
[Fc(η^6-BC$_5$H$_5$)Co(η^5-C$_5$H$_5$)]	CH$_2$Cl$_2$	650	90	—	HRS	1.06	52
FeNi							
[Fc{CH=C(SCH$_2$Ph)C=NCMe$_2$CH$_2$O(*N*,*S*-NiCl$_2$)}]	CHCl$_3$	500	20	4	EFISH	1.34	49,50
[Fc{CH=C(SCH$_2$Ph)C=NCMe$_2$CH$_2$O(*N*,*S*-NiBr$_2$)}]	CHCl$_3$	511	7	3	EFISH	1.34	49,50
[Fc{CH=C{S(O)CH$_2$Ph}C=NCMe$_2$CH$_2$O(*N*,*S*-NiCl$_2$)}]	CHCl$_3$	498	16	6	EFISH	1.34	49,50
[Fc{CH=C{S(O)CH$_2$Ph}C=NCMe$_2$CH$_2$O(*N*,*S*-NiCl$_2$)}]	CHCl$_3$	494	6	2	EFISH	1.34	49,50
FeZn							
[Zn(OAc)$_2${Fc-(*E*)-CH=CH-4-bipy-4′-C$_9$H$_{19}$}]	[b]	508	17	—	EFISH	1.34	43
FeMo							
[Fe(η^5-C$_5$HMe$_4$){η^5-C$_5$Me$_4$-(*E*)-CH=CH-4-C$_5$H$_4$NMoCl(NO)(Tp^{Me,Me})}]	CH$_2$Cl$_2$	555	205	14	HRS	1.06	46
[Fe(η^5-C$_5$HMe$_4$){η^5-C$_5$Me$_4$-(*E*)-CH=CH-4-C$_5$H$_4$NMoCl(NO)(Tp^{Me,Me})}](PF$_6$)	CH$_2$Cl$_2$	589	129	20	HRS	1.06	46
[Fc{NHMoCl(NO)(Tp^{Me,Me})}]	CH$_2$Cl$_2$	737	119	57	HRS	1.06	46
[Fc{(*E*)-CH=CH-4-C$_5$H$_4$NMo(CO)$_5$}]	CHCl$_3$	487	95	12	HRS	1.06	40
[Fc{(*E*)-CH=CH-4-C$_6$H$_4$-(*E*)-CH=CH-4-C$_5$H$_4$NMo(CO)$_5$}]	CHCl$_3$	476	448	—	HRS	1.06	42
[Fc{(*E*)-CH=CHC(O)-4-C$_5$H$_4$NMoCl(NO)(Tp^{Me,Me})}]	CH$_2$Cl$_2$	570	74	8	HRS	1.06	46
[Fc{4-C$_6$H$_4$-(*E*)-CH=CH-4-C$_5$H$_4$NMoCl(NO)(Tp^{Me,Me})}]	CH$_2$Cl$_2$	495	166	17	HRS	1.06	46
[Fc{NHMoBr(NO)(Tp^{Me,Me})}]	CH$_2$Cl$_2$	753	135	67	HRS	1.06	46
[Fc{NHMoI(NO)(Tp^{Me,Me})}]	CH$_2$Cl$_2$	764	197	101	HRS	1.06	46
[Fc{4-C$_6$H$_4$NHMoCl(NO)(Tp^{Me,Me})}]	CH$_2$Cl$_2$	596	297	52	HRS	1.06	46
[Fc{4-C$_6$H$_4$NHMoBr(NO)(Tp^{Me,Me})}]	CH$_2$Cl$_2$	609	289	60	HRS	1.06	46
[Fc{4-C$_6$H$_4$NHMoI(NO)(Tp^{Me,Me})}]	CH$_2$Cl$_2$	627	431	110	HRS	1.06	46
[Fc{4-C$_6$H$_4$-(*E*)-CH=CH-4-C$_6$H$_4$NHMoCl(NO)(Tp^{Me,Me})}]	CH$_2$Cl$_2$	530	975	3	HRS	1.06	46
[Fc{4-C$_6$H$_4$-(*E*)-N=N-4-C$_6$H$_4$NHMoCl(NO)(Tp^{Me,Me})}]	CH$_2$Cl$_2$	479	564	85	HRS	1.06	46

TABLE X
(CONTINUED)

Complex	Solvent	λ_{max} (nm)	β (10^{-30} esu)	β_0 (10^{-30} esu)	Technique	Fundamental (μm)	Ref.
[Fc{(*E*)-CH=CH-4-C_5H_4NMoCl(NO)($Tp^{Me,Me}$)}]	CH_2Cl_2	511	42	3	HRS	1.06	46
[Fc{(*E*)-CH=CH-4-C_5H_4NMoI(NO)($Tp^{Me,Me}$)}]	CH_2Cl_2	516	36	2	HRS	1.06	46
[Fe(η^5-C_5HMe_4){η^5-C_5H_4-(*E*)-CH=CH-4-C_5H_4NMoCl(NO)($Tp^{Me,Me}$)}]	CH_2Cl_2	547	177	8	HRS	1.06	46
FeRu							
[Fc{(1,4-dimethyl-7-isopropylazulene)Ru(η^5-C_5H_5)}]$(PF_6)_2$	CH_2Cl_2	698	326	134	HRS	1.06	29
[Fc{(*E*)-CH=CH-4-C_6H_4C≡NRu$(PPh_3)_2$(η^5-C_5H_5)}](PF_6)	$CHCl_3$	485	186	25	HRS	1.06	115
[Fc{(*E*)-CH=CH-4-C_6H_4C≡NRu$(PPh_3)_2$(η-C_5H_5)}](BF_4)	$CHCl_3$	484	325	44	HRS	1.06	115
[Fc{(η^7-C_7H_6)Ru(η^5-C_5H_5)}]$(PF_6)_2$	$MeNO_2$	614	125	28	HRS	1.06	117
[Fc{(η^7-C_7H_6)Ru(η^5-C_5Me_5)}](PF_6)	$MeNO_2$	605	115	23	HRS	1.06	117
[Ru{C≡C-(*E*)-CH=CHFc}$(PPh_3)_2$(η^5-indenyl)]	CH_2Cl_2	345	273	141	HRS	1.06	100
[Ru{C=CH-(*E*)-CH=CHFc}$(PPh_3)_2$(η^5-indenyl)](BF_4)	CH_2Cl_2	301	117	73	HRS	1.06	100
FeRh							
[Fc{(C≡C-η^6-BC_5H_5)Rh(η^5-C_5Me_5)}](PF_6)	[b]	450	59	—	HRS	1.06	118
FePd							
[Fc{CH=C(SCH_2Ph)C=$NCMe_2CH_2O$(*N,S*-$PdCl_2$)}]	$CHCl_3$	530	12	4	EFISH	1.34	48–50
	$CHCl_3$	530	9	6	EFISH	1.91	49
[Fc{CH=C(S(O)CH_2Ph)C=$NCMe_2CH_2O$(*N,S*-$PdCl_2$)}]	$CHCl_3$	540	30	8	EFISH	1.34	48–50
[Fc{(*E*)-CH=CHCH=C(SCH_2Ph)C=$NCMe_2CH_2O$(*N,S*-$PdCl_2$)}]	$CHCl_3$	567	58	13.5	EFISH	1.34	49,50
	$CHCl_3$	567	18.2	10.8	EFISH	1.91	49,50
[Fc{(*E*)-CH=CHCH=C(S(O)CH_2Ph)C=$NCMe_2CH_2O$(*N,S*-$PdCl_2$)}]	$CHCl_3$	602	123.5	19	EFISH	1.34	49,50
[Fc{C≡CCH=C(SCH_2Ph)C=$NCMe_2CH_2O$(*N,S*-$PdCl_2$)}]	$CHCl_3$	558	33.6	8.5	EFISH	1.34	49,50
FeW							
[Fc{(*E*)-CH=CH-4-C_6H_4-(*E*)-CH=CH-4-C_5H_4NW$(CO)_5$}]	$CHCl_3$	487	535	—	HRS	1.06	42
[Fe(η^5-C_5HMe_4){η^5-C_5Me_4-(*E*)-CH=CH-4-C_5H_4NW$(CO)_5$}]	CH_2Cl_2	552	252	14	HRS	1.06	46
[Fe(η^5-C_5HMe_4){η^5-C_5Me_4-(*E*)-CH=CH-2-C_4H_2S-5-(*E*)-CH=CH-4-C_5H_4NW$(CO)_5$}]	CH_2Cl_2	571	306	34	HRS	1.06	46
[Fc{(*E*)-CH=CHC(OMe)W$(CO)_5$}]	Hexane	559	67	—	HRS	1.06	111
	CH_2Cl_2	—	82	—	HRS	1.06	111
	Acetone	—	102	—	HRS	1.06	111
	MeCN	—	160	—	HRS	1.06	111
[Fc{(*E,E*)-$(CH=CH)_2$C(OMe)W$(CO)_5$}]	Hexane	558	220	—	HRS	1.06	111
	CH_2Cl_2	—	402	—	HRS	1.06	111
	Acetone	—	520	—	HRS	1.06	111

	MeCN	—	600	—	HRS	1.06	111
[Fc{(*E,E,E*)-(CH=CH)$_3$C(OMe)W(CO)$_5$}]	Hexane	551	408	—	HRS	1.06	111
	CH_2Cl_2	—	580	—	HRS	1.06	111
	Acetone	—	840	—	HRS	1.06	111
	MeCN	—	1140	—	HRS	1.06	111
[Fc{(*E,E,E,E*)-(CH=CH)$_4$C(OMe)W(CO)$_5$}]	Hexane	541	780	—	HRS	1.06	111
	CH_2Cl_2	—	960	—	HRS	1.06	111
	Acetone	—	1720	—	HRS	1.06	111
	MeCN	—	2420	—	HRS	1.06	111
[Fc{(*E*)-CH=CH-4-C$_5$H$_4$N}W(CO)$_5$]	$CHCl_3$	491	101	12	HRS	1.06	40
	$MeNO_2$	492	282	—	HRS	1.06	45
[Fc{(*E*)-CH=CH-4-C$_6$H$_4$CN}W(CO)$_5$]	$CHCl_3$	487	375	48	HRS	1.06	40
[Fc{NHWCl(NO)(Tp^{Me,Me})}]	CH_2Cl_2	589	240	38	HRS	1.06	46
[Fc{4-C$_6$H$_4$NHWCl(NO)(Tp^{Me,Me})}]	CH_2Cl_2	510	56	4	HRS	1.06	46
FeRe							
[Fc{(*E*)-CH=CH-4-C$_5$H$_4$NRe(CO)$_5$}](BF$_4$)	$MeNO_2$	516	712	—	HRS	1.06	45
[Fc{(*E,E,E,E,E*)-CH=CHCMe=CHCH=CHCH=CMeCH=CHC$_6$H$_4$N-4}]$_2$ReBr(CO)$_3$	$CHCl_3$	501[c]	24	—	HRS	1.30	164
[ReCl(CO)$_3${Fc-(*E*)-CH=CH-4-bipy-4′-C$_9$H$_{19}$}]	[b]	524	17	—	EFISH	1.34	43
FeIr							
[Fc(C≡C-η^6-BC$_5$H$_5$)Ir(η^5-C$_5$Me$_5$)](PF$_6$)	[b]	442	46	—	HRS	1.06	118
FePt							
[Fc{CH=C(SCH$_2$Ph)C=NCMe$_2$CH$_2$O(*N,S*-PtCl$_2$)}]	$CHCl_3$	523	16	5.2	EFISH	1.34	50
[Fc{CH=C(S(O)CH$_2$Ph)C=NCMe$_2$CH$_2$O(*N,S*-PtCl$_2$)}]	$CHCl_3$	548	28.0	7.7	EFISH	1.34	49,50
ZnRu							
[Zn(OAc)$_2${Ru(η^5-C$_5$H$_5$){η^5-C$_5$H$_4$-(*E*)-CH=CH-4-bipy-4′-Me}]	[b]	441	18	—	EFISH	1.34	43
MoRu							
[Mo{C≡NRu(NH$_3$)$_5$}(CO)$_5$]$^{2+}$	Acetone	693	225	90	HRS	1.06	113
Ru$_2$							
[Ru{N≡CRu(PPh$_3$)$_2$(η^5-C$_5$H$_5$)}(NH$_3$)$_5$](CF$_3$SO$_3$)$_3$	H_2O	720	1080	—	HRS	1.06	109
	$MeNO_2$	716	157	69	HRS	1.06	113
[Ru{C≡C-(*E*)-CH=CH-4-C$_6$H$_4$CNRu(NH$_3$)$_5$}(PPh$_3$)$_2$(η^5-indenyl)](PF$_6$)$_3$	Acetone	442	315	80	HRS	1.06	74
[Ru{C≡CCH=CH-4-C$_6$H$_4$C≡NRu(NH$_3$)$_5$}(PPh$_3$)$_2$(η^5-indenyl)](CF$_3$SO$_3$)$_3$	Acetone	442	315	80	HRS	1.06	100
[Ru{C≡NRu(NH$_3$)$_5$}(PPh$_3$)$_2$(η^5-indenyl)](PF$_6$)$_3$	Acetone	621	108	26	HRS	1.06	74
[Ru{C≡NRu(NH$_3$)$_5$}(PPh$_3$)$_2$(η^5-indenyl)](CF$_3$SO$_3$)$_3$	Acetone	621	108	26	HRS	1.06	100
[Ru(η^5-C$_5$H$_5$)(η^5-C$_5$H$_4$CH=CH-η^6-C$_7$H$_7$)Ru(η^5-C$_5$H$_5$)](PF$_6$)	CH_2Cl_2	423	162	50	HRS	1.06	112
[Ru(η^5-C$_5$H$_5$)(η^5-C$_5$H$_4$CH=CH-η^6-C$_7$H$_7$)Ru(η^5-C$_5$Me$_5$)](PF$_6$)	CH_2Cl_2	424	75	23	HRS	1.06	112
	$MeNO_2$	409	87	30			
[Ru(η^5-C$_5$H$_5$)(η^5-C$_5$H$_4$CH=CH-η^6-C$_7$H$_7$)Ru(η^5-C$_5$H$_5$)](PF$_6$)$_2$	$MeNO_2$	560	358	28	HRS	1.06	112
[Ru(η^5-C$_5$H$_5$)(η^5-C$_5$H$_4$CH=CH-η^6-C$_7$H$_7$)Ru(η^5-C$_5$Me$_5$)](PF$_6$)$_2$	$MeNO_2$	541	701	17	HRS	1.06	112

TABLE X
(CONTINUED)

Complex	Solvent	λ_{max} (nm)	β (10^{-30} esu)	β_0 (10^{-30} esu)	Technique	Fundamental (μm)	Ref.
RuW							
$[Ru\{C{\equiv}NW(CO)_5\}(PPh_3)_2(\eta^5\text{-indenyl})]$	CH_2Cl_2	392	40	15	HRS	1.06	74,100
$[Ru\{C{\equiv}CCH{=}CH\text{-}4\text{-}C_6H_4C{\equiv}NW(CO)_5\}(PPh_3)_2(\eta^5\text{-indenyl})]$	CH_2Cl_2	456	700	150	HRS	1.06	74,100
$[Ru\{C{\equiv}C\text{-}(E)\text{-}CH{=}CH\text{-}4\text{-}C_5H_4NW(CO)_5\}(PPh_3)_2(\eta^5\text{-indenyl})]$	CH_2Cl_2	462	535	71	HRS	1.06	74,100
$[W\{C{\equiv}NRu(NH_3)_5\}(CO)_5]^{2+}$	Acetone	708	130	56	HRS	1.06	113
$[W(CO)_3(phen)(\mu\text{-}Cl)\text{-}fac\text{-}Ru(CO)_3]$	$CHCl_3$	500	−76	—	EFISH	1.9	114
RuRe							
$[ReCl(CO)_3(Ru(\eta^5\text{-}C_5H_5)\{\eta^5\text{-}C_5H_4\text{-}(E)\text{-}CH{=}CH\text{-}4\text{-}bipy\text{-}4'\text{-}Me\})]$	[b]	460	15	—	EFISH	1.34	43
RuOs							
$[Ru\{CNOs(NH_3)_5\}(PPh_3)_2(\eta^5\text{-}C_5H_5)]^{3+}$	DMSO	440	65	16	HRS	1.06	113
RhW							
$[W(CO)_5\{NC_5H_4\text{-}4\text{-}(E)\text{-}CH{=}CH\text{-}4\text{-}C_5H_4N\}\text{-}cis\text{-}RhCl(CO)_2]$	[b]	435	−41	−31	EFISH	1.9	77
$[W(CO)_5(pyz)\text{-}cis\text{-}RhCl(CO)_2]$	[b]	499	−28	−19	EFISH	1.9	77
$[W(CO)_3(phen)(\mu\text{-}Cl)\text{-}cis\text{-}Rh(CO)_2]$	$CHCl_3$	517	−31.7	—	EFISH	1.9	114
WRe							
$[W(CO)_5\{NC_5H_4\text{-}4\text{-}(E)\text{-}CH{=}CH\text{-}4\text{-}C_5H_4N\}\text{-}cis\text{-}ReCl(CO)_4]$	[b]	437	—	−39	EFISH	1.91	77
WOs							
$[W(CO)_3(phen)(\mu\text{-}Cl)\text{-}fac\text{-}Os(CO)_3]$	$CHCl_3$	495	−75.6	—	EFISH	1.9	114
WIr							
$[W(CO)_3(phen)(\mu\text{-}Cl)\text{-}cis\text{-}Ir(CO)_2]$	$CHCl_3$	503	−25.8	—	EFISH	1.9	114
Fe_3							
$[(Fc\{(E)\text{-}CH{=}CH\text{-}4\text{-}C_6H_4\})_3C(OH)]$	CF_3COOH	[b]	1700 ± 250	—	HRS	1.06	39
$[\{Fe(CO)(\eta^5\text{-}C_5H_5)\}_2(\mu\text{-}CO)\{\mu\text{-}C\text{-}(E)\text{-}CH{=}CHFc\}](BF_4)$	CH_2Cl_2	655	113	36	HRS	1.06	119
$[\{Fe(CO)(\eta^5\text{-}C_5H_5)\}_2(\mu\text{-}CO)\{\mu\text{-}C\text{-}(E,E)\text{-}(CH{=}CH)_2Fc\}](BF_4)$	CH_2Cl_2	734	156	74	HRS	1.06	119
$[\{Fe(CO)(\eta^5\text{-}C_5H_5)\}_2(\mu\text{-}CO)\{\mu\text{-}C\text{-}(E,E,E)\text{-}(CH{=}CH)_3Fc\}](BF_4)$	CH_2Cl_2	775	227	120	HRS	1.06	119
$[\{Fe(CO)(\eta^5\text{-}C_5H_5)\}_2(\mu\text{-}CO)\{\mu\text{-}C\text{-}(E,E,E,E)\text{-}(CH{=}CH)_4Fc\}](BF_4)$	CH_2Cl_2	813	562	313	HRS	1.06	119
$[\{Fe(CO)(\eta^5\text{-}C_5H_5)\}_2(\mu\text{-}CO)\{\mu\text{-}C\text{-}(E)\text{-}CH{=}CH\text{-}4\text{-}C_6H_4\text{-}(E)\text{-}CH{=}CHFc\}](BF_4)$	CH_2Cl_2	705	1106	469	HRS	1.06	119
$[\{Fe(CO)(\eta^5\text{-}C_5H_5)\}_2(\mu\text{-}CO)\{\mu\text{-}C\text{-}(E)\text{-}CH{=}CH\text{-}2\text{-}C_4H_2S\text{-}5\text{-}(E)\text{-}CH{=}CHFc\}](BF_4)$	CH_2Cl_2	757	679	343	HRS	1.06	119
$[\{Fe(CO)(\eta^5\text{-}C_5H_5)\}_2(\mu\text{-}CO)\{\mu\text{-}C\text{-}(E)\text{-}CH{=}CH\text{-}2\text{-}C_4H_2O\text{-}5\text{-}(E)\text{-}CH{=}CHFc\}](BF_4)$	CH_2Cl_2	747	352	173	HRS	1.06	119
	CH_2Cl_2	764	1195	615	HRS	1.06	119

[{Fe(CO)(η^5-C_5H_5)}$_2$(μ-CO){μ-C-(*E*)-CH=CH-2-C_4H_2S-5-(*E,E*)-(CH=CH)$_2$Fc}](BF$_4$)							
[{Fe(CO)(η^5-C_5H_5)}$_2$(μ-CO){μ-C-(*E*)-CH=CH-(*E*)-CH=CClFc}](BF$_4$)	CH_2Cl_2	746	459	225	HRS	1.06	119
[{Fe(CO)(η^5-C_5H_5)}$_2$(μ-CO){μ-C-(*E*)-CH=CH-2-C_4H_2S-5-(*E*)-CH=CH-(*E*)-CH=CClFc}](BF$_4$)	CH_2Cl_2	731	1424	674	HRS	1.06	119
[{Fe(CO)(η^5-C_5H_5)}$_2$(μ-CO){μ-C-(*E*)-CH=CH-2,5-C_4H_2S-2,5-C_4H_2S-(*E*)-CH=CHFc}](BF$_4$)	CH_2Cl_2	733	429	202	HRS	1.06	120
[{Fe(CO)(η^5-C_5H_5)}$_2$(μ-CO){μ-C-(*E*)-CH=CH-3-(6-(*E*)-2-ferrocenylethenyldithieno-[3,2-b:2,3-d]-thiophene)}](BF$_4$)	CH_2Cl_2	747	624	307	HRS	1.06	120
[{Fe(CO)(η^5-C_5H_5)}$_2$(μ-CO){μ-C-(*E*)-CH=CH-2,5-C_4H_2S-(*E*)-CH=CH-2-C_4H_2S-5-(*E*)-CH=CHFc}](BF$_4$)	CH_2Cl_2	750	867	430	HRS	1.06	120
[{Fe(CO)(η^5-C_5H_5)}$_2${μ-C-(*E*)-CH=CH-2-(7-{(*E*)-2-ferrocenylethenyl}thieno-[3,2-e][1]-benzothiophene)}](BF$_4$)	CH_2Cl_2	712	625	273	HRS	1.06	120
[{Fe(CO)(η^5-C_5H_5)}$_2${μ-C-(*E*)-CH=CH-2-(2′-{(*E*)-2-ferrocenylethenyl}-4,5,4′,5′-tetrahydro-[6,6′]-bi[cyclopenta[b]thiophenylidene])}](BF$_4$)	CH_2Cl_2	830	1243[d]	697[d]	HRS	1.06	120
[{Fe(CO)(η^5-C_5H_5)}$_2$(μ-CO){μ-C-(*E*)-CH=CH-2,5-C_4H_2S-2,5-C_4H_2S-2,5-C_4H_2S-(*E*)-CH=CHFc}](BF$_4$)	CH_2Cl_2	772	650	—	HRS	1.3	121
[{Fe(CO)(η^5-C_5H_5)}$_2$(μ-CO){μ-C-(*E*)-CH=CH-2,5-C_4H_2S-2,7-C_8H_4S-2,5-C_4H_2S-(*E*)-CH=CHFc}](BF$_4$)	CH_2Cl_2	880	344	—	HRS	1.3	121
FeW_2							
[Fe{η^5-3,4-$PC_4H_2Me_2$-1-W(CO)$_5$}{η^5-3,4-PC_4HMe_2-2-(*N,N′*-diethylbarbiturate)-1-W(CO)$_5$}]	$CHCl_3$	565	−59	−35	EFISH	1.907	59

[a] β_{333} value calculated using the two-level model on the π-π^* absorption.
[b] Not stated.
[c] CH_2Cl_2 solvent.
[d] Fluorescence contribution.

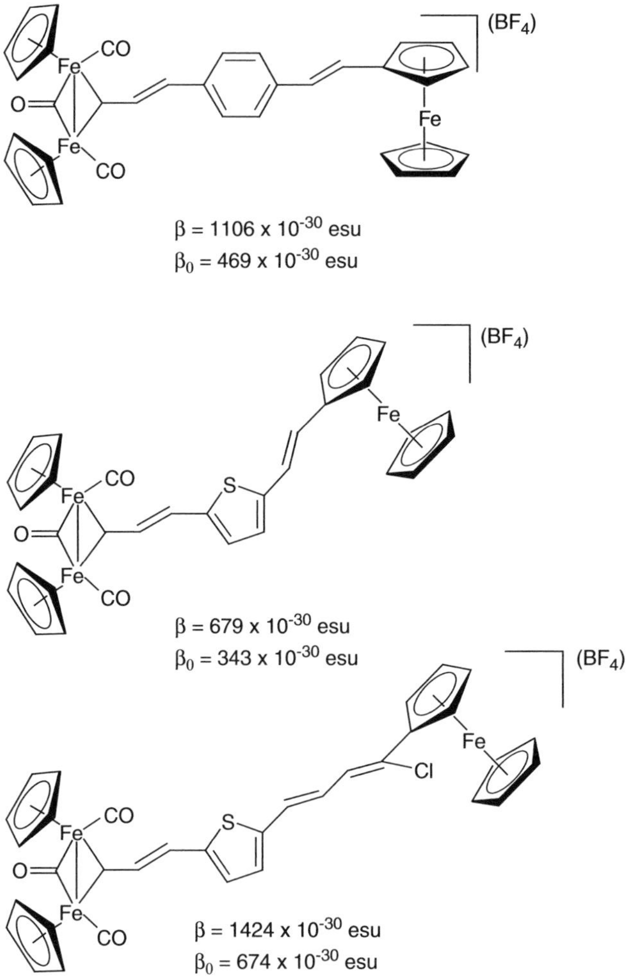

$\beta = 1106 \times 10^{-30}$ esu
$\beta_0 = 469 \times 10^{-30}$ esu

$\beta = 679 \times 10^{-30}$ esu
$\beta_0 = 343 \times 10^{-30}$ esu

$\beta = 1424 \times 10^{-30}$ esu
$\beta_0 = 674 \times 10^{-30}$ esu

FIG. 10. Selected trimetallic complexes with large β values.

remarkably small. Nonlinearities for the pyridine complexes are larger; metal replacement has no effect upon β, but π-bridge lengthening leads to a significant increase.

IV
MOLECULAR THIRD-ORDER MEASUREMENTS

Researchers have used a variety of techniques to measure the third-order hyperpolarizabilities of organometallic complexes, *Z*-scan in particular being very popular over the past few years. Data have been collected at several wavelengths with varying pulse lengths using these various techniques, and in many instances with

TABLE XI
MOLECULAR SECOND-ORDER NLO MEASUREMENTS OF OTHER COMPLEXES

Complex	Solvent	λ_{max}	β (10^{-30} esu)	β_0 (10^{-30} esu)	Technique	Fundamental (μm)	Ref.
[NiCl(PPh$_3$)(η^5-C$_5$H$_5$)]	THF	330	89	45	HRS	1.06	89
[RuCl(PPh$_3$)$_2$(η^5-C$_5$H$_5$)]	THF	357	<7	<4	HRS	1.06	94
[Ru(C≡N)(PPh$_3$)$_2$(η^5-indenyl)]	CH_2Cl_2	396	13	5	HRS	1.06	75,122
[RhCl(η^2-cod)$_2$(4-NC$_5$H$_4$CH=CH-4-C$_6$H$_4$NMe$_2$)]	$CHCl_3$	397	77	—	EFISH	1.34	80
trans-[PdI(4-C$_6$H$_4$NO$_2$)(PEt$_3$)$_2$]	$CHCl_3$	[a]	0.5	—	EFISH	1.91	32
trans-[PdI(4-C$_6$H$_4$NO$_2$)(PPh$_3$)$_2$]	$CHCl_3$	[a]	1.5	—	EFISH	1.91	32
[IrCl(η^2-cot)$_2$(4-NC$_5$H$_4$CH=CH-4-C$_6$H$_4$NMe$_2$)]	$CHCl_3$	413	82	—	EFISH	1.34	80
[IrCl(η^2-cot)$_2${4-NC$_5$H$_4$(CH=CH)$_2$-4-C$_6$H$_4$NMe$_2$}]	$CHCl_3$	430	152	—	EFISH	1.34	80
trans-[PtBr(4-C$_6$H$_4$CHO)(PEt$_3$)$_2$]	$CHCl_3$	[a]	2.1	—	EFISH	1.91	32
trans-[PtBr(4-C$_6$H$_4$NO$_2$)(PEt$_3$)$_2$]	$CHCl_3$	[a]	3.8	—	EFISH	1.91	32
trans-[PtI(4-C$_6$H$_4$NO$_2$)(PEt$_3$)$_2$]	$CHCl_3$	[a]	1.7	—	EFISH	1.91	32

[a]Not stated.

large error margins for the resultant data, all of which serve to render meaningful comparison difficult and development of structure–property relationships particularly challenging. Accordingly, the discussion that follows is of necessity somewhat more cautious than that for molecular second-order measurements.

Ferrocene was the first organotransition metal complex for which third-order NLO properties were reported, and metallocenes and their derivatives remain popular (Tables XII and XIII). Ferrocene itself has been measured several times. Early values obtained from optical power limiting (OPL) measurements[123] were very high compared to later values from DFWM and THG,[124] probably because OPL is carried out on a nanosecond time-scale and results can be enhanced by thermal effects. The results from *Z*-scan studies at 0.53 μm are also very large, probably stemming from resonance enhancement.[125] Large errors with some reports have resulted in difficulties in developing structure–activity relationships.

Cyclopentadienyl ring substitution usually results in increased cubic NLO response. The γ value of [Fc(CH=CHPh)] is larger than those of ferrocene and styrene ($16.9 \pm 0.8 \times 10^{-36}$ esu)[126] combined, suggesting that communication between the components is important for optical nonlinearity. The ferrocenyl unit has been embellished with a variety of cyclopentadienyl substituents, but meaningful comparisons are probably restricted to results emanating from a single laboratory, and the only general comment that can be made is that π-system lengthening usually results in increased nonlinearity. There have been a large number of bimetallic complexes incorporating ferrocene units, selected efficient examples being shown in Fig. 11, and one of which incorporates both the ferrocenyl and the metal alkynyl units, the two most popular moieties in organometallics for nonlinear optics. Incorporation of the second (and in some cases third) metal center generally results in complexes with significantly increased nonlinearity.

While ferrocenyl and, to a lesser extent, ruthenocenyl complexes have been intensively studied, other metallocenyl complexes are little investigated, the group 4 examples being listed in Table XIII, and the most efficient example being shown in Fig. 12. Although comparisons must be cautious, nonlinearities are lower than those for ferrocenyl and ruthenocenyl complexes, possibly related to the fact that the group 4 species are d^0 16 electron complexes. The errors for these low nonlinearities are also low, so development of structure–NLO activity correlations is possible. The increased π-conjugation resultant upon replacing halogen by alkenyl or alkynyl ligand results in enhanced nonlinearity, with the cubic nonlinearity for $[Ti(C{\equiv}CPh)_2(\eta^5\text{-}C_5H_5)_2]$ three times that of $[Ti(C{\equiv}CPh)Cl(\eta^5\text{-}C_5H_5)_2]$. Metal replacement has been examined, increasing γ being seen on proceeding up group 4. This corresponds to an increase in electron-accepting strength of the metal;[124] it is suggested that mixing of titanium orbitals with those of the ligand is more effective than mixing of zirconium or hafnium with the ligand-based orbitals.[124]

The carbonyl complexes in Table XIV are of three types, tricarbonylchromium η^6-arene complexes, pentacarbonyltungsten σ-pyridine complexes, and carbonylruthenium σ-alkenyl complexes, the first two possessing low γ values. For the first two types, lengthening the π-system or replacing acceptor by donor substituent on the tricarbonylchromium-coordinated arene ring results in

TABLE XII

THIRD-ORDER NLO MEASUREMENTS OF FERROCENYL AND RUTHENOCENYL COMPLEXES

Complex	λ_{max} (nm)	Technique	Solvent	Fundamental (μm)	n_2	γ_{real} (10^{-36} esu)	γ_{imag} (10^{-36} esu)	$\lvert\gamma\rvert$(10^{-36} esu)	γ (10^{-45} m^5 V^{-2})	Ref.
[FcH]	[a]	OPL	Molten	1.06	1.1×10^{-17} m^2 W^{-1}	—	—	—	4.4	123
	[a]	OPL	Molten	1.06	1.1×10^{-17} m^2 W^{-1}	—	—	—	2.9	127
	[a]	OPL	Molten	1.06	1.5×10^{-11} esu	—	—	39000	—	128
	[a]	OPL	EtOH	1.06	2.1×10^{-19} m^2 W^{-1}	—	—	—	3.9	123
	[a]	OPL	EtOH	1.06	2.6×10^{-13} esu	—	—	12000	—	128
	440	DFWM	THF	0.60	—	—	—	16.1 ± 1.8	—	126
[Fc(CH=CHPh)]	445	DFWM	THF	0.60	—	—	—	85.5 ± 19.8	—	126
[Fc{C(O)NH-4-C_5H_4N}]	[a]	Z-scan	DMF	0.53	—	—	—	1510000	—	125
[FcC(O)CH_3]	[a]	THG	Dioxane	1.91	—	—	—	27 ± 3	—	32,129
[Fc{C(O)NH-4-C_5H_4N}$_2$]	[a]	Z-scan	DMF	0.53	—	—	—	1530000	—	125
[Fc-4-C_6H_4-(E)-CH=CH-4-$C_6H_4NO_2$]	[a]	THG	Neat	1.91	—	—	—	25	—	124
[Fe(η^5-$C_5H_4Bu^n$)$_2$]	[a]	THG	Neat	1.91	—	—	—	25 ± 4	—	124
[Fe(η^5-$C_5H_4SiMe_3$)$_2$]	[a]	OPL	Neat	1.06	1.3×10^{-17} m^2 W^{-1}	—	—	—	5.5	123,127
	[a]	OPL	Neat	1.06	1.51×10^{-11} esu	—	—	64000	—	128
[Fe{η^5-C_5H_4C(O)NH-4-C_5H_4N}$_2$]	469	THG	$CHCl_3$	1.91	—	605	860	1050	—	131,132
[Fe{η^5-C_5H_4-(E)-4-CH=CHC_6H_4I}$_2$]	468	Z-scan	THF	0.80	—	200 ± 50	100 ± 20	220 ± 50	—	133
[Fe{η^5-C_5H_4-(E)-4-CH=CHC_6H_4C≡$CSiMe_3$}$_2$]	468	Z-scan	THF	0.80	—	0 ± 50	200 ± 40	200 ± 40	—	133
[Fe{η^5-C_5H_4-(E)-4-CH=CHC_6H_4C≡CH}$_2$]	469	Z-scan	THF	0.80	—	−400 ± 250	200 ± 40	450 ± 240	—	133
[Fe(η^5-C_5H_4CH=CHPh)$_2$]	461	DFWM	THF	0.60	—	—	—	270 ± 26	—	126
[Fe(η^5-C_5H_4CHO)(η^5-C_5H_4CH=CHPh)]	462	DFWM	THF	0.60	—	—	—	305 ± 36	—	126
[Fe{η^5-C_5H_4-(E)-CH=CHC_6H_4-4-CN}$_2$]	492	Z-scan	THF	0.80	—	280 ± 150	30 ± 20	280 ± 150	—	134
[Fe{η^5-C_5H_4-(Z)-CH=CHC_6H_4-4-NO_2}$_2$]	481	Z-scan	THF	0.80	—	600 ± 300	0 ± 50	600 ± 300	—	134
[Fe{η^5-C_5H_4-(E)-CH=CHC_6H_4-4-CN}{η^5-C_5H_4-(Z)-CH=CHC_6H_4-4-CN}]	469	Z-scan	THF	0.80	—	310 ± 200	50 ± 30	310 ± 200	—	134
[Fe{η^5-C_5H_4-(E)-CH=CHC_6H_4-4-NO_2}{η^5-C_5H_4-(Z)-CH=CHC_6H_4-4-NO_2}]	488	Z-scan	THF	0.80	—	840 ± 400	770 ± 200	1140 ± 430	—	134

TABLE XII
(CONTINUED)

Complex	λ_{max} (nm)	Technique	Solvent	Fundamental (μm)	n_2	γ_{real} (10^{-36} esu)	γ_{imag} (10^{-36} esu)	$\lvert\gamma\rvert$(10^{-36} esu)	γ (10^{-45} m^5 V^{-2})	Ref.
[Ru(η^5-C_5H_5)$_2$]	[a]	OPL	Molten	1.06	3.3×10^{-11} esu	—	—	—	3.9	128
	[a]	OPL	Molten	1.06	2.2×10^{-17} m^2 W^{-1}	—	—	—	3.9	127
[Ru{η^5-C_5H_4-(*E*)-CH=CH-4-$C_6H_4NO_2$}(η^5-C_5Me_5)]	424	THG	Dioxane	1.91	—	—	—	140±14	—	31,32,129
[Ru{η^5-C_5H_4-(*E*)-CH=CH-4-$C_6H_4NO_2$}(η^5-C_5H_5)]	390	THG	Dioxane	1.91	—	—	—	114±11	—	32,129
[Ru{η^5-C_5H_4-(*E*)-CH=C(CN)-4-$C_6H_4NO_2$}(η^5-C_5Me_5)]	443	THG	Dioxane	1.91	—	—	—	105±11	—	32
[Fc(C≡CC≡CC≡CC≡C)Fc]	[a]	THG	$CHCl_3$	1.91	—	—	—	110	—	130
[Fc(CH=CH-4-C_6H_4CH=CH)Fc]	452	DFWM	THF	0.60	—	—	—	504±52	—	126
[Fe(η^5-C_5H_4CHO)(η^5-C_5H_4CH=CH-4-C_6H_4CH=CH)Fc]	455	DFWM	THF	0.60	—	—	—	925±86	—	126
[Fc-(*E*)-CH=CH-4-C_6H_4-(*E*)-CH=CHFc)]	457	*Z*-scan	THF	0.80	—	640±300	30±20	640±300	—	134
[MoCl(HNC_6H_4-4-Fc)(NO){BH($N_2C_3H_2$-3-C_6H_4-4-OMe)$_3$}]	598	THG	$CHCl_3$	1.91	—	−450	460	640	—	131,132
[MoCl(HNC_6H_4-4-N=N-4-C_6H_4Fc)(NO){BH($N_2C_3H_2$-3-C_6H_4-4-OMe)$_3$}]	537	THG	$CHCl_3$	1.91	—	−3450	3210	4710	—	131,132
[MoCl(HNC_6H_3-3-Me-4-N=NC_6H_4-4-Fc)(NO){BH($N_2C_3H_2$-3-C_6H_4-4-OMe)$_3$}]	540	THG	$CHCl_3$	1.91	—	−467	1320	1400	—	131,132
[MoCl(HNC_6H_3-3-Me-4-N=NC_6H_3-3-Me-4-Fc)(NO){BH($N_2C_3H_2$-3-C_6H_4-4-OMe)$_3$}]	530	THG	$CHCl_3$	1.91	—	−428	5770	5790	—	131,132
[MoCl(HNC_6H_3-3-Me-4-N=NC_6H_3-5-Me-2-Fc)(NO){BH($N_2C_3H_2$-3-C_6H_4-4-OMe)$_3$}]	571	THG	$CHCl_3$	1.91	—	−180	410	448	—	131,132
[Fc-(*E*)-CH=CH-4-{η^6-C_6H_4)-Cr(CO)$_3$}-(*E*)-CH=CH-Fc)]	455	*Z*-scan	THF	0.80	—	850±300	95±30	860±300	—	134
[Zn{NC_5H_4-4-NHC(O)Fc}$_2$(NO_3)$_2$]	[a]	*Z*-scan	DMF	0.53	—	—	—	2460000	—	125
[Ag{1,2-(FcCH=NH)$_2$$CH_2CH_2$}($NO_3$)]	447	*Z*-scan	DMF	0.53	—	—	—	268000	—	135
[HgI_2{1,2-(FcCH=NH)$_2$$CH_2CH_2$}($NO_3$)]	447	*Z*-scan	DMF	0.53	—	—	—	1440000	—	135
[HgI_2(1,2-{(FcCH=)NH}$_2$$CH_2CH_2$)]	447	*Z*-scan	DMF	0.53	—	—	—	0.144×10^6	—	135
[Fe{η^5-C_5H_4-(*E*)-4-CH=CHC_6H_4CH=CRuCl(dppm)$_2$}$_2$](PF_6)$_2$	383	*Z*-scan	THF	0.80	—	−3000±1200	2300±800	3800±1400	—	133

[Fe{η^5-C_5H_4-(*E*)-4-$CH{=}CHC_6H_4C{\equiv}CRuCl(dppm)_2\}_2$]	396	*Z*-scan	THF	0.80	—	-7100 ± 3000	10600 ± 2000	13000 ± 3000	—	133
[Fe{η^5-C_5H_4-(*E*)-4-$CH{=}CHC_6H_4C{\equiv}CAu(PMe_3)\}_2$]	463	*Z*-scan	THF	0.80	—	200 ± 150	0 ± 30	200 ± 150	—	133
[Fe{η^5-C_5H_4-(*E*)-4-$CH{=}CHC_6H_4C{\equiv}CAu(PPh_3)\}_2$]	465	*Z*-scan	THF	0.80	—	-1100 ± 300	300 ± 60	1140 ± 310	—	133
[Fe{η^5-C_5H_4-(*E*)-4-$CH{=}CHC_6H_4C{\equiv}CAu(PCy_3)\}_2$]	468	*Z*-scan	THF	0.80	—	-400 ± 500	500 ± 100	640 ± 390	—	133
[$Cd_2(\{NC_5H_4$-4-NHC(O)-η^5-$C_5H_4\}_2Fe)_2(OAc)_4$]	[a]	*Z*-scan	DMF	0.53	—	—	—	3.1×10^6	—	125
[$Hg_2(\{NC_5H_4$-4-NHC(O)-η^5-$C_5H_4\}_2Fe)_2(OAc)_4$]	[a]	*Z*-scan	DMF	0.53	—	—	—	3.12×10^6	—	125

[a] Not stated.

TABLE XIII
MOLECULAR THIRD-ORDER NLO RESULTS OF OTHER METALLOCENE COMPLEXES

Complex	λ_{max} (nm)	Technique	Solvent	Fundamental (μm)	n_2	$\|\gamma\|(10^{-36}$ esu)	γ (10^{-45} m^5 V^{-2})	Ref.
$[TiF_2(\eta^5\text{-}C_5H_4Me)_2]$	414	THG	$CHCl_3$	1.907	—	<3	—	124,136,137
$[TiCl_2(\eta^5\text{-}C_5H_4Me)_2]$	518	THG	$CHCl_3$	1.907	—	<5	—	124,136,137
$[TiBr_2(\eta^5\text{-}C_5H_4Me)_2]$	568	THG	$CHCl_3$	1.907	—	<5	—	124,136,137
$[Ti(C{\equiv}CBu^n)_2(\eta^5\text{-}C_5H_5)_2]$	390	THG	$CHCl_3$	1.91	—	12±2	—	136
$[Ti(C{\equiv}CBu^n)_2(\eta^5\text{-}C_5H_4Me)_2]$	[a]	THG	$CHCl_3$	1.91	—	15±2	—	137
$[Ti(C{\equiv}CPh)_2(\eta^5\text{-}C_5H_5)_2]$	416	THG	THF	1.907	—	92±14	—	124,136
$[Ti(C{\equiv}CPh)Cl(\eta^5\text{-}C_5H_5)_2]$	[a]	THG	THF	1.907	—	31±5	—	124,136
$[ZrCl_2(\eta^5\text{-}C_5H_5)_2]$	[a]	OPL	Molten	1.06	1.0×10^{-11} esu	—	—	128
	[a]	OPL	Molten	1.06	0.8×10^{-11} esu	—	1.9	127
	338	THG	$CHCl_3$	1.91	—	<5	—	124,136,137
$[\{ZrCl(\eta^5\text{-}C_5H_5)_2\}_2O]$	282	THG	THF	1.91	—	10±2	—	136
$[Zr(C{\equiv}CPh)_2(\eta^5\text{-}C_5H_5)_2]$	390	THG	THF	1.91	—	58±9	—	124,136
$[Zr(\overline{CPh{=}CPhCPh{=}CPh})(\eta^5\text{-}C_5H_5)_2]$	474	THG	THF	1.91	—	47±7	—	136
$[ZrCl(CH{=}CHC_6H_4Me)(\eta^5\text{-}C_5H_5)_2]$	356	THG	THF	1.91	—	24±4	—	136
$[\{ZrCl(\eta^5\text{-}C_5H_5)_2\}_2\{\mu\text{-}(E)\text{-}CH{=}CH\text{-}3\text{-}C_6H_4\text{-}(Z)\text{-}CH{=}CH\}]$	356	THG	THF	1.91	—	68±10	—	136
$[\{ZrCl(\eta^5\text{-}C_5H_5)_2\}_2\{\mu\text{-}(E)\text{-}CH{=}CH\text{-}4\text{-}C_6H_4\text{-}(E)\text{-}CH{=}CH\}]$	380	THG	THF	1.91	—	154±23	—	136
$[HfCl_2(\eta^5\text{-}C_5H_5)_2]$	[a]	OPL	Molten	1.06	1.3×10^{-11} esu	—	—	128
	[a]	OPL	Molten	1.06	1.0×10^{-17} m^2 W^{-1}	—	2.4	127
	306	THG	$CHCl_3$	1.91	—	—	<5	136
$[Hf(C{\equiv}CPh)_2(\eta^5\text{-}C_5H_5)_2]$	390	THG	THF	1.91	—	51±8	—	124,136

[a]Not stated.

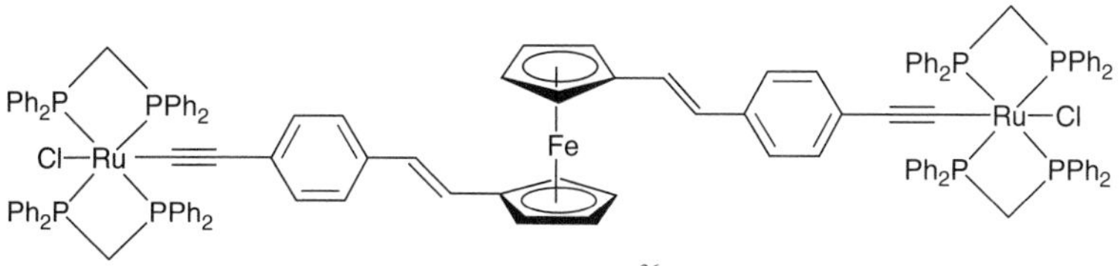

γ_{real} = -7100 +/– 3000 x 10^{-36} esu
γ_{imag} = 10600 +/– 2000 x 10^{-36} esu
$|\gamma|$ = 13000 +/– 3000 x 10^{-36} esu
σ_2 = 2500 x 10^{-50} cm^4 s

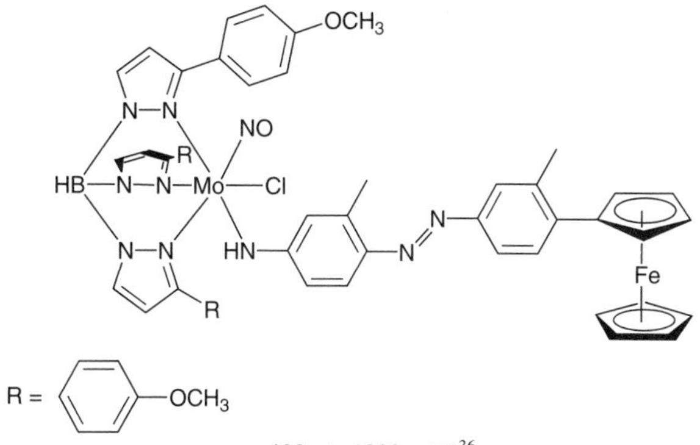

R = C_6H_4–OCH_3

γ_{real} = -428 +/– 1200 x 10^{-36} esu
γ_{imag} = 5770 +/– 800 x 10^{-36} esu
$|\gamma|$ = 5790 +/– 1400 x 10^{-36} esu

FIG. 11. Selected bi- and trimetallic ferrocene complexes with large γ values.

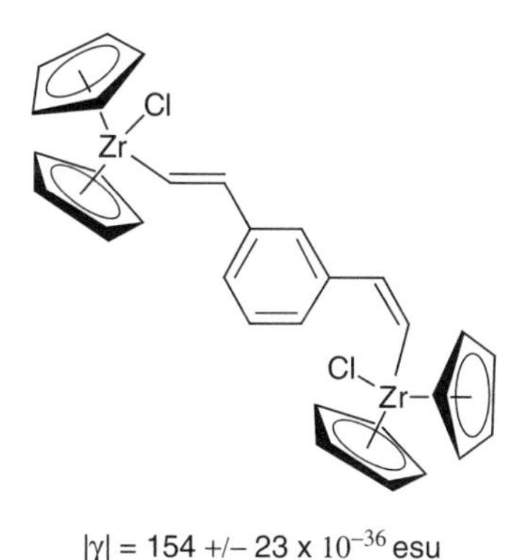

$|\gamma|$ = 154 +/– 23 x 10^{-36} esu

FIG. 12. A metallocene with a large γ value.

increased nonlinearity. Cubic nonlinearities for the alkenyl complexes increase on carbonyl replacement by phosphine and introduction of nitro substituent, as can be seen in Fig. 13.

Data for phosphine-substituted molybdenum carbonyl complexes reported as either $\langle \gamma \rangle$ or as a ratio of γ_{CS_2} are collected in Table XV. These studies were carried

TABLE XIV

THIRD-ORDER NLO MEASUREMENTS OF CARBONYL COMPLEXES

Complex	λ_{max} (nm)	Technique	Solvent	Fundamental (μm)	γ_{real} (10^{-36} esu)	γ_{imag} (10^{-36} esu)	Ref.
$[Cr(CO)_3(\eta^6-C_6H_6)]$	310	THG	$CHCl_3$	1.91	2	—	32
$[Cr(CO)_3(\eta^6-C_6H_5OMe)]$	310	THG	$CHCl_3$	1.91	3	—	32
$[Cr(CO)_3(\eta^6-C_6H_5NH_2)]$	313	THG	$CHCl_3$	1.91	12	—	32
$[Cr(CO)_3(\eta^6-C_6H_5NMe_2)]$	318	THG	$CHCl_3$	1.91	10	—	32
$[Cr(CO)_3\{\eta^6-C_6H_5C(O)OMe\}]$	318	THG	$CHCl_3$	1.91	6	—	32
$[Cr(CO)_3\{\eta^6-C_6H_5-(E)-CH{=}CHPh\}]$	410	THG	$CHCl_3$	1.91	21	—	32
$[Ru\{(Z)-HC{=}CHPh\}(CO)_2(\eta^5-C_5Ph_5)]$	289	Z-scan	CH_2Cl_2	0.80	10 ± 50	100 ± 40	73
$[Ru\{(Z)-HC{=}CH-4-C_6H_4NO_2\}(CO)_2(\eta^5-C_5Ph_5)]$	364	Z-scan	CH_2Cl_2	0.80	-180 ± 80	5 ± 5	73
$[Ru\{(Z)-HC{=}CH-4-C_6H_4NO_2\}(CO)(PMe_2Ph)(\eta^5-C_5Ph_5)]$	434	Z-scan	CH_2Cl_2	0.80	-420 ± 150	25 ± 15	73
$[W(CO)_5(NC_5H_4-4-NH_2)]$	290	THG	DMSO	1.91	15	—	32
$[W(CO)_5(NC_5H_4-4-Bu^n)]$	328	THG	Dioxane	1.91	15	—	32
$[W(CO)_5(NC_5H_5)]$	332	THG	Toluene	1.91	8	—	32
$[W(CO)_5(NC_5H_4-4-Ph)]$	340	THG	$CHCl_3$	1.91	12	—	32
$[W(CO)_5\{NC_5H_4-4-C(O)Me\}]$	440	THG	$CHCl_3$	1.91	14	—	32

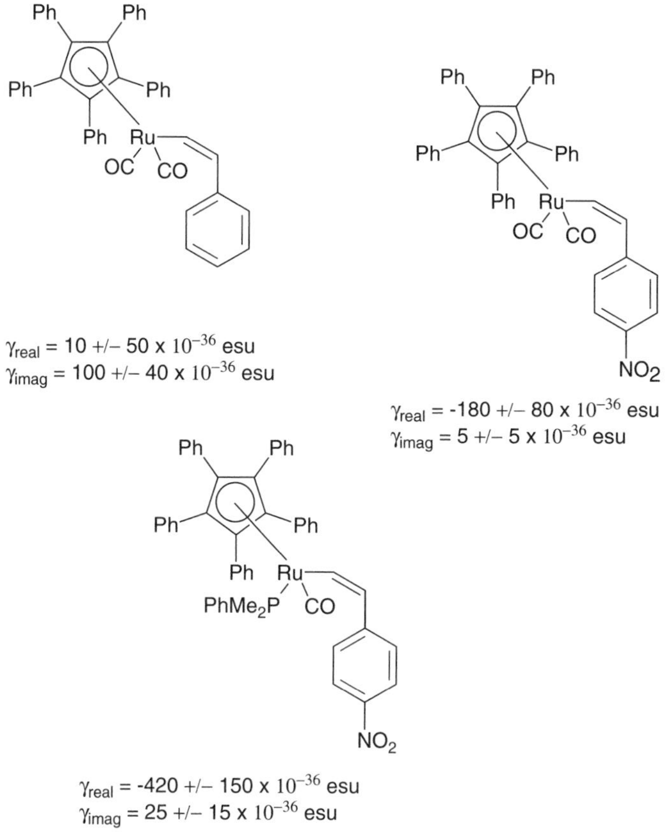
Ph
Ph
Ph
Ph
Ph
Ru
OC
CO
γreal = 10 +/– 50 x 10−36 esu
γimag = 100 +/– 40 x 10−36 esu
Ph
Ph
Ph
Ph
Ph
Ru
OC
CO
NO2
γreal = -180 +/– 80 x 10−36 esu
γimag = 5 +/– 5 x 10−36 esu
Ph
Ph
Ph
Ph
Ph
Ru
PhMe2P
CO
NO2
γreal = -420 +/– 150 x 10−36 esu
γimag = 25 +/– 15 x 10−36 esu

FIG. 13. Selected examples of metal carbonyl complexes with large γ values.

out at a laser fundamental wavelength of 0.53 μm, and the significant nonlinearities for some of the bis-phosphine complexes may not reflect off-resonance nonlinearities. The most extensively studied organometallic compounds are alkynyl and vinylidene complexes, the results of which are listed in Table XVI, and efficient examples of which are depicted in Figs. 14–17. Almost all studies have been carried out at 0.800 μm using the *Z*-scan technique and in the same laboratory, permitting a cautious comparison of results. While recent studies have emphasized ruthenium alkynyl complexes, earlier studies contrasted ruthenium, nickel, and gold alkynyl complexes containing the same systematically varied alkynyl ligands. Investigations of (cyclopentadienyl)ruthenium alkynyl complexes established the importance of a number of molecular modifications for cubic nonlinearity: phosphine ligand substitution (replacing a triarylphosphine by a trialkylphosphine resulted in minimal change), variation in the 4-phenylalkynyl substituent (although replacing 4-H by 4-Br resulted in no appreciable change, introduction of the strong acceptor NO_2 resulted in a significant increase), and chain lengthening (progression from phenyl to biphenyl to stilbenyl and tolane-linked to an imino-linked two-ring-ligand resulted in progressively increasing nonlinearity). For this series of complexes,

TABLE XV

MOLECULAR THIRD-ORDER NLO RESULTS OF MOLYBDENUM CARBONYL COMPLEXES

Complex	Technique	Solvent	Fundamental (μm)	γ/γ_{CS_2} (10^{-34} esu)	$\lvert\gamma\rvert/\gamma_{CS_2}$ (10^{-34} esu)	$\langle\gamma\rangle$ (10^{-34} esu)	Ref.
cis-[Mo(CO)$_4$(PPh$_3$)$_2$]	DFWM	THF	0.53	—	—	1700	138
	Z-scan	THF	0.53	6750	—	—	139
	DFWM	THF	0.53	—	12593	—	139
trans-[Mo(CO)$_4$(PPh$_3$)$_2$]	DFWM	THF	0.53	—	—	330	138
	Z-scan	THF	0.53	2125	—	—	139
	DFWM	THF	0.53	—	2444	—	139
cis-[Mo(CO)$_4$(PPh$_2$OMe)$_2$]	DFWM	THF	0.53	—	—	44.0	138
	Z-scan	THF	0.53	375	—	—	139
	DFWM	THF	0.53	—	326	—	139
cis-[Mo(CO)$_4$(PPh$_2$Me)$_2$]	DFWM	THF	0.53	—	—	4.0	138
	Z-scan	THF	0.53	250	—	—	139
	DFWM	THF	0.53	—	296	—	139
[Mo(CO)$_5$(PPh$_3$)]	DFWM	THF	0.53	—	—	1.5	138
	Z-scan	THF	0.53	120	—	—	139
	DFWM	THF	0.53	—	11	—	139
[Mo(CO)$_5$(PPh$_2$NHMe)]	DFWM	THF	0.53	—	—	6.0	138
	Z-scan	THF	0.53	73	—	—	139
	DFWM	THF	0.53	—	44	—	139
[Mo(CO)$_5$(PPh$_2$NH$_2$)]	DFWM	THF	0.53	—	—	8.6	138
cis-[Mo(CO)$_4$(PPh$_2$NHMe)$_2$]	DFWM	THF	0.53	—	—	270.0	138
cis-[Mo(CO)$_4${PPh$_2$C(O)Me}$_2$]	DFWM	THF	0.53	—	—	260.0	138
cis-[Mo(CO)$_4$(PPh$_2$Cl)$_2$]	DFWM	THF	0.53	—	—	14.0	138
cis-[Mo(CO)$_4$(PPh$_2$CH$_2$Ph)$_2$]	DFWM	THF	0.53	—	—	630.0	138
cis-[Mo(CO)$_4${PPh(C$_4$H$_3$S)$_2$}$_2$]	DFWM	THF	0.53	—	—	350.0	138
cis-[Mo(CO)$_4${(PPh$_2$O)$_2$SiBu^{t}Me}]	DFWM	THF	0.53	—	—	39.0	138

TABLE XVI

MOLECULAR THIRD-ORDER NLO MEASUREMENTS OF ALKYNYL AND VINYLIDENE COMPLEXES

Complex	λ_{max} (nm)	Technique	Solvent	Fundamental (μm)	γ_{real} (10^{-36} esu)	γ_{imag} (10^{-36} esu)	$\lvert\gamma\rvert$ (10^{-36} esu)	σ_2 (10^{-50} cm^4 s)	Ref.
[Ti(C≡CPh)Cl(η^5-C_5H_5)$_2$]	510	THG	THF	1.91	—	—	31±5	—	124,136
[Ti(C≡CBu^n)$_2$(η^5-C_5H_5)$_2$]	390	THG	$CHCl_3$	1.91	—	—	12±2	—	136
[Ti(C≡CBu^n)$_2$(η^5-C_5H_4Me)$_2$]	[a]	THG	$CHCl_3$	1.91	—	—	15±2	—	137
[Ti(C≡CPh)$_2$(η^5-C_5H_5)$_2$]	410	THG	THF	1.91	—	—	92±14	—	124,136
[Fe(4-C≡C$C_6H_4NO_2$)(dppe)(η^5-C_5H_5)]	497	*Z*-scan	THF	0.80	−410±200	580±200	710±280	—	83
[Fe{C≡C-4-C_6H_4-(*E*)-CH=CH-4-$C_6H_4NO_2$}(dppe)(η^5-C_5H_5)]	499	*Z*-scan	THF	0.80	−2200±600	1200±300	2500±670	—	83
[Fe(C≡CPh)(dppe)(η^5-C_5Me_5)]	350	*Z*-scan	THF	0.69	110±100	17±10	110±100	6±3	144
[Fe(C≡C-4-C_6H_4C≡CPh)(dppe)(η^5-C_5Me_5)]	436	*Z*-scan	THF	0.69	−1500±1200	200±40	1500±1200	66±15	144
[1,3,5-{(η^5-C_5Me_5)(dppe)Fe(C≡C-4-C_6H_4C≡C)}$_3C_6H_3$]	460	*Z*-scan	THF	0.69	−3300±800	2800±700	4300±1100	920±250	144
[1,3,5-{(η^5-C_5Me_5)(dppe)Fe(C≡C-4-C_6H_4C≡C)}$_3C_6H_3$]$^{3+}$	1864	*Z*-scan	THF	0.69	−2000±1000	−3300±1000	3900±1400	−1100±330	144
[Ni(C≡CPh)$_2$(PEt$_3$)$_2$]	370	DFWM	$CHCl_3$	1.06	−280	150	310	—	145
[Ni(C≡CPh)(PPh$_3$)(η^5-C_5H_5)]	307	*Z*-scan	THF	0.80	15±10	<10	—	—	89
[Ni(C≡C-4-$C_6H_4NO_2$)(PPh$_3$)(η-C_5H_5)]	439	*Z*-scan	THF	0.80	−270±100	70±50	—	—	89
[Ni(C≡C-4-C_6H_4-4-$C_6H_4NO_2$)(PPh$_3$)(η^5-C_5H_5)]	413	*Z*-scan	THF	0.80	−580±200	300±60	—	—	89
[Ni{C≡C-4-C_6H_4-(*E*)-CH=CH-4-$C_6H_4NO_2$}(PPh$_3$)(η^5-C_5H_5)]	437	*Z*-scan	THF	0.80	−420±100	480±150	—	—	89
[Ni{C≡C-4-C_6H_4-(*Z*)-CH=CH-4-$C_6H_4NO_2$}(PPh$_3$)(η^5-C_5H_5)]	417	*Z*-scan	THF	0.80	−230±50	160±80	—	—	89
[Ni(C≡C-4-C_6H_4C≡C-4-$C_6H_4NO_2$)(PPh$_3$)(η^5-C_5H_5)]	417	*Z*-scan	THF	0.80	−640±300	720±300	—	—	89
[Ni(C≡C-4-C_6H_4N=CH-4-$C_6H_4NO_2$)(PPh$_3$)(η^5-C_5H_5)]	448	*Z*-scan	THF	0.80	<120	360±100	—	—	89
[Zr(C≡CPh)$_2$(η^5-C_5H_5)$_2$]	390	THG	THF	1.91	—	—	58±9	—	124,136
[Ru(C≡CPh)(PPh$_3$)$_2$(η^5-C_5H_5)]	310	*Z*-scan	THF	0.80	⩽150	0	—	—	92,140
[Ru(C≡C-4-C_6H_4CH{OC(O)Me}$_2$)(PPh$_3$)$_2$(η^5-C_5H_5)]	326	*Z*-scan	THF	0.80	100±100	0	100±100	—	93
[Ru(C≡C-4-C_6H_4CHO)(PPh$_3$)$_2$(η^5-C_5H_5)]	400	*Z*-scan	THF	0.80	−75±50	210±50	220±60	—	93
[Ru(C≡C-4-C_6H_4Br)(PPh$_3$)$_2$(η^5-C_5H_5)]	325	*Z*-scan	THF	0.80	⩽150	0	—	—	92,140
[Ru(C≡C-4-$C_6H_4NO_2$)(PMe$_3$)$_2$(η^5-C_5H_5)]	480	*Z*-scan	THF	0.80	−230±70	74±30	—	—	92,140
[Ru(C≡C-4-$C_6H_4NO_2$)(PPh$_3$)$_2$(η^5-C_5H_5)]	461	DFWM	THF	0.80	—	—	−260±60	—	92,140
		Z-scan	THF	0.80	−210±50	⩽10	—		
[Ru(C≡C-4-C_6H_4-4-$C_6H_4NO_2$)(PPh$_3$)$_2$(η^5-C_5H_5)]	448	*Z*-scan	THF	0.80	−380±200	320±160	—	—	92
[Ru{C≡C-4-C_6H_4-(*E*)-CH=CH-4-$C_6H_4NO_2$}(PPh$_3$)$_2$(η^5-C_5H_5)]	476	*Z*-scan	THF	0.80	−450±100	210±100	—	—	92,140
[Ru(C≡C-4-C_6H_4C≡C-4-$C_6H_4NO_2$)(PPh$_3$)$_2$(η^5-C_5H_5)]	446	*Z*-scan	THF	0.80	−450±100	⩽20	—	—	92,140
[Ru(C≡C-4-C_6H_4N=CH-4-$C_6H_4NO_2$)(PPh$_3$)$_2$(η^5-C_5H_5)]	496	*Z*-scan	THF	0.80	−850±300	360±200	—	—	92
trans-[Ru(C=CHPh)Cl(dppm)$_2$](PF$_6$)	320	*Z*-scan	THF	0.80	<440	<50	<440	—	101
trans-[Ru(C≡CPh)Cl(dppm)$_2$]	308	*Z*-scan	THF	0.80	<120	0	<120	—	101
trans-[Ru(C≡CPh)Cl(dppm)$_2$](PF$_6$)	833	*Z*-scan	CH_2Cl_2	0.80	1300±500	−2200±1000	2600±1000	—	142,147
trans-[Ru(C≡CPh)Cl(dppm)$_2$]	318	*Z*-scan	CH_2Cl_2	0.80	<300	<200	≈0	—	142,147
trans-[Ru{C=CH-4-$C_6H_4$$\overline{\text{CHO(CH}_2)_3}$O}Cl(dppm)$_2$](PF$_6$)	317	*Z*-scan	THF	0.80	75±75	0	75±75	—	93

negative nonlinearities were observed for all of the nitro-containing

TABLE XVI
(CONTINUED)

Complex	λ_{max} (nm)	Technique	Solvent	Fundamental (μm)	γ_{real} (10^{-36} esu)	γ_{imag} (10^{-36} esu)	$\lvert\gamma\rvert$(10^{-36} esu)	σ_2 (10^{-50} cm^4 s)	Ref.
trans-[Ru{C≡C-4-$C_6H_4\overline{CHO(CH_2)_3O}$}Cl(dppm)$_2$]	320	*Z*-scan	THF	0.80	50±50	0	50±50	—	93
trans-[Ru{C=CH-2-C_6H_4CHO}Cl(dppm)$_2$](PF$_6$)	555	*Z*-scan	THF	0.80	450±150	150±60	470±160	—	93
trans-[Ru{C=CH-3-C_6H_4CHO}Cl(dppm)$_2$](PF$_6$)	320	*Z*-scan	THF	0.80	200±200	0	200±200	—	93
trans-[Ru{C≡C-3-C_6H_4CHO}Cl(dppm)$_2$]	321	*Z*-scan	THF	0.80	150±150	0	150±150	—	93
trans-[Ru(C=CH-4-C_6H_4C≡CPh)Cl(dppm)$_2$](PF$_6$)	380	*Z*-scan	THF	0.80	<500	0	<500	—	101
trans-[Ru(C≡C-4-C_6H_4C≡CPh)Cl(dppm)$_2$]	381	*Z*-scan	THF	0.80	65±40	520±200	520±200	—	101
trans-[Ru(C=CH-4-C_6H_4CHO)Cl(dppm)$_2$](PF$_6$)	403	*Z*-scan	THF	0.80	0	<20	<20	—	101
trans-[Ru(C≡C-4-C_6H_4CHO)Cl(dppm)$_2$]	405	*Z*-scan	THF	0.80	<120	210±60	210±60	—	101
trans-[Ru(C=CH-4-$C_6H_4NO_2$)Cl(dppm)$_2$](PF$_6$)	470	*Z*-scan	THF	0.80	<50	<30	<50	—	101
trans-[Ru(C≡C-4-$C_6H_4NO_2$)Cl(dppm)$_2$]	466[b]	*Z*-scan	CH_2Cl_2	0.80	170±34	230±46	290±60	—	148,149
trans-[Ru(C≡C-4-C_6H_4-4-$C_6H_4NO_2$)Cl(dppm)$_2$]	448[b]	*Z*-scan	CH_2Cl_2	0.80	140±28	64±13	—	—	149
trans-[Ru{C≡C-4-C_6H_4-(*E*)-CH=CH-4-$C_6H_4NO_2$}Cl(dppm)$_2$]	471[b]	*Z*-scan	CH_2Cl_2	0.80	200±40	1100±220	—	—	149
trans-[Ru{C=CH-4-C_6H_4-(*E*)-CH=CHPh}Cl(dppm)$_2$](PF$_6$)	360	*Z*-scan	THF	0.80	<200	90±50	90±50	21±12	107
trans-[Ru{C≡C-4-C_6H_4-(*E*)-CH=CHPh}Cl(dppm)$_2$]	397	*Z*-scan	THF	0.80	−600±400	700±400	920±600	170±100	107
trans-[Ru(C=CH-4-C_6H_4C≡C-4-$C_6H_4NO_2$)Cl(dppm)$_2$](PF$_6$)	326	*Z*-scan	THF	0.80	<500	420±60	420±60	—	101
trans-[Ru(C≡C-4-C_6H_4C≡C-4-$C_6H_4NO_2$)Cl(dppm)$_2$]	464	*Z*-scan	THF	0.80	−160±80	160±60	230±100	—	101
trans-[Ru(C≡C-4-C_6H_4C≡C-4-C_6H_4C≡C-4-$C_6H_4NO_2$)Cl(dppm)$_2$]	439	*Z*-scan	THF	0.80	−920±200	970±200	1300±300	—	101
trans-[Ru{C≡C-4-C_6H_4-(*E*)-CH=CH-4-$C_6H_4NO_2$}Cl(dppm)$_2$]	490	*Z*-scan	THF	0.80	200±40	1100±220	1100±220	—	148
trans-[Ru(C≡C-4-$C_6H_4NO_2$)$_2$(dppm)$_2$]	474[b]	*Z*-scan	CH_2Cl_2	0.80	300±60	490±98	—	—	149
trans-[Ru(C≡C-4-C_6H_4-4-$C_6H_4NO_2$)$_2$(dppm)$_2$]	453[b]	*Z*-scan	CH_2Cl_2	0.80	⩽800	2500±500	—	—	149
trans-[Ru{C≡C-4-C_6H_4-(*E*)-CH=CH$C_6H_4NO_2$}$_2$(dppm)$_2$]	367[b]	*Z*-scan	CH_2Cl_2	0.80	⩽1100	3400±680	—	—	149
[Fe{η^5-C_5H_4-(*E*)-4-CH=CHC_6H_4CH=CRuCl(dppm)$_2$}$_2$](PF$_6$)$_2$	383	*Z*-scan	THF	0.80	−3000±1200	2300±800	3800±1400	—	133
[Fe{η^5-C_5H_4-(*E*)-4-CH=CHC_6H_4C≡CRuCl(dppm)$_2$}$_2$]	396	*Z*-scan	THF	0.80	−7100±3000	10600±2000	13000±3000	2500	133
trans-[Ru(C=CHPh)Cl(dppe)$_2$](PF$_6$)	317	*Z*-scan	THF	0.80	380±400	<50	380±400	—	101
trans-[Ru(C≡CPh)Cl(dppe)$_2$]	319	*Z*-scan	THF	0.80	−170±40	71±20	180±45	—	11,104
trans-[Ru(C=CH-4-C_6H_4CHO)Cl(dppe)$_2$](PF$_6$)	412	*Z*-scan	THF	0.80	<260	0	<260		101
trans-[Ru(C≡C-4-C_6H_4CHO)Cl(dppe)$_2$]	413	*Z*-scan	THF	0.80	−300±500	<200	300±500	—	101
trans-[Ru(C=CH-4-$C_6H_4NO_2$)Cl(dppe)$_2$](PF$_6$)	476	*Z*-scan	THF	0.80	250±300	<50	250±300	—	101
trans-[Ru(C≡C-4-$C_6H_4NO_2$)Cl(dppe)$_2$]	477	*Z*-scan	THF	0.80	320±55	<50	320±55	—	101
trans-[Ru{C≡C-4-C_6H_4-(*E*)-CH=CHPh}Cl(dppe)$_2$]	404	*Z*-scan	THF	0.80	300±400	300±100	420±350	70±30	107

trans-[Ru{C=CH-4-C$_6$H$_4$-(*E*)-CH=CH-4-C$_6$H$_4$NO$_2$}Cl(dppe)$_2$](PF$_6$)	473	*Z*-scan	THF	0.80	650 ± 500	< 50	650 ± 500	—	101
trans-[Ru{C≡C-4-C$_6$H$_4$-(*E*)-CH=CH-4-C$_6$H$_4$NO$_2$}Cl(dppe)$_2$]	489	*Z*-scan	THF	0.80	40 ± 200	< 100	40 ± 200	—	101
trans-[Ru(C=CH-4-C$_6$H$_4$C≡CPh)Cl(dppe)$_2$](PF$_6$)	893	*Z*-scan	CH$_2$Cl$_2$	0.80	2900 ± 1000	−1200 ± 600	3100 ± 1000	—	142,147
trans-[Ru(C≡C-4-C$_6$H$_4$C≡CPh)Cl(dppe)$_2$]	387	*Z*-scan	CH$_2$Cl$_2$	0.80	−100 ± 100	450 ± 200	460 ± 200	—	142,147
trans-[Ru(C≡C-4-C$_6$H$_4$C≡CPh)(C≡CPh)(dppe)$_2$]	383	*Z*-scan	CH$_2$Cl$_2$	0.80	−670 ± 300	1300 ± 300	1500 ± 500	—	104,142,147
[1,3,5-(*trans*-[(dppm)$_2$ClRu{(*E*)-4-C=CHC$_6$H$_4$CH=CH}])$_3$C$_6$H$_3$](PF$_6$)$_3$	396	*Z*-scan	THF	0.80	−900 ± 500	700 ± 400	1100 ± 700	170 ± 100	107,151
[1,3,5-(*trans*-[(dppm)$_2$ClRu{(*E*)-4-C≡CC$_6$H$_4$CH=CH}])$_3$C$_6$H$_3$]	415	*Z*-scan	THF	0.80	−640 ± 500	2000 ± 500	2100 ± 600	480 ± 120	107,151
[1,3,5-(*trans*-[(dppe)$_2$ClRu{(*E*)-4-C≡CC$_6$H$_4$CH=CH}])$_3$C$_6$H$_3$]	426	*Z*-scan	THF	0.80	−4600 ± 2000	4200 ± 800	6200 ± 2000	1000 ± 200	107,151
[1,3,5-(*trans*-[(dppe)$_2$(PhC≡C)Ru{(*E*)-4-C≡CC$_6$H$_4$CH=CH}])$_3$C$_6$H$_3$]	421	*Z*-scan	THF	0.80	−11200 ± 3000	8600 ± 2000	14000 ± 4000	2100 ± 500	107,151
[1,3,5-(*trans*-[(dppe)$_2$(PhC≡C)Ru(4-C≡CC$_6$H$_4$C≡C)])$_3$C$_6$H$_3$]	411	*Z*-scan	THF	0.80	−600 ± 200	2900 ± 500	3000 ± 600	690 ± 120	104,107,151
[1,3,5-(*trans*-[(dppe)$_2$ClRu(4-C≡CC$_6$H$_4$C≡C])$_3$C$_6$H$_3$]	414	*Z*-scan	CH$_2$Cl$_2$	0.80	−330 ± 100	2200 ± 500	2200 ± 600	—	107,147
		DFWM	CH$_2$Cl$_2$	0.80	—	—	2000[c]	—	143
		Z-scan	THF	0.80	−330 ± 100	2200 ± 500	2200 ± 600	530 ± 120	104
[1,3,5-(*trans*-[(dppe)$_2$ClRu(4-C=CHC$_6$H$_4$C≡C])$_3$C$_6$H$_3$](PF$_6$)$_3$	893	*Z*-scan	CH$_2$Cl$_2$	0.80	13500 ± 3000	−4700 ± 500	14000 ± 3000	—	147
		DFWM	CH$_2$Cl$_2$	0.80	—	—	20000	—	143
[1-(Me$_3$SiC≡C)C$_6$H$_3$-3,5-{4-C≡CC$_6$H$_4$C≡C-*trans*-[RuCl(dppe)$_2$]}$_2$]	411	*Z*-scan	THF	0.80	−510 ± 500	4700 ± 1500	4700 ± 2000	1100 ± 360	150
[1-(Me$_3$SiC≡C)C$_6$H$_3$-3,5-{4-C≡CC$_6$H$_4$C≡C-*trans*-[Ru(C≡CPh)(dppe)$_2$]}$_2$]	407	*Z*-scan	THF	0.80	−700 ± 100	2270 ± 300	2400 ± 300	550 ± 70	150
[1-(HC≡C)C$_6$H$_3$-3,5-{4-C≡CC$_6$H$_4$C≡C-*trans*-[Ru(C≡CPh)(dppe)$_2$]}$_2$]	408	*Z*-scan	THF	0.80	−830 ± 100	2200 ± 300	2400 ± 300	530 ± 70	150
[1,3,5-C$_6$H$_3$(4-C≡CC$_6$H$_4$C≡C-*trans*-[Ru(dppe)$_2$]C≡C-3,5-C$_6$H$_3${4-C≡CC$_6$H$_4$C≡C-*trans*-[Ru(C≡CPh)(dppe)$_2$]}$_2$)$_3$]	402	*Z*-scan	THF	0.80	−5050 ± 500	20100 ± 2000	20700 ± 2000	4800 ± 500	150
[Pd(C≡CPh)$_2$(PEt$_3$)$_2$]	370	DFWM	CHCl$_3$	1.06	−210	34	210	—	145
trans-[Pd(C≡CPh)$_2$(PBu$^n_3$)$_2$]	[a]	[a]	[a]	[a]	0.63	110		—	152
[Ag(C≡CPh)]$_7$[d]	260	OHD-OKE	1:1 DMSO/CHCl$_3$	0.65	—	—	90700[c]	—	153
	271	OHD-OKE	1:1:1 DMSO/CHCl$_3$/CH$_2$Cl$_2$	0.65	—	—	−105900[c]	—	153
[AgC≡CPh · AgS(Bu^t)]$_7$[d]	264	OHD-OKE	1:1 DMSO/CHCl$_3$	0.65	—	—	74400	—	153
	271	OHD-OKE	1:1:1 DMSO/CHCl$_3$/CH$_2$Cl$_2$	0.65	—	—	−64900	—	153
cis-[PtCl(PBu$^n_3$)$_2$(C≡C-4-C$_6$H$_4$C≡C)PtCl(PBu$^n_3$)$_2$]	[a]	OKG/IDS	THF	1.06/0.53	11 ± 3	224 ± 56	—	—	154
trans-[PtCl(PBu$^n_3$)$_2$(C≡C-4-C$_6$H$_4$C≡C)PtCl(PBu$^n_3$)$_2$]	[a]	OKG/IDS	THF	1.06/0.53	19 ± 5	827 ± 207	—	—	154

TABLE XVI
(CONTINUED)

Complex	λ_{max} (nm)	Technique	Solvent	Fundamental (μm)	γ_{real} (10^{-36} esu)	γ_{imag} (10^{-36} esu)	$\lvert\gamma\rvert$ (10^{-36} esu)	σ_2 (10^{-50} cm^4 s)	Ref.
trans-$[PtCl(PBu^n_3)_2(C{\equiv}C\text{-}4\text{-}C_6H_4C{\equiv}C)Pt(PBu^n_3)_2(C{\equiv}C\text{-}4\text{-}C_6H_4C{\equiv}C)PtCl(PBu^n_3)_2]$	[a]	OKG/IDS	THF	1.06/0.53	45 ± 11	1196 ± 300	—	—	154
trans-$[PtCl(PBu^n_3)_2(C{\equiv}C\text{-}4\text{-}C_6H_4C{\equiv}CC{\equiv}C\text{-}4\text{-}C_6H_4C{\equiv}C)PtCl(PBu^n_3)_2]$	[a]	OKG/IDS	THF	1.06/0.53	88 ± 22	2167 ± 542	—	—	154
cis-$[Pt(C{\equiv}C\text{-}4\text{-}C_6H_4C{\equiv}CH)_2(PBu^n_3)_2]$	[a]	FWM	[a]	0.63	230[c]	260[c]	290[c]	—	152
trans-$[Pt(C{\equiv}C\text{-}4\text{-}C_6H_4C{\equiv}CH)(PBu^n_3)_2(C{\equiv}C\text{-}4\text{-}C_6H_4C{\equiv}C)Pt(C{\equiv}C\text{-}4\text{-}C_6H_4C{\equiv}CH)(PBu^n_3)_2]$	[a]	OKG/IDS	THF	1.06/0.53	66 ± 17	1328 ± 332	—	—	154
trans-$[Pt(NCS)(PBu^n_3)_2(C{\equiv}C\text{-}4\text{-}C_6H_4C{\equiv}C)Pt(NCS)(PBu^n_3)_2]$	[a]	OKG/IDS	THF	1.06/0.53	30 ± 8	1134 ± 284	—	—	154
$(PPN)[Au(C{\equiv}C\text{-}4\text{-}C_6H_4NO_2)_2]$	376	*Z*-scan	CH_2Cl_2	0.80	−800 ± 400	115 ± 50	—	—	155
$(NPr_4)[Au(C{\equiv}C\text{-}4\text{-}C_6H_4NO_2)_2]$	374	*Z*-scan	CH_2Cl_2	0.80	90 ± 150	190 ± 50	—	—	155
$[Au(C{\equiv}CPh)(PPh_3)]$	296	*Z*-scan	THF	0.80	39 ± 20	—	—	—	141,155,156
$[Au(C{\equiv}C\text{-}4\text{-}C_6H_4CHO)(PMe_3)]$	322	*Z*-scan	THF	0.80	35 ± 20	45 ± 30	60 ± 35	—	93
$[Au(C{\equiv}C\text{-}4\text{-}C_6H_4CHO)(PPh_3)]$	322	*Z*-scan	THF	0.80	300 ± 150	0	300 ± 150	—	93
$[Au(4\text{-}C{\equiv}CC_6H_4\overline{CHO\{CH_2\}_3}O)(PPh_3)]$	296	*Z*-scan	THF	0.80	210 ± 100	0	210 ± 100	—	93
$[Au(C{\equiv}C\text{-}4\text{-}C_6H_4NO_2)(PPh_3)]$	338	*Z*-scan	THF	0.80	120 ± 40	20 ± 15	—	—	141,155
$[Au(C{\equiv}C\text{-}4\text{-}C_6H_4\text{-}4\text{-}C_6H_4NO_2)(PPh_3)]$	350	*Z*-scan	THF	0.80	540 ± 150	120 ± 50	—	—	141,155
$[Au(C{\equiv}C\text{-}4\text{-}C_6H_4C{\equiv}CPh)(PPh_3)]$	336	*Z*-scan	THF	0.80	−900 ± 400	0 ± 100	900 ± 400	0 ± 24	107
$[Au(C{\equiv}C\text{-}4\text{-}C_6H_4C{\equiv}CPh)(PMe_3)]$	335	*Z*-scan	THF	0.80	−200 ± 150	0 ± 50	200 ± 150	0 ± 12	107
$[Au(C{\equiv}C\text{-}4\text{-}C_6H_4C{\equiv}C\text{-}4\text{-}C_6H_4NO_2)(PPh_3)]$	362	*Z*-scan	THF	0.80	1300 ± 400	560 ± 150	—	—	141,155
$[Au\{C{\equiv}C\text{-}4\text{-}C_6H_4\text{-}(E)\text{-}CH{=}CHPh\}(PPh_3)]$	338	*Z*-scan	THF	0.80	0 ± 300	0 ± 50	0	0 ± 12	107
$[Au\{C{\equiv}C\text{-}4\text{-}C_6H_4\text{-}(E)\text{-}CH{=}CH\text{-}4\text{-}C_6H_4NO_2\}(PPh_3)]$	386	*Z*-scan	THF	0.80	1200 ± 200	470 ± 150	—	—	141,155
$[Au\{C{\equiv}C\text{-}4\text{-}C_6H_4\text{-}(Z)\text{-}CH{=}CH\text{-}4\text{-}C_6H_4NO_2\}(PPh_3)]$	362	*Z*-scan	THF	0.80	420 ± 150	92 ± 30	—	—	141,155
$[Au\{C{\equiv}C\text{-}4\text{-}C_6H_4\text{-}(E)\text{-}N{=}CH\text{-}4\text{-}C_6H_4NO_2\}(PPh_3)]$	392	*Z*-scan	THF	0.80	130 ± 30	330 ± 60	—	—	141,155
$[Au(C{\equiv}C\text{-}4\text{-}C_6H_4NO_2)(CNBu^t)]$	332	*Z*-scan	CH_2Cl_2	0.80	≤130	≤50	—	—	155
$[Au(C{\equiv}C\text{-}4\text{-}C_6H_4\text{-}4\text{-}C_6H_4NO_2)(CNBu^t)]$	343	*Z*-scan	CH_2Cl_2	0.80	20 ± 100	70 ± 50	—	—	155
$[Au\{C{\equiv}C\text{-}4\text{-}C_6H_4\text{-}(E)\text{-}CH{=}CH\text{-}4\text{-}C_6H_4NO_2\}(CNBu^t)]$	381	*Z*-scan	CH_2Cl_2	0.80	390 ± 200	1050 ± 300	—	—	155
$[Au(C{\equiv}C\text{-}4\text{-}C_6H_4\text{-}4\text{-}C_6H_4NO_2)\{C(NHBu^t)(NEt_2)\}]$	354	*Z*-scan	CH_2Cl_2	0.80	10 ± 100	160 ± 40	—	—	155
$[Au\{C{\equiv}C\text{-}4\text{-}C_6H_4\text{-}(E)\text{-}CH{=}CH\text{-}4\text{-}C_6H_4NO_2\}\{C(NHBu^t)(NEt_2)\}]$	389	*Z*-scan	CH_2Cl_2	0.80	−200 ± 80	610 ± 200	—	—	155
$[Fe\{\eta^5\text{-}C_5H_4\text{-}(E)\text{-}4\text{-}CH{=}CHC_6H_4C{\equiv}CAu(PMe_3)\}_2]$	463	*Z*-scan	THF	0.80	200 ± 150	0 ± 30	200 ± 150	—	133
$[Fe\{\eta^5\text{-}C_5H_4\text{-}(E)\text{-}4\text{-}CH{=}CHC_6H_4C{\equiv}CAu(PCy_3)\}_2]$	468	*Z*-scan	THF	0.80	−400 ± 500	500 ± 100	640 ± 390	—	133
$[Fe\{\eta^5\text{-}C_5H_4\text{-}(E)\text{-}4\text{-}CH{=}CHC_6H_4C{\equiv}CAu(PPh_3)\}_2]$	465	*Z*-scan	THF	0.80	−1100 ± 300	300 ± 60	1140 ± 310	—	133

[a]Not stated.
[b]Measured in MeCN.
[c]Error not stated.
[d]Assumed polymer length.

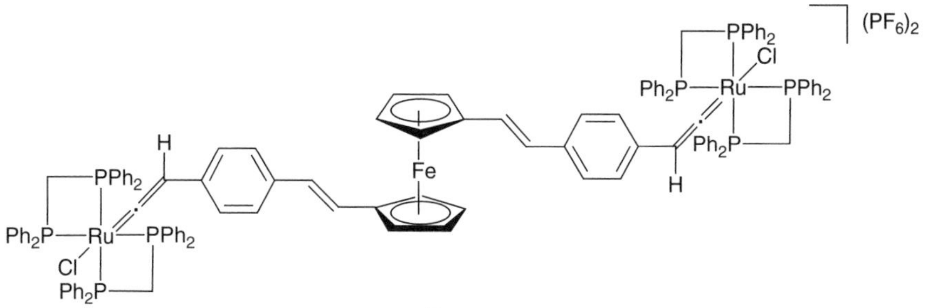

γ_{real} = -3000 +/– 1200 x 10^{-36} esu
γ_{imag} = 2300 +/– 800 x 10^{-36} esu
$|\gamma|$ = 3800 +/– 1400 x 10^{-36} esu

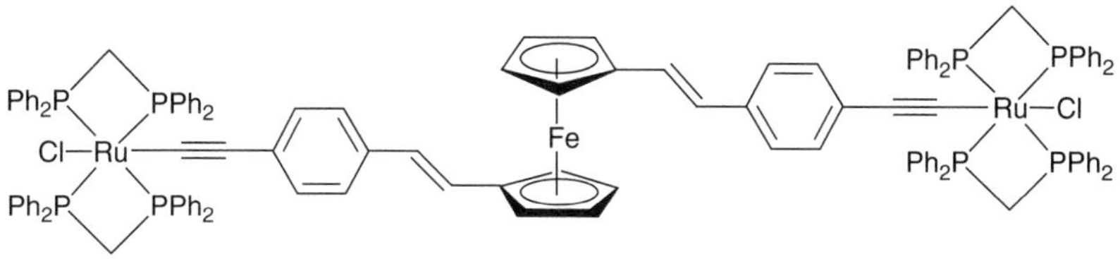

γ_{real} = -7100 +/– 3000 x 10^{-36} esu
γ_{imag} = 10600 +/– 2000 x 10^{-36} esu
$|\gamma|$ = 13000 +/– 3000 x 10^{-36} esu
σ_2 = 2500 x 10^{-50} cm^4 s

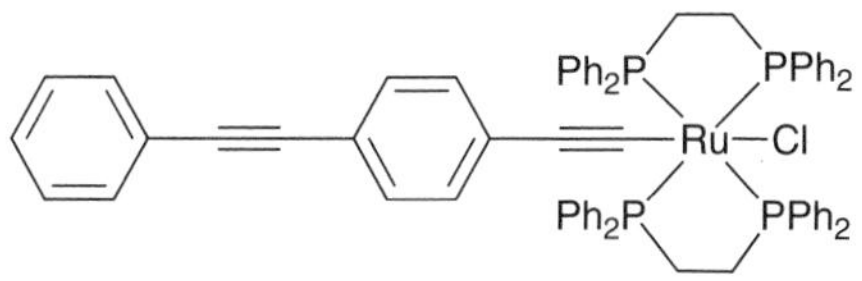

γ_{real} = -100 +/– 100 x 10^{-36} esu
γ_{imag} = 450 +/– 200 x 10^{-36} esu
$|\gamma|$ = 460 +/– 200 x 10^{-36} esu

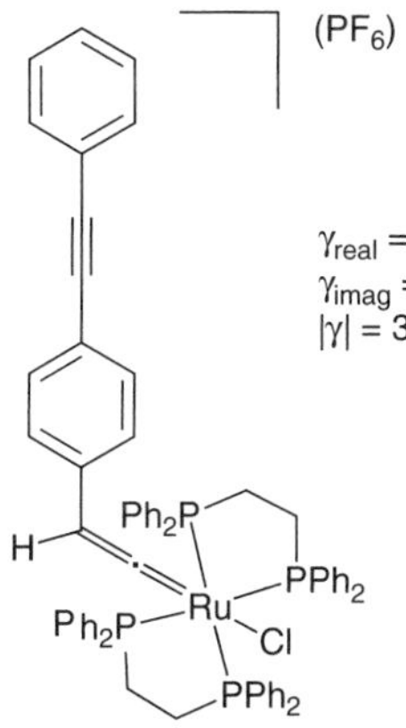

γ_{real} = 2900 +/– 1000 x 10^{-36} esu
γ_{imag} = -1200 +/– 600 x 10^{-36} esu
$|\gamma|$ = 3100 +/– 1000 x 10^{-36} esu

FIG. 14. Selected examples of alkynyl and vinylidene complexes with large γ values.

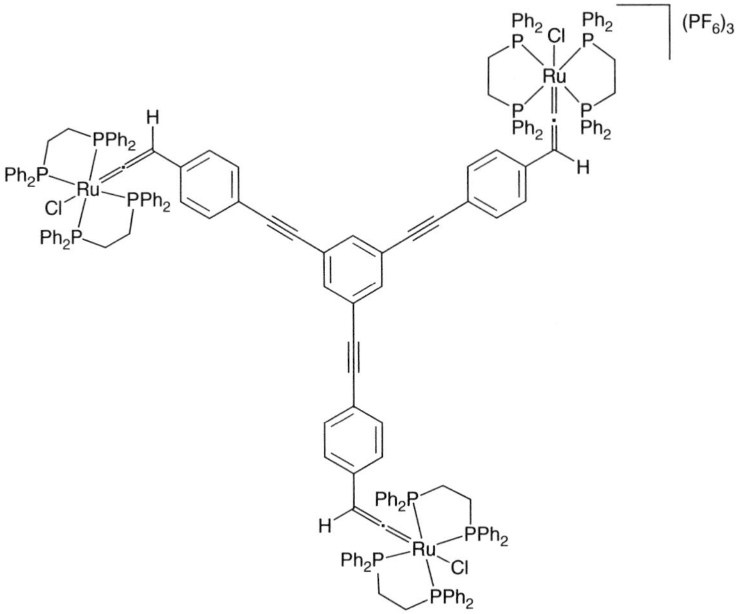

γ_{real} = 13500 +/– 3000 x 10^{-36} esu
γ_{imag} = -4700 +/– 500 x 10^{-36} esu
$|\gamma|$ = 14000 +/– 3000 x 10^{-36} esu

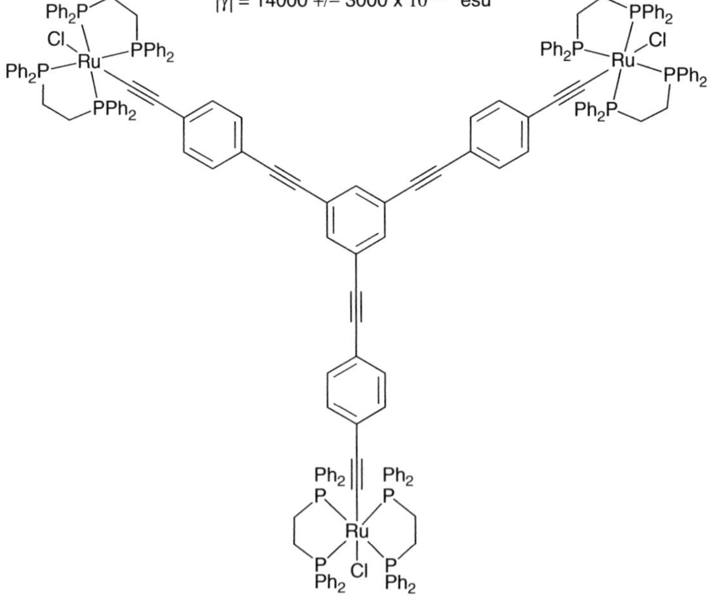

γ_{real} = -330 +/– 100 x 10^{-36} esu
γ_{imag} = 2200 +/– 500 x 10^{-36} esu
$|\gamma|$ = 2200 +/– 600 x 10^{-36} esu

FIG. 15. Selected examples of alkynyl and vinylidene complexes with large γ values.

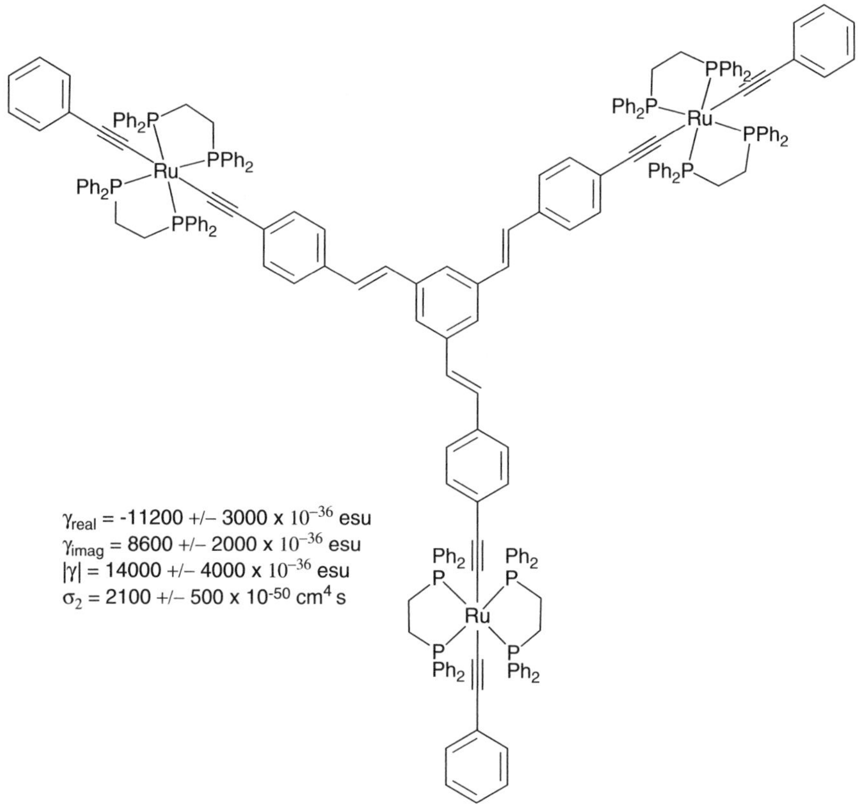

γ_{real} = -11200 +/− 3000 x 10^{-36} esu
γ_{imag} = 8600 +/− 2000 x 10^{-36} esu
$|\gamma|$ = 14000 +/− 4000 x 10^{-36} esu
σ_2 = 2100 +/− 500 x 10^{-50} cm^4 s

FIG. 16. A zero-generation alkynylruthenium dendrimer with large γ values.

examples. There are several possible explanations for this. Thermal lensing was rejected as the reason for the negative γ values, and although two-photon dispersion is the most likely cause, a negative static hyperpolarizability was not ruled out.[92,140]

Surprisingly, when these same alkynyl ligands are coupled to ligated gold centers, γ values for some complexes are larger than those of their ruthenium analogues; this is the opposite trend to that observed with β. γ values for the nickel complexes are the same as those of the ruthenium analogues, despite the ruthenium complexes possessing greater delocalization possibilities with the extra phosphine co-ligand. The later transition metal complexes in which the ligated metal acts as a donor group have larger nonlinearities than early transition metal examples in which the metal acts as an acceptor. Nonlinearity increases significantly upon installation of a nitro group; complexes lacking this substituent have much lower nonlinearities than those containing this group.

More recent research has emphasized dimensional progression from 1D linear complexes to 2D octupolar complexes, and subsequently dendrimers. Nonlinearities increase for this progression, with values for the dendrimers extremely large (Figs. 14–17). Another recent focus has been the possibility of "switching" cubic

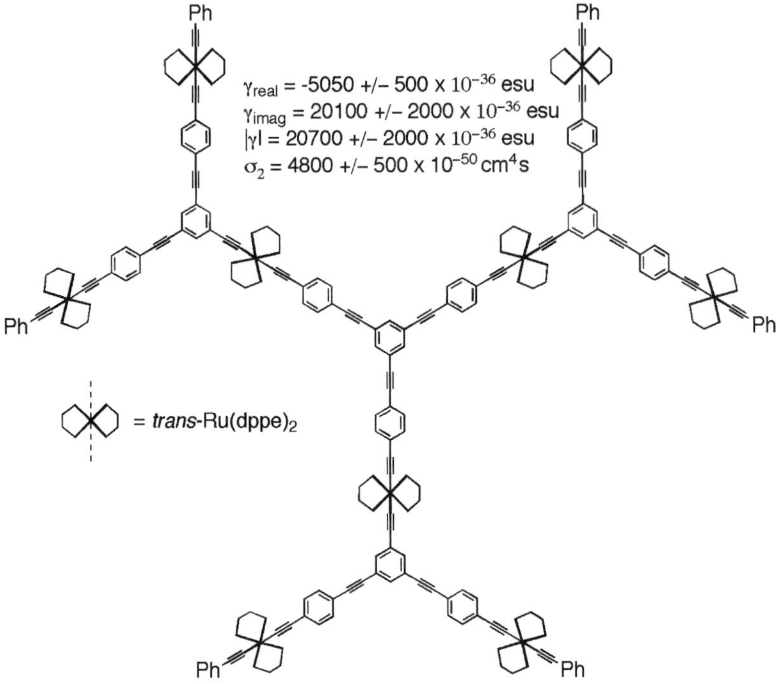

FIG. 17. A first-generation alkynylruthenium dendrimer with large γ values.

nonlinearity. The reversible protonation/deprotonation sequence for interconverting alkynyl and vinylidene complexes potentially affords protonically switchable materials, an idea that has been pursued for ruthenium complexes,[101] with differences of up to an order of magnitude being observed. Reversible oxidation/reduction processes have been exploited to examine iron and ruthenium alkynyl complexes as electrochemically switchable materials[81,82,107,133,142–144] with changes of sign and magnitude of both γ_{real} and γ_{imag} observed, and a potential on/off switch identified.[142]

Cubic NLO data for selected group 10 metal bis(alkynyl) complexes are listed in Table XVII. Hyperpolarizability decreases progressing down the group for phenylalkynyl examples, the same observation as with group 4 metal alkynyl complexes (see above). For these complexes λ_{max} is, in all cases, close to 3ω, and a high-order intensity dependence was observed; this is characteristic of multiphoton resonant enhancement, so it is possible that three-photon effects exist.

$[Pt(\eta^2\text{-}C_{60})(PPh_3)_2]$ was measured by *Z*-scan using both circular and linear polarized light (Table XVIII).[159] The difference in nonlinearity for the two polarizations confirms that the two-photon absorption contribution is significant. A broad absorption band extending to ~800 nm rendered extraction of off-resonance nonlinearity impossible with the fundamental frequency employed.

TABLE XVII
THIRD-ORDER NLO MEASUREMENTS OF SELECTED GROUP 10 BIS-ALKYNYL COMPLEXES

Complex	λ_{max} (nm)	Technique	Solvent	Fundamental (μm)	γ_{real} (10^{-45} m^5 V^{-2})	γ_{imag} (10^{-45} m^5 V^{-2})	$\lvert\gamma\rvert$ (10^{-45} m^5 V^{-2})	n_2 (10^{-18} m^2 W^{-1})	Ref.
trans-$[Ni(C{\equiv}CPh)_2(PEt_3)_2]$	370	DFWM	$CHCl_3$	1.06	−280	150	311		160,161
	370	DFWM	$CHCl_3$	1.06	−0.28	0.15	0.31		145
trans-$[Ni(C{\equiv}CC{\equiv}CH)_2(PEt_3)_2]$	336	DFWM	$CHCl_3$	1.06	−78.7	172	189		160,161
	336	DFWM	$CHCl_3$	1.06	−0.079	0.17	0.19		145
trans-$[Ni(C{\equiv}CPh)_2(PBu^n_3)_2]$	[a]	*Z*-scan	THF	0.53	—	—	—	-16 ± 5	162
trans-$[Pd(C{\equiv}CPh)_2(PEt_3)_2]$	370	DFWM	$CHCl_3$	1.06	−210	34	213		160,161
	370	DFWM	$CHCl_3$	1.06	−0.21	0.034	0.21		145
trans-$[Pd(C{\equiv}CC{\equiv}CH)_2(PEt_3)_2]$	290	DFWM	$CHCl_3$	1.06	−38.5	9.19	3.96		145,160,161
	290	DFWM	$CHCl_3$	1.06	−0.039	0.0092	0.040		145
trans-$[Pd(C{\equiv}CPh)_2(PBu^n_3)_2]$	[a]	*Z*-scan	THF	0.53	—	—	—	-0.5 ± 0.1	162
trans-$[Pd(C{\equiv}C\text{-}4\text{-}C_6H_4C{\equiv}CPh)_2(PBu^n_3)_2]$	[a]	*Z*-scan	THF	0.53	—	—	—	-25 ± 3	162
trans-$[Pt(C{\equiv}CPh)_2(PEt_3)_2]$	332	DFWM	$CHCl_3$	1.06	−110	22	110		160,161
	332	DFWM	$CHCl_3$	1.06	−0.11	0.022	0.11		145
trans-$[Pt(C{\equiv}CC{\equiv}CH)_2(PEt_3)_2]$	318	DFWM	$CHCl_3$	1.06	−19.3	7.71	20.8		145,160,161
	318	DFWM	$CHCl_3$	1.06	−0.019	0.0077	0.021		145
trans-$[Pt(C{\equiv}CPh)_2(PBu^n_3)_2]$	[a]	*Z*-scan	THF	0.53	—	—	—	-3.0 ± 0.1	162
trans-$[Pt(C{\equiv}C\text{-}4\text{-}C_6H_4C{\equiv}CPh)_2(PBu^n_3)_2]$	[a]	*Z*-scan	THF	0.53	—	—	—	-209 ± 27	162

[a]Not stated.

TABLE XVIII

MOLECULAR THIRD-ORDER NLO MEASUREMENTS OF A FULLERENE DERIVATIVE BY Z-SCAN AT 0.53 μM

Complex	Solvent	γ_{real} (10^{-31} esu)	γ_{imag} (10^{-31} esu)	Polarization	τ (ps)	Ref.
$[Pt(\eta^2\text{-}C_{60})(PPh_3)_2]$	Toluene	6.9[a]	13	Linear	10	159
		14[a]	34	Circular	10	159

[a]Authors note that values are less than or comparable to the total experimental error and may not be reliable.

TABLE XIX

THIRD-ORDER NLO MEASUREMENTS OF SQUARE-PLANAR GROUP 10 COMPLEXES

Complex	λ_{max} (nm)	Technique	Solvent	Fundamental (μm)	γ_{real} (10^{-48} m^5 V^{-2})	γ_{imag} (10^{-48} m^5 V^{-2})	$\lvert\gamma\rvert$(10^{-48} m^5 V^{-2})	Ref.
$[Ni(2\text{-}C_4H_3S)_2(PEt_3)_2]$	374	DFWM	$CHCl_3$	1.06	−510	460	680	145,157,158
$[Ni(2\text{-}C_8H_5S)_2(PBu^n_3)_2]$	388	DFWM	$CHCl_3$	1.06	−200	1000	1000	145,157,158
$[Ni(2\text{-}C_8H_5S)_2(PEt_3)_2]$	370	DFWM	$CHCl_3$	1.06	−280	150	310	145
$[Ni(2\text{-}C_4H_2S)(PBu^n_3)_2]_n$	515	DFWM	$CHCl_3$	1.06	−17000	20000	26000	145,157,158
$[Pd(2\text{-}C_4H_3S)_2(PEt_3)_2]$	330	DFWM	$CHCl_3$	1.06	−69	130	150	145,157,158
$[Pd(2\text{-}C_8H_5S)_2(PEt_3)_2]$	370	DFWM	$CHCl_3$	1.06	−210	34	210	145
$[Pd_2(\mu\text{-dppm})_2Me_2]$	[a]	DFWM	$CHCl_3$	1.06	−100	200	200	165
$[Pt(2\text{-}C_4H_3S)_2(PEt_3)_2]$	320	DFWM	$CHCl_3$	1.06	−25	60	65	145,157,158
$[Pt(2\text{-}C_8H_5S)_2(PEt_3)_2]$	332	DFWM	$CHCl_3$	1.06	−110	22	110	145

Complex	λ_{max} (nm)	Technique	Solvent	Fundamental (μm)	$\langle\gamma\rangle$ (10^{-34} esu)	γ (10^{-36} esu)	Ref.
cis-$[PtCl_2\{PPh_2(C_5H_3S)\}_2]$	[a]	DFWM	CH_2Cl_2	0.53	1900.0	—	138
trans-$[PdI(4\text{-}C_6H_4NO_2)(PEt_3)_2]$	[a]	THG	$CHCl_3$	1.91	—	36	32
trans-$[PdI(4\text{-}C_6H_4NO_2)(PPh_3)_2]$	[a]	THG	$CHCl_3$	1.91	—	50	32
trans-$[PtBr(4\text{-}C_6H_4CHO)(PEt_3)_2]$	[a]	THG	$CHCl_3$	1.91	—	37	32
trans-$[PtBr(4\text{-}C_6H_4NO_2)(PEt_3)_2]$	[a]	THG	$CHCl_3$	1.91	—	55	32
trans-$[PtI(4\text{-}C_6H_4NO_2)(PEt_3)_2]$	[a]	THG	$CHCl_3$	1.91	—	36	32

[a]Not stated.

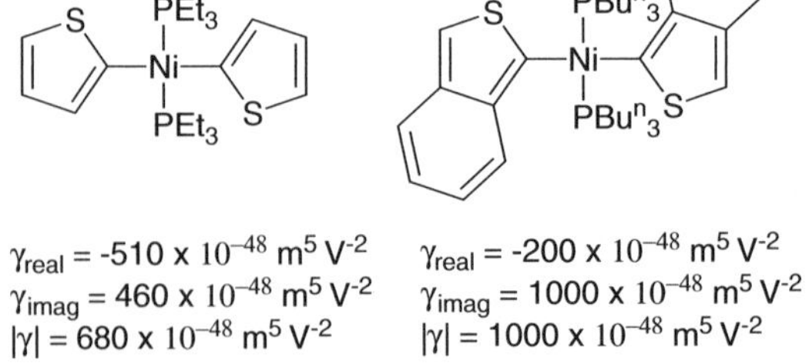

FIG. 18. Examples of thienyl complexes with large γ values.

Several thiophenyl and isobenzothiophenyl complexes have been subjected to degenerate four-wave mixing at 1.06 μm (see Table XIX),[145,157,158] efficient examples being depicted in Fig. 18. The second hyperpolarizability decreases on proceeding down the group;[158] the same trend was observed with group 10 metal alkynyl complexes. Replacing thiophenyl by isobenzothiophenyl ligand results in increased third-order nonlinearity for palladium and platinum complexes, consistent with the response being dominated by the MLCT transition, but the opposite is seen with nickel.

Several square-planar arylpalladium and arylplatinum complexes have been examined, and results are also collected in Table XIX. These THG studies at 1.91 μm revealed that metal replacement in *trans*-$[MI(C_6H_4\text{-}4\text{-}NO_2)(PEt_3)_2]$ (M = Pd, Pt) had no effect on nonlinearity; γ values are larger than in related carbonyl complexes, although data are low in absolute terms. Replacing triethylphosphine by triphenylphosphine, replacing 4-formyl by 4-nitro substituent, and proceeding from iodo to bromo, all result in a 50% increase in nonlinearity.

V
CONCLUDING REMARKS

The studies summarized above have established structure–NLO property relationships for organometallic systems, and both quadratic and cubic molecular nonlinearities obtained for specific complexes are extremely large, suggesting that the potential for application of organometallics remains. The most popular complexes subjected to NLO study are ferrocenyl and alkynyl complexes – it is reassuring that these are amongst the most stable of organometallics, thereby satisfying an important materials requirement for putative applications.

Whether or not organometallics will be used in functioning devices remains unclear. Although organometallic complexes can have similar NLO merit to the best organic molecules, the possibility of their use may well depend on identifying aspects of their performance that are clearly superior to that offered by organic compounds. For example, the facile reversible redox processes at transition metal complexes are often accompanied by marked changes in the linear optical and NLO properties, and so electrochromic switching has commanded significant recent attention. This is an area that is likely to remain topical in the near future.

One problem that has stymied comparison of NLO merit of complexes is that data are frequently reported at one wavelength only. The first wavelength-dependence study of both nonlinear refraction and nonlinear absorption of an organometallic complex appeared recently.[146] Although these data were collected point-by-point, the complete spectrum requiring over a week of data collection, the recent development of techniques such as white light continuum *Z*-scan that permit rapid acquisition of an NLO "spectrum" should ensure that wavelength dependence studies become much more common in the near future.

Acknowledgment

We thank the Australian Research Council for support. M.G.H. holds an ARC Australian Professorial Fellowship and thanks the Alexander von Humboldt-Stiftung for a Fellowship during which part of this review was written.

APPENDIX: ABBREVIATIONS

bipy	2,2′-bipyridyl
Bu	butyl
CA	*p*-chloranil
c.c.	complex conjugate
CT	charge transfer
dc	direct current
DDQ	2,3-dichloro-5,6-dicyano-1,4-benzoquinone
DFWM	degenerate four-wave mixing
DMF	dimethyl formamide
DMSO	dimethyl sulfoxide
diop	2,3-*O*-isopropylidene-2,3-dihydroxy-1,4-bis (diphenylphosphino)butane
diph	1,2-bis(methylphenylphosphino)benzene
dppe	1,2-bis(diphenylphosphino)ethane
dppm	bis(diphenylphosphino)methane
EFISH	electric field-induced second-harmonic generation
Et	ethyl
Fc	ferrocenyl
FWM	four-wave mixing
HRS	hyper-Rayleigh scattering
Im	imaginary component
Me	methyl
mes	mesityl
MLCT	metal-to-ligand charge transfer
NLO	nonlinear optical
OAc	acetate
OHD-OKE	optical heterodyne-detected optical Kerr effect
OKG	optical Kerr gate
OPL	optical power limiting
Ph	phenyl
phen	1,10-phenanthroline
Pr	propyl
prophos	1,2-bis(diphenylphosphino)propane
pyz	pyrazine
Re	real component
SHG	second-harmonic generation

TCNE	tetracyanoethylene
TCNQ	7,7′,8,8′-tetracyanoquinodimethane
THF	tetrahydrofuran
THG	third-harmonic generation
$Tp^{Me,Me}$	tris(3,5-dimethylpyrazolyl)borate

References

(1) Whittall, I. R.; McDonagh, A. M.; Humphrey, M. G.; Samoc, M. *Adv. Organomet. Chem.* **1998**, *42*, 291.
(2) Whittall, I. R.; McDonagh, A. M.; Humphrey, M. G.; Samoc, M. *Adv. Organomet. Chem.* **1999**, *43*, 349.
(3) Williams, D. J., Ed., *Nonlinear Optical Properties of Organic and Polymeric Materials*, American Chemical Society, Washington, DC, 1983.
(4) Kirtman, B.; Champagne, B. *Int. Rev. Phys. Chem.* **1997**, *16*, 389.
(5) Hann, R. A., Bloor, D., Eds., *Organic Materials for Nonlinear Optics*, Royal Society of Chemistry, London, UK, 1989.
(6) Hann, R. A., Bloor, D., Eds., *Organic Materials for Nonlinear Optics II*, Royal Society of Chemistry, London, UK, 1991.
(7) Verbiest, T.; Houbrechts, S.; Kauranen, M.; Clays, K.; Persoons, A. *J. Mater. Chem.* **1997**, *7*, 2175.
(8) Long, N. J. *Angew. Chem.-Int. Ed. Engl.* **1995**, *34*, 21.
(9) Di Bella, S. *Chem. Soc. Rev.* **2001**, *30*, 355.
(10) Chelma, D. S., Zyss, J., Eds., *Nonlinear Optical Properties of Organic Molecules and Crystals I*, Academic Press, Orlando, FL, 1987.
(11) Chelma, D. S., Zyss, J., Eds., *Nonlinear Optical Properties of Organic Molecules and Crystals II*, Academic Press, Orlando, FL, 1987.
(12) Marder, S. R. (D. W. Bruce, D. O'Hare, Eds.), *Inorganic Materials*, John Wiley, Chichester, UK, **1992**, p. 115.
(13) Nie, W. *Adv. Mater.* **1993**, *5*, 520.
(14) Allen, S. D. *New Scientist* **1989**, 1 July, p. 31.
(15) Heeger, A. J., Orenstein, J., Ulrich, D., Eds., *Nonlinear Optical Properties of Polymers*, Materials Research Society, Pittsburgh, PA, 1988.
(16) Ashwell, G. J., Bloor, D., Eds., *Organic Materials for Nonlinear Optics III*, Royal Society of Chemistry, Cambridge, UK, 1993.
(17) Messier, J., Kajzar, F., Prasad, P., Ulrich, D., Eds., *Nonlinear Optical Effects in Organic Polymers*, Kluwer Academic Publishers, Dordrecht, The Netherlands, 1989.
(18) Messier, J., Kajzar, F., Prasad, P., Eds., *Organic Molecules for Nonlinear Optics and Photonics*, Springer-Verlag, Berlin, Germany, 1991.
(19) Kobayashi, T., Ed., *Nonlinear Optics of Organics and Semiconductors*, Springer-Verlag, Berlin, Germany, 1989.
(20) Lyons, M. H., Ed., *Materials for Nonlinear and Electro-optics*, Institute of Physics, Bristol, UK, 1989.
(21) Marder, S. R., Sohn, J. E., Stucky, G. D., Eds., *Materials for Nonlinear Optics, Chemical Perspectives*, American Chemical Society, Washington, DC, 1991.
(22) Powell, C. E.; Humphrey, M. G. *Coord. Chem. Rev.* **2004**, *248*, 725.
(23) Cifuentes, M. P.; Humphrey, M. G. *J. Organomet. Chem.* **2004**, *689*, 3968.
(24) Zyss, J.; Chemla, D. S. (D. S. Chemla, J. Zyss, Eds.), *Nonlinear Optical Properties of Organic Molecules and Crystals I*, Academic Press, Orlando, FL, **1987**, p. 23.
(25) Chemla, D. S.; Oudar, J. L.; Jerphagnon, J. *Phys. Rev. B* **1975**, *12*, 4534.
(26) Clays, K.; Persoons, A. *Rev. Sci. Instrum.* **1992**, *63*, 3285.
(27) Sheik-Bahae, M.; Said, A. A.; Wei, T.; Hagan, D. J.; Van Stryland, E. W. *IEEE J. Quantum Electron.* **1990**, *26*, 760.
(28) Sutherland, R. L. *Handbook of Nonlinear Optics*, Marcel Dekker, New York, NY, 1996.

(29) Farrell, T.; Meyer-Friedrichsen, T.; Malessa, M.; Haase, D.; Saak, W.; Asselberghs, I.; Wostyn, K.; Clays, K.; Persoons, A.; Heck, J.; Manning, A. R. *J. Chem. Soc., Dalton Trans.* **2001**, 29.
(30) Sharma, H. K.; Pannell, K. H.; Ledoux, I.; Zyss, J.; Ceccanti, A.; Zanello, P. *Organometallics* **2000**, *19*, 770.
(31) Calabrese, J. C.; Cheng, L. T.; Green, J. C.; Marder, S. R.; Tam, W. *J Am. Chem. Soc.* **1991**, *113*, 7227.
(32) Cheng, L. T.; Tam, W.; Meredith, G. R. *Mol. Cryst. Liq. Cryst.* **1990**, *189*, 137.
(33) Pedersen, B.; Wagner, G.; Herrmann, R.; Scherer, W.; Meerholz, K.; Schmalzlin, E.; Brauchle, C. *J. Organomet. Chem.* **1999**, *590*, 129.
(34) Pal, S. K.; Krishnan, A.; Das, P. K.; Samuelson, A. G. *J. Organomet. Chem.* **2000**, *604*, 248.
(35) Marder, S. R.; Beratan, D. N.; Tiemann, B. G.; Cheng, L.-T.; Tam, W. (R. A. Hann, D. Bloor, Eds.), *Organic Materials for Nonlinear Optics II*, **1991**, Royal Society of Chemistry, London, UK.
(36) Balavoine, G. G. A.; Daran, J. C.; Iftime, G.; Lacroix, P. G.; Manoury, E.; Delaire, J. A.; Maltey-Fanton, I.; Nakatani, K.; Di Bella, S. *Organometallics* **1999**, *18*, 21.
(37) Cheng, L.-T. (J. Messier, F. Kajzar, P. Prasad, Eds.), *Organic Molecules for Nonlinear Optics and Photonics*, Springer, Berlin, **1991**, p. 121.
(38) Tiemann, B. G.; Marder, S. R.; Perry, J. W.; Cheng, L.-T. *Chem. Mater.* **1990**, *2*, 690.
(39) Arbez-Gindre, C.; Steele, B. R.; Heropoulos, G. A.; Screttas, C. G.; Communal, J.-E.; Blau, W.; Ledoux-Rak, I. *J. Organomet. Chem.* **2005**, *690*, 1620.
(40) Mata, J.; Uriel, S.; Peris, E.; Llusar, R.; Houbrechts, S.; Persoons, A. *J. Organomet. Chem.* **1998**, *562*, 197.
(41) Mata, J. A.; Peris, E.; Asselberghs, I.; Van Boxel, R.; Persoons, A. *New J. Chem.* **2001**, *25*, 299.
(42) Mata, J. A.; Peris, E.; Asselberghs, I.; Van Boxel, R.; Persoons, A. *New J. Chem.* **2001**, *25*, 1043.
(43) Bourgault, M.; Baum, K.; Le Bozec, H.; Pucetti, G.; Ledoux, I.; Zyss, J. *New J. Chem.* **1998**, *22*, 517.
(44) Krishnan, A.; Pal, S. K.; Nandakumar, P.; Samuelson, A. G.; Das, P. K. *Chem. Phys.* **2001**, *265*, 313.
(45) Lee, I. S.; Lee, S. S.; Chung, Y. K.; Kim, D.; Song, N. W. *Inorg. Chim. Acta* **1998**, *279*, 243.
(46) Malaun, M.; Kowallick, R.; McDonagh, A. M.; Marcaccio, M.; Paul, R. L.; Asselberghs, I.; Clays, K.; Persoons, A.; Bildstein, B.; Fiorini, C.; Nunzi, J. M.; Ward, M. D.; McCleverty, J. A. *J. Chem. Soc., Dalton Trans.* **2001**, 3025.
(47) Yuan, Z.; Taylor, N. J.; Sun, Y.; Marder, T. B. *J. Organomet. Chem.* **1993**, *449*, 27.
(48) Doisneau, G.; Balavoine, G.; Fillebeenkhan, T.; Clinet, J. C.; Delaire, J.; Ledoux, I.; Loucif, R.; Puccetti, G. *J. Organomet. Chem.* **1991**, *421*, 299.
(49) Loucif-Saibi, R.; Delaire, J. A.; Bonazzola, L.; Doisneau, G.; Balavoine, G.; Fillebeenkhan, T.; Ledoux, I.; Puccetti, G. *Chem. Phys.* **1992**, *167*, 369.
(50) Ledoux, I. *Synth. Met.* **1993**, *54*, 123.
(51) Behrens, U.; Brussaard, H.; Hagenau, U.; Heck, J.; Hendrickx, E.; Koernich, J.; van der Linden, J. G. M.; Persoons, A.; Spek, A. L. *Chem.-Eur. J.* **1996**, *2*, 98.
(52) Hagenau, U.; Heck, J.; Hendrickx, E.; Persoons, A.; Schuld, T.; Wong, H. *Inorg. Chem.* **1996**, *35*, 7863.
(53) Malaun, M.; Reeves, Z. R.; Paul, R. L.; Jeffery, J. C.; McCleverty, J. A.; Ward, M. D.; Asselberghs, I.; Clays, K.; Persoons, A. E. *Chem. Commun.* **2001**, *49* .
(54) Lee, I. S.; Choi, D. S.; Shin, D. M.; Chung, Y. K.; Choi, C. H. *Organometallics* **2004**, *23*, 1875.
(55) Heck, J.; Dabek, S.; Meyer-Friedrichsen, T.; Wong, H. *Coord. Chem. Rev.* **1999**, *192*, 1217.
(56) Wong, H.; Meyer-Friedrichsen, T.; Farrell, T.; Mecker, C.; Heck, J. *Eur. J. Inorg. Chem.* **2000**, 631.
(57) Blanchard-Desce, M.; Runser, C.; Fort, A.; Barzoukas, M.; Lehn, J. M.; Bloy, V.; Alain, V. *Chem. Phys.* **1995**, *199*, 253.
(58) Moore, A. J.; Chesney, A.; Bryce, M. R.; Batsanov, A. S.; Kelly, J. F.; Howard, J. A. K.; Perepichka, I. F.; Perepichka, D. F.; Meshulam, G.; Berkovuc, G.; Kotler, Z.; Mazor, R.; Khodorkovsky, V. *Eur. J. Org. Chem.* **2001**, *14*, 2671.
(59) Klys, A.; Zakrzewski, J.; Nakatani, K.; Delaire, J. A. *Inorg. Chem. Commun.* **2001**, *4*, 205.
(60) Tam, W.; Cheng, L.-T.; Bierlein, J.; Cheng, L. K.; Wang, Y.; Feiring, A. E.; Meredith, G. R.; Eaton, D. F.; Calabrese, J. C.; Rikken, G. (S. R. Marder, J. E. Sohn, G. D. Stucky, Eds.),

Materials for Nonlinear Optics: Chemical Perspectives, American Chemical Society, Washington, DC, **1991**, p. 158.
(61) Cheng, L.-T.; Tam, W.; Eaton, D. F. *Organometallics* **1990**, *9*, 2856.
(62) Muller, T. J. J.; Netz, A.; Ansorge, M.; Schmälzlin, E.; Bräuchle, C.; Meerholz, K. *Organometallics* **1999**, *18*, 5066.
(63) Garcia, M. H.; Royer, S.; Robalo, M. P.; Dias, A. R.; Tranchier, J.-P.; Chavignon, R.; Prim, D.; Auffrant, A.; Rose-Munch, F.; Rose, E.; Vaissermann, J.; Persoons, A.; Asselberghs, I. *Eur. J. Inorg. Chem.* **2003**, 3895.
(64) Maiorana, S.; Papagni, A.; Licandro, E.; Persoons, A.; Clay, K.; Houbrechts, S.; Porzio, W. *Gazz. Chim. Ital.* **1995**, *125*, 377.
(65) Roth, G.; Fischer, H.; Meyer-Friedrichsen, T.; Heck, J.; Houbrechts, S.; Persoons, A. *Organometallics* **1998**, *17*, 1511.
(66) Fischer, H.; Podschadly, O.; Roth, G.; Herminghaus, S.; Klewitz, S.; Heck, J.; Houbrechts, S.; Meyer, T. *J. Organomet. Chem.* **1997**, *541*, 321.
(67) Farrell, T.; Manning, A. R.; Mitchell, G.; Heck, J.; Meyer-Friedrichsen, T.; Malessa, M.; Wittenburg, C.; Prosenc, M. H.; Cunningham, D.; McArdle, P. *Eur. J. Inorg. Chem.* **2002**, 1677.
(68) Farrell, T.; Meyer-Friedrichsen, T.; Malessa, M.; Wittenburg, C.; Heck, J.; Manning, A. R. *J. Organomet. Chem.* **2001**, *625*, 32.
(69) Farrell, T.; Meyer-Friedrichsen, T.; Heck, J.; Manning, A. R. *Organometallics* **2000**, *19*, 3410.
(70) Tamm, M.; Grzegorzewski, A.; Steiner, T.; Jentzsch, T.; Werncke, W. *Organometallics* **1996**, *15*, 4984.
(71) Tamm, M.; Bannenberg, T.; Baum, K.; Frohlich, R.; Steiner, T.; Meyer-Friedrichsen, T.; Heck, J. *Eur. J. Inorg. Chem.* **2000**, 1161.
(72) Lee, I. S.; Seo, H. M.; Chung, Y. K. *Organometallics* **1999**, *18*, 5194.
(73) Humphrey, P. A.; Turner, P.; Masters, A. F.; Field, L. D.; Cifuentes, M. P.; Humphrey, M. G.; Asselberghs, I.; Persoons, A.; Samoc, M. *Inorg. Chim. Acta* **2005**, *358*, 1663.
(74) Houbrechts, S.; Clays, K.; Persoons, A.; Cadierno, V.; Gamasa, M. P.; Gimeno, J. *Organometallics* **1996**, *15*, 5266.
(75) Houbrechts, S.; Clays, K.; Persoons, A.; Cadierno, V.; Gamasa, M. P.; Gimeno, J.; Whittall, I. R.; Humphrey, M. G. *Proc. SPIE-Int. Soc. Opt. Eng.* **1996**, *2852*, 98.
(76) Bruce, D. W.; Thornton, A. *Mol. Cryst. Liq. Cryst.* **1993**, *231*, 253.
(77) Pizzotti, M.; Ugo, R.; Roberto, D.; Bruni, S.; Fantucci, P.; Rovizzi, C. *Organometallics* **2002**, *21*, 5830.
(78) Lee, I. S.; Seo, H.; Chung, Y. K. *Organometallics* **1999**, *18*, 1091.
(79) Wu, Z.; Ortiz, R.; Fort, A.; Barzoukas, M.; Marder, S. R. *J. Organomet. Chem.* **1997**, *528*, 217.
(80) Roberto, D.; Ugo, R.; Bruni, S.; Cariati, E.; Cariati, F.; Fantucci, P.; Invernizzi, I.; Quici, S.; Ledoux, I.; Zyss, J. *Organometallics* **2000**, *19*, 1775.
(81) Powell, C. E.; Cifuentes, M. P.; McDonagh, A. M.; Hurst, S. K.; Lucas, N. T.; Delfs, C. D.; Stranger, R.; Humphrey, M. G.; Houbrechts, S.; Asselberghs, I.; Persoons, A.; Hockless, D. C. R. *Inorg. Chim. Acta* **2003**, *352*, 9.
(82) Weyland, T.; Ledoux, I.; Brasselet, S.; Zyss, J.; Lapinte, C. *Organometallics* **2000**, *19*, 5235.
(83) Garcia, M. H.; Robalo, M. P.; Dias, A. R.; Duarte, M. T.; Wenseleers, W.; Aerts, G.; Goovaerts, E.; Cifuentes, M. P.; Hurst, S.; Humphrey, M. G.; Samoc, M.; Luther-Davies, B. *Organometallics* **2002**, *21*, 2107.
(84) McDonagh, A. M.; Cifuentes, M. P.; Humphrey, M. G.; Houbrechts, S.; Maes, J.; Persoons, A.; Samoc, M.; Luther-Davies, B. *J. Organomet. Chem.* **2000**, *610*, 71.
(85) Garcia, M. H.; Robalo, M. P.; Dias, A. R.; Piedade, M. F. M.; Galvao, A.; Wenseleers, W.; Goovaerts, E. *J. Organomet. Chem.* **2001**, *619*, 252.
(86) Wenseleers, W.; Gerbrandij, A. W.; Goovaerts, E.; Garcia, M. H.; Robalo, M. P.; Mendes, P. J.; Rodrigues, J. C.; Dias, A. R. *J. Mater. Chem.* **1998**, *8*, 925.
(87) Wenseleers, W.; Goovaerts, E.; Hepp, P.; Garcia, M. H.; Robalo, M. P.; Dias, A. R.; Piedade, M. F. M.; Duarte, M. T. *Chem. Phys. Lett.* **2003**, *367*, 390.
(88) Whittall, I. R.; Humphrey, M. G.; Houbrechts, S.; Maes, J.; Persoons, A.; Schmid, S.; Hockless, D. C. R. *J. Organomet. Chem.* **1997**, *544*, 277.

(89) Whittall, I. R.; Cifuentes, M. P.; Humphrey, M. G.; Luther-Davies, B.; Samoc, M.; Houbrechts, S.; Persoons, A.; Heath, G. A.; Bogsanyi, D. *Organometallics* **1997**, *16*, 2631.
(90) Naulty, R. H.; Cifuentes, M. P.; Humphrey, M. G.; Houbrechts, S.; Boutton, C.; Persoons, A.; Heath, G. A.; Hockless, D. C. R.; Luther-Davies, B.; Samoc, M. *J. Chem. Soc., Dalton Trans.* **1997**, 4167.
(91) Dias, A. R.; Garcia, M. H.; Mendes, P.; Piedade, M. F. M.; Duarte, M. T.; Calhorda, M. J.; Mealli, C.; Wenseleers, W.; Gerbrandij, A. W.; Goovaerts, E. *J. Organomet. Chem.* **1998**, *553*, 115.
(92) Whittall, I. R.; Cifuentes, M. P.; Humphrey, M. G.; Luther-Davies, B.; Samoc, M.; Houbrechts, S.; Persoons, A.; Heath, G. A.; Hockless, D. C. R. *J. Organomet. Chem.* **1997**, *549*, 127.
(93) Hurst, S. K.; Lucas, N. T.; Cifuentes, M. P.; Humphrey, M. G.; Samoc, M.; Luther-Davies, B.; Asselberghs, I.; Van Boxel, R.; Persoons, A. *J. Organomet. Chem.* **2001**, *633*, 114.
(94) Whittall, I. R.; Humphrey, M. G.; Persoons, A.; Houbrechts, S. *Organometallics* **1996**, *15*, 1935.
(95) Wu, I. Y.; Lin, J. T.; Luo, J.; Sun, S. S.; Li, C.-S.; Lin, K. J.; Tsai, C.; Hsu, C. C.; Lin, J.-L. *Organometallics* **1997**, *16*, 2038.
(96) Wu, I. Y.; Lin, J. T.; Luo, J.; Li, C. S.; Tsai, C.; Wen, Y. S.; Hsu, C. C.; Yeh, F. F.; Liou, S. *Organometallics* **1998**, *17*, 2188.
(97) McDonagh, A. M.; Cifuentes, M. P.; Lucas, N. T.; Humphrey, M. G.; Houbrechts, S.; Persoons, A. *J. Organomet. Chem.* **2000**, *605*, 184.
(98) Cifuentes, M. P.; Driver, J.; Humphrey, M. G.; Asselberghs, I.; Persoons, A.; Samoc, M.; Luther-Davies, B. *J. Organomet. Chem.* **2000**, *607*, 72.
(99) McDonagh, A. M.; Lucas, N. T.; Cifuentes, M. P.; Humphrey, M. G.; Houbrechts, S.; Persoons, A. *J. Organomet. Chem.* **2000**, *605*, 193.
(100) Cadierno, V.; Conejero, S.; Gamasa, M. P.; Gimeno, J.; Asselberghs, I.; Houbrechts, S.; Clays, K.; Persoons, A.; Borge, J.; Garcia-Granda, S. *Organometallics* **1999**, *18*, 582.
(101) Hurst, S. K.; Cifuentes, M. P.; Morrall, J. P. L.; Lucas, N. T.; Whittall, I. R.; Humphrey, M. G.; Asselberghs, I.; Persoons, A.; Samoc, M.; Luther-Davies, B.; Willis, A. C. *Organometallics* **2001**, *20*, 4664.
(102) Naulty, R. H.; McDonagh, A. M.; Whittall, I. R.; Cifuentes, M. P.; Humphrey, M. G.; Houbrechts, S.; Maes, J.; Persoons, A.; Heath, G. A.; Hockless, D. C. R. *J. Organomet. Chem.* **1998**, *563*, 137.
(103) Hurst, S. K.; Lucas, N. T.; Humphrey, M. G.; Asselberghs, I.; Van Boxel, R.; Persoons, A. *Aust. J. Chem.* **2001**, *54*, 447.
(104) McDonagh, A. M.; Humphrey, M. G.; Samoc, M.; Luther-Davies, B.; Houbrechts, S.; Wada, T.; Sasabe, H.; Persoons, A. *J. Am. Chem. Soc.* **1999**, *121*, 1405.
(105) Whittall, I. R.; Humphrey, M. G.; Houbrechts, S.; Persoons, A.; Hockless, D. C. R. *Organometallics* **1996**, *15*, 5738.
(106) Hurst, S. K.; Cifuentes, M. P.; McDonagh, A. M.; Humphrey, M. G.; Samoc, M.; Luther-Davies, B.; Asselberghs, I.; Persoons, A. *J. Organomet. Chem.* **2002**, *642*, 259.
(107) Hurst, S. K.; Lucas, N. T.; Humphrey, M. G.; Isoshima, T.; Wostyn, K.; Asselberghs, I.; Clays, K.; Persoons, A.; Samoc, M.; Luther-Davies, B. *Inorg. Chim. Acta* **2003**, *350*, 62.
(108) Paul, F.; Costuas, K.; Ledoux, I.; Deveau, S.; Zyss, J.; Halet, J.-F.; Lapinte, C. *Organometallics* **2002**, *21*, 5229.
(109) Laidlaw, W. M.; Denning, R. G.; Verbiest, T.; Chauchard, E.; Persoons, A. *Nature* **1993**, *363*, 58.
(110) Morrison, I. D.; Denning, R. G.; Laidlaw, W. M.; Stammers, M. A. *Rev. Sci. Instrum.* **1996**, *67*, 1445.
(111) Jayaprakash, K. N.; Ray, P. C.; Matsuoka, I.; Bhadbhade, M. M.; Puranik, V. G.; Das, P. K.; Nishihara, H.; Sarkar, A. *Organometallics* **1999**, *18*, 3851.
(112) Meyer-Friedrichsen, T.; Wong, H.; Prosenc, M. H.; Heck, J. *Eur. J. Inorg. Chem.* **2003**, 936.
(113) Laidlaw, W. M.; Denning, R. G.; Verbiest, T.; Chauchard, E.; Persoons, A. *Proc. SPIE-Int. Soc. Opt. Eng.* **1994**, *2143*, 14.
(114) Pizzotti, M.; Ugo, R.; Dragonetti, C.; Annoni, E.; Demartin, F.; Mussini, P. *Organometallics* **2003**, *22*, 4001.
(115) Mata, J. A.; Peris, E.; Uriel, S.; Llusar, R.; Asselberghs, I.; Persoons, A. *Polyhedron* **2001**, *20*, 2083.
(116) Hendrickx, E.; Persoons, A.; Samson, S.; Stephenson, G. R. *J. Organomet. Chem.* **1997**, *542*, 295.

(117) Meyer-Friedrichsen, T.; Mecker, C.; Prosenc, M. H.; Heck, J. *Eur. J. Inorg. Chem.* **2002**, 239.
(118) Behrens, U.; Meyer-Friedrichsen, T.; Heck, J. *Z. Anorg. Allg. Chem.* **2003**, *629*, 1421.
(119) Farrell, T.; Manning, A. R.; Murphy, T. C.; Meyer-Friedrichsen, T.; Heck, J.; Asselberghs, I.; Persoons, A. *Eur. J. Inorg. Chem.* **2001**, 2365.
(120) Hudson, R. D. A.; Manning, A. R.; Nolan, D. F.; Asselberghs, I.; Van Boxel, R.; Persoons, A.; Gallagher, J. F. *J. Organomet. Chem.* **2001**, *619*, 141.
(121) Hudson, R. D. A.; Asselberghs, I.; Clays, K.; Cuffe, L. P.; Gallagher, J. F.; Manning, A. R.; Persoons, A.; Wostyn, K. *J. Organomet. Chem.* **2001**, *637*, 435.
(122) Houbrechts, S.; Clays, K.; Persoons, A.; Pikramenou, Z.; Lehn, J. M. *Chem. Phys. Lett.* **1996**, *258*, 485.
(123) Winter, C. S.; Oliver, S. N.; Rush, J. D. *Opt. Commun.* **1988**, *69*, 45.
(124) Myers, L. K.; Langhoff, C.; Thompson, M. E. *J. Am. Chem. Soc.* **1992**, *114*, 7560.
(125) Li, G.; Song, Y.; Hou, H.; Li, L.; Fan, Y.; Zhu, Y.; Meng, X.; Mi, L. *Inorg. Chem.* **2003**, *42*, 913.
(126) Ghosal, S.; Samoc, M.; Prasad, P. N.; Tufariello, J. J. *J. Phys. Chem.* **1990**, *94*, 2847.
(127) Winter, C. S.; Oliver, S. N.; Rush, J. D. (R. A. Hann, D. Bloor, Eds.), *Organic Materials for Nonlinear Optics*, Royal Society of Chemistry, London, UK, **1989**, p. 232.
(128) Winter, C. S.; Oliver, S. N.; Rush, J. D. (J. Messier, F. Kajzar, P. Prasad, D. Ulrich, Eds.), *Nonlinear Optical Effects in Organic Polymers*, Kluwer, Dordrecht, The Netherlands, **1989**, p. 247.
(129) Calabrese, J. C.; Tam, W. *Chem. Phys. Lett.* **1987**, *133*, 244.
(130) Yuan, Z.; Stringer, G.; Jobe, I. R.; Kreller, D.; Scott, K.; Koch, L.; Taylor, N. J.; Marder, T. B. *J. Organomet. Chem.* **1993**, *452*, 115.
(131) Rojo, G.; Agullo-Lopez, F.; Campo, J. A.; Heras, J. V.; Cano, M. *J. Phys. Chem. B* **1999**, *103*, 11016.
(132) Rojo, G.; Agullo-Lopez, F.; Campo, J. A.; Cano, M.; Lagunas, M. C.; Heras, J. V. *Synth. Met.* **2001**, *124*, 201.
(133) Hurst, S. K.; Humphrey, M. G.; Morrall, J. P.; Cifuentes, M. P.; Samoc, M.; Luther-Davies, B.; Heath, G. A.; Willis, A. C. *J. Organomet. Chem.* **2003**, *670*, 56.
(134) Mata, J. A.; Peris, E.; Llusar, R.; Uriel, S.; Cifuentes, M. P.; Humphrey, M. G.; Samoc, M.; Luther-Davies, B. *Eur. J. Inorg. Chem.* **2001**, 2113.
(135) Hou, H.; Li, G.; Song, Y.; Fan, Y.; Zhu, Y.; Zhu, L. *Eur. J. Inorg. Chem.* **2003**, 2325.
(136) Myers, L. K.; Ho, D. M.; Thompson, M. E.; Langhoff, C. *Polyhedron* **1995**, *14*, 57.
(137) Thompson, M. E.; Chiang, W.; Myers, L. K.; Langhoff, C. *Proc. SPIE-Int. Soc. Opt. Eng.* **1991**, *1497*, 423.
(138) Zhai, T.; Lawson, C. M.; Gale, D. C.; Gray, G. M. *Opt. Mater.* **1995**, *4*, 455.
(139) Fargin, E.; Berthereau, A.; Cardinal, T.; Videau, J. J.; Villesuzzanne, A.; Le Flem, G. *Ann. Chim.-Sci. Mater.* **1998**, *23*, 27.
(140) Whittall, I. R.; Humphrey, M. G.; Samoc, M.; Swiatkiewicz, J.; Luther-Davies, B. *Organometallics* **1995**, *14*, 5493.
(141) Whittall, I. R.; Humphrey, M. G.; Samoc, M.; Luther-Davies, B. *Angew. Chem.-Int. Ed. Engl.* **1997**, *36*, 370.
(142) Powell, C. E.; Cifuentes, M. P.; Morrall, J. P.; Stranger, R.; Humphrey, M. G.; Samoc, M.; Luther-Davies, B.; Heath, G. A. *J. Am. Chem. Soc.* **2003**, *125*, 602.
(143) Powell, C. E.; Humphrey, M. G.; Cifuentes, M. P.; Morrall, J. P.; Samoc, M.; Luther-Davies, B. *J. Phys. Chem. A* **2003**, *107*, 11264.
(144) Cifuentes, M. P.; Humphrey, M. G.; Morrall, J. P.; Samoc, M.; Paul, F.; Lapinte, C.; Roisnel, T. *Organometallics* **2005**, *24*, 4280.
(145) Davey, A. P.; Page, H.; Blau, W.; Byrne, H. J.; Cardin, D. J. *Synth. Met.* **1993**, *57*, 3980.
(146) Powell, C. E.; Morrall, J. P.; Ward, S. A.; Cifuentes, M. P.; Notaras, E. G. A.; Samoc, M.; Humphrey, M. G. *J. Am. Chem. Soc.* **2004**, *126*, 12234.
(147) Cifuentes, M. P.; Powell, C. E.; Humphrey, M. G.; Heath, G. A.; Samoc, M.; Luther-Davies, B. *J. Phys. Chem. A* **2001**, *105*, 9625.
(148) McDonagh, A. M.; Whittall, I. R.; Humphrey, M. G.; Skelton, B. W.; White, A. H. *J. Organomet. Chem.* **1996**, *519*, 229.
(149) McDonagh, A. M.; Cifuentes, M. P.; Whittall, I. R.; Humphrey, M. G.; Samoc, M.; Luther-Davies, B.; Hockless, D. C. R. *J. Organomet. Chem.* **1996**, *526*, 99.

(150) McDonagh, A. M.; Humphrey, M. G.; Samoc, M.; Luther-Davies, B. *Organometallics* **1999**, *18*, 5195.
(151) Hurst, S. K.; Humphrey, M. G.; Isoshima, T.; Wostyn, K.; Asselberghs, I.; Clays, K.; Persoons, A.; Samoc, M.; Luther-Davies, B. *Organometallics* **2002**, *21*, 2024.
(152) Frazier, C. C.; Chauchard, E.; Cockerham, M. P.; Porter, P. L. *Mater. Res. Soc. Symp. Proc.* **1988**, *109*, 323.
(153) Teo, B. K.; Xu, Y. H.; Zhong, B. Y.; He, Y. K.; Chen, H. Y.; Qian, W.; Deng, Y. J.; Zou, Y. H. *Inorg. Chem.* **2001**, *40*, 6794.
(154) Porter, P. L.; Guha, S.; Kang, K.; Frazier, C. C. *Polymer* **1991**, *32*, 1756.
(155) Vicente, J.; Chicote, M. T.; Abrisqueta, M. D.; de Arellano, M. C. R.; Jones, P. G.; Humphrey, M. G.; Cifuentes, M. P.; Samoc, M.; Luther-Davies, B. *Organometallics* **2000**, *19*, 2968.
(156) Hudson, R. D. A.; Manning, A. R.; Gallagher, J. F.; Garcia, M. H.; Lopes, N.; Asselberghs, I.; Van Boxel, R.; Persoons, A.; Lough, A. J. *J. Organomet. Chem.* **2002**, *655*, 70.
(157) Blau, W. J.; Cardin, D. J.; Cardin, C. J.; Davey, A. (G. J. Ashwell, D. Bloor, Eds.), *Organic Materials for Nonlinear Optics III*, Royal Society of Chemistry, Cambridge, UK, **1993**, p. 124.
(158) Davey, A. P.; Byrne, H. J.; Page, H.; Blau, W.; Cardin, D. J. *Synth. Met.* **1993**, *58*, 161.
(159) Ergorov, A. N.; Marivtsky, O. B.; Petrovsky, A. N.; Yakubrovsky, K. V. *Laser Phys.* **1995**, *5*, 1006.
(160) Blau, W. J.; Byrne, H. J.; Cardin, D. J.; Davey, A. P. *J. Mater. Chem.* **1991**, *1*, 245.
(161) Davey, A.; Cardin, D. J.; Byrne, H. J.; Blau, W. (J. Messier, F. Kajzar, P. Prasad, Eds.), *Organic Molecules for Nonlinear Optics and Photonics*, Springer, Berlin, Germany, **1991**, p. 391.
(162) Haub, J.; Johnson, M.; Orr, B.; Woodruff, M.; Crisp, G. T. *CLEO/QUELS'91*, Baltimore, MA, **1991**.
(163) Alain, V.; Blanchard-Desce, M.; Chen, C.-T.; Marder, S. R.; Fort, A.; Barzoukas, M. *Synth. Met.* **1996**, *81*, 133.
(164) Briel, O.; Sünkel, K.; Krossing, I.; Nöth, H.; Schmälzlin, E.; Meerholz, K.; Bräuchle, C.; Beck, W. *Eur. J. Inorg. Chem.* **1999**, 483.
(165) Page, H.; Blau, W.; Davey, A. P.; Lou, X.; Cardin, D. J. *Synth. Met.* **1994**, *63*, 179.

Pentadienyl Complexes of the Group 4 Transition Metals

LOTHAR STAHL[a] and RICHARD D. ERNST[b,*]

[a]*Department of Chemistry, University of North Dakota, Grand Forks, ND 58202-9024, USA*
[b]*Department of Chemistry, University of Utah, 315 S. 1400 E., Salt Lake City, UT 84112-0850, USA*

I
INTRODUCTION

The utilization of allyl[1] and cyclopentadienyl ligands[2] has been vigorously pursued since the 1950s and 1960s. However, it was not until nearly 1980 before pentadienyl ligands attracted more than cursory attention, even though Fe(*cyclo*-C_6H_7)$(CO)_3^+$ and Fe(C_5H_7)$(CO)_3^+$ had been reported in 1960 and 1962, respectively.[3] While most metal pentadienyl studies continued at first to focus on later transition metals,[4] extensions to earlier metals did ultimately follow,[5] and in many respects the chemistry appeared to become more intriguing with every step to the left. Much of this no doubt derives from the unique properties of pentadienyl ligands, especially their great steric demands and strong δ acid character.[6] The metals themselves play a further role by promoting coupling reactions between unsaturated organic molecules and pentadienyl ligands, much as had already been observed for early metal diene complexes.[7] As the group 4 transition metals

*Corresponding author
E-mail: ernst@chem.utah.edu (R.D. Ernst).

ADVANCES IN ORGANOMETALLIC CHEMISTRY
VOLUME 55 ISSN 0065-3055/DOI 10.1016/S0065-3055(07)55003-3

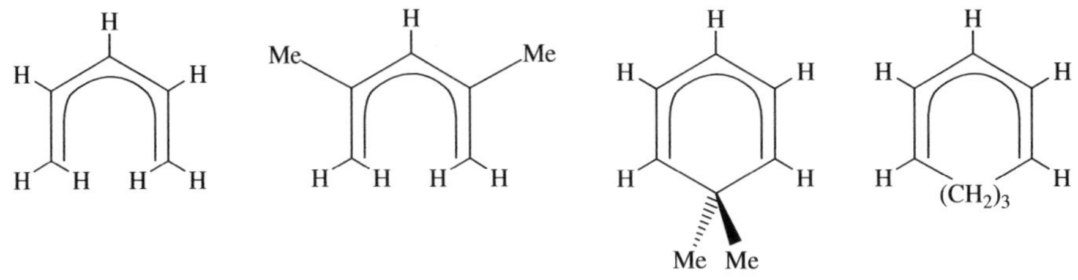

C_5H_7	2,4-C_7H_{11}	6,6-dmch	c-C_8H_{11}
pentadienyl	2,4-dimethylpentadienyl	6,6-dimethyl-cyclohexadienyl	cyclooctadienyl

CHART 1.

titanium, zirconium, and hafnium have generated especially rich coupling chemistry, and as their metallocenes also have proven to be remarkably unique,[8] this review will focus on the advances made in the pentadienyl chemistry of these three elements. Some of the most commonly used pentadienyl ligands are illustrated in Chart 1. In general, a numbering scheme will be utilized in which the dienyl termini are defined as C1 and C5, with C2–C4 appearing between C1 and C5. Formally, any charge delocalized over the dienyl fragment will be shared by C1, C3, and C5, leaving C2 and C4 uncharged.[5,6]

II
BIS(η^5-PENTADIENYL)METAL COMPLEXES

A. *Open Metallocenes*

While simple metallocenes, $M(\eta\text{-}C_5H_5)_2$ (**1**), are very stable thermally, all the way from the 15 electron vanadocene to the 20 electron nickelocene,[9] it has long been recognized that $Ti(C_5H_5)_2$ itself is quite unstable. Numerous attempts to isolate this species have led instead to dinitrogen adducts and to a wide variety of species in which C–H bond activations have taken place.[10] Better results were obtained with decamethyltitanocene, $Ti(C_5Me_5)_2$. Although this compound is also readily coordinated by nitrogen, its isolation as a 14 electron species can be accomplished using an argon atmosphere.[11] Nevertheless, it too decomposes at room temperature *via* C–H activation. All indications are that 14 electron zirconocenes would be even more unstable than titanocenes, paralleling the dimerization processes that have been observed for heavier metal analogues of vanadocene, chromocene, manganocene, cobaltocene, and nickelocene.[12] In any event, the profoundly differing properties of pentadienyl as compared to cyclopentadienyl ligands suggest that at least species such as "open titanocenes" (bis(pentadienyl)titanium compounds) could perhaps be isolable. In particular, the fact that pentadienyl ligands are even more sterically demanding than C_5Me_5 could inhibit coordination by other ligands, and their wider nature should also serve to prevent C–H activation reactions.[6]

Additionally, since pentadienyl ligands may also bond more strongly to metals than cyclopentadienyl,[5] this could lead to further stabilization of an open titanocene. Although not fully appreciated at the time, δ backbonding interactions, generally considered negligible for C_5H_5, can be quite significant for pentadienyl ligands.[6,13] This would also be expected to greatly reduce the chances that dinitrogen would coordinate to an open titanocene.

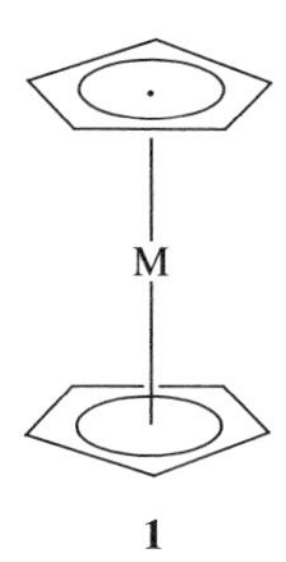

1

Indeed, in 1982 it was reported that the 14 electron $Ti(2,4\text{-}C_7H_{11})_2$ could be isolated as a deep green liquid under a nitrogen atmosphere.[14] Heating to at least 120 °C did not lead to observable decomposition, although the compound reacts very vigorously with air. While not noticeably coordinated by dinitrogen (at least not at 1 atm pressure), $Ti(2,4\text{-}C_7H_{11})_2$ can be reversibly coordinated by a variety of phosphines (Section IIB).[13,15] Titanocenes, however, display the reverse preference, which is consistent with the strongly δ acidic character of pentadienyl ligands. To date the structure of this open titanocene has not been determined, but electronic arguments clearly favor the adoption of an ideally staggered geometry (**2**), as observed for its vanadium analogue.[16]

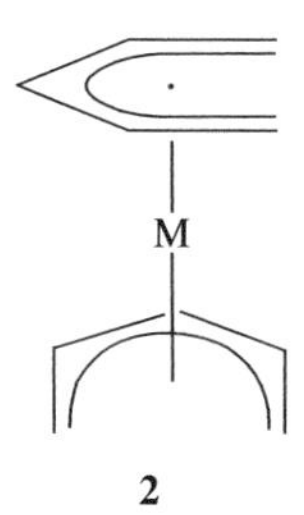

2

Subsequently, several other open titanocenes were observed.[17] Two such isolable species utilized the $2,3\text{-}C_7H_{11}$ or 6,6-dmch ligands, while the existence of $Ti(3\text{-}C_6H_9)_2$ as at least an intermediate in solution was established based on the observation of reversible dissociation of PEt_3 by $Ti(3\text{-}C_6H_9)_2(PEt_3)$.[15a] Furthermore, deep green solutions containing the presumed $Ti(2\text{-}C_6H_9)_2$ complex have been reported,[18] the formulation of which was based upon the isolation of a carbonyl adduct upon exposure to CO. None of these 14 electron complexes has yet been characterized structurally, although at least some are solids at room temperature. NMR spectroscopic data do establish, however, that $Ti(2,3\text{-}C_7H_{11})_2$ exists as a pair of diastereomeric species, represented for simplicity as **3a** and **3b**, although their actual conformations are expected to be staggered (cf., **2**).[17b] As in the case of the iron and ruthenium analogues,[6,19] the $Ti(2,3\text{-}C_7H_{11})_2$ isomers have been observed to interconvert, through a process proposed to involve species related to **4** as a key intermediate. Although the magnitude of the barrier to the process has not yet been established, it appears to be greater than that for $Fe(2,3\text{-}C_7H_{11})_2$ (ca. 22 kcal/mol).[6]

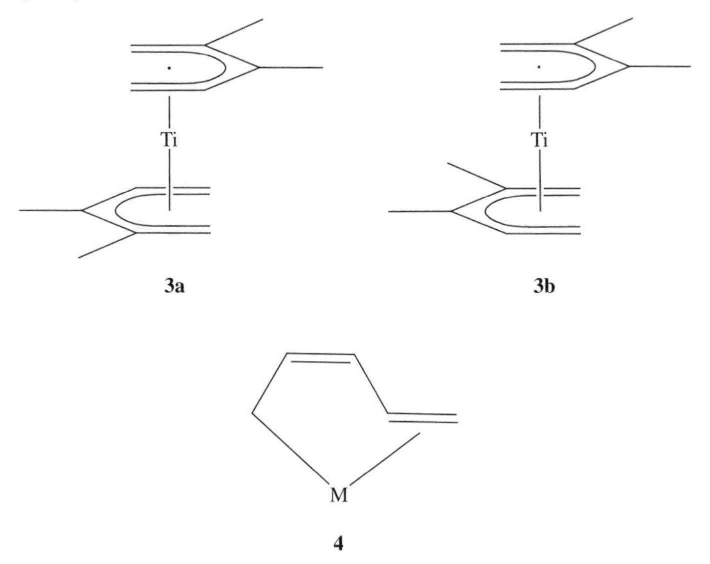

The appropriate incorporation of Me_3C[20] or Me_3Si[21] substituents has been found to lead to the solid open titanocenes, $Ti[2,4\text{-}(Me_3C)_2C_5H_5]_2$ and $Ti[1,5\text{-}(Me_3Si)_2C_5H_5]_2$. The latter has been structurally characterized (Fig. 1) and found to adopt something close to the expected staggered conformation (**5c**). A conformation angle, χ, of 82.5° was observed, defined relative to values of 0° and 180°, respectively, for the *syn*-eclipsed and *anti*-eclipsed conformations, **5a** and **5b**.

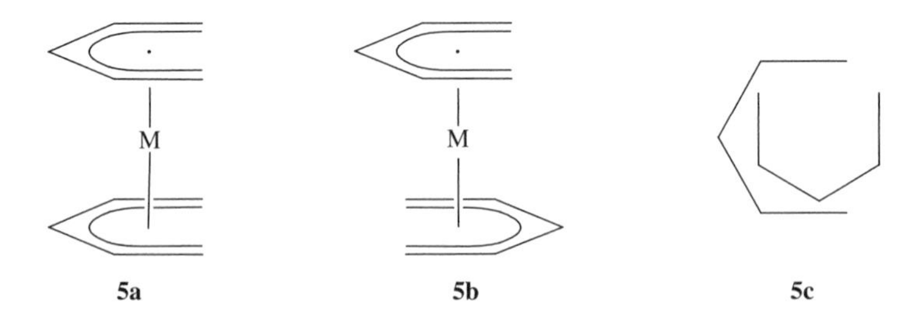

Subsequently, another solid open titanocene, $Ti(c\text{-}C_8H_{11})_2$, was isolated.[22] A structural study also revealed something close to a staggered conformation ($\chi = 88.4°$, Fig. 2), as expected.

In this species, as in $Ti[1,5\text{-}(Me_3Si)_2C_5H_5]_2$, the M–C(1,5) bonds (C1 and C5 being the dienyl termini) are significantly shorter than the others, whereas for open

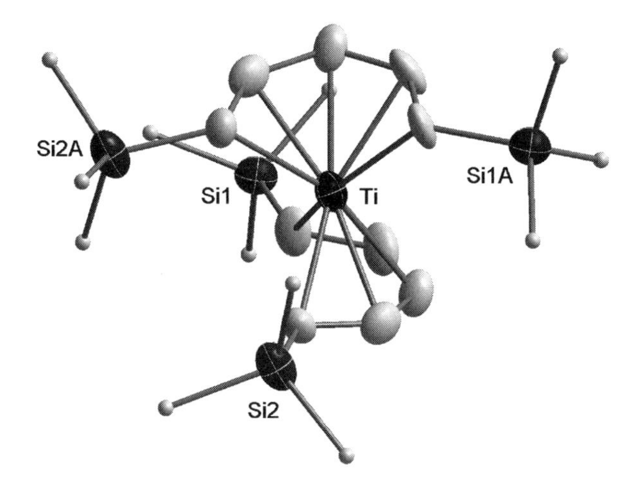

FIG. 1. Molecular structure of $Ti[(1,5\text{-}Me_3Si)_2C_5H_5]_2$ (hydrogen atoms omitted; methyl groups simplified).

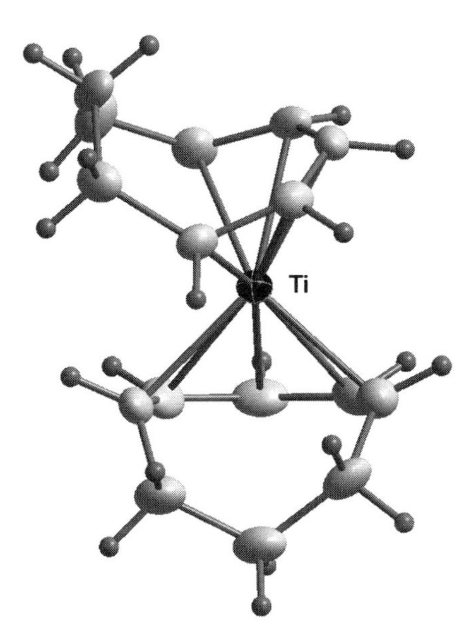

FIG. 2. Molecular structure of $Ti(c\text{-}C_8H_{11})_2$.

ferrocenes these bonds tended to be the longest. Overall, the average Ti–C bond distances in $Ti(c\text{-}C_8H_{11})_2$ and $Ti[1,5\text{-}(Me_3Si)_2C_5H_5]_2$ are 2.245 and 2.275 Å, respectively. While a comparison with titanocene itself cannot be made, the recent isolation and structural characterization of the stable, highly substituted 14 electron titanocenes, $Ti[C_5Me_4(SiMe_2R)]_2$ (R = Me, ^{i}Pr, ^{t}Bu), has been achieved.[23] The average Ti–C distances for these species have been found to be 2.352, 2.345, and 2.352 Å, respectively, quite close to what had been predicted.[9] Structures of the vanadium, chromium, and iron $M(c\text{-}C_8H_{11})_2$ complexes were also determined, at present comprising the only fully structurally characterized series of open metallocenes for these four metals.

Given the existence of Ti(6,6-dmch)$_2$ and $Ti(c\text{-}C_8H_{11})_2$, one could easily consider it possible to prepare an analogous, intermediate cycloheptadienyl complex, $Ti(c\text{-}C_7H_9)_2$. However, attempts to prepare this species have instead led to the previously reported[24] $Ti(\eta\text{-}C_7H_7)(\eta^5\text{-}C_7H_9)$ (Section IIID), apparently as a result of the much greater susceptibility of $c\text{-}C_7H_9$ vs. $c\text{-}C_8H_{11}$ towards C–H activation reactions.[25] In fact, the simple cyclohexadienyl ligand itself is quite prone to such reactions, with even the 18 electron $Fe(c\text{-}C_6H_7)_2$ complex isomerizing at room temperature to $Fe(\eta\text{-}C_6H_6)(\eta^4\text{-}c\text{-}C_6H_8)$, while no evidence of the existence of a related titanium compound was observed.[17a] Clearly, the methyl substituents on the 6,6-dmch ligand play a key role in stabilizing its complexes, although a C–Me activation reaction has been promoted thermally in a ruthenium complex.[26]

Even more surprising than the isolation of stable 14 electron open titanocenes has been the existence of the 14 electron $Zr[1,5\text{-}(Me_3Si)_2C_5H_5]_2$[21] (Fig. 3). This complex, like its titanium analogue, was isolated *via* sublimation at ca. 120 °C and exists in a similar arrangement ($\chi = 82.2°$). The average Zr–C distance in this compound is 2.396 Å, which may be implicated as the reason this species is much more air-sensitive than its titanium analogue. To date it is not known whether a hafnium analogue can be prepared.

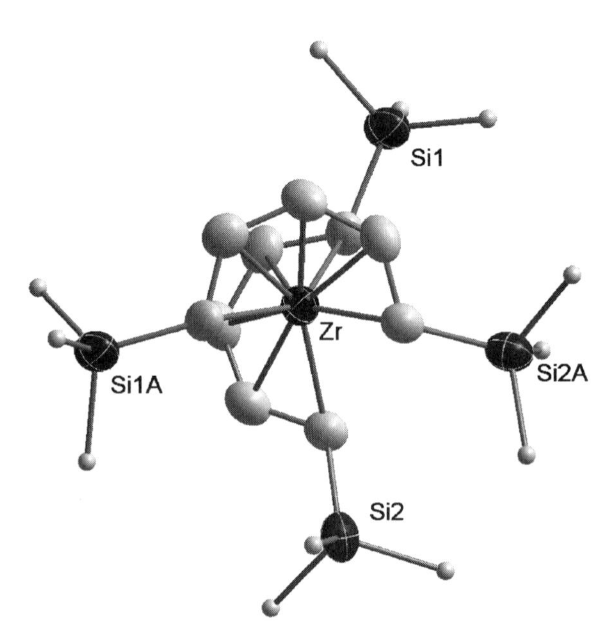

FIG. 3. Molecular structure of $Zr[1,5\text{-}(Me_3Si)_2C_5H_5]_2$ (hydrogen atoms omitted; methyl groups simplified).

B. *Ligand Complexes of the Open Metallocenes*

The first such species were isolated by the reaction of $Ti(2,4\text{-}C_7H_{11})_2$ with CO, various phosphines or phosphites [Eq. (1)].[13,14,27]

$$
\begin{gathered}
Ti(2,4\text{-}C_7H_{11})_2 + L \rightleftharpoons Ti(2,4\text{-}C_7H_{11})_2(L) \\
L = CO, PF_3, PMe_3, PEt_3, PMe_2Ph, P(OMe)_3, \\
P(OEt)_3, P(OCH_2)_3CEt, PO_3C_4H_7
\end{gathered}
\tag{1}
$$

Two interesting points may be made concerning this reaction. The first is that only one ligand is incorporated into the complex, despite the fact that two carbonyls can coordinate to $Ti(C_5Me_5)_2$.[28] Additionally, even the single ligand incorporated undergoes partial dissociation in solution. The greatly diminished Lewis acidity of the open titanocenes relative to titanocene fragments may readily be attributed to the great steric demands of the pentadienyl ligands, a good part of which derives from the geometric requirement for a much shorter separation between the metal atom and the dienyl ligand plane in order to maintain similar M–C bond lengths. The fact that M–pentadienyl bond lengths can be significantly shorter than corresponding M–C_5H_5 bond lengths further enhances the steric differences between the open and closed dienyl ligands. A good illustration of this may be seen from a comparison of the structural arrangements in these ligand adducts. While the attachment of an additional ligand to a metallocene leads to a dramatic tilt between the ligand planes (**6a**), the shorter M–open dienyl plane separations generally lead to these ligands already being close to or within a van der Waals separation of each other, thereby preventing such tilting and requiring the adoption of a *syn*-eclipsed conformation in order to allow ligand coordination by the open edges of the dienyl ligands (**6b**, Fig. 4).

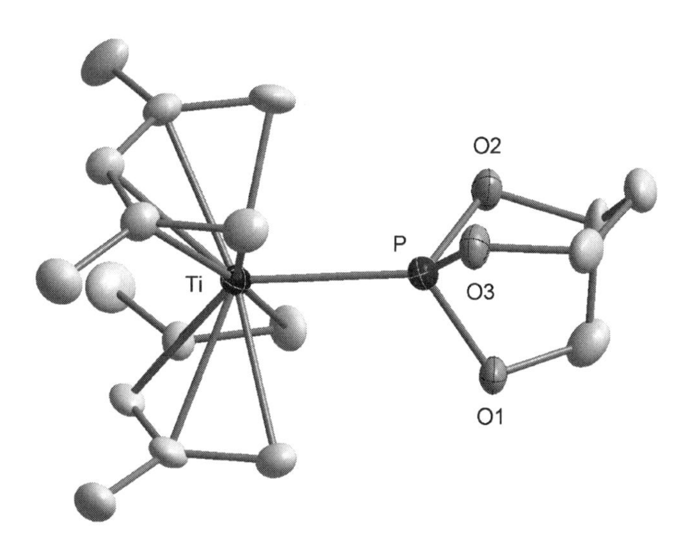

FIG. 4. Molecular structure of $Ti(2,4\text{-}C_7H_{11})_2(PO_3C_4H_7)$ (hydrogen atoms omitted).

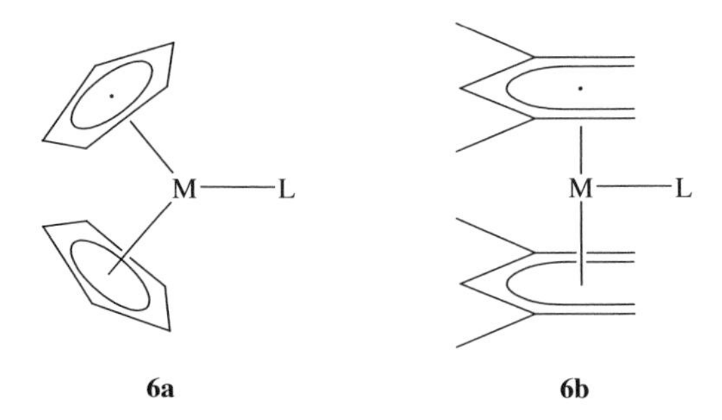

The trifluorophosphine adducts of the isomeric $Ti(2,3\text{-}C_7H_{11})_2$ complexes may also be prepared by direct interaction.[29] While spectroscopic data clearly indicate the presence of the two expected isomers, the isolation of $Ti(2,3\text{-}C_7H_{11})_2(CO)$ has been reported to yield a single isomer.[18] Given the observations above regarding the PF_3 complexes, it seems likely that two carbonyl isomers are also formed, but one may preferentially crystallize, and/or undergoes significant isomerization to the other during manipulation.

Related CO and PEt_3 complexes of the $Ti(C_5H_7)_2$ and $Ti(3\text{-}C_6H_9)_2$ fragments have been isolated by the formation of the otherwise apparently unstable fragments in the presence of the appropriate ligand [Eq. (2)].[17b] A structural study of $Ti(3\text{-}C_6H_9)_2(CO)$ has revealed the expected *syn*-eclipsed geometry.[18]

$$\begin{gathered} \text{“TiCl}_2\text{”} + 2\text{Pdl}^- \xrightarrow{\text{L}} \text{Ti(Pdl)}_2\text{(L)} \\ \text{Pdl} = \text{C}_5\text{H}_7, 3\text{-C}_6\text{H}_9; \text{L} = \text{CO, PEt}_3 \end{gathered} \tag{2}$$

The thermodynamics of the ligand dissociation processes of several of these species have been studied by variable temperature ^{31}P NMR spectroscopy.[14] These studies have found that the favorability of coordination of $Ti(2,4\text{-}C_7H_{11})_2$ by added ligands falls in the order: PF_3 (17.4 ± 0.8) > $P(OCH_2)_3CEt$ (16.3 ± 0.9) > PMe_3 (14.5 ± 0.8) > PMe_2Ph (12.9 ± 0.5) > $P(OMe)_3$ (11.4 ± 0.9) > $P(OEt)_3$ (10.6 ± 0.6) > PEt_3 (10.0 ± 0.1), with the binding energies being provided in kcal/mol in parentheses. Nearly all the data support the proposal that steric influences dominate the binding preferences, with the phosphites being the most notable exceptions. However, cone angles originally estimated for phosphite ligands were based on unrealistic, non-existent conformations, having all three substituents bent directly backwards. Revision of these estimates to account for actual phosphite conformations[30] led to excellent agreement between binding and steric factors.[14] The greatly enhanced binding by the cage phosphite, required to direct all three substituents back, relative to $P(OMe)_3$ and $P(OEt)_3$, provided additional confirmation for the claim of prior underestimations of phosphite cone angles. A slight preference for the binding of PF_3 (104°) over $P(OCH_2)_3CEt$ (101°) was interpreted as arising from a secondary electronic preference, favoring acceptors over donors.

Structural studies of the PF_3, $P(OEt)_3$, and PMe_3 complexes revealed Ti–P bond distances of 2.324(1), 2.472(4), and 2.550(2) Å, respectively.[31] The Ti–P bond distances thus do not correlate with their respective strengths. Instead, the bond lengths seem to be correlated with the electronegativities of the attached substituents. As the substituent electronegativities increase, there could be contraction of the phosphorus orbitals, or an increase in *s* orbital character of the phosphorus lone pair, either or both then resulting in a shorter Ti–P bond.

Open titanocenes incorporating the edge-bridged 6,6-dmch or c-C_8H_{11} ligands have also been observed to form ligand adducts. Generally, the additional steric interactions brought about by these bridges have led to weaker binding. Thus, while $Ti(6,6\text{-dmch})_2$ forms a readily isolable monocarbonyl, binding by PF_3 had to be established by low-temperature NMR spectroscopy, while PMe_3 and $P(OMe)_3$ bindings were supported by the observation of reversible color changes at low temperatures.[17a] Parallel behavior has been displayed by $Ti(c\text{-}C_8H_{11})_2$, which does form readily isolable $Ti(c\text{-}C_8H_{11})_2(L)$ complexes for L = CO or $P(OCH_2)_3CEt$.[22,32] For all these edge-bridged complexes, a *syn*-eclipsed arrangement has been adopted, with the additional ligand located by the electronically open dienyl edges. As will be described later, however, this is not what is observed for related half-open complexes. Spectroscopic studies of the carbonyl complexes have revealed that the edge-bridged pentadienyl ligands are actually electronically intermediate between other pentadienyls and C_5H_5.[17a]

The related chemistry of zirconium and hafnium has not been so deeply developed. As the 14 electron open metallocenes themselves are not generally stable, the preparations of ligand adducts have required the presence of a coordinating ligand in the initial preparative stages [Eq. (3)].[33] In addition to the $MCl_4(L)_2$ complexes, $[MCl_3(PEt_3)_2]_2$ complexes may also be used as starting materials in similar reactions. Of initial note is the spontaneous reduction of Zr(IV) and even Hf(IV) to their divalent states, the facility of which can be attributed largely to the strongly δ-acidic character of pentadienyl ligands, leading to a dramatic favorability for binding to metals in lower oxidation states. The wide girth of the open pentadienyl edges could also be involved, as it would not likely allow for favorable overlap with the contracted orbitals of a metal in a higher oxidation state (see Sections IIC, IIIE, and IIIF).

$$\begin{gathered} MCl_4(PR_3)_2 + 4K(2,4\text{-}C_7H_{11}) \rightarrow M(2,4\text{-}C_7H_{11})_2(L) \\ M = Zr,\ Hf;\ L = PMe_3,\ PMe_2Ph,\ PEt_3 \end{gathered} \tag{3}$$

As PEt_3 is relatively weakly bound in the above complexes, it may be readily replaced by smaller phosphines, or alternatively, by $P(OMe)_3$ or $P(OCH_2)_3CMe$ for at least the zirconium complex [Eq. (4)].

$$\begin{gathered} Zr(2,4\text{-}C_7H_{11})_2(PEt_3) + L \rightleftharpoons Zr(2,4\text{-}C_7H_{11})_2L + PEt_3 \\ L = PMe_3, P(OMe)_3, P(OCH_2)_3CMe \end{gathered} \tag{4}$$

$$\begin{gathered} M(2,4\text{-}C_7H_{11})_2(PMe_2Ph) \overset{CO}{\rightleftharpoons} M(2,4\text{-}C_7H_{11})_2(CO) \overset{CO}{\rightleftharpoons} M(2,4\text{-}C_7H_{11})_2(CO)_2 \\ M = Zr,\ Hf \end{gathered} \tag{5}$$

Additionally, the phosphines may be replaced by CO [Eq. (5)].[33b] Notably, the reaction in Eq. (5) has provided the first examples of bis(ligand) adducts of open metallocenes. The structure of the dicarbonyl zirconium complex is rather unsymmetric, and is presented in Fig. 5. The complex, as well as its hafnium analogue,

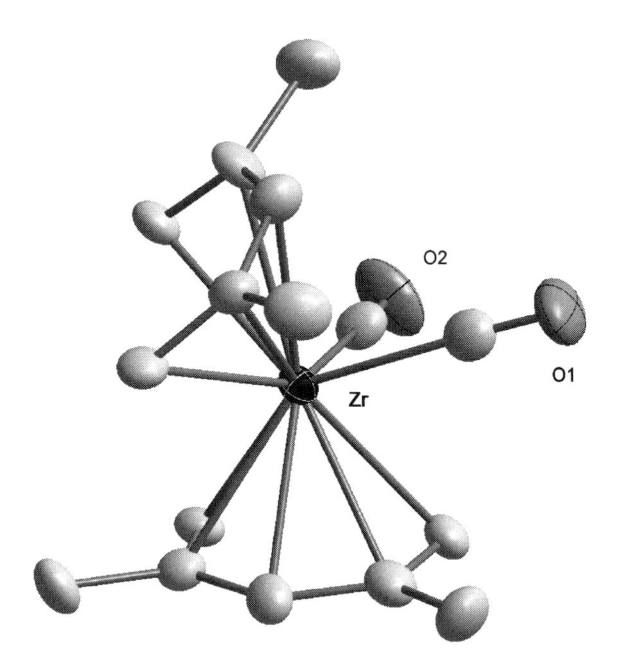

FIG. 5. Molecular structure of $Zr(2,4\text{-}C_7H_{11})_2(CO)_2$ (hydrogen atoms omitted).

undergoes fluxional behavior in solution, rendering the two dienyl ligands equivalent, with $\Delta G^{\neq} = 12.0(2)$ and $13.1(2)$ kcal/mol, respectively. Placing the red zirconium dicarbonyl complex in solution under a static vacuum leads to loss of one CO ligand, thereby forming the green 16 electron monocarbonyl complex, which adopts the expected *syn*-eclipsed geometry (**6b**).

The isolation of related zirconium complexes incorporating the C_5H_7 ligand has also been achieved [Eqs. (6) and (7)].[34]

$$ZrCl_4(dmpe)_2 + 4K(C_5H_7) \rightarrow Zr(C_5H_7)_2(dmpe) \tag{6}$$

$$ZrCl_4 + 4K(C_5H_7) + PMe_2Ph \rightarrow Zr(C_5H_7)_2(PMe_2Ph) \tag{7}$$

In contrast to the unsymmetric structure for $Zr(2,4\text{-}C_7H_{11})_2(CO)_2$, the dmpe complex adopts a more symmetric structure in which the phosphine donor sites each coordinate by an open edge of one of the C_5H_7 ligands (Fig. 6).

The reaction in Eq. (7) leads to the isolation of the green $Zr(C_5H_7)_2(PMe_2Ph)$ complex, although its adoption of a red color in solutions containing excess PMe_2Ph could be an indication that the 18 electron bis(PMe_2Ph) complex is formed as part of an equilibrium [Eq. (8)]. Based on spectroscopic data, it would appear that $Zr[1,5\text{-}(Me_3Si)_2C_5H_5]_2$ reacts with CO, although the nature of the product is unclear.[35]

$$Zr(C_5H_7)_2(PMe_2Ph) + PMe_2Ph \overset{?}{\rightleftharpoons} Zr(C_5H_7)_2(PMe_2Ph)_2 \tag{8}$$

Edge-bridged open zirconocenes have been isolated as well [Eq. (9)].[36] The $Zr(6,6\text{-dmch})_2(PMe_3)_2$ complex (**7**; $L = PMe_3$) is symmetric, so that the dienyl

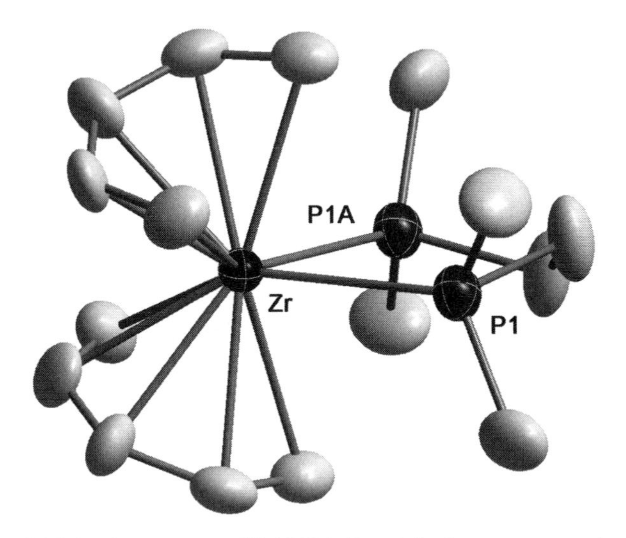

FIG. 6. Molecular structure of $Zr(C_5H_7)_2(dmpe)$ (hydrogen atoms omitted).

ligands are equivalent, but as opposed to $Zr(C_5H_7)_2(dmpe)$ and most other related species, the PMe_3 ligands are not situated by the electronically open dienyl edges, but instead by the central (C3) positions of the dienyl ligands (Fig. 7).

$$ZrCl_4 + 4K(6,6\text{-dmch}) + 2PMe_3 \rightarrow Zr(6,6\text{-dmch})_2(PMe_3)_2 \quad (9)$$

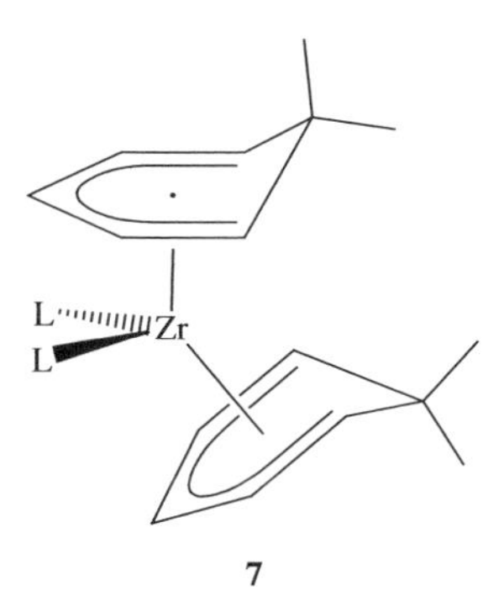

7

One of the PMe_3 ligands is very labile, and reactions with CO or PhC_2SiMe_3 lead to its rapid replacement [Eq. (10)].[37] The crystal structures of each product have been determined. The carbonyl complex adopts a different structure than the bis(phosphine) complex, such that the PMe_3 ligand remains located by the central dienyl carbon atom of one 6,6-dmch ligand (Fig. 8), but the smaller carbonyl resides by an edge bridge. The isolation of an alkyne complex was unexpected, as in previous reactions of group 4 metal pentadienyl complexes with alkynes (Section VB), dienyl–alkyne coupling was always observed (*vide infra*). The structure of the alkyne complex was actually disordered, with a second image appearing to reflect a slow-to-form coupling product. The reduced reactivity of 6,6-dmch to these

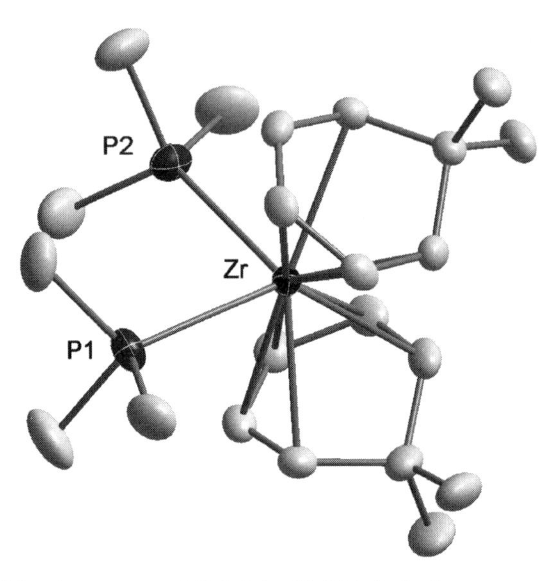

FIG. 7. Molecular structure of $Zr(6,6\text{-dmch})_2(PMe_3)_2$ (hydrogen atoms omitted).

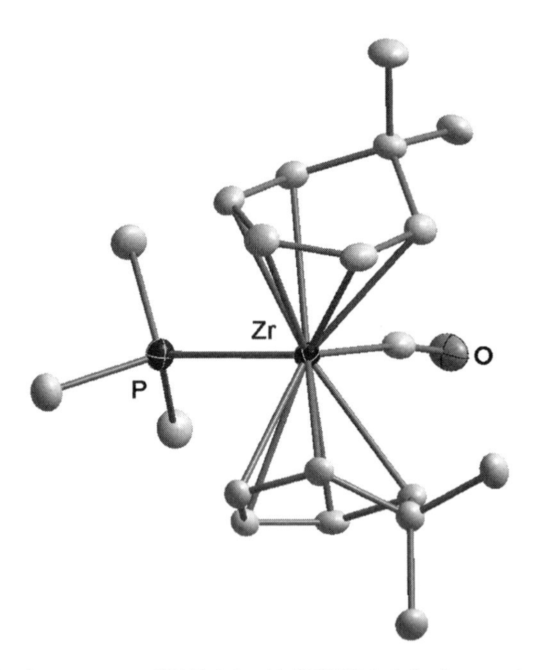

FIG. 8. Molecular structure of $Zr(6,6\text{-dmch})_2(CO)(PMe_3)$ (hydrogen atoms omitted).

coupling reactions was attributed to the short C1–C5 separation, which would lead to improved metal–dienyl overlap, thereby stabilizing the dienyl compound.

$$\begin{gathered} Zr(6,6\text{-dmch})_2(PMe_3)_2 + L \rightarrow Zr(6,6\text{-dmch})_2(PMe_3)(L) \\ L = CO, PhC_2SiMe_3 \end{gathered} \tag{10}$$

It is noteworthy that the formation of the 18 electron Zr(6,6-dmch)$_2$(PMe$_3$)$_2$ complex, in contrast to an analogue of the 16 electron Zr(2,4-C$_7$H$_{11}$)$_2$(PMe$_3$), suggests that the 6,6-dmch ligand is sterically less demanding than 2,4-C$_7$H$_{11}$. Most other data, however, including the weaker binding of ligands by Ti(6,6-dmch)$_2$ as compared to Ti(2,4-C$_7$H$_{11}$)$_2$, suggest the opposite. An explanation for this seemingly contradictory behavior has been proposed based on the fact that addition of the first ligand to a M(6,6-dmch)$_2$ fragment brings about more steric distortion than would occur for a M(2,4-C$_7$H$_{11}$)$_2$ fragment. This greater distortion then opens up the metal coordination sphere aiding the accommodation of a second ligand, especially for the larger zirconium center.[36]

C. *Open Metallocene Complexes with Halide and Related Ligands*

Despite the tremendous importance of M(C$_5$H$_5$)$_2$X$_2$ compounds in many chemical applications,[8,10,38] related pentadienyl-containing species have until recently been conspicuously absent. It has been recognized that the wide girth of pentadienyl ligands would more likely result in poor overlap with the contracted orbitals of a metal in a higher ($> +3$) oxidation state and, furthermore, the high δ acidity of pentadienyls would also favor bonding to metals in lower oxidation states. Nevertheless, pentadienyl analogues of the M(C$_5$H$_5$)$_2$X$_2$ species have finally been accessed, utilizing zirconium, which may be regarded as the largest transition metal. Thus, reaction of Zr(6,6-dmch)$_2$(PMe$_3$)$_2$ with 1,2-C$_2$H$_4$XX′ (X = X′ = Cl, Br, I; X = Cl, X′ = OMe) leads to the stable, brightly colored (orange-red to dark red) Zr(6,6-dmch)$_2$X$_2$ complexes [Eq. (11), **8a–c**], as well as to Zr(6,6-dmch)$_2$(Cl)(OMe) (**8d**).[39] Furthermore, an oxidative addition reaction [Eq. (12)] could be used to prepare the thermally unstable Zr(6,6-dmch)$_2$(Me)(Br). The reaction of the dihalides with two equivalents of methyl lithium appears to lead only to a paler, oily dimethyl compound, which was not characterized. Reactions designed to provide access to fluoro analogues of the above species have thus far led to insoluble products, possibly promoted by the high favorability for fluorides to bridge zirconium centers. The bright colors of species **8** contrast markedly with the colorless Zr(C$_5$H$_5$)$_2$X$_2$ analogues. Based upon theoretical studies of Zr(C$_5$H$_5$)(6,6-dmch)X$_2$ complexes (Section IIIF), these colors may be attributed to ligand-to-metal charge transfer (LMCT) transitions, which are shifted to the visible region as a consequence of the HOMO on the 6,6-dmch ligand being destabilized by ca. 1 eV relative to the HOMOs of the C$_5$H$_5$ ligand.

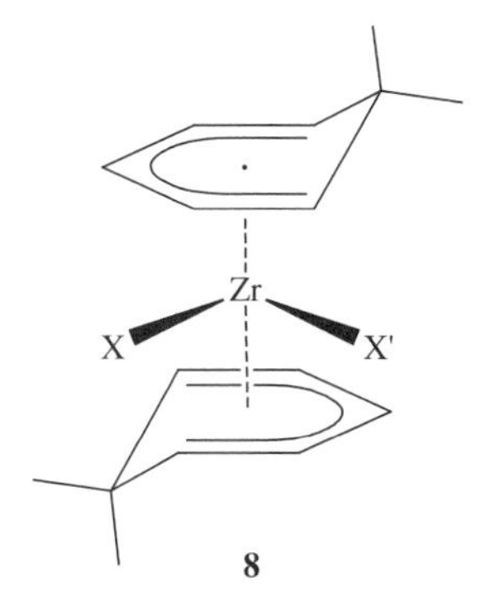

8

$$\mathrm{Zr(6,6\text{-}dmch)_2(PMe_3)_2 + 1,2\text{-}C_2H_4XX' \rightarrow Zr(6,6\text{-}dmch)_2XX'} \quad X = X' = Cl, Br, I \quad X = Cl, X' = OMe \tag{11}$$

$$\mathrm{Zr(6,6\text{-}dmch)_2(PMe_3)_2 + MeBr \rightarrow Zr(6,6\text{-}dmch)_2(Me)(Br)} \tag{12}$$

The structural data for these species (Fig. 9) provide confirmation for the previous proposals that the wide natures of the pentadienyl ligands lead to a loss in metal–ligand overlap, which would be particularly severe for higher valent metals. Thus, for each of these compounds, there is a single, short Zr–C bond to the central dienyl carbon atom (C3), averaging 2.457(4) Å, while the Zr–C bonds for the C(2,4) and C(1,5) atoms become progressively longer, and less precisely defined at 2.536(5) and 2.695(11) Å. These trends suggest the apparent onset of a slipping of the η^5-dienyl ligands from the metal coordination sphere, consistent with the expected loss in metal–dienyl overlap due to the presence of Zr(IV).

While the relative orientations of the 6,6-dmch ligands differ significantly for some of these complexes, it seems likely that the differences arise more from steric than electronic considerations, and would seem relatively unimportant. One item of interest, however, is the Zr–O–C angle of 153.2(1)° in Zr(6,6-dmch)$_2$(Cl)(OMe), which suggests at least partial donation of both oxygen lone pairs to the metal center, in accord with other results which have established the ability of alkoxide ligands to serve as formal 5 electron donors.[40] In this case, there would certainly be competition between the alkoxide and 6,6-dmch ligands (if not also the halides) for π donation, in order to avoid a 20 electron count. Indeed, support that the alkoxide ligand competes with the 6,6-dmch ligands can be seen in the average Zr–C bond distances of 2.483, 2.552, and 2.732 Å, respectively, for the C(3), C(2,4), and C(1,5) atoms, all of which are longer than their counterparts in the dihalide compounds.

Besides these higher valent open metallocenes, there are some boratabenzene complexes of Zr(IV) (e.g., **9**)[41] which bear at least some similarity to the dienyl compounds, by virtue of the resonance forms indicated. The location of the boron atom near the "dienyl" ligand plane, as opposed to the much greater deviations found for other edge-bridged dienyls such as 6,6-dmch, seems to reveal a greater contribution from the η^6 rather than the η^5 bonding mode, although the trends in Zr–(C,B) bond distances generally parallel expectations for η^5 coordination. Thus, in Zr(C_5H_5BOEt)$_2$Cl$_2$, the average Zr–C distances were found to be 2.497(3), 2.582(2), and 2.656(3) Å, respectively, for the C(3), C(2,4), and C(1,5) positions, with the Zr–B distance being 2.826(2) Å. This provides support for the participation of both resonance structures above. However, in Zr(4-^{t}BuC_5H_4BPh)$_2$Cl$_2$,[41d] the Zr–C distances are quite similar for the C(3), C(2,4), and C(1,5) positions, averaging 2.66(1), 2.61(2), and 2.62(3) Å, respectively.

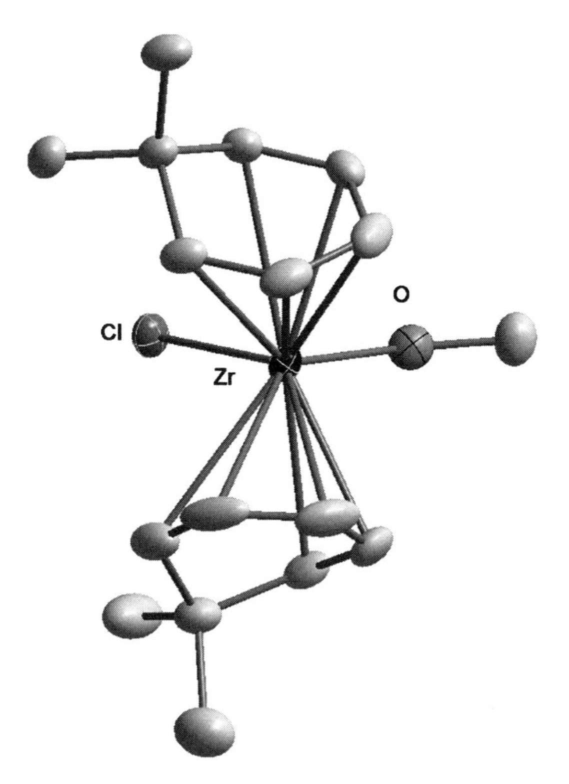

FIG. 9. Molecular structure of $Zr(6,6\text{-dmch})_2(Cl)(OMe)$ (hydrogen atoms omitted).

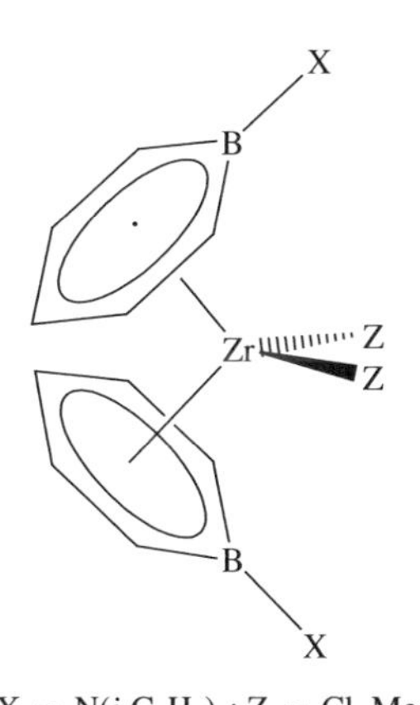

X = $N(i\text{-}C_3H_7)_2$; Z = Cl, Me
X = OEt; Z = Cl
X = Me, Ph; Z = Cl, CO

9

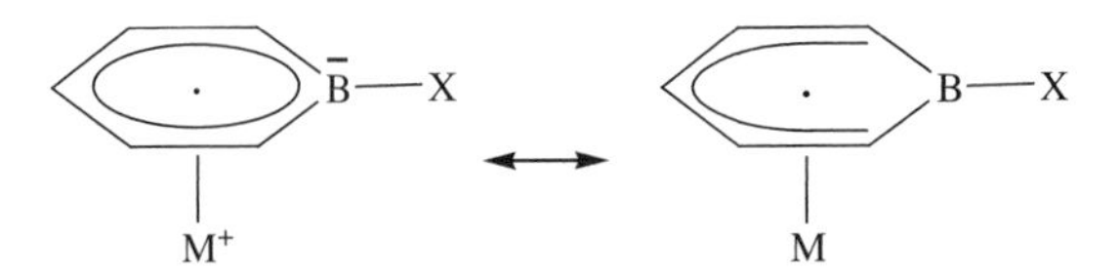

Electrochemical reductions of the dichlorides having R = NMe_2, OEt, Me, and Ph have been carried out and a good correlation of reduction potential was observed with a two-parameter Hammett function.[42] The dichlorides could also be reduced under a CO atmosphere by magnesium metal, yielding $Zr(C_5H_5BR)_2(CO)_2$ complexes.

III
MONO(η^5-PENTADIENYL)METAL COMPLEXES

A. *Half-Open Metallocenes*

Perhaps in large part due to the smaller cone angle of C_5H_5 and even C_5Me_5 relative to pentadienyl ligands (ca. 136°, 165°, and 180°, respectively),[15a] there are far fewer unligated half-open metallocenes than open metallocenes, which is a general trend for at least groups 4 and 5. Apparently, the only unligated examples for group 4, to date, involve titanium,[43] which forms $Ti(C_5H_5)[\eta^5\text{-}c\text{-}1,2,5,6\text{-}(Me_3Si)_4\text{-}4\text{-}R\text{–}C_6H]$ (**10**) through a coupling reaction [Eq. (13)] involving three alkynes. Clearly, the steric bulk imposed by the highly substituted cyclohexadienyl ligand is the key to the isolation of these unligated complexes. Indeed, these species (e.g., Fig. 10) show no coordination by CO or a variety of other ligands under typical conditions.

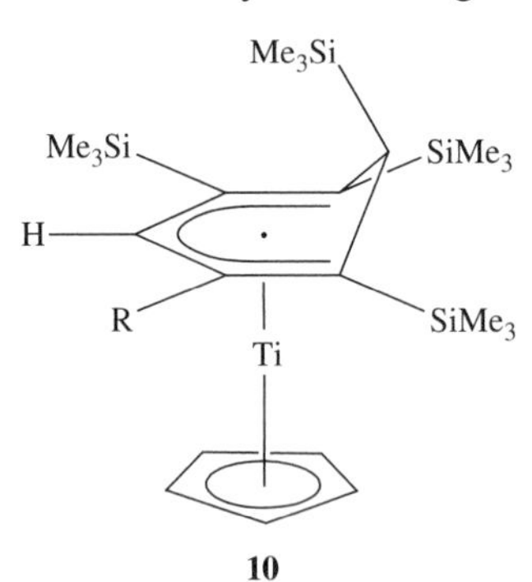

$$\begin{aligned} &(C_5H_5)Ti(\mu\text{-}Me_3Si_2C_2)_2Mg(C_5H_5) + \text{excess } RC_2H \\ &\rightarrow Ti(C_5H_5)[\eta^5\text{-}c\text{-}1,2,5,6\text{-}(Me_3Si)_4\text{-}4\text{-}R\text{-}C_6H] \\ &R = Me_3Si, Ph, {}^nBu, {}^tBu, Cy, Fe(C_5H_5)(C_5H_4) \end{aligned} \tag{13}$$

These compounds have been shown to be low spin (diamagnetic), as is the case for their open metallocene analogues, but not for titanocenes themselves.[11] This is consistent with observations on $V(C_5H_5)[1,5\text{-}(Me_3Si)_2C_5H_5]$[44] which, like $V(2,4\text{-}C_7H_{11})_2$,[16] is low spin, despite vanadocene being high spin.[45] Of interest is the fact

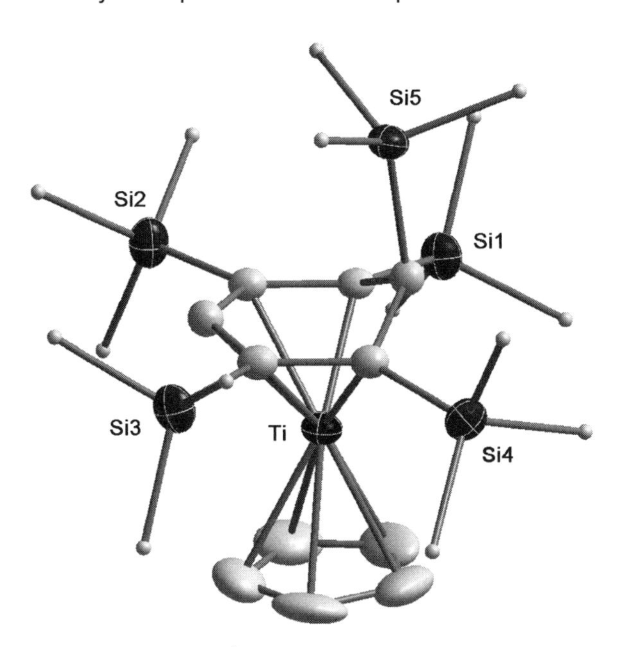

FIG. 10. Molecular structure of $Ti(C_5H_5)[\eta^5$-c-1,2,4,5,6-$(Me_3Si)_5$-$C_6H]$ (hydrogen atoms omitted; methyl groups simplified).

that the Ti–C distances for the electronically open dienyl ligands average ca. 2.15 Å, vs. 2.36 Å for the C_5H_5 ligands. The presence of shorter M–C bonds for the pentadienyl ligands vs. cyclopentadienyl ligands is typical of early metal half-open metallocenes, and reflects stronger bonding for the electronically open, non-aromatic dienyl ligands (Section IIIB).[5,6] However, the differences observed for the 16 electron mono(adducts) have tended to be ca. 0.1 Å, whereas the shortening of some 0.2 Å in the 14 electron species must be regarded as very remarkable.

The isolation of the 14 electron complex $Ti(C_5Me_5)(2,4$-$C_7H_{11})$, as a deep black crystalline solid, has been reported, and supported through spectroscopic data.[43c] The deep black color contrasts markedly with the generally green colors of **10**, which could be a result of the Me_3Si substituents of **10**, or perhaps might reflect dinitrogen coordination to $Ti(C_5Me_5)(2,4$-$C_7H_{11})$. The potentially analogous $Ti(Ph_2CH)_2$ and $Ti(Ph_2CSiMe_3)_2$ derived complexes were also described. Each complex was prepared from the reaction of lower valent titanium chloride complexes with the appropriate anion. Their activities as polymerization catalysts will be discussed in Section VC.

B. *Ligand Complexes of the Half-Open Metallocenes*

The half-open metallocenes of titanium and zirconium have proven of great value in allowing for direct comparisons of the metal–ligand bonding involving pentadienyl and cyclopentadienyl ligands, and they have also exhibited very rich reaction chemistry (Section V).

$Ti(C_5H_5)Cl_2(THF)_2$ has been found to be very effective for preparing these complexes, whether with C_5H_7,[46] 2,4-C_7H_{11},[46] 6,6-dmch,[47] or *c*-C_8H_{11}[48] ligands. Variations of this procedure are also possible, utilizing C_5H_4Me[49] or 1,3-$C_5H_3(^tBu)_2$[50] in place of C_5H_5, e.g., Eqs. (14)–(16).

$$\begin{gathered}Ti(C_5H_5)Cl_2(THF)_2 + 2Pdl^- + PR_3 \rightarrow Ti(C_5H_5)(Pdl)(PR_3)\\ Pdl = C_5H_7, R = Me, Et\\ Pdl = 3\text{-}C_6H_9, R = Me\\ Pdl = 2,4\text{-}C_7H_{11}, R = Me, Et, n\text{-}C_4H_9\\ Pdl = 6,6\text{-dmch}, R = Me\\ Pdl = c\text{-}C_8H_{11}, R = Me, Et\end{gathered} \tag{14}$$

$$\begin{gathered}Ti[1,3\text{-}C_5H_3^tBu_2]Cl_2(THF)_2 + 2KPdl + PR_3\\ \rightarrow Ti[1,3\text{-}C_5H_3^tBu_2](Pdl)(PR_3)\\ Pdl = C_5H_7, R = Me, Et\\ Pdl = 2,4\text{-}C_7H_{11} \text{ or } 6,6\text{-dmch}, R = Me\end{gathered} \tag{15}$$

$$\begin{gathered}Ti[1,3\text{-}C_5H_3(SiMe_3)_2]Cl_2(THF)_2 + 2K(2,4\text{-}C_7H_{11}) + PMe_3\\ \rightarrow Ti[1,3\text{-}C_5H_3(SiMe_3)_2](2,4\text{-}C_7H_{11})(PMe_3)\end{gathered} \tag{16}$$

$Ti(C_5H_4Me)(C_5H_7)(PMe_3)$ can be prepared similarly. Replacement of PEt_3 by phosphites occurs readily [Eq. (17)],[46] Likewise, reaction with excess CO is very rapid [Eq. (18)], leading to the thermally unstable 18 electron dicarbonyl complex.[46b] Removal of the CO atmosphere using a static vacuum leads to a loss of one of the carbonyl ligands, yielding the likewise thermally unstable 16 electron $Ti(C_5H_5)(2,4\text{-}C_7H_{11})(CO)$.

$$\begin{gathered}Ti(C_5H_5)(2,4\text{-}C_7H_{11})(PEt_3) + P(OR)_3 \rightarrow Ti(C_5H_5)(2,4\text{-}C_7H_{11})[P(OR)_3]\\ R = Me, Et\\ P(OR)_3 = PO_3C_4H_7\end{gathered} \tag{17}$$

$$\begin{gathered}Ti(C_5H_5)(2,4\text{-}C_7H_{11})(PEt_3) \underset{PEt_3}{\overset{CO}{\rightleftharpoons}} Ti(C_5H_5)(2,4\text{-}C_7H_{11})(CO)\\ \underset{-CO}{\overset{CO}{\rightleftharpoons}} Ti(C_5H_5)(2,4\text{-}C_7H_{11})(CO)_2\end{gathered} \tag{18}$$

As $Ti(2,4\text{-}C_7H_{11})_2$ was found only to form a monocarbonyl complex, it is clear that there needs to be some relief of steric crowding to allow for the second CO to bind. As noted above, this can be accomplished also by replacing titanium by zirconium. The C–O stretching frequencies for the dicarbonyl are observed at 1932 and 1984 cm^{-1}, while that for the monocarbonyl appears at 1959 cm^{-1}. For comparison, the C–O stretching frequencies in $Ti(C_5H_5)_2(CO)_2$,[51] appear at 1897 and 1975 cm^{-1}, consistent with the enhanced δ acidity of the pentadienyl ligands. While the structure of the monocarbonyl complex can be presumed to be symmetric, and similar to that of $Ti(C_5H_5)(2,4\text{-}C_7H_{11})(PEt_3)$, having the additional ligand by the open dienyl edge, the dicarbonyl compound is less symmetric. ^{1}H and ^{13}C NMR spectra reveal that the carbonyl ligands are non-equivalent in the ground state, as are the two ends of the 2,4-dimethylpentadienyl ligand. The barrier to the process by which equivalence is established was found to be 12.7 ± 0.2 kcal/mol.

Structural data have been obtained for several half-open titanocenes, including Ti$(C_5H_5)(2,4\text{-}C_7H_{11})(PEt_3)$,[46] Ti$[1,5\text{-}C_5H_3({}^tBu)_2](C_5H_7)(PMe_3)$,[50] Ti$(C_5H_5)$(6,6-dmch)$(PMe_3)$,[52] Ti$[1,3\text{-}C_5H_3({}^tBu)_2]$(6,6-dmch)$(PMe_3)$,[53] Ti$(C_5H_5)(c\text{-}C_8H_{11})(PMe_3)$,[54] and Ti$(C_5H_4Me)(c\text{-}C_8H_{11})(PEt_3)$. For the open-edge complexes, the phosphine ligands reside by the open dienyl edge (e.g., **11**, Fig. 11), whereas for the $c\text{-}C_8H_{11}$ and 6,6-dmch complexes, the phosphine is located by the central dienyl carbon atom (e.g., **12**, Fig. 12). The change in orientation does lead to some lengthening of the Ti–C bond for central carbon atoms of the 6,6-dmch ligands but in all cases the average Ti–C distances for the electronically open dienyl ligand are significantly shorter than those for their cyclic counterparts. The values for the six listed compounds are 2.24 Å vs. 2.35 Å, 2.24 Å vs. 2.39 Å, 2.35 Å vs. 2.38 Å, 2.26 Å vs. 2.39 Å, 2.24 Å vs. 2.38 Å, and 2.25 Å vs. 2.40 Å, respectively. The observation of

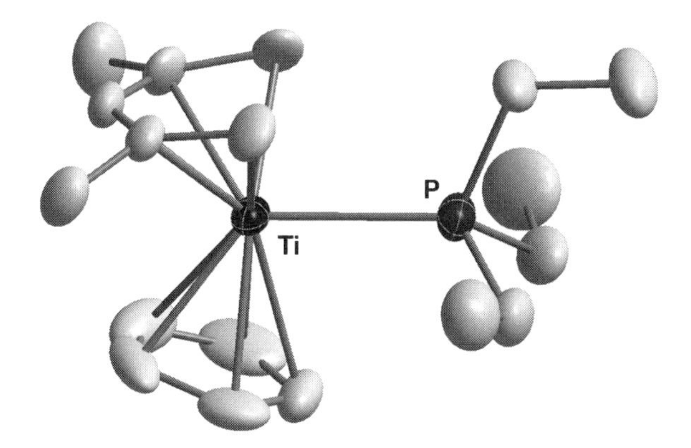

FIG. 11. Molecular structure of Ti$(C_5H_5)(2,4\text{-}C_7H_{11})(PEt_3)$ (hydrogen atoms omitted).

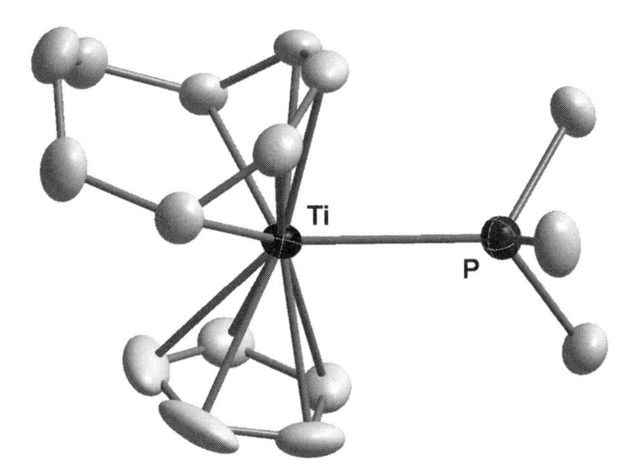

FIG. 12. Molecular structure of Ti$(C_5H_5)(c\text{-}C_8H_{11})(PMe_3)$ (hydrogen atoms omitted).

a longer Ti–P bond in $Ti[C_5H_3(^tBu)_2](6,6\text{-dmch})(PMe_3)$ relative to $Ti[C_5H_3(^tBu)_2](C_5H_7)(PMe_3)$, 2.632(1) Å vs. 2.527(1) Å, is consistent with the proposal that the 6,6-dmch ligand is, at least for the mono(ligand) complexes, more sterically demanding than typical (non-edge-bridged) pentadienyl ligands.

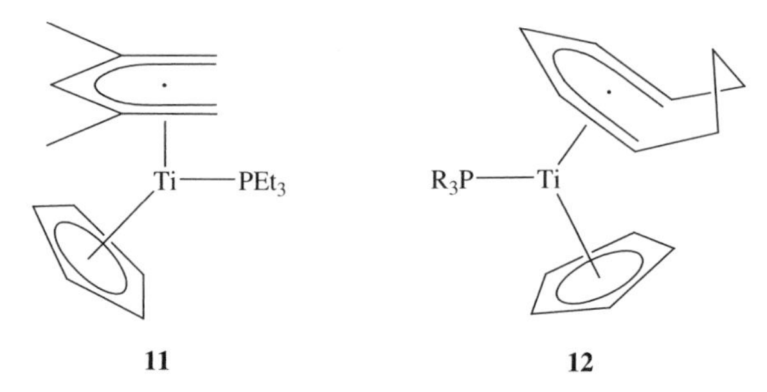

That the shortening of the Ti–C distances for pentadienyl vs. cyclopentadienyl ligands is an indication of stronger bonding for the former, non-aromatic ligand, was substantiated by *ab initio* and extended Hückel MO studies. The *ab initio* studies reproduced quite well the general structural features of $Ti(C_5H_5)(2,4\text{-}C_7H_{11})(PEt_3)$. The extended Hückel studies revealed the total Ti–C bond overlap population for the open dienyl ligand carbon atoms to be 0.35 vs. only 0.11 for the C_5H_5 ligand. A calculation in which the Ti–C bond distances for the two ligands were reversed did not lead to a noticeable change for the pentadienyl ligand, but did actually lead to a *decrease* in the Ti–C bond overlap population for the C_5H_5 ligand to 0.03. The calculations also estimated the net charges on the C_5H_7 and C_5H_5 ligands to be ca. −0.39 and −0.18, consistent with the δ acidity of open pentadienyl ligands.

The isolation of half-open zirconocenes was not as straightforward as for titanium. The main problem appeared to be that in the preparation of $Zr(C_5H_5)Cl_3$ from $Zr(C_5H_5)_2Cl_2$, over-chlorination could occur,[55] leading to significant contamination by Zr(IV) side products. Subsequently, however, it was found that $Zr(C_5H_5)Cl_2Br$[56] could be isolated more readily, although traces of species such as $[Zr_3(C_5H_5)_3X_3(\mu_3\text{-}X)_2(\mu\text{-}X)_3]^+[Zr_2X_9]^-$ (X = Cl, Br) could be isolated.[57] The monometallic starting material has allowed for the isolation of a variety of half-open zirconocenes, incorporating the $2,4\text{-}C_7H_{11}$, 6,6-dmch, and the $c\text{-}C_8H_{11}$ ligands [Eqs. (19)–(21))]. As in the case of $Zr(6,6\text{-dmch})_2(PMe_3)_2$, $Zr(C_5H_5)(6,6\text{-dmch})(PMe_3)_2$ reacts readily with CO to replace one phosphine ligand [Eq. (22)].[37]

$$Zr(C_5H_5)Cl_2Br + 3K(2,4\text{-}C_7H_{11}) + \text{dmpe} \rightarrow Zr(C_5H_5)(2,4\text{-}C_7H_{11})(\text{dmpe}) \quad (19)$$

$$\begin{gathered} Zr(C_5H_5)Cl_2Br + 3K\text{Pdl} + 2PMe_3 \rightarrow Zr(C_5H_5)(\text{Pdl})(PMe_3)_2 \\ \text{Pdl} = 6,6\text{-dmch},\ 2,6,6\text{-tmch} \end{gathered} \quad (20)$$

$$\begin{gathered} Zr(C_5H_5)Cl_2Br + 3K(c\text{-}C_8H_{11}) + PR_3 \rightarrow Zr(C_5H_5)(c\text{-}C_8H_{11})(PR_3) \\ R = \text{Me, Et} \end{gathered} \quad (21)$$

$$Zr(C_5H_5)(6,6\text{-dmch})(PMe_3)_2 + CO \rightarrow Zr(C_5H_5)(6,6\text{-dmch})(CO)(PMe_3) \quad (22)$$

The molecular structures of $Zr(C_5H_5)(2,4\text{-}C_7H_{11})(\text{dmpe})$[58] (Fig. 13), $Zr(C_5H_5)(2,6,6\text{-tmch})(PMe_3)_2$[59] (Fig. 14), $Zr(C_5H_5)(6,6\text{-dmch})(CO)(PMe_3)$,[37] $Zr(C_5H_5)$

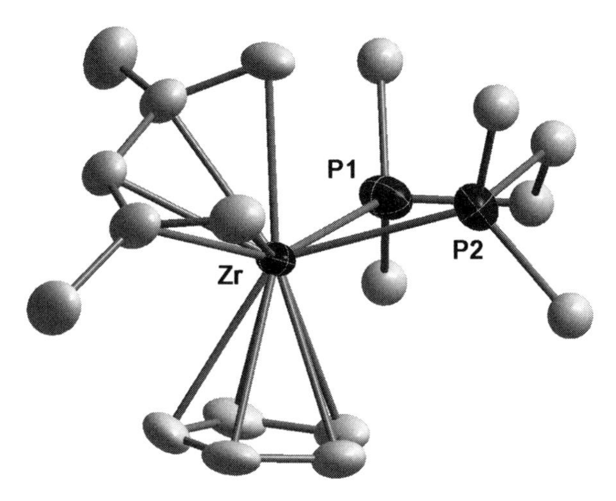

FIG. 13. Molecular structure of $Zr(C_5H_5)(2,4\text{-}C_7H_{11})(dmpe)$ (hydrogen atoms omitted).

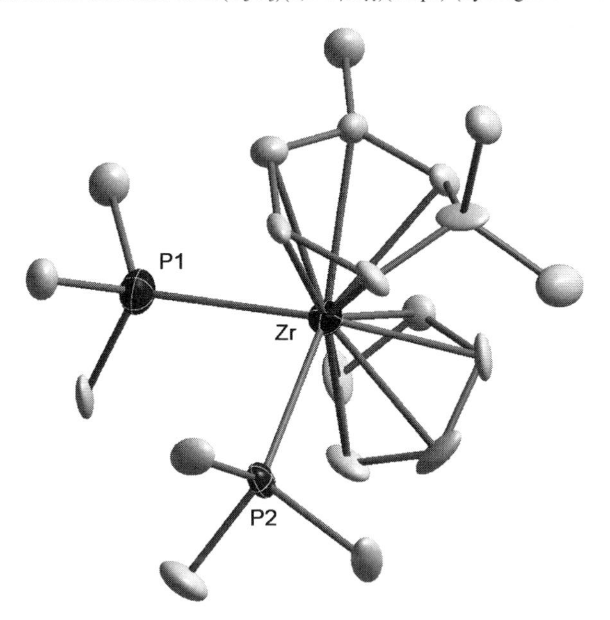

FIG. 14. Molecular structure of $Zr(C_5H_5)(2,6,6\text{-tmch})(PMe_3)_2$ (hydrogen atoms omitted).

$(c\text{-}C_8H_{11})(PMe_3)$, and $Zr(C_5H_5)(c\text{-}C_8H_{11})(PEt_3)$[54] (Fig. 15) have been determined. Most surprising is the observation that only one phosphine center could coordinate for the $c\text{-}C_8H_{11}$ complex, revealing a subtle distinction between this ligand and 6,6-dmch, likely related to the C1–C5 and M–dienyl plane separations.

Interestingly, the $c\text{-}C_8H_{11}$ complexes are a beautiful deep blue color, while the other complexes are red. As with the half-open titanocenes, the Zr–C bond distances for the electronically open dienyl ligands are shorter than those for the C_5H_5 ligands, the values for the five compounds being 2.42 Å vs. 2.54 Å, 2.47 Å vs. 2.53 Å, 2.47 Å vs. 2.51 Å, 2.24 Å vs. 2.38 Å, and 2.34 Å vs. 2.51 Å, respectively.

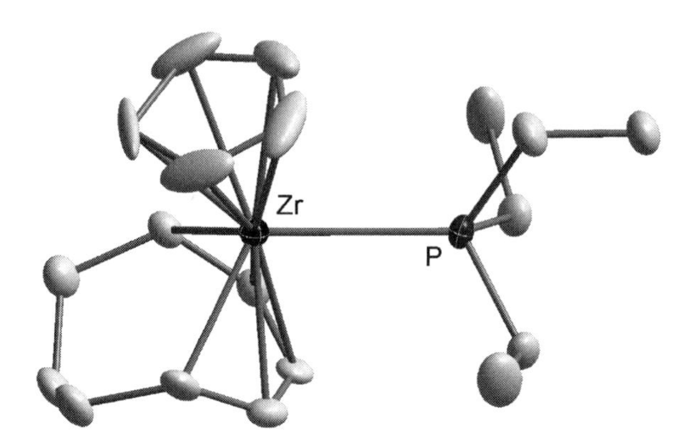

FIG. 15. Molecular structure of $Zr(C_5H_5)(c\text{-}C_8H_{11})(PEt_3)$ (hydrogen atoms omitted).

A noteworthy point is the observation that on replacing one PMe_3 ligand in $Zr(6,6\text{-dmch})_2(PMe_3)_2$ by CO, the degree of shortening has decreased, as has been observed in related cases.[60] This is consistent with the proposal that the strong π-acceptor CO competes more with pentadienyl than C_5H_5, in accord with the δ-acidic nature of pentadienyls. Also of note is the fact that greater electron deficiencies seem to lead to greater favorabilities for the open ligand. Thus, the shortening of the M–C bonds to the electronically open dienyl ligand relative to those for the Cp ligand was ca. 0.21 Å for the 14 electron **10**, but dropped to an average of ca. 0.12 Å for the six 16 electron half-open titanocene complexes. For the 16 electron $Zr(C_5H_5)(c\text{-}C_8H_{11})(PR_3)$ complexes (R = Me, Et), shortenings of ca. 0.14–0.17 Å were observed, relative to ca. 0.05–0.12 Å for the two 18 electron half-open zirconocenes, $Zr(C_5H_5)(2,4\text{-}C_7H_{11})(dmpe)$ and $Zr(C_5H_5)(2,6,6\text{-tmch})(PMe_3)_2$. This appears more likely to be an electronic influence of some sort, although one could propose that the greater steric crowding induced by the incorporation of an additional phosphine donor site could affect the more sterically demanding pentadienyl ligands more than the generally smaller Cp ligands.

The availability of the $Zr(C_5H_5)(6,6\text{-dmch})X_2$ and $Zr(6,6\text{-dmch})_2X_2$ complexes, and the fact that 6,6-dmch is electronically intermediate between C_5H_5 and typical pentadienyl ligands, have provided an opportunity to attempt the preparation of N_2 complexes of an open or half-open zirconocene complex. In fact, reduction of these complexes with sodium amalgam under a nitrogen atmosphere was observed to lead to deeply colored (blue to purple) solutions, which changed color under vacuum.[61] While crystalline products were not readily isolated, this at least provides an indication that N_2 coordination to metal pentadienyl complexes can occur.

At the present time no half-open hafnocene complexes have been reported, but their syntheses should be feasible.

C. *Half-Open Metallocenes with Halide and Related Ligands*

Just as $Zr(6,6\text{-dmch})_2(PMe_3)_2$ reacts with 1,2-dihaloalkanes, the half-open complex $Zr(C_5H_5)(6,6\text{-dmch})(PMe_3)_2$ can likewise be converted to higher valent complexes [Eq. (23)].[62]

$$Zr(C_5H_5)(6,6\text{-dmch})(PMe_3)_2 + 1,2\text{-}C_2H_4X_2 \rightarrow Zr(C_5H_5)(6,6\text{-dmch})X_2 \quad X = Cl, Br, I \tag{23}$$

Similar to the $Zr(6,6\text{-dmch})_2X_2$ analogues, these species are orange (X = Cl) to red (X = I) in color. Each has been structurally characterized, as has been a 3-Me_3Si-substituted diiodide analogue[61] (e.g., Fig 16), and found to adopt similar arrangements (**13**), with the two halides oriented by one of the external dienyl C–C bonds (C4–C5). The Zr–dmch bonding follows the pattern found for the Zr(6,6-dmch)$_2X_2$ compounds, having a short Zr–C3 bond, with the subsequent atoms becoming further removed. This pattern may again be traced to the likely poor overlap generated between the contracted Zr(IV) orbitals and the π molecular orbitals of the 6,6-dmch ligand.

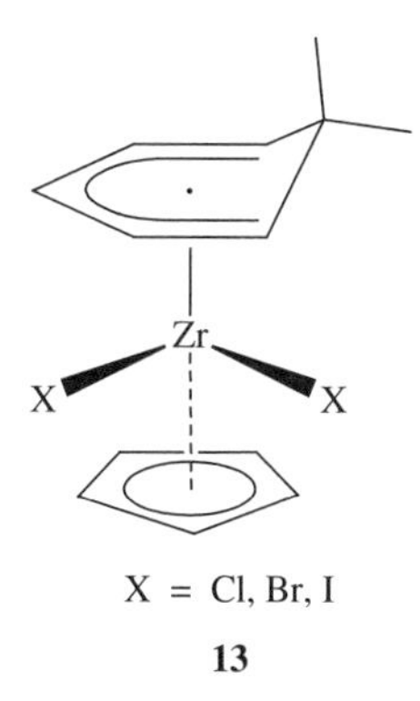

13

Of greatest interest for these compounds is that they provide the first opportunity to compare the bonding of a higher valent metal center with (electronically) open

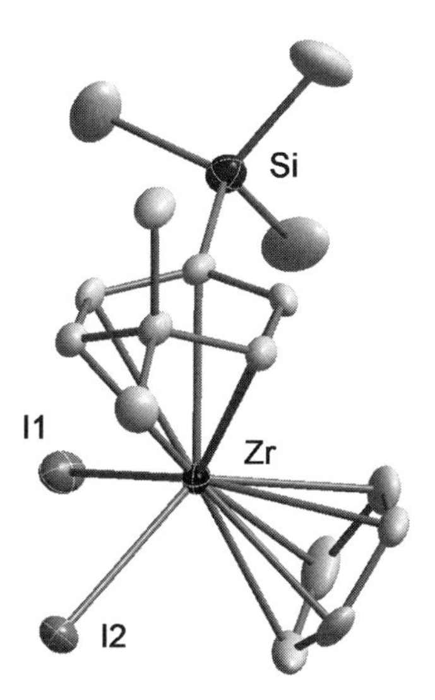

FIG. 16. Molecular structure of $Zr(C_5H_5)(3\text{-}Me_3Si\text{-}6,6\text{-dmch})I_2$ (hydrogen atoms omitted).

and closed dienyl ligands. While all lower valent (+2) half-open titanocenes and zirconocenes have significantly shorter M–C distances for their electronically open dienyl ligands, the shortening being as large as 0.2 Å or as small as 0.06 Å for $Zr(C_5H_5)(2,6,6\text{-tmch})(PMe_3)_2$, the $Zr(C_5H_5)(6,6\text{-dmch})X_2$ compounds all show exactly the opposite pattern, with the Zr–C (C_5H_5) bonds being shorter, by ca. 0.09 Å. As pentadienyl ligands should be both better donors and acceptors than C_5H_5,[5] the explanation for this dramatic reversal in bonding favorability would seem to lie in the contracted nature of the higher valent metal orbitals, particularly given that even with the lower valent metals there is already ample evidence of somewhat ineffective overlap.[5] These data have therefore confirmed earlier suspicions that the reason for the apparent inability to isolate the long sought after higher valent species had to do with orbital overlap issues, as well as to the loss in δ backbonding.

Accompanying PES and MO studies have yielded useful additional insight into the bonding of these species. The PES data revealed that in these higher valent d^0 complexes, for which δ backbonding should be negligible, the 6,6-dmch and C_5H_5 ligands appeared nearly equal in their electronegativities. Each dihalide compound has a similar first ionization energy, which MO results indicate arises from a 6,6-dmch-based HOMO. These dmch-based HOMOs are ca. 0.5 eV higher in energy than the C_5H_5 HOMOs, leading to the orange-to-red colors of the dmch-containing compounds. The LUMO is primarily metal-localized, having the appearance of being able to lead to a δ bonding interaction.[5] In fact, for $Zr(C_5H_5)(2,6,6\text{-tmch})(PMe_3)_2$, the HOMO (below) was found to involve a prominent δ-bonding interaction, in accord with the above considerations. Mayer bond order analyses for $Zr(C_5H_5)(6,6\text{-dmch})(PMe_3)_2$ and $Zr(C_5H_5)(6,6\text{-dmch})I_2$ found respective Zr-Cp and Zr-dmch bond orders of 1.33 vs. 2.26 for the phosphine complex and 1.60 vs. 1.68 for the iodide complex,[63] in line with the structural data, and also revealing opposite Cp and dmch preferences for Zr(II) and Zr(IV).

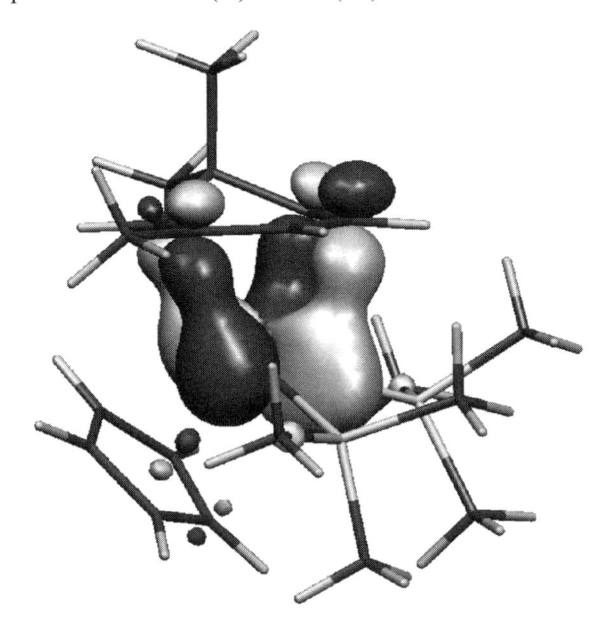

Another interesting example of a higher valent half-open metallocene was obtained from the reaction of $Zr(C_5H_5)(6,6\text{-dmch})(PMe_3)_2$ with two equivalents of PhC_2SiMe_3.[59] As in the reaction of $Zr(6,6\text{-dmch})_2(PMe_3)_2$ with one equivalent of alkyne, the 6,6-dmch ligand in the isolated product did not undergo any coupling reaction. In this case, the two alkynes coupled together, yielding a zirconacyclopentadiene fragment (**14**, Fig. 17). The product also contained a small amount of a second isomer, shown spectroscopically to have the orientation of one alkyne reversed. Structural data reveal much the same as had been found for the other high-valent half-open zirconocenes. Thus, the Zr–C (C_5H_5) bonds were shorter than those for the 6,6-dmch ligand, 2.521 Å vs. 2.585 Å, the average Zr–C(3), Zr–C(2,4), and Zr–C(1,5) bond distances show an increasing trend from 2.505(2) to 2.538(2) to 2.672(1) Å, and there is a short–long–long–short pattern in delocalized C–C bond lengths of the 6,6-dmch ligand.

$SiMe_3$
Ph
Zr
Ph
$SiMe_3$

14

While boratabenzene analogues of open metallocenes were considered in Section IIC, a series of half-open analogues has also been reported.[42] Thus, reactions of $Zr(C_5Me_5)Me_2Cl$ with C_5H_5BR anions (R = N^iPr_2, OEt, Ph) were found to lead to $Zr(C_5Me_5)(C_5H_5BR)Me_2$ complexes. All three complexes reacted quantitatively

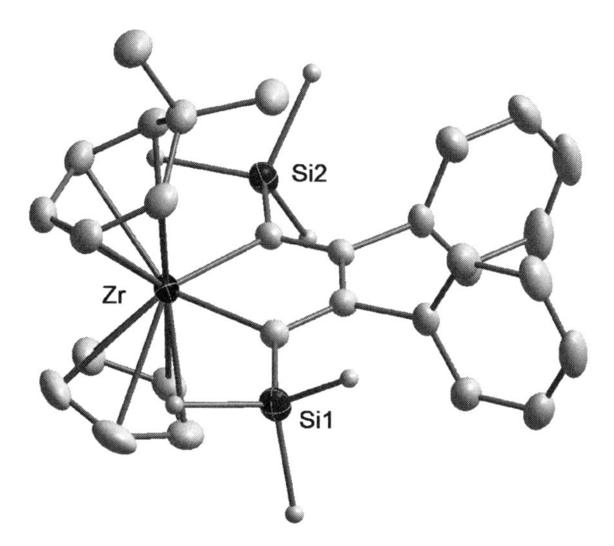

FIG. 17. Molecular structure of a zirconacyclopentadiene complex (hydrogen atoms omitted, methyl groups simplified).

with $B(C_6F_5)_3$ to yield the appropriate $[Zr(C_5Me_5)(C_5H_5BR)Me]^+[MeB(C_6F_5)_3]^-$ "salts." Structural data for two of the compounds were obtained, each revealing some residual interaction between the metal center and the abstracted methyl group, with the Zr–C distances being ca. 2.55 Å, vs. ca. 2.27 Å for the remaining methyl group. It was also reported that reduction of the $Zr(C_5Me_5)(C_5H_5BR)Cl_2$ complexes (R = N^iPr_2, OEt, Me, Ph) by magnesium in the presence of CO led to the formation of the dicarbonyl complexes $Zr(C_5Me_5)(C_5H_5BR)(CO)_2$, which, like simple pentadienyl analogues, displayed higher C–O stretching frequencies than their C_5H_5 or C_5Me_5 analogues.

D. *Other Complexes with Additional Delocalized π Ligands*

The 16 electron complex Ti(η^5-*c*-C_7H_9)(η-C_7H_7) has been isolated from several different synthetic approaches, including the reduction of $TiCl_3$ in the presence of cycloheptatriene, the reaction of titanium vapor with cycloheptatriene, and the reaction of "$TiCl_2$" with the cycloheptadienyl anion.[24,25] Although a structural study was carried out, it was subject to severe disorder in the dienyl ligand. The fact that several different reactions led to the same complex was interpreted as an indication that this, and a number of similar reactions, are subject to an element of thermodynamic control, through which the formation of at least one aromatic η-C_nH_n ligand is favored.[25] These reactions seem less common for *c*-C_8H_{11} ligands, although the preparation of their half-open titanocenes can lead to varying amounts of accompanying $Ti(C_5H_5)(C_8H_8)$.[64]

Coupling reactions involving one dienyl ligand in open metallocenes have led to complexes retaining one pentadienyl ligand, together with an allyl or diene group. These species will be described in Section VB, with related coupling products.

A quite bizarre complex has been isolated from the reaction of $Zr(C_5H_4R)_2(Me)(NMe_2Ph)^+$ species (R = H, Me_3Si, Me_3C) with two equivalents of 2-butyne.[65] The resulting cationic complexes, e.g., **15** (Fig. 18), incorporate two cyclopentadienyl ligands, in addition to an unusual η^5-*S*-pentadienyl ligand. Generally, η^5-*S* coordination has been favored electronically for 18 electron complexes when the metal center had a d^4 configuration.[33b,44,66] It is possible in this case, as well as at least one other,[67] that the η^5-*S* coordination is favored by steric effects. Unlike the other η^5-*S* species, however, **15** displays a very short Zr–C bond for C(1) 2.315(7) Å, while the other four bond lengths range from 2.60 to 2.76 Å. Together with a very pronounced long–short–long–short alternation in the C–C bonds, this suggests that valence bond contributor **15** plays a dominant role in the bonding.

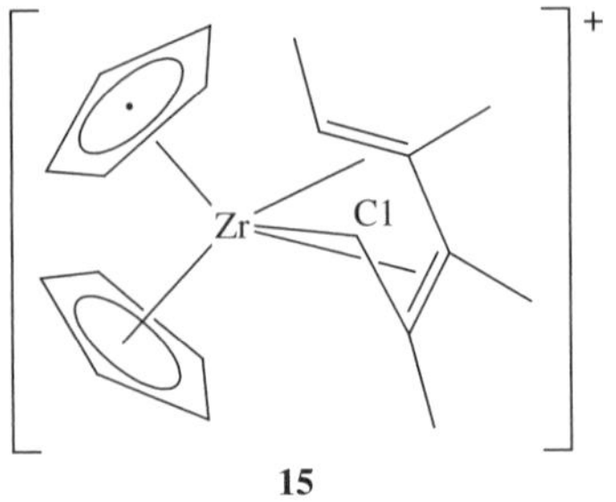

15

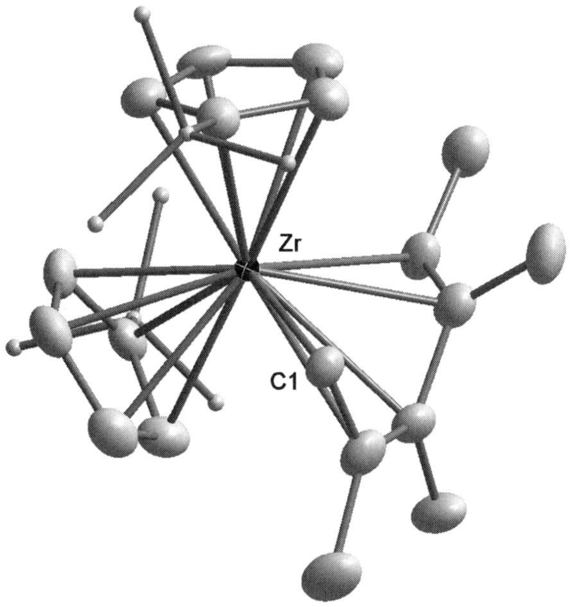

FIG. 18. Molecular structure of the $Zr[C_5H_4(CMe_3)]_2$(2,3,4-trimethylhexadienyl) cation, **15** (hydrogen atoms omitted; CMe_3 groups simplified).

E. *Other Complexes with Additional Neutral Ligands*

This category of complex is currently quite limited in number, although one could expect such species to display very rich reaction chemistry. A potentially useful route has been developed [Eq. (24)] which yields (η^5-dienyl)$Ti(CO)_4^-$ complexes, **16** (R = H, OMe).[68] However, while **16** is depicted as a pentadienyl complex, the small 14.9° deviation of C2 from the dienyl plane for the R = H complex seems to implicate a more than negligible contribution from an η^6-arene resonance form, with negative charge then being placed on C1, and various locations on the two uncoordinated arene rings. Quite possibly, the replacement of these two phenyl groups by H or simply alkyl substituents would lead to a greater contribution of the η^5 bonding mode.

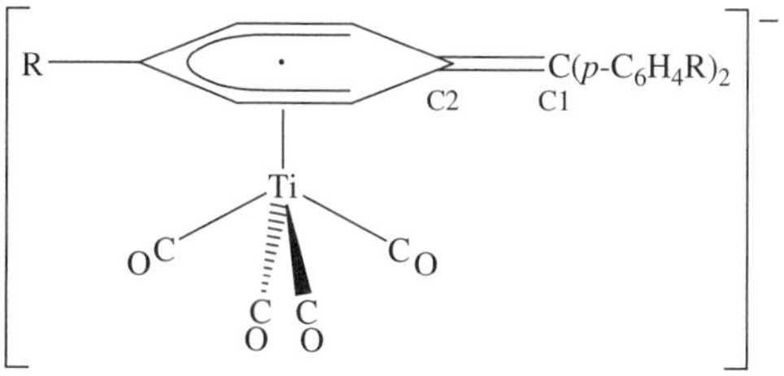

16

$$Ti(CO)_6^{2-} + (4\text{-R-}C_6H_4)_3CCl \rightarrow Ti[\eta^5\text{-}C(4\text{-}C_6H_4R)_3](CO)_4^- \quad R = H, OMe \qquad (24)$$

An analogous complex, $Ti(2,4\text{-}C_7H_{11})(CO)_4^-$, appears to have been prepared through the reaction of $Ti(CO)_5$(dmpe) and $2,4\text{-}C_7H_{11}^-$. Some similar C–O stretching frequencies (1785, 1914 cm^{-1}) to those of the known C_5H_5 analogue were observed,[69] but full characterization of this species has not been achieved. Given the rich chemistry observed for $Mo(2,4\text{-}C_7H_{11})(CO)_3^-$, it can be expected that these related titanium, and perhaps ultimately zirconium and hafnium complexes, would also display interesting chemistry.[70]

F. *Other Complexes with Halide and Related Ligands*

As an outgrowth of the great interest in Cp-containing constrained geometry catalysts, several related cyclohexadienyl-containing analogues, **17**, have been prepared. These species are of further note in that they are currently the only η^5-pentadienyl complexes of Ti(IV).[71] Such complexes would generally be expected to undergo loss of a pentadienyl ligand, leading to a lower metal oxidation state. However, as noted earlier, the 6,6-dmch ligand is actually intermediate electronically between Cp and pentadienyl, and its short C1–C5 separation would tend to offset problems of maintaining reasonable overlap with the contracted orbitals of a higher valent metal center. Furthermore, the fact that the dmch ligand is tethered to an amide ligand may prevent the loss of the 6,6-dmch ligand, and the tether may help to inhibit intermolecular couplings between dienyl ligands. Finally, the strong π donor ability of the amide ligand can reduce the net charge on the titanium center, thereby improving orbital overlap and perhaps even allowing for a small degree of δ backbonding. Both chloride complexes have been structurally characterized (e.g., Fig. 19), and reveal the patterns of M–C and C–C bond distances typical of these higher valent complexes.

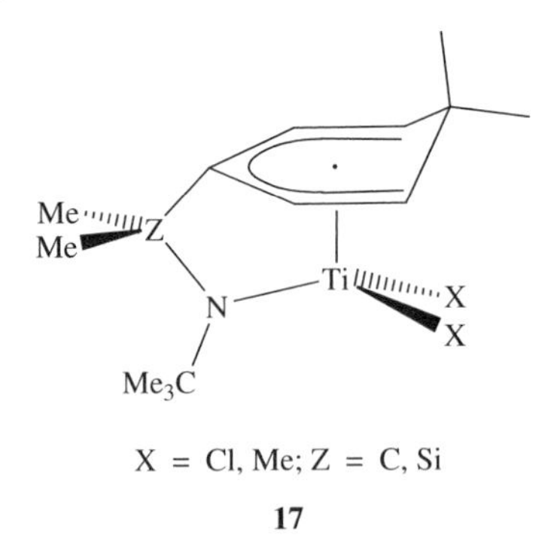

X = Cl, Me; Z = C, Si

17

While $Zr(6,6\text{-dmch})_2(PMe_3)_2$ complexes react with dihaloalkanes to yield Zr(IV) dienyl complexes, $Ti(6,6\text{-dmch})_2$ yields a Ti(III) complex [Eq. (25)].[72] A structural study (Fig. 20) revealed the compound to crystallize as a one-dimensional chloride-bridged polymer. Even though the oxidation state of +3 is not especially high, one again sees the M–C and C–C bonding patterns typical of the higher oxidation state species. It seems quite possible that $[Ti(C_5H_5)Cl_2]_n$[73] adopts a similar arrangement.

$$Ti(6,6\text{-dmch})_2 + 1,2\text{-}C_2H_4Cl_2 \rightarrow [Ti(6,6\text{-dmch})Cl_2]_\infty \quad (25)$$

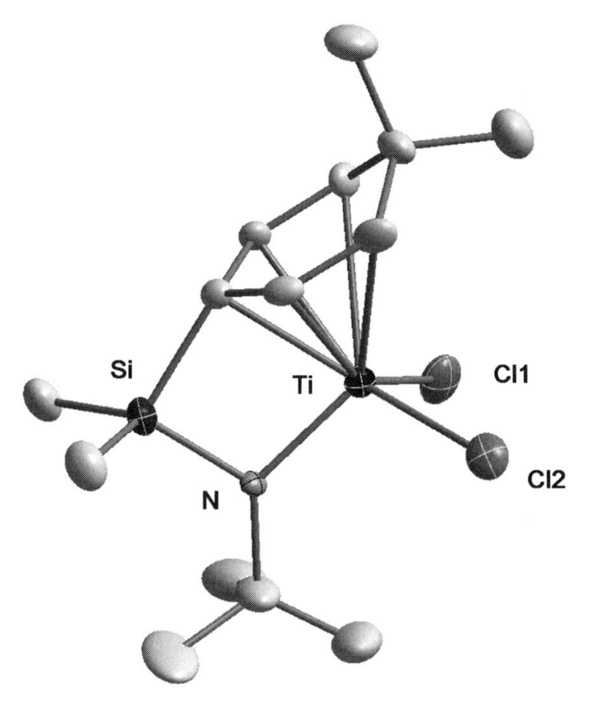

FIG. 19. Molecular structure of the X = Cl, Z = Si constrained geometry catalyst precursor **17**.

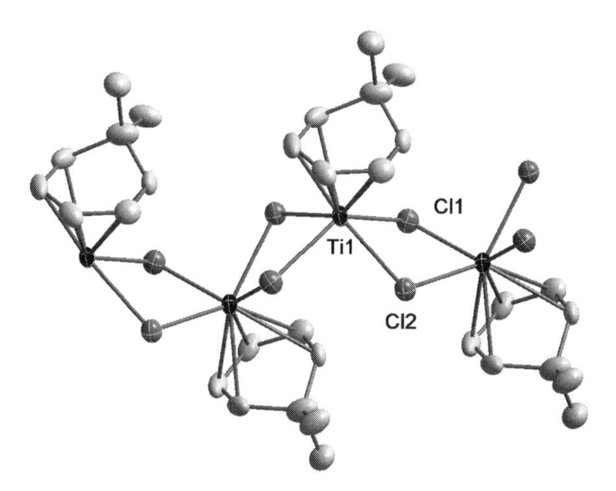

FIG. 20. Molecular structure of $[Ti(6,6\text{-dmch})Cl_2]_\infty$ (hydrogen atoms omitted).

The reaction of $Ti(1,3\text{-}C_5H_3\,^tBu_2)(6,6\text{-dmch})(PMe_3)$ with SF_6 has been reported to yield $\{Ti(1,3\text{-}C_5H_3\,^tBu_2)F_2\}_4$, presumably *via* a $Ti(1,3\text{-}C_5H_3\,^tBu_2](6,6\text{-dmch})F_2$ intermediate.[53] Given the generally low stability of higher valent metal–pentadienyl complexes, it is not surprising that the dmch ligand would be lost from the titanium center, even though Zr(IV) dmch complexes are known, as well as a Ti(IV) dmch complex, the stabilization of which was discussed above.

G. *Other Complexes with Neutral and Halide or Related Ligands*

The reduction of $ZrCl_4(dmpe)_2$ by Na/Hg in the presence of 1,3-cyclohexadiene[74] or 1,3-cyclooctadiene[75] leads to Zr(dienyl)(H)$(dmpe)_2$ complexes [Eq. (26)]. A structural determination for the cyclooctadienyl complex was reported (Fig. 21). The structure may be described as pseudooctahedral, with the c-C_8H_{11} ligand opposite to the hydride ligand (d(Zr–H)$\sim$1.67 Å).

$$\begin{aligned} &ZrCl_4(dmpe)_2 + Na/Hg + diene \rightarrow Zr(dienyl)(H)(dmpe)_2 \\ &diene = 1,3\text{-}C_6H_8, 1,3\text{-}C_8H_{12}; dienyl = c\text{-}C_6H_7, c\text{-}C_8H_{11} \end{aligned} \quad (26)$$

Another pseudooctahedral complex results from the reaction of Zr(6,6-dmch)$_2$Br$_2$ with dmpe [Eq. (27)].[72]

$$Zr(6,6\text{-dmch})_2Br_2 + dmpe \rightarrow Zr(6,6\text{-dmch})Br_3(dmpe) + ? \quad (27)$$

The reaction [Eq. (27)] occurs immediately on mixing, which would not be expected at all for a Cp-like complex, but is quite consistent with the structural data for high-valent metal–pentadienyl complexes, which have revealed the open dienyl ligands to be in something of a state of slippage from the metal coordination sphere. This seems to represent an intermediate stage in the usual reduction process that takes place on the addition of pentadienyl anions to higher valent metal species [e.g., Eqs. (3), (6), (7), (9), (14)–(16), and (19)–(21)]. A deep purple intermediate is initially observed, which might be a simple adduct of dmpe, with η^1 coordination by the 6,6-dmch ligand. Subsequently, either loss of the 6,6-dmch ligand or a ligand redistribution reaction occurs, leading to the formation of Zr(6,6-dmch)Br$_3$

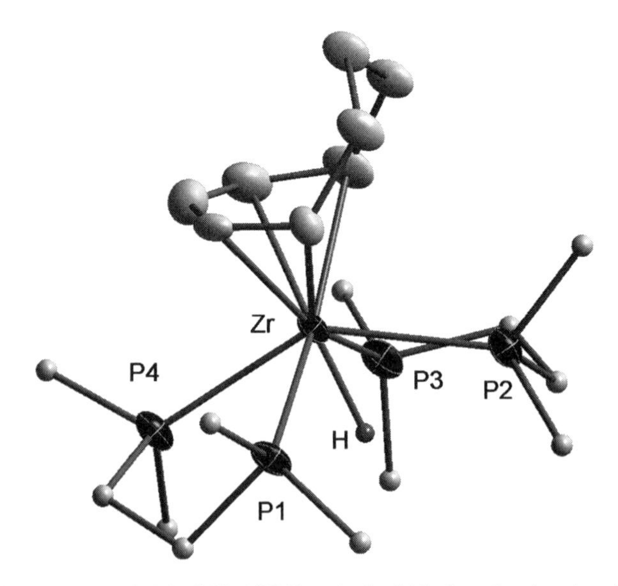

FIG. 21. Molecular structure of Zr(c-C_8H_{11})(H)$(dmpe)_2$ (hydride ligand reduced and phosphine substituents simplified).

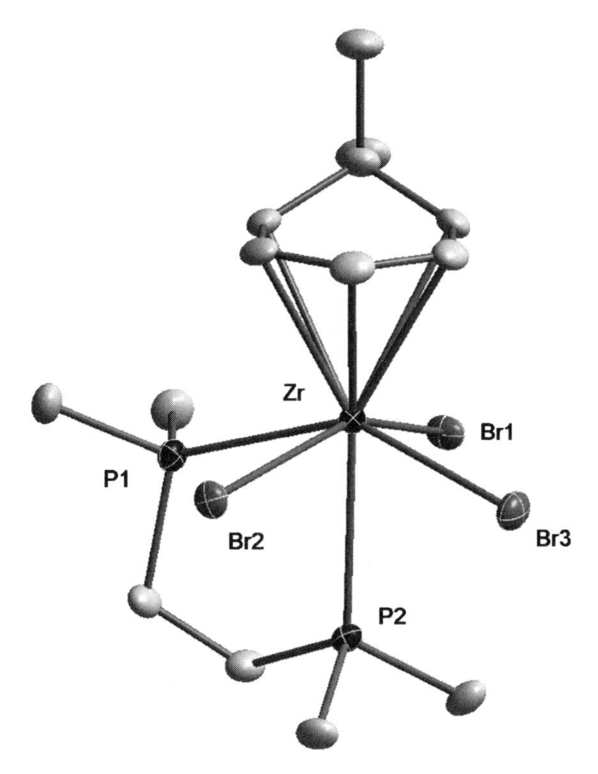

FIG. 22. Molecular structure of $Zr(6,6\text{-dmch})Br_3(dmpe)$ (hydrogen atoms omitted).

(dmpe) as one of the products. A structural study (Fig. 22) reveals this product to have adopted the relatively common *mer*, *cis* arrangement. The Zr–dmch bonding parameters follow the now well-established norms for higher valent species.

A related complex, $Zr(6,6\text{-dmch})I_3(PMe_3)$, was actually isolated as a byproduct during the synthesis of $Zr(6,6\text{-dmch})_2I_2$ from $Zr(6,6\text{-dmch})_2(PMe_3)_2$ and 1,2-diiodoethane [Eqs. (11) and (28) [39]]. This may well have been formed from a reaction of the desired diiodide complex with the liberated PMe_3, analogous to the process involved in Eq. (27). A structural study was carried out, but suffered from a severe disorder. Nonetheless, it appeared to reveal a four-legged piano stool geometry. Given the facility of these reactions, it is possible that they could account for the relatively modest yields of $Zr(6,6\text{-dmch})_2X_2$ and $Zr(C_5H_5)(6,6\text{-dmch})X_2$ complexes obtained as described by Eqs. (11) and (23). Conceivably, the appropriate dmch ligand sources could lead to the selective introductions of 6,6-dmch ligands with ZrX_4 or $Zr(C_5H_5)X_3$ complexes, yielding the desired higher valent species more directly and in higher yields.[39]

$$Zr(6,6\text{-dmch})_2(PMe_3)_2 + 1,2\text{-}C_2H_4I_2 \rightarrow Zr(6,6\text{-dmch})_2I_2 + Zr(6,6\text{-dmch})I_3(PMe_3) \quad (28)$$

IV
MISCELLANEOUS COMPLEXES

A. *Heteroatom-Containing Pentadienyl Complexes*

While there are numerous heteroatom-containing analogues of pentadienyl,[76] acac being an especially well-utilized example, the electropositive nature of the group 4 transition metals would seemingly greatly favor σ coordination through one or two heteroatoms rather than through the π system. Only carbon-bound complexes fall in the scope of this review, and a variety of these complexes has been reported. In some cases, especially when the heterodienyl fragment is part of a larger macrocycle, as in **18**,[77] it may not be clear to what extent the macrocycle's framework may contribute to the metal center's approach to the carbon atoms. In other cases, the mode of coordination often appears related to the steric environment about the metal center, as well as to the relative donating properties of the additional ligands. Many of the observed cases of at least partial π coordination have thus far involved NacNac types of ligands, **19**. In the first structural study of such a complex, $Zr(NacNac)Cl_3$ (one R = ^{t}Bu, the other Ph; R′ = $SiMe_3$), Zr–N distances of 2.138(5) and 2.187(5) Å were observed, along with relatively short Zr–C distances, 2.61(1), 2.54(1), and 2.60(1) Å, reflecting some degree of interaction with the π system.[78] Other structural studies have shown that the $Zr(NacNac)_2X_2$ complexes (R = Me, R′ = Ph, X = NMe_2 or Cl) do not display any significant π interaction, as the metal center lies nearly in the ligand plane for X = NMe_2, and only deviates slightly for X = Cl. In contrast, however, a structural study of $Zr(C_9H_7)(NacNac)Cl_2$ (R = CH_3, R′ = Ph, C_9H_7 = indenyl) did reveal a significant deviation,[79] leading to Zr–C distances ranging from 2.703(2) to 2.748(2) Å. A somewhat similar coordination mode was subsequently found for $\{[Zr(C_5H_5)(NacNac)]_2(\mu\text{-}Cl)_2\}^{2+}$, with the Zr–C distances ranging from 2.651(3) to 2.706(3) Å.[80] A much bulkier NacNac ligand, with R = ^{t}Bu and R′ = 2,6-$^i Pr_2C_6H_3$ has also been found to lead to complexes having some degree of π coordination. Included in this group are $Zr(NacNac)Cl_3$ (Zr–C, 2.540(3)–2.867(4) Å), $Zr(NacNac)(OTf)_3$ (Zr–C, 2.508(7)–2.770(6) Å), $Zr(NacNac)(CH_2CMe_3)_2Cl$ (Zr–C, 2.612(7)–2.936(8) Å), and $Zr(NacNac)(CH_2SiMe_3)_2Cl$ (Zr–C, 2.582(5)–2.913(6) Å).[81a]

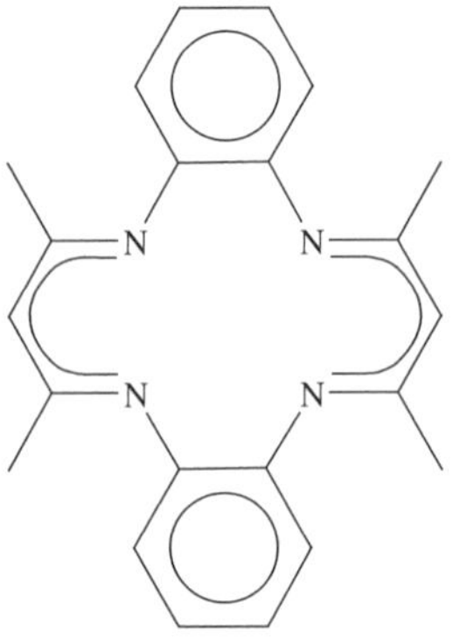

N N N N

18

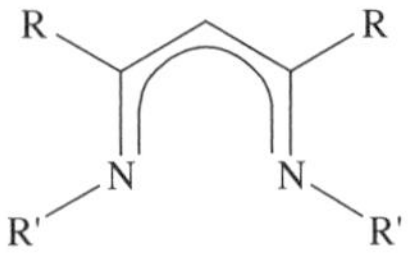

R R N N R′ R′

19

Interesting C–N bond activations have been observed in a number of cases when NacNac ligands are bound to group 4 metal centers.[81] While the precursors generally involve $\sigma(N),\sigma(N')$ coordination of the heterodienyl, intermediates of increased hapticity are naturally implicated, leading *via* C–N cleavage to the formation of imido ligands and metallapyrrole arrangements. Thus, from the reactions of $Zr(NacNac)Cl_3$ with two equivalents of KC_8 or of $Zr(NacNac)(OTf)_3$ with two equivalents of $LiCH_2CMe_3$ or $LiCH_2SiMe_3$, each product was found to contain a terminal imido ligand as well as the "Nac^{t}Bu" ligand, **20**.[81a] The chloro complex was converted to Zr(Nac^{t}Bu)(NR′)Cl(THF), while the triflato complex was converted to $[Zr(Nac^tBu)(NR')]_2(\mu\text{-}OTf)_2$. In a similar manner, the more sterically encumbered "TbtNacNacMes" (Tbt = 2,4,6-tri(bis(trimethylsilyl)methylphenyl, Mes = 2,4,6-trimethylphenyl) substituted NacNac ligand forms the complexes $TiCl_3$(TbtNacNacMes) or MCl_3(THF)(TbtNacNacMes) (M = Zr, Hf), which on reduction with KC_8 provide [Li(tmeda)][MCl_2(=NTbt)(NMesCMeCHCMe)] (M = Ti, Zr, Hf) and Hf(=NTbt)(THF)(NMesCMeCHCMe). Remarkably, hydrolysis results in liberation of the reconstituted HTbtNacNacMes pro-ligand.[81b] In a similar manner, $TiCl_2$(RNacNacR) (R = 2,6-$^iPr_2C_6H_3$) upon reduction (Na/K) provides the titanium(III) binuclear hydride bridged complex $K_2[Ti_2(\mu\text{-}H)_2(=NR)_2(RNCMeCHCMe)_2]$ in which the RNCMeCHCMe ligand adopts a tetrahapto coordination mode.[81c]

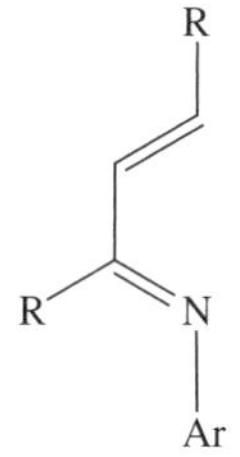

R = $t\text{-}C_4H_9$
Ar = $2,6\text{-}(i\text{-}C_3H_7)_2C_6H_3$

20

Other complexes have been reported in which part of the heterodienyl ligand is incorporated into an aromatic ring. When the heteroatom is part of the ring, as in various pyridine derivatives,[82] this generally leads to non-planarity of the heterodienyl ligand and a close approach of the metal atom to some of the carbon atoms. When the heteroatoms are not included in the aromatic ring, as in **21**, the ligand tends to be planar and the derived zirconium complexes have exhibited varying degrees of interaction with the heterodienyl π system.[83]

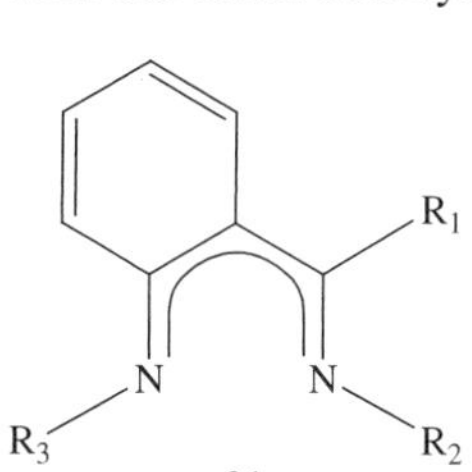

21

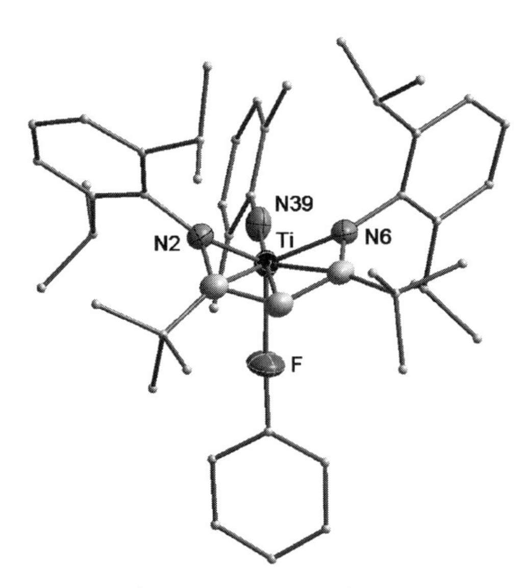

FIG. 23. Molecular structure of Ti(η^5-NacNac)(NAr)(FPh)$^+$ (hydrogen atoms omitted; organic substituent carbon atoms simplified).

More recently, related titanium complexes have received attention. Structural studies of various Ti(NacNac) complexes generally show modest interactions between the metal center and the π system.[84] Included in this group is an interesting example incorporating coordinated fluorobenzene [Eq. (29)].[85] The product incorporates an imido ligand, a NacNac ligand [ArNC(^{t}Bu)]$_2$CH (Ar = 2,6-^{i}Pr$_2$C$_6$H$_3$), and coordinated fluorobenzene (**22**, Fig. 23). The complex has been shown to be an active catalyst for various carbo-amination reactions. Some related complexes have been prepared, in which one of the terminal nitrogen atoms has been replaced by a carbon atom of a carbene ligand.[86] In one of these species, Ti(azadienyl)(NAr)(κ^2-OTf), containing an azadienyl ligand **23**, coordination has been described as occurring through the terminal carbon and nitrogen atoms, which can be regarded as entailing σ Ti–C and N:$\rightarrow$Ti interactions (distances, 2.263(1) and 2.015(1) Å, respectively). However, relatively close approaches are also made to the interior carbon atoms. Interestingly, one equivalent of ether can be incorporated into the complex, leading to η^1-azadienyl coordination through the nitrogen atom and to monodentate triflate coordination.

R = t-C$_4$H$_9$; Ar = 2,6-(i-C$_3$H$_7$)$_2$C$_6$H$_3$

22

23

$$\text{Ti(NacNac)(NAr)Cl} \xrightarrow{-\text{Cl}^-} \xrightarrow{\text{C}_6\text{H}_5\text{F}} \text{Ti}(\eta^5\text{-NacNac})(\text{NAr})(\text{FPh})^+ \qquad (29)$$

B. *Complexes with η^1 or η^3 Coordination*

A bis(η^1-pentadienyl) complex of zirconium has been prepared as shown in Eq. (30).[87] The η^1-pentadienyl ligands are attached to the zirconium center selectively in the product through the end furthest from the methyl substituents. There are also several more complicated species known in which what might be considered as highly modified η^1-pentadienyl coordination is present.[88]

$$Zr(C_5H_5)_2Cl_2 + 2K(2\text{-}MeC_5H_6) \rightarrow Zr(C_5H_5)_2(\eta^1\text{-}4\text{-}MeC_5H_6)_2 \qquad (30)$$

V
REACTION CHEMISTRY

A variety of reaction chemistry has already been covered, such as Lewis base coordination and dissociation, and redox chemistry. In those specific cases, one isolates new products that retain pentadienyl ligands. This section will therefore focus on reactions in which pentadienyl ligands are altered, as may occur *via* a coupling reaction, and on applications of metal pentadienyl compounds in materials and catalytic processes.

A. *Pentadienyl–Pentadienyl Couplings*

While the reaction of $Zr(C_5H_5)Cl_2Br$ with three equivalents of $K(2,4\text{-}C_7H_{11})$ in the presence of dmpe led to the formation of a half-open zirconocene (Section IIIB), a similar reaction employing $Mg(2,4\text{-}C_7H_{11})Br$ but without dmpe was found to lead to a more complicated half-open zirconocene, **24**, in which the role of dmpe is replaced by a diene ligand (Fig. 24).[58] In actuality, the diene is really present as an enediyl ligand, thus currently at least making this the only higher valent group 4 complex with a non-edge-bridged pentadienyl ligand. In accord with this formulation, the Zr–C bonds for the C_5H_5 ligand are on average ca. 0.1 Å shorter than the bonds for the pentadienyl ligand.

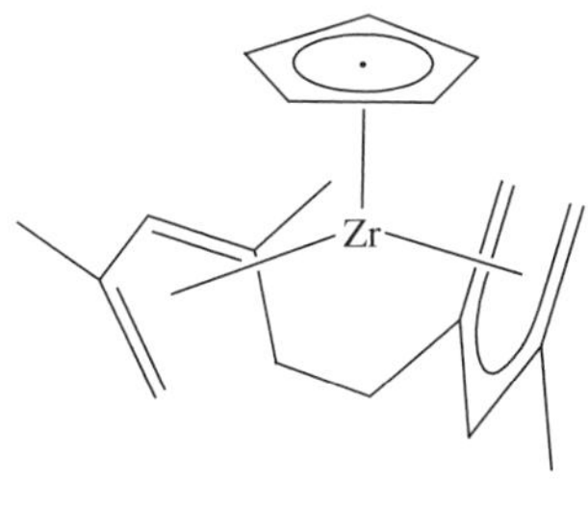

24

Attempts to prepare a $Hf(6,6\text{-dmch})_2(PMe_3)_2$ complex by the route employed for its zirconium analogue [Eq. (31)] have instead led to a complex of stoichiometry "$Hf(6,6\text{-dmch})_4(PMe_3)$" (Fig. 25), **25**.[89] In this complex, one 6,6-dmch ligand has

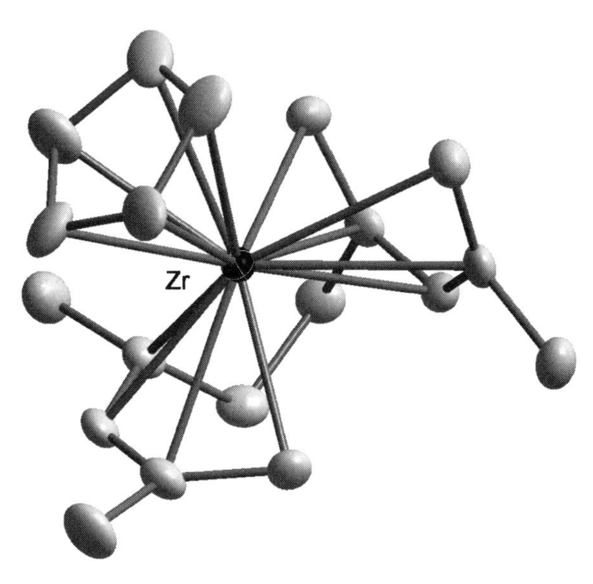

FIG. 24. Molecular structure of the half-open zirconocene diene complex, **24** (hydrogen atoms omitted).

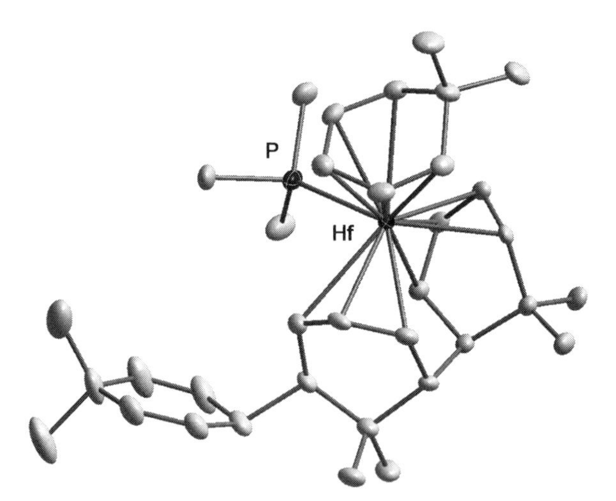

FIG. 25. Molecular structure of "Hf(6,6-dmch)$_4$(PMe$_3$)," **25** (hydrogen atoms omitted).

undergone coupling with two others, converting the latter to dienes, one of which coordinated to the hafnium center, while the central dmch fragment has become a coordinated allyl. The fourth dmch remained uncoupled, and fully coordinated to the metal center, yielding an 18 electron species. Although the approach of Eq. (31) was unsuccessful, it seems likely that more selective incorporation of the dmch ligands should be possible, leading to the desired Hf(6,6-dmch)$_2$(PR$_3$)$_n$ complexes.

$$HfCl_4 + 4K(6,6\text{-dmch}) + 2PMe_3 \rightarrow \text{“}Hf(6,6\text{-dmch})_4(PMe_3)\text{”} \qquad (31)$$

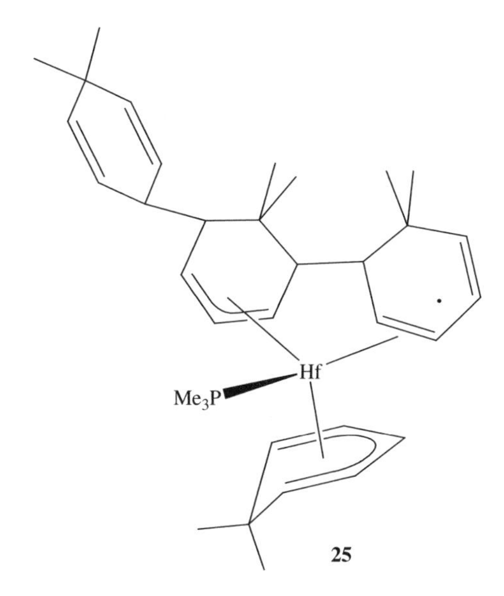

B. *Coupling Reactions with Unsaturated Organic Molecules*

Group 4 transition metal pentadienyl compounds undergo a rich variety of coupling reactions with just about any unsaturated organic molecule, with the apparent exception of olefins. This section will first examine the coupling reactions for non-edge-bridged pentadienyl ligands such as C_5H_7 and 2,4-C_7H_{11}, followed by their edge-bridged analogues, *c*-C_8H_{11} and 6,6-dmch. The reactions of imines, ketones, aldehydes, and nitriles are generally straightforward, and will be considered first for each class of dienyl ligand. Subsequently, reactions with isonitriles and alkynes will be detailed, which tend to be more complicated. To date, while reactions have been observed with esters, the resulting products have yet to be identified.

The first coupling reaction to be reported for a group 4 transition metal pentadienyl complex was that shown in Eq. (32).[46] The product, **26** (Fig 26), reveals that the more strongly bound open dienyl (2,4-C_7H_{11}) ligand is also the more reactive, and undergoes coupling through one of its two dienyl termini, followed by dimerization as a result of the nitrogen centers subsequently bridging a second metal center. It can be noted that the coupling process has led to the rearrangement of the original *U*-dienyl framework into an *S* (sickle) geometry, and also that structural and spectroscopic data suggest that the diene ligand is actually serving more as an enediyl. Although not appreciated at the time, the product is unusual in that it appears relatively resistant to further coupling reactions.

$$\mathrm{Ti(C_5H_5)(2,4\text{-}C_7H_{11})(PEt_3) + MeCN \rightarrow \tfrac{1}{2}\{Ti(C_5H_5)[2,4\text{-}C_7H_{11}\text{-}MeCN]\}_2} \quad (32)$$

26

In contrast to the nitrile coupling results, pentadienyl ligands tend to undergo couplings with ketones at both dienyl termini. Attempts to isolate mono(ketone) coupling products have generally not succeeded, suggesting that the second coupling step is faster than the first. The R = C_2H_5 product of Eq. (33), **27**, has been structurally characterized (Fig. 27), revealing that the generated allyl fragment is

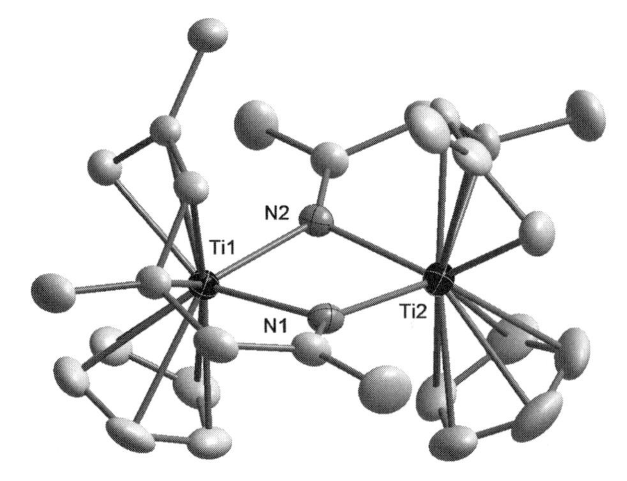

FIG. 26. Molecular structure of $\{Ti(C_5H_5)[2,4\text{-}C_7H_{11}\text{-}CH_3CN]\}_2$ (hydrogen atoms omitted).

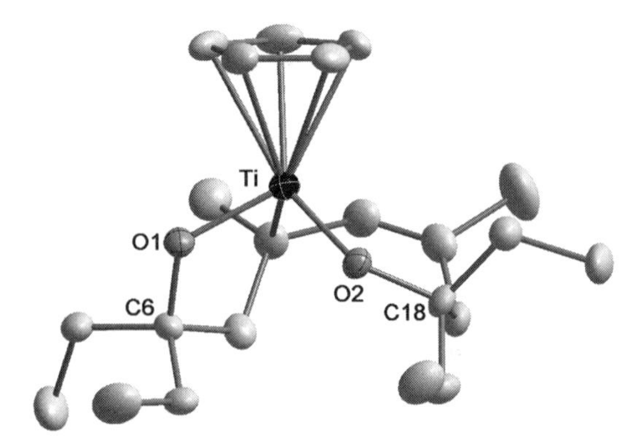

FIG. 27. Molecular structure of the bis(3-pentanone)/$Ti(C_5H_5)(2,4\text{-}C_7H_{11})$ coupling product, **27** (hydrogen atoms omitted).

best described as being bound to the metal center in σ (η^1) fashion.[40a] This feature and some other geometric considerations (Ti–O1 = 1.846(3) Å, Ti–O2 = 1.809(3) Å, Ti–O1–C6 = 116.5(3)°, Ti–O2–C18 = 142.0(3)°) led to the proposal that at least partial π donation was occurring from both lone pairs of electrons on O2, tantamount to some contribution as a 5 electron donor.

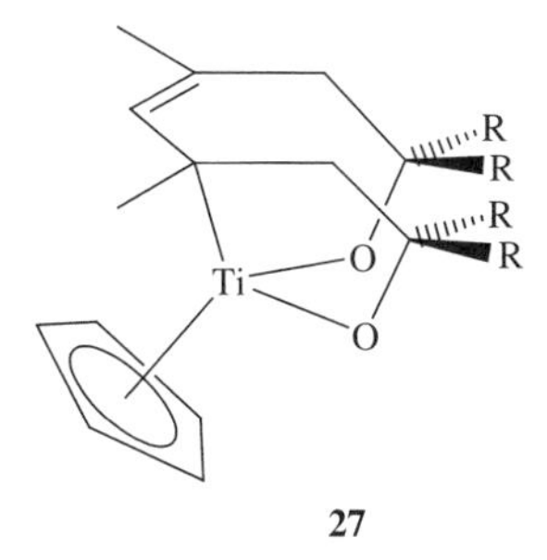

27

$$\mathrm{Ti(C_5H_5)(2,4\text{-}C_7H_{11})(PEt_3) + 2R_2CO \rightarrow Ti(C_5H_5)[2,4\text{-}C_7H_{11}\text{-}(CR_2O)_2]} \qquad (33)$$
$$\mathrm{R = Me, Et}$$

An interesting difference is observed for the coupling reaction of acetone with $Ti(C_5H_4R)(C_5H_7)(PMe_3)$ complexes [Eq. (34)].[49] A crystallographic study (Fig. 28) of the R = Me product revealed that again both pentadienyl termini had coupled to acetone molecules, but this time, the product dimerized, accompanied by the conversion of the original dienyl fragment from a *U* to *S* conformation (**28**). This is in accord with the fact that the 2,4-dimethyl substitution pattern favors the *U* conformation, whereas for an unsubstituted pentadienyl, the *W* and *S* conformers are preferable.[90] As can be seen, there is again σ-allyl coordination present, and perhaps

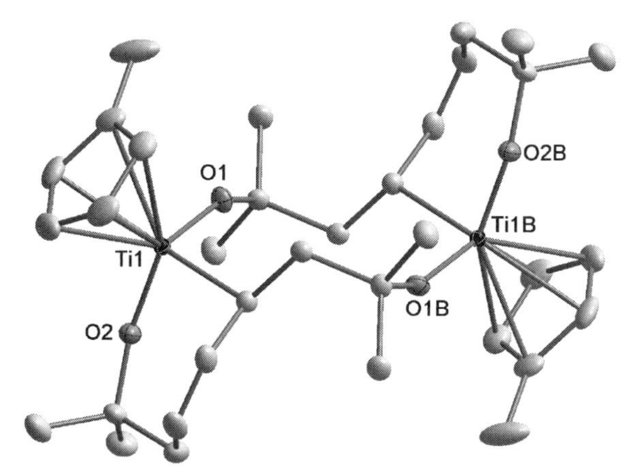

FIG. 28. Molecular structure of the dimeric bis(acetone)/$Ti(MeC_5H_4)(C_5H_7)$ coupling product, **28** (hydrogen atoms omitted).

due to there being fewer constraints on the geometric parameters for the alkoxide coordination, larger Ti–O–C angles and shorter Ti–O distances were found, again suggesting at least a partial contribution of the alkoxide ligands as 5 electron donors.

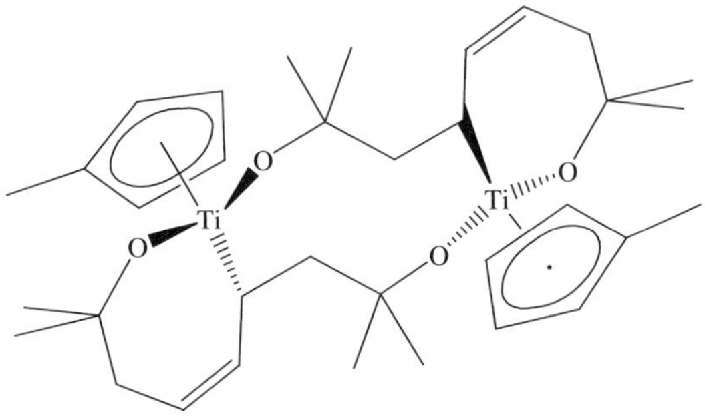

28

$$Ti(C_5H_4R)(C_5H_7)(PMe_3) + 2Me_2CO \rightarrow \tfrac{1}{2}[Ti(C_5H_4R)(OCMe_2\text{-}C_5H_7\text{-}CMe_2O)]_2 \quad R = H,\ Me \qquad (34)$$

The use of imines as coupling partners has led to even more diverse opportunities. Whether due to their greater steric bulk relative to acetone or to the lower electronegativity of nitrogen compared to oxygen, only one equivalent of PhCH=NMe reacts rapidly with $Ti(C_5H_5)(2,4\text{-}C_7H_{11})(PEt_3)$, yielding a red product **29** (R = Me),[40a] although a second equivalent can incorporate more slowly, yielding a green product, **30**.[40b] For the bulkier PhCH=N(i-C_3H_7), only one equivalent of imine was observed to incorporate.[40b] Diffraction studies have been carried out for the mono(coupling) products of PhCH=NR (R = Ph, ^{i}Pr)[89,91] and the bis(coupling) product of PhCH=NMe,[40b] revealing the structures shown, with η^4-diene and η^3-allyl coordination, respectively, and with the hydrogen substituent of the imine carbon atom directed toward the open dienyl edge. An exception is observed for the product resulting from the reaction of $Ti[1,3\text{-}C_5H_3{}^tBu_2](2,4\text{-}C_7H_{11})(PEt_3)$ with PhCH=NMe, which was shown from a structural study to couple with the phenyl substituent of the imine carbon atom directed toward the open dienyl edge.[50]

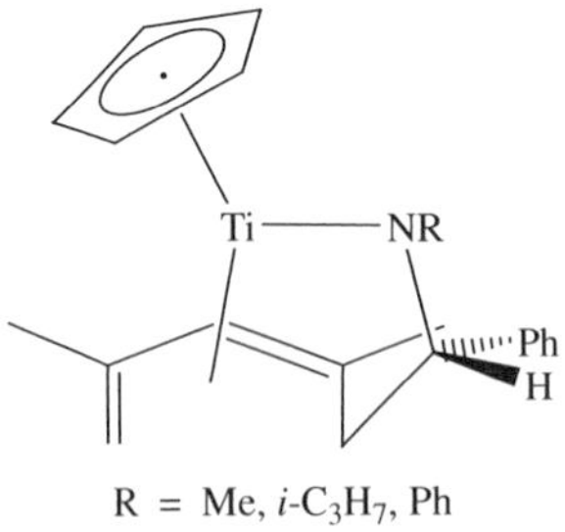

R = Me, i-C_3H_7, Ph

29

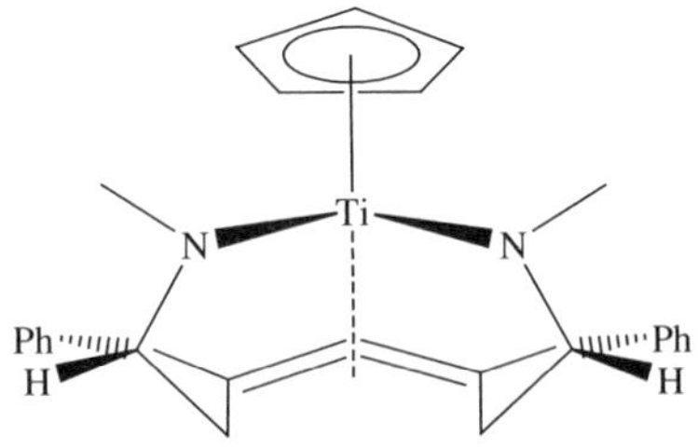

30

As a result of the second imine coordination occurring much more slowly than the first, the opportunity was afforded to bring about tandem couplings involving imines and additional unsaturated organic molecules. Thus, even though **29** (for R = ^{i}Pr) does not readily incorporate a second imine, it will readily incorporate acetone[40b] or 4-tolunitrile,[50] yielding **31** and **32** (R = 4-MeC_6H_4), respectively, each of which has been structurally characterized. The mixed imine/ketone product is notable in having σ-allyl coordination, like the bis(ketone) products, but unlike the bis(imine) products. This was attributed to the ability of the alkoxide ligands to serve as formal 5 electron donors. The mixed imine/nitrile product is notable in that the mono(MeCN) product described earlier would not readily undergo further coupling reactions. Hence, in general, mixed imine/nitrile products may need to be accessed by initial reaction with imine, rather than nitrile.

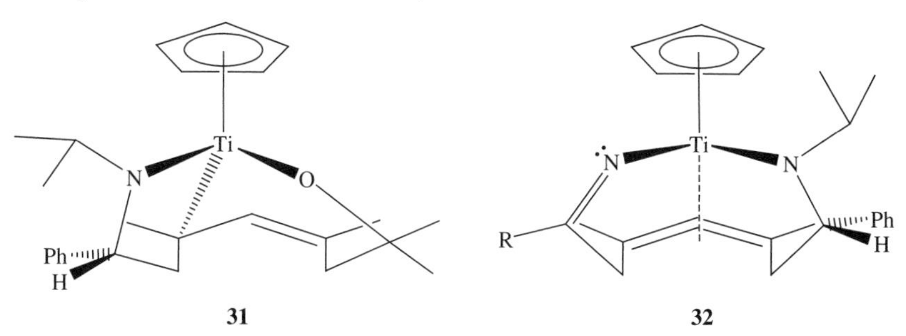

Coupling reactions of lower valent metal–pentadienyl compounds with isonitriles have generally been found to be very complicated. Thus, $Ti(C_5H_5)(2,4\text{-}C_7H_{11})(PEt_3)$ was shown to incorporate four equivalents of aryl isocyanide, RNC (R = 4-XC_6H_4NC; X = H, Me, Et), yielding **33**, in which two isonitriles can be seen to have coupled between the two dienyl termini, leading to a diazadiene ligand, while the two others have formed an indole coupled to the 2 position of the dienyl framework.[40a] The product may then be formulated as a tris(amide) complex, although weak coordination by the C=C bond in the diazadiene fragment might be present, somewhat reminiscent of the bonding situation for the NacNac ligands (Section IVA).

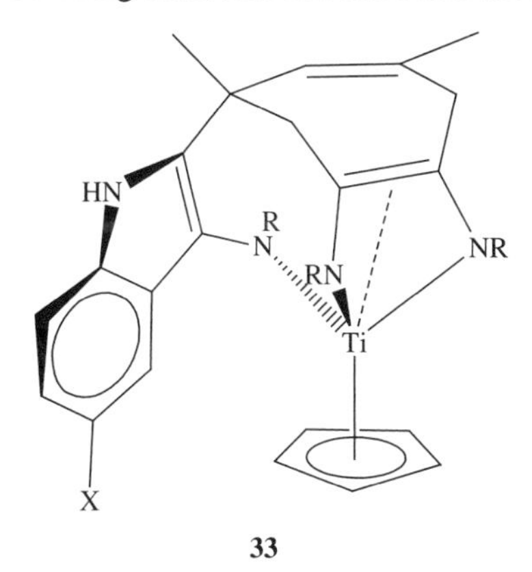

33

Several products have been reported from various tandem imine/isonitrile coupling reactions. The simplest involves the hindered CN^{t}Bu which reacts with the PhCH=N^{i}Pr/$Ti(C_5H_5)(2,4\text{-}C_7H_{11})$ coupling product (**29**) to yield **34**.[40b] Two equivalents of CN^{t}Bu can be seen to have incorporated, but only one having undergone coupling. Thus, the coupling of the imine to one end of the dienyl ligand converted it to a diene ligand, and the unsaturated carbon atom of the subsequently added isonitrile coupled to both ends of the diene, yielding the cyclopentene ring.

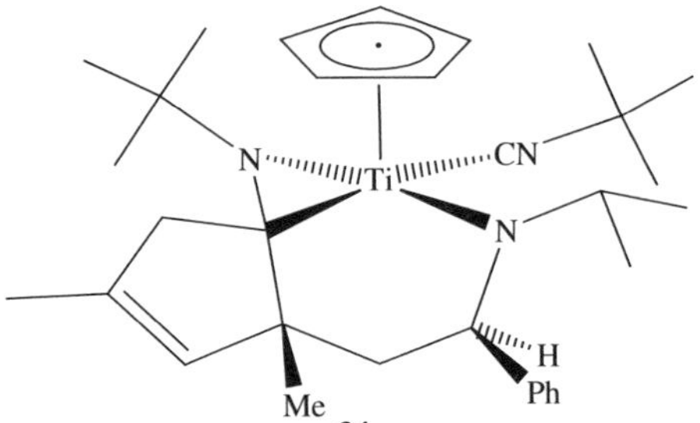

N Ti CN N H Ph Me

34

More complicated tandem imine/isonitrile couplings are observed for aryl isonitriles[50]. Thus, subsequent to coupling of the $Ti(C_5H_5)(2,4\text{-}C_7H_{11})$ fragment with $PhCH{=}N^iPr$ four equivalents of $4\text{-}MeC_6H_4NC$ also incorporate and couple, yielding **35** ($R = {}^iPr$, $R' = 4\text{-}MeC_6H_4$). As in the tandem imine/CN^tBu reaction, the first isonitrile to incorporate coupled across the diene ligand termini to yield a five-membered ring. Addition of the three final isonitriles led to the observed complex, with a coordination environment similar to that of **33**, although no indole ring was constructed. Interestingly, if the isonitrile in the above reaction has not been completely freed of ^tBuOH used in its preparation, only three equivalents of isonitrile are incorporated, and a different product, **36** (R = 4-tolyl), is isolated.[40b] The proposed mechanisms for the two reactions bear a number of steps in common, such that one does again form a five-membered ring from the diene and one isonitrile. However, once the alcohol is incorporated, leading to protonation and decoordination of one of the amide ligands, the mechanism changes significantly, leading to the formation of an indole and coordination of the metal center by C_5H_5, the alkoxide, and two amide ligands. In this complex, a very short Ti–O distance (1.760(10) Å) and a very wide Ti–O–C angle (169.0(11)°) were observed. These values correlated well with those of related compounds, and together with other arguments, provided further support for the proposition that alkoxide ligands could serve as formal 5 electron donors.

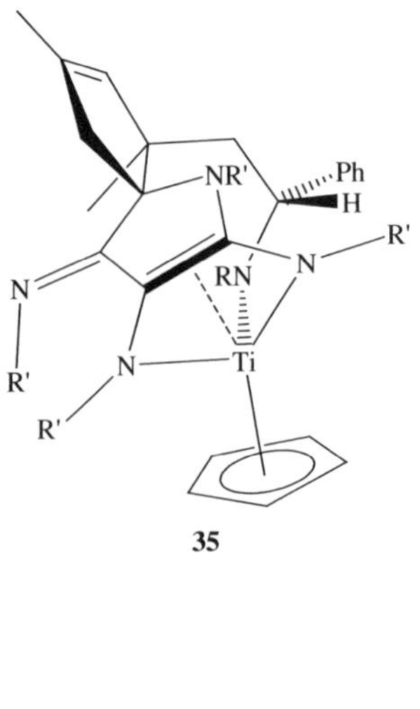

NR' Ph H R' N RN N Ti R' N R'

35

H N Ph R H N N Ti N O R

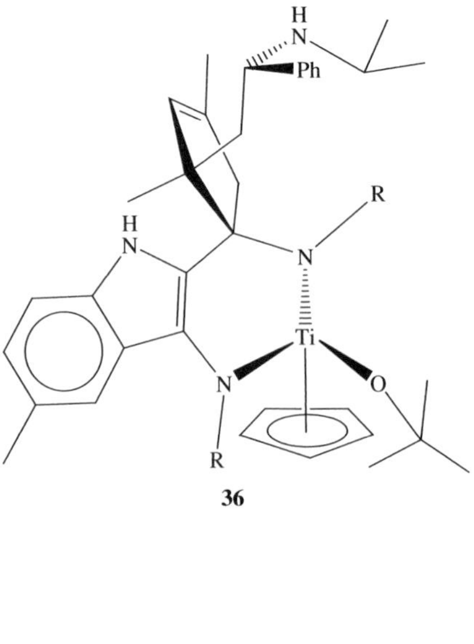

36

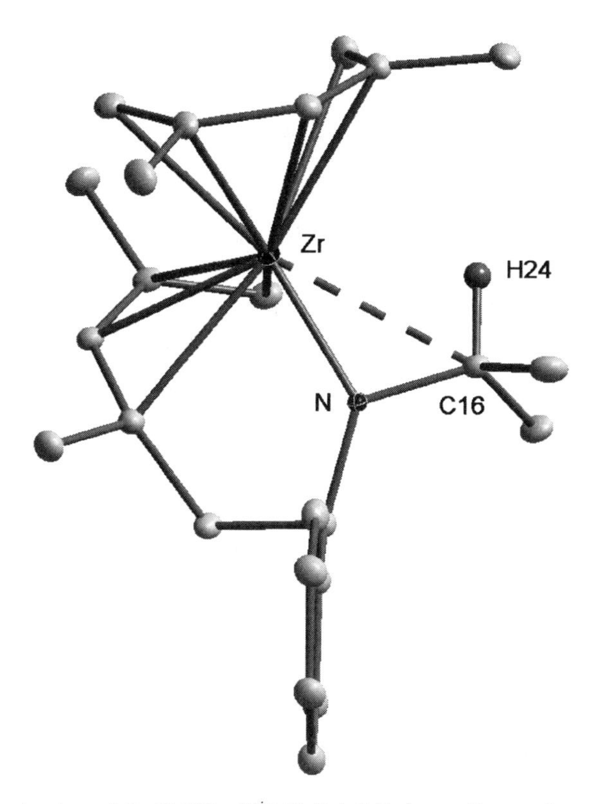

FIG. 29. Molecular structure of the PhCH=N^{i}Pr/Zr(2,4-C_7H_{11})$_2$ coupling product, **37** (most hydrogen atoms omitted).

The open zirconocene Zr(2,4-C_7H_{11})$_2$(PMe_3) has been found to undergo a 1:1 reaction with PhCH=N^{i}Pr leading to **37** (Fig. 29).[92] At first glance, one observes η^5-2,4-C_7H_{11}, diene, and π-amide coordination. Perhaps due to constraints imposed by the tethering between the diene and amide ligands, there are some geometric deformations that tend to obscure the natures of the bonding interactions. Nevertheless, it is clear that there is at least a significant component of enediyl character in the "diene" coordination, thereby leading to some Zr(IV) character. The Zr–C and C–C bond distances for the 2,4-C_7H_{11} ligand are consistent with this, with the Zr–C3 bond being shortest. However, the lengthening of the following Zr–C bonds is not as pronounced as generally seen for a higher oxidation state complex.

H Me Me Zr N Ph H

37

Considering the dienyl, diene, and π-amide coordinations, **37** would seem to be a 16 electron complex. There is, however, clear evidence of an additional interaction. The Zr–N–C16 angle was found to be 102.86(4)°, quite smaller than would be expected of sp^2 hybridization, and suggesting the presence of an agostic interaction. While it would be expected that the interaction would involve the C16–H bond, its $^1J(^{13}C^1H)$ value of 128 Hz provides no support of this. A careful electron density study at 16 K was carried out, but not only was there no bond path for any agostic interaction, but bond paths were missing as well for most Zr–C interactions. Although this study did not reveal the nature of the agostic interaction, a later NMR study revealed a significantly smaller $J(^{13}C^{15}N)$ value for the N–C16 bond relative to such a bond in a titanium analogue,[91] thereby implicating the N–C16 bond as being mainly responsible for the agostic interaction.[6] Additionally, the compound has been the subject of a study designed to develop a general approach to choosing the optimal set of orbital coordinate axes for a low-symmetry compound.[93] Not surprisingly, this complex reacts readily with two equivalents of CN^{t}Bu leading to the same overall conversion as observed for the Ti(C_5H_5)(2,4-C_7H_{11}) fragment (cf., **34**).[94] Thus, one CN^{t}Bu has coupled to the imine-coupled diene ligand, leaving the 2,4-C_7H_{11} ligand intact, while the second equivalent of isonitrile simply coordinates to the metal center, as in **38**. In contrast to the above, Ti(2,4-C_7H_{11})$_2$ has been found to undergo a very unusual coupling reaction with two equivalents of imine, yielding **39**.[91]

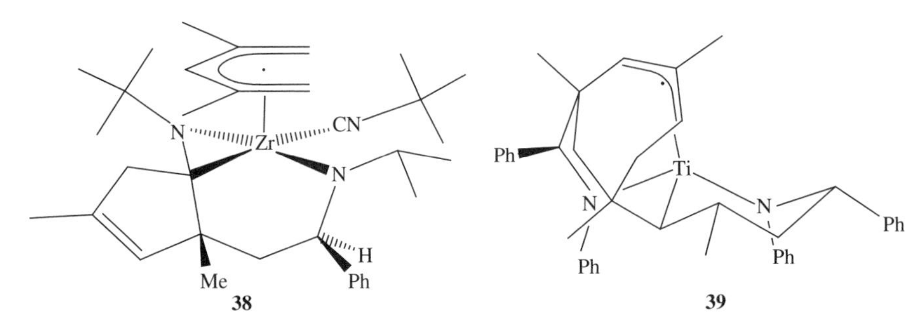

Reactions with alkynes can also follow complicated paths. Thus, reaction of Ti(C_5H_5)(2,4-C_7H_{11})(PEt_3) with PhC_2SiMe_3 led to the incorporation of two equivalents of alkyne, generating **40**.[95] Attempts to limit the alkyne content only led to lower yields of **40**. Based on previous pentadienyl coupling reactions, it can be expected that the first alkyne coupled to both dienyl termini (C1, C5). A subsequent hydride shift from C5 to C2 could then be followed by incorporation of the second alkyne, and coupling of its phenyl-substituted end to C3. Hydride transfer from C1 to C8, together with coupling of the silyl substituted end of the second alkyne to C1, then complete the process. The presence of single Cp, diene, and alkyl coordinate interactions would provide this complex with a 14 electron configuration. Spectroscopic and structural data indicate the presence of at least one additional agostic interaction, involving the C8–H bond.

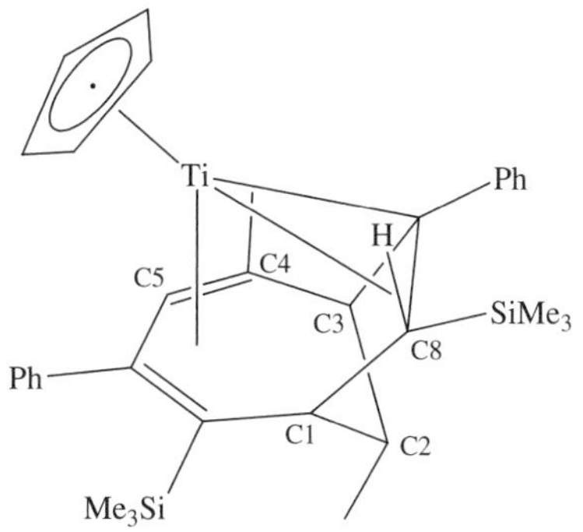

40

The selective incorporation of two alkynes in the above reaction led to the exploration of similar reactions with diynes.[95] Interestingly, a much simpler, and potentially more useful, path was followed for reactions with α,ω-hexadiynes, heptadiynes, and octadiynes, leading to deep green, bicyclic products **41**,[95] having a nine-membered ring fused to a four-, five-, or six-membered ring. That significant functionalization of the diynes is possible despite the presence of the electropositive titanium center suggests that these reactions could be of some synthetic utility.[96] Indeed, the organic fragments may be removed from the metal center, and are typically released as conjugated trienes.[97] The six-membered ring complex, however, decomposes rapidly at room temperature, perhaps due to the ability of the larger ring to approach more closely the electron deficient (16 electron) metal center. As would be expected, longer diynes also lead to thermally unstable products. In fact, even the smaller fused ring species undergo rearrangements, *via* C–C bond cleavage, to much more complicated species, e.g., **42**. Furthermore, it has been found that other dienyl ligands such as C_5H_7 also yield analogous products, some of which undergo as yet unclear rearrangements through one or more observable intermediates.

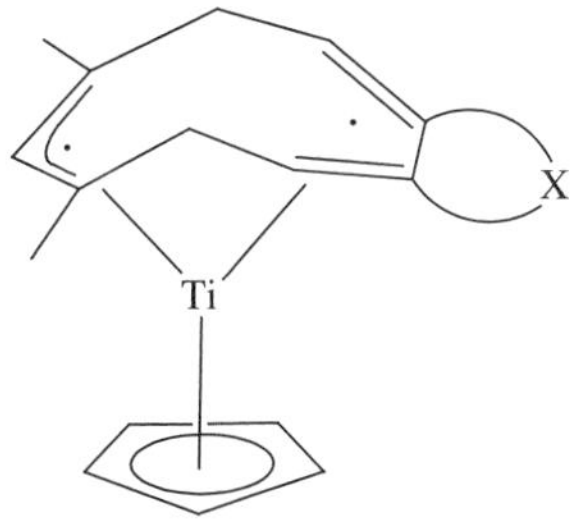

X = C_2H_4, CH_2ZCH_2, C_4H_8
Z = O, S, NMe, N(n-C_4H_9), CH_2, $C(CO_2Me)_2$, $C(CO_2Et)_2$

41

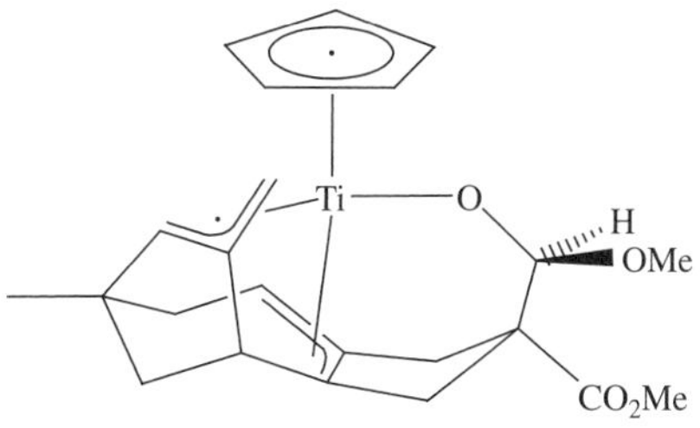

42

The presence of an edge-bridge in a pentadienyl ligand more often than not leads to dramatic differences in reaction chemistry. Having said that, we will begin with coupling reactions involving the one partner, specifically an imine, that thus far has been found not to have its chemistry affected dramatically by the presence of an edge-bridge. Thus, $Ti(C_5H_5)(6,6\text{-dmch})(PMe_3)$, $Zr(C_5H_5)(6,6\text{-dmch})(PMe_3)_2$,[61,98] and $Zr(6,6\text{-dmch})_2(PMe_3)_2$[36] react readily with PhCH=NPh, leading to the incorporation and coupling of two equivalents of the imine. In the last case, given the presence of two 6,6-dmch ligands, the couplings could either both occur on a single dmch ligand, or an imine could couple to each of the dmch ligands. The observed product, **43** (Fig. 30), involves the former possibility, which is reasonable given that the first coupling converts a dienyl ligand to a diene, which should entail substantial enediyl character, and therefore more negative charge on its termini. The resulting product is another example of a higher valent metal–pentadienyl complex and shows the dramatic trends in dienyl M–C and C–C bond distances expected for such a species.

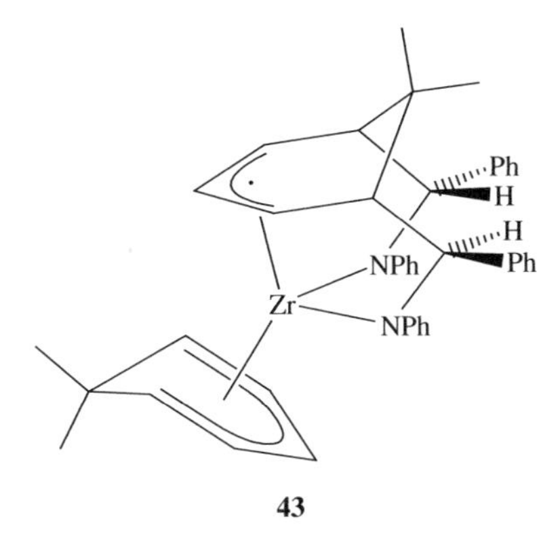

Some particularly interesting results have come from similar reactions of $Zr(6,6\text{-dmch})_2(PMe_3)_2$ with N-tethered diimines, $(CH_2)_n[NC(H)Ph]_2$.[61,98] For $n = 2$, the tether is so short that only one imine can couple with the $Ti(C_5H_5)(6,6\text{-dmch})$ fragment. As a result, the second end coordinates without coupling, leading to an 18

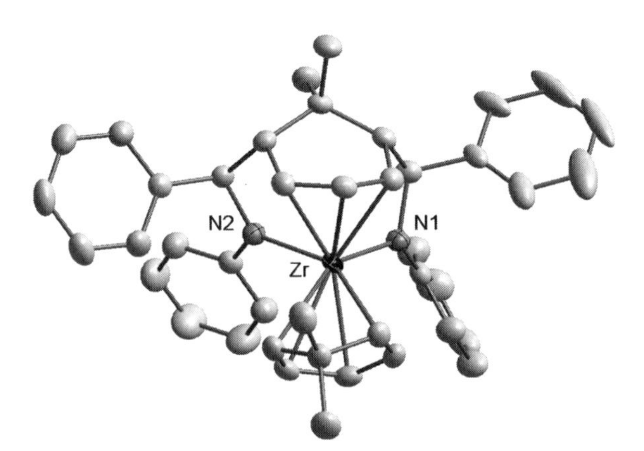

FIG. 30. Molecular structure of the bis(PhCH=NPh)/$Zr(6,6\text{-dmch})_2$ coupling product, **43**.

electron complex with C_5H_5, diene, amide, and κ^1-imine coordination. Through a 2:1 reaction of $Zr(C_5H_5)(6,6\text{-dmch})(PMe_3)_2$ with the C_2H_4-bridged diimine, it is possible to isolate a complex in which the two imine carbon atoms each couple to a 6,6-dmch terminus in different metal coordination spheres, with one PMe_3 ligand remaining bound to each 18 electron metal center. For $n = 3$ or 4, the diimine can effectively span the two dienyl termini of a $Zr(6,6\text{-dmch})_2$ fragment, leading to the formation of the desired 12- and 13-membered rings (**44**; Fig. 31). For $n = 6$, however, an intermolecular coupling has been observed for either the $Ti(C_5H_5)(6,6\text{-dmch})$ or the $Zr(6,6\text{-dmch})_2$ fragment, involving two dmch ligands and two diimines and generating a 30-membered ring.

44

In contrast to all the coupling reactions cited above, coupling reactions of 6,6-dmch ligands with ketones and aldehydes have led quite unexpectedly to 1,4-regiochemistry, rather than the usual 1,5.[47,52] While reactions of $Ti(C_5H_5)$

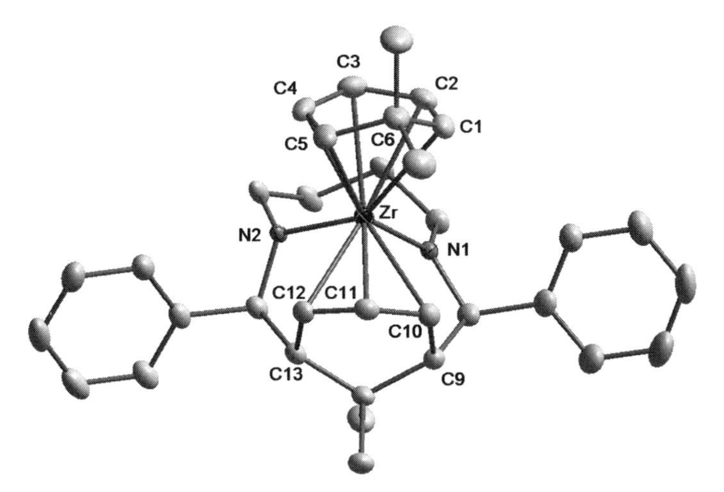

FIG. 31. Molecular structure of a diimine ($n = 4$) coupling product (**44**) with the $Zr(6,6\text{-dmch})_2$ fragment.

(6,6-dmch)(PMe_3) with ketones (Me_2CO, Et_2CO, $(CH_2)_4CO$, $(CH_2)_5CO$) or aldehydes (R = ^{i}Pr, ^{t}Bu) did not readily lead to crystalline metal-containing products, protonation of the metal complexes led to release of the enediols **45**, or in cases in which oxidation preceded hydrolysis, enetriols **46**. As would have to be expected, the two ketones have coupled to the same side of the dienyl ligand, thereby setting relative configurations at two carbon atoms. In the case of the enetriol, a third configuration is set, but surprisingly the oxygen introduced *via* the oxidation step added on the side of the ligand opposite to the metal center, indicating direct attack by oxygen on the ligand rather than at the metal center. This results in formal inversion of configuration at the new chiral center. To what extent this may play a role in other M–C bond oxidations is unclear. An interesting difference is observed with benzaldehyde, which leads only to a mono(coupling) product. A monocoupling byproduct was also observed for the reactions with bulky aldehydes (R = ^{i}Pr, ^{t}Bu). The aldehyde coupling processes lead to two additional chiral centers, and for bulky aldehydes, high stereoselectivities could be observed.

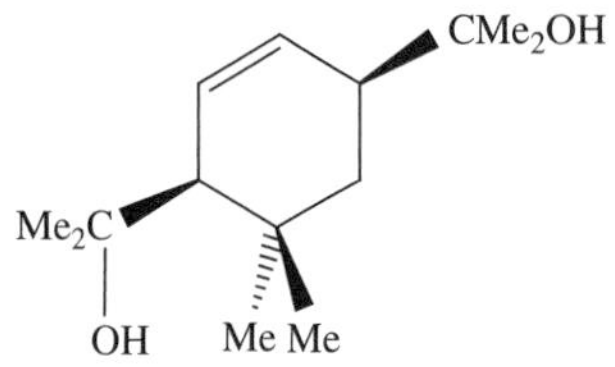

45 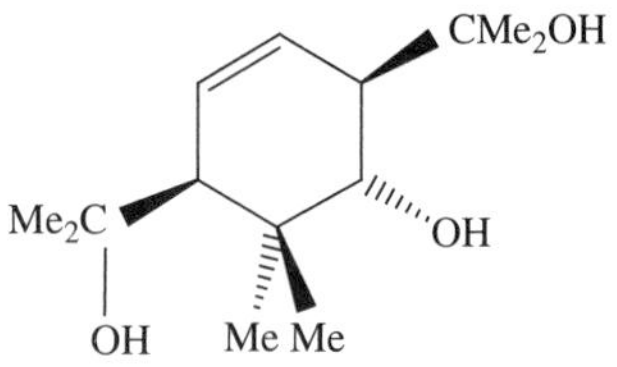 **46**

An important point relevant to mechanistic considerations is that couplings involving cyclopropyl ketones did not lead to ring opening, thereby ruling out an electron transfer process.[97] The mechanism responsible for the 1,4-regiochemistry was further probed by way of an analogous, more crystalline zirconium complex, **47**, prepared from the reaction of $Zr(C_5H_5)$(6,6-dmch)$(PMe_3)_2$ and acetone.[99] The product was isolated as a dimer, through formation of a Zr_2O_2 ring, and revealed that 1,4-regiochemistry is established at the time of incorporation of the two ketone equivalents, in contradiction to the initially proposed mechanism. The alternative mechanism was supported by the isolation of a mono(benzophenone) coupling product, **48**, which showed the expected coupling to occur at a dienyl terminus.[99] Apparently due to the ability of the resulting alkoxide ligand to serve as a 5 electron donor (Ti–O–C = 140.68(11)°), the metal center could not bond to the entire diene unit, and instead is only coordinated by one of the C=C bonds. Presumably the steric bulk of the edge-bridge is then responsible for directing the second coupling to C4 rather than C5, accounting for the observed regiochemistry. Conceivably, appropriately substituted diones could lead to 1,5 dicouplings, perhaps even yielding structures bearing some similarity to Taxol®.

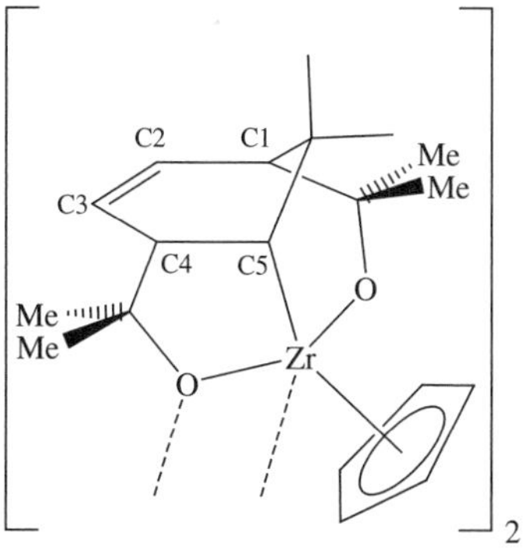

47

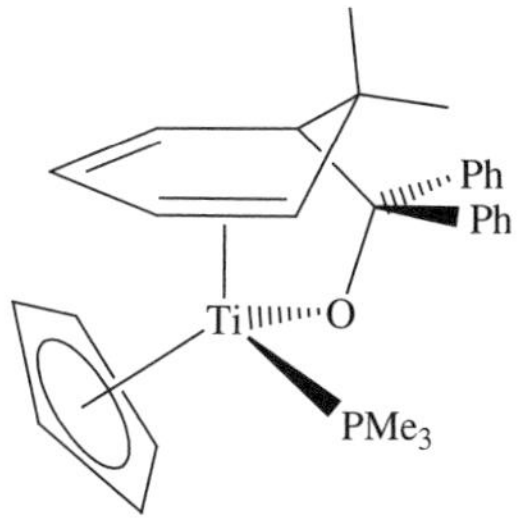

48

Reactions of edge-bridged complexes with isocyanides have only begun to be examined, and to date there is only one report which has appeared.[100] From the reaction of the thermally unstable Zr(6,6-dmch)$_2$(Me)(Br) with 2-MeC$_6$H$_4$NC, a small quantity of **49** was isolated, and characterized structurally. The product is clearly quite different from other isonitrile coupling products, especially in that only one dienyl site was involved, and it was the central carbon atom, not a terminus. As all previous dienyl/isonitrile coupling reactions involved low oxidation state metals and open edge dienyl ligands, it can be expected that many other modes of coupling should be realizable for the higher valent compounds.

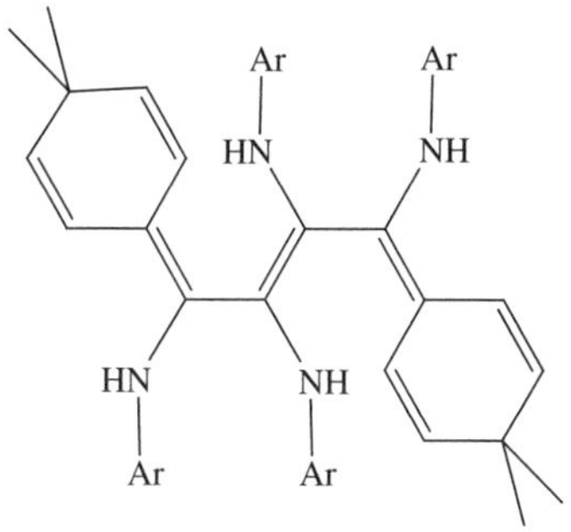

Ar = 2-MeC$_6$H$_4$

49

As noted in Sections IIB and IIIB, alkynes can be incorporated into Zr(6,6-dmch)$_2$ or Zr(C$_5$H$_5$)(6,6-dmch) complexes without dienyl/alkyne coupling taking place, in contrast to all observations to date for other pentadienyl ligand complexes. It nonetheless appears that subsequent coupling reactions, specifically 5+2 and 5+2+2 ring constructions, can take place under the appropriate conditions. Thus, some of the alkyne adducts undergo transformations with time, perhaps as a result of these reactions. In fact, while addition of two alkyne equivalents into a half-open Zr(II) dmch complex led to a zirconacyclopentadiene complex (Fig. 17), a similar reaction for the sterically more crowded $Zr(6,6\text{-dmch})_2$ unit did indeed rapidly lead to a 5 + 2 + 2 ring construction, yielding **50** (Fig. 32).[89,91] The mode of coupling is similar to that which took place for **41**, but unlike any other for simple alkynes. It is not yet clear if **50** is subject to similar rearrangements as observed for **41**. Overall, it can be seen that by allowing for alkynes to be incorporated without their undergoing almost immediate dienyl/alkyne coupling, the 6,6-dmch ligand can provide more control over any subsequent coupling, leading to potentially useful ring constructions.

50

In contrast to some of the results for 6,6-dmch ligand complexes, reactions of group 4 metal cyclooctadienyl complexes with alkynes have all led to

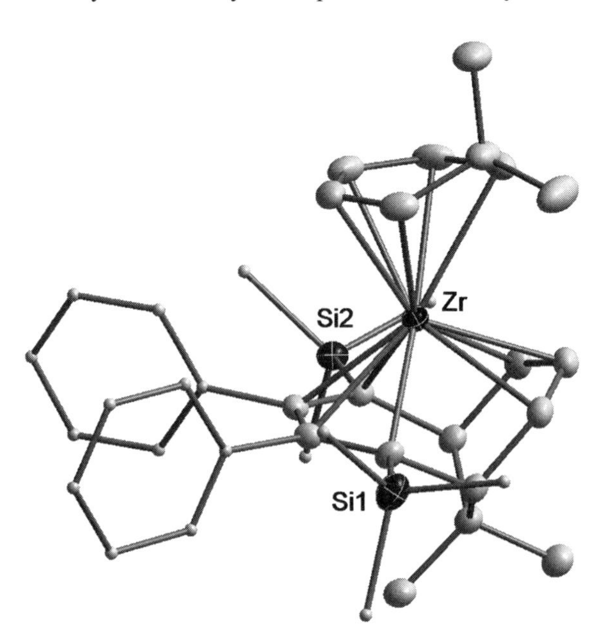

FIG. 32. Molecular structure of the bis(PhC_2SiMe_3)/$Zr(6,6\text{-dmch})_2$ coupling product, **50** (hydrogen atoms omitted, phenyl and Me_3Si groups simplified).

dienyl/alkyne couplings. The first product to be characterized was **51**, formed by the reaction of $Ti(C_5H_5)(c\text{-}C_8H_{11})(PEt_3)$ with three equivalents of PhC_2SiMe_3.[48] A mechanism of its formation was proposed through which the first alkyne underwent coupling between the dienyl termini (C1, C9), thereby generating an allyl ligand, one end of which (C2) then coupled to the phenyl-substituted end of the second alkyne ligand, whose other end coupled to the phenyl-substituted end of the third alkyne. The silyl-substituted end of the third alkyne then coupled to what had been the central carbon atom (C7) of the original dienyl ligand. The resulting complex, like **40**, then has Cp, diene, and alkyl ligands, leading to a 14 electron count. Unlike **40**, however, there are no C–H bonds in proximity to the metal center (primarily due to the edge-bridge), and the metal then was proposed to engage in agostic interactions with four C–C bonds. This was established through a combination of structural, MO, and spectroscopic methods. In particular, remarkably low $^1J(^{13}C^{13}C)$ values (17.9–29.6 Hz) were observed for the C–C bonds (formed between the five atoms in bold) proposed to engage in the interactions, while notably high $^1J(^{13}C^1H)$ values (e.g., 149 Hz) were observed for some C–H bonds. These trends parallel those generally observed for C–H agostic interactions.[101]

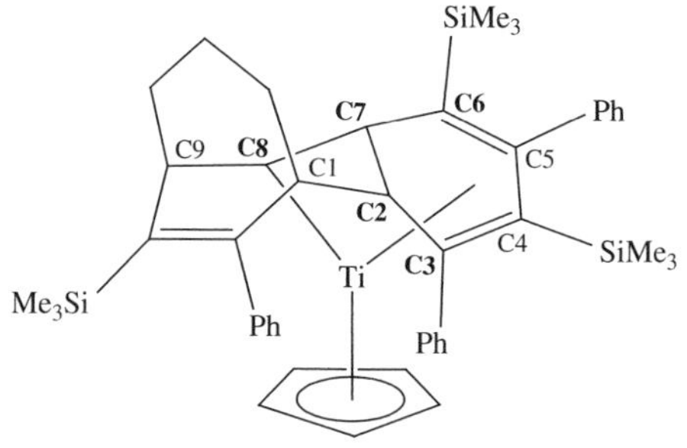

SiMe3 C7 C6 Ph C9 C8 C1 C5 C2 C4 SiMe3 Ti C3 Me3Si Ph Ph

51

In what would seem to be the similar reaction of $Ti(C_5H_5)(c\text{-}C_8H_{11})(PMe_3)$ with PhC_2SiMe_3, a product, **52**, incorporating only two alkyne equivalents was isolated.[102] In this case, one alkyne was, quite unexpectedly, incorporated into the edge-bridge, while the second took part in the expected coupling between the dienyl termini. With C_5H_5, alkyl (σ-allyl), olefin, and phosphine coordination, the complex attains a 14 electron configuration. Given the electron deficiency of the complex, the question arose as to why the allyl ligand appeared to be bound primarily through a σ, rather than π, mode. As in complex **51**, spectroscopic and structural data established the presence of at least one (C–C)→Ti agostic interaction. The comparable or greater favorability of titanium to engage in agostic coordination by a C–C bond (C5–C10) relative to olefin coordination was cited as a possible reason for many early transition metal complexes being active polymerization catalysts. A similarity was noted between **52** and electron deficient metallacyclobutanes, which were also demonstrated to possess agostic interactions. These interactions were proposed to play a role in olefin metathesis reactions.

52

Interestingly, accompanying the bis(alkyne) coupling product is a smaller amount of a tetra(alkyne) coupling product, **53**.[54b] Although this species is clearly far more complicated, once again one finds lengthened C–C bonds (C8–C9, 1.570(3) Å; C9–C10, 1.562(2) Å) close to the metal center, suggestive of (C–C)→Ti agostic interactions. The framework geometry of **53** is similar to that of **51**, and it indeed appears to result from the reaction of **51** with alkyne. The structural data also suggested agostic interactions with the metallacyclobutene fragment, and comparisons of structural and spectroscopic data for previously reported metallacyclobutenes seemed to demonstrate the presence of such interactions in other electron deficient species.

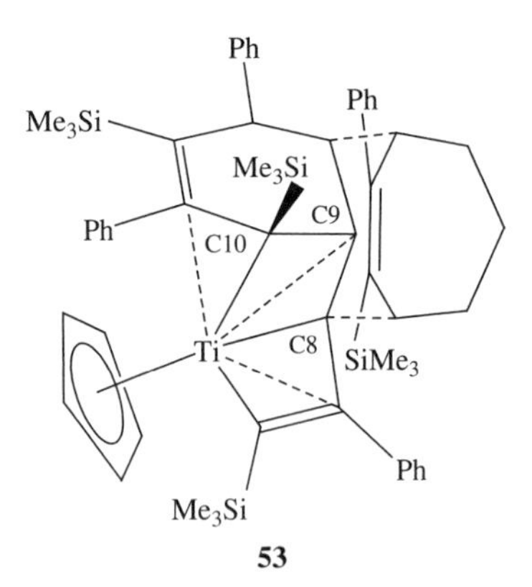

53

While the reactions of the Ti(C_5H_5)(*c*-C_8H_{11}) unit with alkynes have led to three compounds with (C–C)→Ti agostic interactions, the reactions of zirconium analogues have led instead to C–Si or C–C bond cleavage.[54c] These reactions, however, appear less selective, and give rise to a host of products. One species isolated from the reaction of Zr(C_5H_5)(*c*-C_8H_{11})(PMe_3) with PhC_2SiMe_3 is **54** (Fig. 33), in which a metal-bound silyl group is clearly present, the silyl substituent having been displaced from C25 by a hydrogen atom. The formation of this complex would seem to have involved the expected initial coupling of a phenyl-substituted alkyne carbon atom (C26) to a dienyl terminus (C10), followed by coupling of this alkyne's other end to the phenyl-substituted end of the second alkyne.

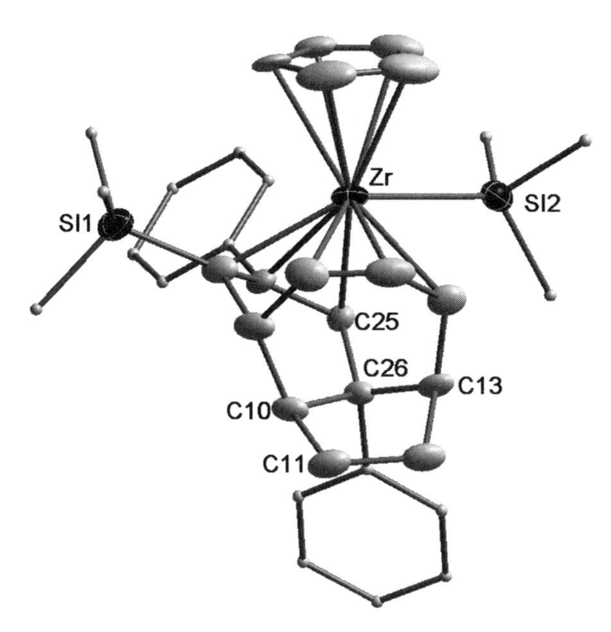

FIG. 33. Molecular structure of the bis(PhC_2SiMe_3) coupling product (**54**) with the $Zr(C_5H_5)(c\text{-}C_8H_{11})$ fragment (hydrogen atoms omitted, phenyl and Me_3Si groups simplified).

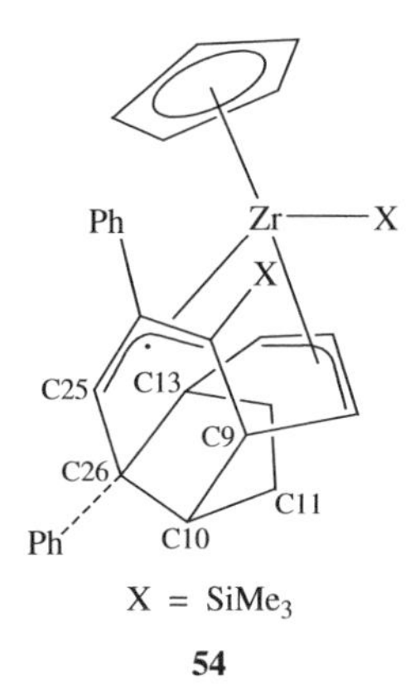

Ultimately, the silyl-substituted end of the second alkyne undergoes a very unusual coupling to the 2 position (C9), rather than to the other terminus, of the original dienyl fragment. Activation of a C–H bond at a methylene bridge position (C13), followed by a second coupling of C26 (to C13), and an exchange of the activated hydrogen atom for the silyl substituent on C25 then leads to the observed product. A related course appears to occur for a product isolated from the analogous reaction with diphenylacetylene, although the lack of C–Si bonds appears to lead to C–C cleavage (**55**, Fig. 34).[89,91] Ironically, the C–C bond appearing to be cleaved is the one formed by coupling of the two alkynes together. While one could propose that such a bond had never really formed, the similar 1,2 dicoupling

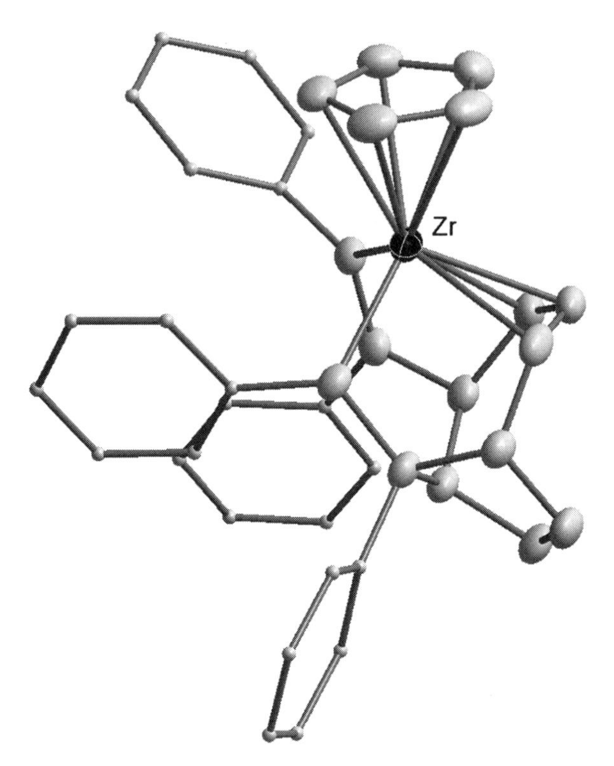

FIG. 34. Molecular structure of the bis(PhC_2Ph) coupling product (**55**) with the $Zr(C_5H_5)(c\text{-}C_8H_{11})$ fragment (hydrogen atoms omitted, phenyl groups simplified).

processes occurring for the dienyl fragments leading to the formations of **54** and **55** leave open the possibility of a common mechanistic pathway, involving the construction of the bicyclic cage in the early stages, followed ultimately by either the C–Si or C–C bond cleavage reaction.

Ph Zr Ph Ph Ph

55

A second product of the PhC_2SiMe_3 coupling reaction is even more unusual.[89] Complex **56** (Fig. 35) can be seen to contain four silicon atoms and three phenyl substituents, the combination clearly derived from three equivalents of alkyne, and

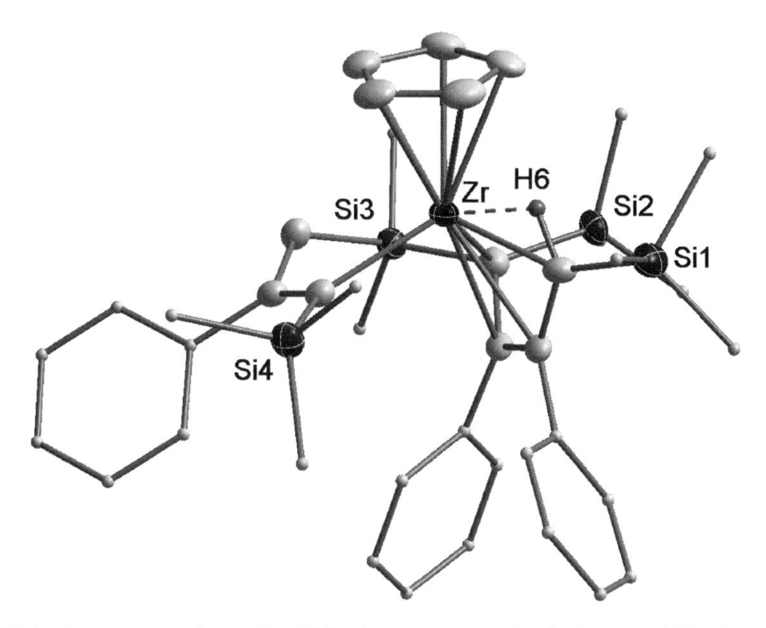

FIG. 35. Molecular structure of complex **56** (hydrogen atoms omitted, phenyl and Me_3Si groups simplified).

one extra silyl substituent. However, in this case, the *c*-C_8H_{11} fragment is missing, and one methyl substituent on Si3 has been converted to a methylene group, with the abstracted hydrogen atom apparently being transferred to C6. It is possible, but by no means certain, that this product was formed from the reaction of **54** with additional alkyne, perhaps *via* an alkyne coupling to the two allyl groups of **54**, making it possible for the cage to be subsequently expelled as a neutral fragment.

Me SiMe$_3$
C9 H SiMe$_3$
Si3 Zr
C10 C6
Me
C11 C12 C8 C7
Ph Ph
SiMe$_3$ Ph

56

The bonding in the new complex is especially unusual. At first glance one can observe apparent C_5H_5, σ-vinyl (through C12), and diene (C6–C9) coordination, and an agostic interaction with the C6–H6 bond. These would lead to an 18 electron configuration. However, the generation of the agostic interaction required some extremely severe deformations of the formal diene ligand. First, it can be seen that the substituents on "diene" carbon atoms C6–C8 all lie close to the "diene" plane, while those of C9 are twisted dramatically, leading to a loss in π overlap, causing the Zr–"diene" bonding to be better described as a combination of σ-alkyl (C9) and

π-allyl (C6–C8) interactions. This also leads to a long C8–C9 bond, 1.514(4) Å. There is still, however, something more unusual here. One can note that the allyl fragment does not adopt its usual geometry, but rather has the metal atom nearly in the plane of the three allyl carbon atoms. This is reminiscent of alkynes which function as 4 electron donors, in which the twist of the second π system results in a change from the usual σ-donor and π-acceptor functions of a single π system to π-donor and δ-acceptor functions.[103] A combination of the two distortions observed, a loss of π overlap and the twist of the allyl fragment, allows for the "diene" fragment to serve as a 6 electron donor, thereby yielding an 18 electron configuration for the zirconium center.

Some applications of group 4 metals in dienylation reactions have been reported. While coupling reactions of $Mg(C_5H_7)_2(THF)_2$ with allyl bromide led primarily to the straight chain triene, the use of $M(C_5H_5)_2Cl_2$ catalysts (M = Ti, Zr) led primarily (ca. 90%) to the branched triene, *via* coupling of the allyl fragment to the 3 position of the dienyl ligand.[104] In another report, the couplings of aldehydes or acetone with zirconium(IV)-bound pentadienyl ligands were also found to occur selectively at the central (3) dienyl position,[105] in contrast to results for the reactions of lower valent compounds (*vide supra*). The presumed $Zr(C_5H_5)_2(Pdl)(OR)$ intermediates (Pdl = (4 or 5)-C_6H_9, 5-PhC_5H_6, or 5,5-C_7H_{11}; R = CH_3, CH_2Ph, Si^tBuMe_2, Si^tBuPh_2) were generated from $Zr(C_5H_5)_2Cl_2$, two equivalents ^{n}BuLi and the appropriate (1 or 3)-dienyl ether. For the aldehyde reactions, a high selectivity for *anti* addition was observed (ca. 84–95%), but in most cases this could be reversed through the addition of $BF_3(OEt_2)$. In one case, involving benzaldehyde and the 4-C_6H_9 ligand, the $BF_3(OEt_2)$ also led to a significant (29%) amount of straight chain product, with the coupling taking place at the dienyl terminus closer to the single methyl substituent.

C. *Materials and Catalytic Applications*

The substantial preference for η^5-bound pentadienyl ligands to bind to transition metals in low oxidation states has led to new and/or improved applications in materials syntheses relative to metal cyclopentadienyl compounds. In one strategy, the addition of potential ligands in oxidizing states can lead to their incorporation with the simultaneous expulsion of the pentadienyl ligands, whether as radicals or as dimers (decatetraenes). Thus, reactions of $M(2,4\text{-}C_7H_{11})_2$ complexes (M = Ti, V, Cr) with phosphine chalcogenides lead to a variety of metal chalcogenides and chalcogenide complexes, some of which had not been previously isolated.[106] A related opportunity is provided by the ability of open metallocene compounds to undergo "naked metal" reactions, analogous to those undergone by metal allyl compounds.[1] While these types of reactions have been demonstrated capable of leading to molecular species from open ferrocenes and open chromocenes,[107] to date the only titanium applications to have been reported involve chemical vapor depositions. Several reports have appeared describing the use of $Ti(2,4\text{-}C_7H_{11})_2$ in the thermally or photochemically promoted depositions of titanium, titanium carbide, and in the presence of nitrogen-containing coreactants, titanium nitride.[108]

Just as metallocenes and metallocene halides have attracted significant interest in olefin polymerization applications, so too have their open pentadienyl analogues. Although the higher reactivities of pentadienyl ligands, particularly toward coupling reactions with unsaturated organic molecules, might cause some concern as to possible side reactions the pentadienyl ligands could undergo, to date metal pentadienyl compounds have not been observed to couple directly to either alkyl or olefin ligands. The fact that dmch ligands can even tolerate, for at least temporary periods, the presence of alkyne ligands, suggests that the dmch ligands could be especially useful. Of course, appropriate substitution of an open edged pentadienyl ligand at or near the 1, 3, and 5 positions could also lead to greater robustness for these species.

In early reports, $Ti(2,4\text{-}C_7H_{11})_2$ was found to lead to highly active catalysts once combined with appropriate supports, such as silica, alumina, or aluminum phosphate,[109] which likely leads to the expulsion of one ligand as a neutral diene *via* protonation. The resulting catalysts led to ultrahigh molecular weight polyethylene, and showed a low response to added hydrogen. Catalysts could also be obtained with other pentadienyl ligands, e.g., $1,5\text{-}(Me_3Si)_2C_5H_5$, and other metals, such as zirconium. $Ti(2,4\text{-}C_7H_{11})_2$ has also been employed as an α-olefin polymerization catalyst precursor when used in combination with various organochromium complexes.[110] These combinations led to polymers with broad molecular weight distributions, which could be adjusted by varying the chromium content, or *via* the addition of hydrogen. Others have reported that the combination of titanium or zirconium halides with various pentadienyl anions, in widely varying ratios, also give rise to active α-olefin polymerization catalysts,[111] utilizing a variety of metal oxide supports, and an appropriate activator such as methylaluminoxane (MAO).

Open or partially open analogues of *ansa*-metallocene polymerization catalysts have also been reported to lead to active catalysts for various α-olefins and dienes.[112] These catalysts employ either two pentadienyl ligands (edge-bridged or not) or a pentadienyl and an aromatic cyclopentadienyl ligand. Analogues of the commercially important "constrained geometry catalysts" have also been developed,[71,113] having pentadienyl and amide ligands connected by a bridging unit (e.g., Fig. 19). One of these reports also included mention of $Ti(C_5H_5)_2$(pentadienyl)- and $Ti(2,4\text{-}C_7H_{11})Me_3$-derived catalysts.[113a] The second presumed complex is especially unusual, given the formal $+4$ metal oxidation state.

The presumed 14 electron $Ti(C_5Me_5)(2,4\text{-}C_7H_{11})$, $Ti(Ph_2CH)_2$, and $Ti(Ph_2C\text{-}SiMe_3)_2$ complexes (Section IIIA), as well as a variety of other open and half-open titanocenes, have been reported to yield, upon activation with MAO and/or poorly coordinating borate ions, catalysts for the polymerization of ethylene/α-olefin mixtures, styrene, vinylcyclohexane, vinylcyclohexene, butadiene, and related species.[43c] The polystyrene was isolated predominately in the syndiotactic form.

A number of open and half-open titanocenes, zirconocenes, and hafnocenes have been reported to yield catalysts for the hydrogenation of unsaturated polymers such as polybutadiene.[114] The hydrogenations were carried out in solution phase, and generally were very efficient (84–89%).

Several of the previously described (Sections IIC and IIIC) boratabenzene complexes have also been found to lead to active olefin oligomerization or polymerization catalysts.[41] Thus, one example of **9**, $Zr(C_5H_5BOEt)_2Cl_2$, on activation with MAO yields a selective catalyst for the formation of α-olefins, having on average some 6.5–8 inserted ethylene units. In contrast, $Zr(C_5H_5BPh)_2Cl_2$ is less selective, yielding various oligomers or polymer, while $Zr(4\text{-}RC_5H_4BPh)_2Cl_2$ (R = H, *t*-C_4H_9) complexes selectively yield 2-alkyl-1-alkenes. From amido-substituted boratabenzene complexes such as $Zr(C_5Me_5)(C_5H_5BN^iPr_2)Cl_2$ and $Zr(C_5H_5BN^iPr_2)_2Cl_2$, again with MAO activation, one obtains α-olefin polymerization catalysts, the activities of which rival that of $Zr(C_5H_5)_2Cl_2$, at least for ethylene polymerization.

Boratabenzene analogues of *ansa*-metallocene catalysts have also been reported. These can have two amido-substituted boratabenzene rings linked *via* either Me_2Si or C_2H_4 bridges, or one amido-substituted boratabenzene ligand linked to a cyclopentadienyl ligand *via* a Me_2C bridge.[41c] These complexes also become active α-olefin polymerization catalysts upon activation with MAO, and can allow for small degrees (ca. 1%) of 1-octene incorporation into polyethylene.

Some of the reported heteroatom-containing pentadienyl ligand complexes (Section IVA) also can yield active polymerization catalysts, especially on activation with MAO. Thus, the fairly simple $Zr(NacNac)Cl_3$ complexes show relatively low polymerization activity, whereas higher activities are displayed by $Zr(NacNac)_2Cl_2$ and $Zr(C_5H_5)(NacNac)Cl_2$, both of which behave as single-site catalysts.[80] Some active catalysts have been observed for the NacNac-types of ligands which incorporate an aromatic ring fused with one of the delocalized NacNac bonds. While low activities were found for an (alkyliminomethyl)phenylamido complex,[83] moderate activities were observed for complexes derived from pyridyl-1-azaallyl[82a] and (isoquinolinyl)naphthyl amides.[82c] Complexes with macrocyclic ligands such as **18** did not yield highly active catalysts.[77b]

VI
FUTURE PROSPECTS

Given that pentadienyl ligands have been shown capable of simultaneously being more strongly bound and more reactive than their cyclic, aromatic counterparts, at least in low-valent complexes, a great deal of chemistry remains to be investigated. The results to date suggest many promising future directions, both for the group 4 transition metals, and for those in at least groups 5 and 6. While it has been demonstrated that N-tethered diimines can undergo dicoupling reactions with pentadienyl ligands in their low-valent metal complexes, leading to medium and large ring systems, there should also be significant promise for developing such reactions for C-tethered diimines, diketones, etc. In some cases, however, appropriate steric control would need to be employed to ensure that the second functional group would be oriented toward the other dienyl terminus. The alkyne (and imine) coupling reactions with edge-bridged titanium and zirconium pentadienyl compounds have led to a number of compounds with (C–C)→M (and one with a (C–N)→M) agostic interactions, and even C–C bond activations. Given that hafnium tends to give rise to even stronger bonding interactions, there should be much to be gained by an extension of these studies to hafnium. Furthermore, the recent access to higher valent group 4 metal–pentadienyl complexes has opened up new avenues for exploration. It is already clear that coupling reactions of unsaturated organic molecules with high-valent pentadienyl complexes differ dramatically from related reactions for the much more investigated low-valent complexes, leading to additional opportunities. Even for the low-valent half-open metallocenes, however, it seems that much remains to be gained from coupling reactions involving complexes which incorporate sterically demanding cyclopentadienyl ligands, which could result in entirely new coupling processes. Finally, the isolation of long sought pentadienyl analogues of the ubiquitous $M(C_5H_5)_2X_2$ species suggests that related Hf, Nb, Ta, Mo, and W analogues can be accessed, which would no doubt lead to further rewarding research directions. Notably, higher valent mono(pentadienyl) complexes of tantalum,[115] tungsten,[116] and rhenium[117] have been reported, although in each case π donor ligands were present, presumably aiding in their stabilizations.

Acknowledgment

Much of the work summarized herein would not have been possible without the dedicated efforts of the coworkers and collaborators listed in some of the accompanying citations below, and their contributions are most gratefully acknowledged. We also thank Prof. Dennis Lichtenberger and Dr. Asha Rajapakshe for providing the delta bonding graphic.

References

(1) (a) Wilke, G.; Bogdanović, B.; Hardt, P.; Heimbach, P.; Keim, W.; Kröner, M.; Oberkirch, W.; Tanaka, K.; Steinrücke, E.; Walter, D.; Zimmermann, H. *Angew. Chem. Int. Ed. Engl.* **1966**, *5*, 151. (b) Green, M. L. H.; Nagy, P. L. I. *Adv. Organomet. Chem.* **1964**, *2*, 325.

(2) For numerous accounts of the history and importance of cyclopentadienyl ligands see: *J. Organomet. Chem.* **2001**, *637–639*.

(3) (a) Fischer, E. O.; Fischer, R. D. *Angew. Chem.* **1960**, *72*, 919. (b) Mahler, J. E.; Pettit, R. *J. Am. Chem. Soc.* **1962**, *84*, 1511.

(4) (a) Powell, P. (West, R., Stone, F. G. A. Eds.), *Adv. Organomet. Chem.* **1986**, *26*, Academic, New York, p. 125. (b) Yasuda, H.; Nakamura, A. *J. Organomet. Chem.* **1985**, *285*, 15.

(5) (a) Ernst, R. D. *Acc. Chem. Res.* **1985**, *18*, 56. (b) Ernst, R. D. *Struct. Bond.* **1984**, *57*, 1. (c) Ernst, R. D. *Chem. Rev.* **1988**, *88*, 1255.

(6) Ernst, R. D. *Comments Inorg. Chem.* **1999**, *21*, 285.

(7) (a) Yasuda, H.; Tatsumi, K.; Nakamura, A. *Acc. Chem. Res.* **1985**, *18*, 120. (b) Erker, G.; Krüger, C.; Muller, G. (West, R., Stone, F. G. A. Eds.), *Adv. Organomet. Chem.* **1985**, *24*, Academic Press, New York, p. 1. (c) Christensen, N. L.; Legzdins, P.; Einstein, F. W. B.; Jones, R. H. *Organometallics* **1991**, *10*, 3070.

(8) Cardin, D. J.; Lappert, M. F.; Raston, C. *Chemistry of Organo-Zirconium and Hafnium Compounds*, Halsted Press, New York, 1986.

(9) Haaland, A. *Acc. Chem. Res.* **1979**, *12*, 415.

(10) (a) Pez, G. P.; Armor, J. N. (West, R., Stone, F. G. A. Eds.), *Adv. Organomet. Chem.* **1981**, *19*, Academic Press, New York, p. 1. (b) Troyanov, S. I.; Antropiusová, H.; Mach, K. *J. Organomet. Chem.* **1992**, *427*, 49.

(11) Bercaw, J. E. *J. Am. Chem. Soc.* **1974**, *96*, 5087.

(12) (a) Guggenberger, L. J. *Inorg. Chem.* **1973**, *12*, 294. (b) Egan, J. W. Jr.; Petersen, J. L. *Organometallics* **1986**, *5*, 906. (c) Couldwell, C.; Prout, K. *Acta Crystallogr. B* **1979**, *35*, 335. (d) Cheung, K. K.; Cross, R. J.; Forrest, K. P.; Wardle, R.; Mercer, M. *J. Chem. Soc., Chem. Commun.* **1971**,

875. (e) Pasman, P.; Snel, J. J. M. *J. Organomet. Chem.* **1984**, *276*, 387. (f) Fischer, E. O.; Wawersik, H. *J. Organomet. Chem.* **1966**, *5*, 559.
(13) Ernst, R. D.; Liu, J.-Z.; Wilson, D. R. *J. Organomet. Chem.* **1983**, *250*, 257.
(14) Liu, J.-Z.; Ernst, R. D. *J. Am. Chem. Soc.* **1982**, *104*, 3737.
(15) (a) Stahl, L.; Ernst, R. D. *J. Am. Chem. Soc.* **1987**, *109*, 5673. (b) Stahl, L.; Trakarnpruk, W.; Freeman, J. W.; Arif, A. M.; Ernst, R. D. *Inorg. Chem.* **1995**, *34*, 1810.
(16) Campana, C. F.; Ernst, R. D.; Wilson, D. R.; Liu, J.-Z. *Inorg. Chem.* **1984**, *23*, 2732.
(17) (a) DiMauro, P. T.; Wolczanski, P. T. *Organometallics* **1987**, *6*, 1947. (b) Newbound, T. D.; Rheingold, A. L.; Ernst, R. D. *Organometallics* **1992**, *11*, 1693.
(18) Wang, Y.; Liang, Y.-Q.; Liu, Y.-J.; Liu, J.-Z. *J. Organomet. Chem.* **1993**, *443*, 185.
(19) Turpin, G. C.; Ernst, R. D. Unpublished results.
(20) Ernst, R. D.; Freeman, J. W.; Swepston, P. N.; Wilson, D. R. *J. Organomet. Chem.* **1991**, *402*, 17.
(21) Gedridge, R. W.; Arif, A. M.; Ernst, R. D. *J. Organomet. Chem.* **1995**, *501*, 95.
(22) Kulsomphob, V.; Tomaszewski, R.; Yap, G. P. A.; Liable-Sands, L. M.; Rheingold, A. L.; Ernst, R. D. *J. Chem. Soc., Dalton Trans.* **1999**, 3995.
(23) (a) Horácek, M.; Kupfer, V.; Thewalt, U.; Štepnicka, P.; Polášek, M.; Mach, K. *Organometallics* **1999**, *18*, 3572. (b) Lukešová, L.; Horácek, M.; Štepnicka, P.; Fejfarová, K.; Gyepes, R.; Císařová, I.; Kubišta, J.; Mach, K. *J. Organomet. Chem.* **2002**, *663*, 134.
(24) (a) van Oven, H. O.; de Liefde Meijer, J. H. *J. Organomet. Chem.* **1971**, *31*, 71. (b) Timms, P. L.; Turney, T. W. *J. Chem. Soc., Dalton Trans.* **1976**, 2021.
(25) Basta, R.; Arif, A. M.; Ernst, R. D. *Organometallics* **2003**, *22*, 812.
(26) DiMauro, P. T.; Wolczanski, P. T. *Polyhedron* **1995**, *14*, 149.
(27) Vos, D.; Arif, A. M.; Ernst, R. D. *J. Organomet. Chem.* **1998**, *553*, 277.
(28) (a) Brintzinger, H.; Bercaw, J. E. *J. Am. Chem. Soc.* **1971**, *93*, 2045. (b) Use of very bulky cyclopentadienyl ligands can also lead to the stabilization of titanocene monocarbonyls.[28c] (c) Hanna, T. E.; Lobkovsky, E.; Chirik, P. J. *J. Am. Chem. Soc.* **2006**, *128*, 6018.
(29) Newbound, T. D.; Ernst, R. D. Unpublished results.
(30) (a) Tolman, C. A. *Chem. Rev.* **1977**, *77*, 313. (b) Bart, J. C. J.; Favini, G.; Todeschini, R. *Phosphorus Sulfur* **1983**, *17*, 205.
(31) Ernst, R. D.; Freeman, J. W.; Stahl, L.; Wilson, D. R.; Arif, A. M.; Nuber, B.; Ziegler, M. L. *J. Am. Chem. Soc.* **1995**, *117*, 5075.
(32) Kulsomphob, V.; Tomaszewski, R.; Rheingold, A. L.; Arif, A. M.; Ernst, R. D. *J. Organomet. Chem.* **2002**, *655*, 158.
(33) (a) Stahl, L.; Hutchinson, J. P.; Wilson, D. R.; Ernst, R. D. *J. Am. Chem. Soc.* **1985**, *107*, 5016. (b) Waldman, T. E.; Stahl, L.; Wilson, D. R.; Arif, A. M.; Hutchinson, J. P.; Ernst, R. D. *Organometallics* **1993**, *12*, 1543. (c) Harvey, B. G.; Basta, R.; Arif, A. M.; Ernst, R. D. *Dalton Trans.* **2004**, 1221.
(34) Waldman, T. E.; Rheingold, A. L.; Ernst, R. D. *J. Organomet. Chem.* **1995**, *503*, 29.
(35) Gedridge, R. W.; Ernst, R. D. Unpublished results.
(36) Basta, R.; Ernst, R. D.; Arif, A. M. *J. Organomet. Chem.* **2003**, *683*, 64.
(37) Basta, R.; Harvey, B. G.; Arif, A. M.; Ernst, R. D. *Inorg. Chim. Acta* **2004**, *357*, 3883.
(38) (a) Togni, A.; Halterman, R. Eds., *Metallocenes: Synthesis, Reactivity, Applications*, Vols. 1 and 2, **1998**, Wiley-VCH, Weinheim, New York. (b) Marek, I. Ed., *Titanium and Zirconium in Organic Synthesis*, **2002**, Wiley-VCH, Weinheim, Cambridge, England.
(39) Basta, R.; Arif, A. M.; Ernst, R. D. *Organometallics* **2005**, *24*, 3974.
(40) (a) Waldman, T. E.; Wilson, A. M.; Rheingold, A. L.; Meléndez, E. M.; Ernst, R. D. *Organometallics* **1992**, *11*, 3201. (b) Tomaszewski, R.; Arif, A. M.; Ernst, R. D. *J. Chem. Soc., Dalton Trans.* **1999**, 1883. (c) Dobado, J. A.; Molina, J. M.; Uggla, R.; Sundberg, M. R. *Inorg. Chem.* **2000**, *39*, 2831.
(41) (a) Bazan, G. C.; Rodriguez, G.; Ashe, A. J. III; Al-Ahmad, S.; Müller, C. *J. Am. Chem. Soc.* **1996**, *118*, 2291. (b) Rogers, J. S.; Bazan, G. C.; Sperry, C. K. *J. Am. Chem. Soc.* **1997**, *119*, 9305. (c) Ashe, A. J. III.; Al-Ahmad, S.; Fang, X.; Kampf, J. W. *Organometallics* **1998**, *17*, 3883.

(d) Bazan, G. C.; Rodriguez, G.; III. Ashe, J. R.; Al-Ahmad, S.; Kampf, J. W. *Organometallics* **1997**, *16*, 2492. (e) Fu, G. C. *Adv. Organomet. Chem.* **2001**, *47*, 101.
(42) Bazan, G. C.; Cotter, W. D.; Komon, Z. J. A.; Lee, R. A.; Lachicotte, R. J. *J. Am. Chem. Soc.* **2000**, *122*, 1371.
(43) (a) Varga, V.; Polášek, M.; Hiller, J.; Thewalt, U.; Sedmera, P.; Mach, K. *Organometallics* **1996**, *15*, 1268. (b) Štepnicka, P.; Císacová, I.; Horácek, M.; Mach, K. *Acta Crystallogr.* **2000**, C56, 1204. (c) Wilson, D. R.; Schmidt, G. F. PCT Int. Appl. WO 96 16,094.
(44) Freeman, J. W.; Hallinan, N. C.; Arif, A. M.; Gedridge, R. W.; Ernst, R. D.; Basolo, F. *J. Am. Chem. Soc.* **1991**, *113*, 6509.
(45) (a) Fritz, H. P.; Schwartzhans, K.-E. *J. Organomet. Chem.* **1964**, *1*, 208. (b) Gordon, K. R.; Warren, K. D. *Inorg. Chem.* **1978**, *17*, 987.
(46) (a) Meléndez, E.; Arif, A. M.; Ziegler, M. L.; Ernst, R. D. *Angew. Chem. Int. Ed.* **1988**, *27*, 1099. (b) Hyla-Kryspin, I.; Waldman, T. E.; Meléndez, E.; Trakarnpruk, W.; Arif, A. M.; Ziegler, M. L.; Ernst, R. D.; Gleiter, R. *Organometallics* **1995**, *14*, 5030.
(47) Wilson, A. M.; West, F. G.; Arif, A. M.; Ernst, R. D. *J. Am. Chem. Soc.* **1995**, *117*, 8490.
(48) Tomaszewski, R.; Hyla-Kryspin, I.; Mayne, C. L.; Arif, A. M.; Gleiter, R.; Ernst, R. D. *J. Am. Chem. Soc.* **1998**, *120*, 2959.
(49) Tomaszewski, R.; Wilson, A. M.; Liable-Sands, L.; Rheingold, A. L.; Ernst, R. D. *J. Organomet. Chem.* **1999**, *583*, 42.
(50) Tomaszewski, R.; Lam, K.-C.; Rheingold, A. L.; Ernst, R. D. *Organometallics* **1999**, *18*, 4174.
(51) Calderazzo, F.; Salzmann, J. J.; Mosimann, P. *Inorg. Chim. Acta* **1967**, *1*, 65.
(52) Wilson, A. M.; West, F. G.; Rheingold, A. L.; Ernst, R. D. *Inorg. Chim. Acta* **2000**, *300–302*, 65.
(53) Basta, R.; Harvey, B. G.; Arif, A. M.; Ernst, R. D. *J. Am. Chem. Soc.* **2005**, *127*, 11924.
(54) (a) Harvey, B. G.; Tomaszewski, R.; Arif, A. M.; Rheingold, A. L.; Ernst, R. D. Unpublished results. (b) Harvey, B. G.; Arif, A. M.; Ernst, R. D. *J. Organomet. Chem.* **2006**, *691*, 5211. (c) Harvey, B. G.; Kulsomphob, V.; Arif, A. M.; Ernst, R. D. *J. Organomet. Chem.* **2007**, in press.
(55) Casey, C. P.; Nief, F. *Organometallics* **2002**, *21*, 3182.
(56) Gorsich, R. D. U.S. Patent 3,080,305, **1963**.
(57) Li, C.; Arif, A. M.; Ernst, R. D. Unpublished results.
(58) Kulsomphob, V.; Arif, A. M.; Ernst, R. D. *Organometallics* **2002**, *21*, 3182.
(59) Kulsomphob, V.; Harvey, B. G.; Arif, A. M.; Ernst, R. D. *Inorg. Chim. Acta* **2002**, *334*, 17.
(60) Gedridge, R. W.; Hutchinson, J. P.; Rheingold, A. L.; Ernst, R. D. *Organometallics* **1993**, *12*, 1553.
(61) Basta, R.; Arif, A. M.; Ernst, R. D. Unpublished results.
(62) Rajapakshe, A.; Gruhn, N. E.; Lichtenberger, D. L.; Basta, R.; Arif, A. M.; Ernst, R. D. *J. Am. Chem. Soc.* **2004**, *126*, 14105.
(63) Rajapakshe, A.; Basta, R.; Arif, A. M.; Ernst, R. D.; Lichtenberger, D. L. *Organometallics* **2007**, *26*, 2867.
(64) Tomaszewski, R.; Harvey, B. G.; Arif, A. M.; Ernst, R. D. Unpublished results.
(65) Horton, A. D.; Orpen, A. G. *Organometallics* **1992**, *11*, 8.
(66) (a) Wooster, T. T.; Geiger, W. E. Jr; Ernst, R. D. *Organometallics* **1995**, *14*, 3455. (b) Stahl, L.; Zahn, T.; Ziegler, M. L.; Ernst, R. D. *Inorg. Chim. Acta* **1999**, *288*, 154.
(67) Sorensen, T. S.; Jablonski, C. R. *J. Organomet. Chem.* **1970**, *25*, C62.
(68) Fischer, P. J.; Ahrendt, K. A.; Young, V. G. Jr.; Ellis, J. E. *Organometallics* **1998**, *17*, 13.
(69) Gedridge, R. W.; Ernst, R. D. Unpublished results.
(70) Kralik, M. S.; Rheingold, A. L.; Hutchinson, J. P.; Freeman, J. W.; Ernst, R. D. *Organometallics* **1996**, *15*, 551.
(71) Feng, S.; Klosin, J.; Kruper, W. J. Jr.; McAdon, M. H.; Neithamer, D. R.; Nickias, P. N.; Patton, J. T.; Wilson, D. R.; Abboud, K. A.; Stern, C. L. *Organometallics* **1999**, *18*, 1159.
(72) Arif, A. M.; Basta, R.; Ernst, R. D. *Polyhedron* **2006**, *25*, 876.
(73) Bartlett, P. D.; Seidel, B. *J. Am. Chem. Soc.* **1961**, *83*, 581.
(74) Datta, S.; Wreford, S. S. *J. Am. Chem. Soc.* **1979**, *101*, 1053.
(75) Fischer, M. B.; James, E. J.; McNeese, T. J.; Nyburg, S. C.; Posin, B.; Wong-Ng, W.; Wreford, S. S. *J. Am. Chem. Soc.* **1980**, *102*, 4941.

(76) (a) Paz-Sandoval, M. A.; Rangel-Salas, I. I. *Coord. Chem. Rev.* **2006**, *250*, 1071. (b) Bleeke, J. R. *Organometallics* **2005**, *24*, 5190.
(77) (a) Giannini, L.; Solari, E.; de Angelis, S.; Ward, T. R.; Floriani, C.; Chiesi-Villa, A.; Rizzoli, C. *J. Am. Chem. Soc.* **1995**, *117*, 5801. (b) Martin, A.; Uhrhammer, R.; Gardner, T. G.; Jordan, R. F.; Rogers, R. D. *Organometallics* **1998**, *17*, 382.
(78) Hitchcock, P. B.; Lappert, M. F.; Liu, D.-S. *J. Chem. Soc., Chem. Commun.* **1994**, 2637.
(79) Rahim, M.; Taylor, N. J.; Xin, S.; Collins, S. *Organometallics* **1998**, *17*, 1315.
(80) Vollmerhaus, R.; Rahim, M.; Tomaszewski, R.; Xin, S.; Taylor, N. J.; Collins, S. *Organometallics* **2000**, *19*, 2161.
(81) (a) Basuli, F.; Kilgore, U. J.; Brown, D.; Huffman, J. C.; Mindiola, D. J. *Organometallics* **2004**, *23*, 6166. (b) Hamaki, H.; Takeda, N.; Tokitoh, N. *Organometallics* **2006**, *25*, 2457. (c) Bai, G.; Wei, P.; Stephan, D. W. *Organometallics* **2006**, *25*, 2649. (d) Bailey, B. C.; Basuli, F.; Huffman, J. C.; Mindiola, D. J. *Organometallics* **2006**, *25*, 3963. (e) Zhao, G.; Basuli, F.; Kilgore, U. J.; Fan, H.; Aneetha, H.; Huffman, J. C.; Wu, G.; Mindiola, D. J. *J. Am. Chem. Soc.* **2006**, *128*, 13575.
(82) (a) Deelman, B.-J.; Hitchcock, P. B.; Lappert, M. F.; Leung, W.-P.; Lee, H.-K.; Mak, T. C. W. *Organometallics* **1999**, *18*, 1444. (b) Cortright, S. B.; Johnston, J. N. *Angew. Chem. Int. Ed.* **2002**, *41*, 345. (c) Cortright, S. B.; Huffman, J. C.; Yoder, R. A.; Coalter, J. N. III; Johnston, J. N. *Organometallics* **2004**, *23*, 2238.
(83) Porter, R. M. *J. Chem. Soc., Dalton Trans.* **2002**, 3290.
(84) (a) Basuli, F.; Bailey, B. C.; Watson, L. A.; Tomaszewski, J.; Huffman, J. C.; Mindiola, D. J. *Organometallics* **2005**, *24*, 1886. (b) Basuli, F.; Clark, R. L.; Bailey, B. C.; Brown, D.; Huffman, J. C.; Mindiola, D. J. *Chem. Commun.* **2005**, 2250.
(85) Basuli, F.; Aneetha, H.; Huffman, J. C.; Mindiola, D. J. *J. Am. Chem. Soc.* **2005**, *127*, 17992.
(86) Basuli, F.; Bailey, B. C.; Tomaszewski, J.; Huffman, J. C.; Mindiola, D. J. *J. Am. Chem. Soc.* **2003**, *125*, 6052.
(87) Yasuda, H.; Nagasuna, K.; Akita, M.; Lee, K.; Nakamura, A. *Organometallics* **1984**, *3*, 1470.
(88) (a) Herberich, G. E.; Kreuder, C.; Englert, U. *Angew. Chem. Int. Ed.* **1994**, *33*, 2465. (b) Mach, K.; Kubista, J.; Gyepes, R.; Trojan, L.; Stepnicka, P. *Inorg. Chem. Commun.* **2003**, *6*, 352. (c) Erker, G.; Petrenz, R.; Krüger, C.; Nolte, M. *J. Organomet. Chem.* **1992**, *431*, 297.
(89) Harvey, B. G.; Arif, A. M.; Ernst, R. D. Unpublished results.
(90) Schlosser, M.; Rauchschwalbe, G. *J. Am. Chem. Soc.* **1978**, *100*, 3258.
(91) Harvey, B. G. Ph.D. Thesis, University of Utah, **2005**.
(92) Pillet, S.; Wu, G.; Kulsomphob, V.; Harvey, B. G.; Ernst, R. D.; Coppens, P. *J. Am. Chem. Soc.* **2003**, *125*, 1937.
(93) Sabino, J. R.; Coppens, P. *Acta Crystallogr.* **2003**, A59, 127.
(94) Ernst, R. D.; Kulsomphob, V.; Arif, A. M. *Z. Kristallogr.-New Cryst. Struct.* **2006**, *221*, 293.
(95) Wilson, A. M.; Waldman, T. E.; Rheingold, A. L.; Ernst, R. D. *J. Am. Chem. Soc.* **1992**, *114*, 6252.
(96) Wilson, A. M.; Klein, M.; Ernst, R. D. Unpublished results.
(97) Wilson, A. M. Ph.D. Thesis, University of Utah, **1994**.
(98) Basta, R. Ph.D. Thesis, University of Utah, **2004**.
(99) Basta, R.; Arif, A. M.; Ernst, R. D. *Organometallics* **2005**, *24*, 3982.
(100) Ernst, R. D.; Basta, R.; Arif, A. M. *Z. Kristallogr.—New Cryst. Struct.* **2004**, *219*, 403.
(101) Brookhart, M.; Green, M. L. H.; Wong, L.-L. (Lippard, S. J. Ed.), *Prog. Inorg. Chem.* **1988**, *36*, Interscience, New York, p. 1.
(102) Harvey, B. G.; Mayne, C. L.; Arif, A. M.; Tomaszewski, R.; Ernst, R. D. *J. Am. Chem. Soc.* **2005**, *127*, 16426 (corrn: **2006**, *128*, 1770).
(103) (a) Otsuka, S.; Nakamura, A. (West, R., Stone, F. G. A. Eds.), *Adv. Organomet. Chem.* **1976**, *14*, 245. (b) Templeton, J. (R. West, F. G. A. Stone, Eds.), *Adv. Organomet. Chem.* **1989**, *29*, 1.
(104) Yasuda, H.; Yamauchi, M.; Nakamura, A.; Sei, T.; Kai, Y.; Yasuoka, N.; Kasai, N. *Bull. Chem. Soc. Jpn.* **1980**, *53*, 1089.
(105) Bertus, P.; Drouin, L.; Laroche, C.; Szymoniak, J. *Tetrahedron* **2004**, *60*, 1375.
(106) (a) Hessen, B.; Stuczynski, S. M.; Steigerwald, M. L. Presented at the 205th National ACS Meeting, Division of Inorganic Chemistry, Denver, CO, March 28, 1993, #18. (b) Hessen, B.; Seigrist, T.; Palstra, T.; Tanzler, S. M.; Steigerwald, M. L. *Inorg. Chem.* **1993**, *32*, 5165.

(107) (a) Severson, S. J.; Cymbaluk, T. H.; Ernst, R. D.; Higashi, J. M.; Parry, R. W. *Inorg. Chem.* **1983**, *22*, 3833. (b) Elschenbroich, Ch.; Nowotny, M.; Behrendt, A.; Harms, K.; Wocadlo, S.; Pebler, J. *J. Am. Chem. Soc.* **1994**, *116*, 6217. (c) Newbound, T. D.; Freeman, J. W.; Wilson, D. R.; Kralik, M. S.; Patton, A. T.; Campana, C. F.; Ernst, R. D. *Organometallics* **1986**, *6*, 2432.

(108) (a) Groshens, T. J.; Lowe-Ma, C. K.; Scheri, R. C.; Dalbey, R. Z. *Mater. Res. Soc. Symp. Proc.* **1993**, *282* (Chemical Perspectives of Microelectronic Materials III), p. 299. (b) Wanner, B. D. U.S. Patent **1993**, 5,273,783. (c) Spencer, J. T.; Ernst, R. D. U.S. Patent **1994**, 5,352,488. (d) Westmoreland, D. L.; Wanner, B. D.; Atwell, D. R. U.S. Patent **1997**, 5,693,377. (e) Westmoreland, D. L.; Wanner, B. D.; Atwell, D. R. U.S. Patent **1999**, 5,902,651. (f) Dando, R. S.; Carpenter, C. M.; Mardian, A. P.; Derderian, G. J.; Gealy, D. U.S. Patent **2004**, 6,797,337.

(109) (a) Smith, P. D.; McDaniel, M. P. *J. Polym. Sci.* **1989**, *27A*, 2695. (b) McDaniel, M. P.; Smith, P. D. U.S. Patent **1991**, 5,075,394.

(110) Dawkins, G. M. Eur. Pat. Appl. EP **1992**, 501,672.

(111) (a) Kohara, T.; Satoshi, K. U.S. Patent **1989**, 4,871,704. (b) Kohara, T.; Satoshi, K. U.S. Patent **1990**, 4,926,002.

(112) (a) Wilson, D. R.; Nickias, P. N.; Neithamer, D. R.; Ernst, R. D. U.S. Patent **1998**, 5,817,849. (b) Wilson, D. R.; Nickias, P. N.; Neithamer, D. R.; Ernst, R. D. U.S. Patent **2000**, 6,034,021. (c) Ewen, J. A.; Strozier, R. W. U.S. Patent **2000**, 6,069,237.

(113) (a) Rodriguez, G.; Speca, A. N.; Kuchta, M. C.; McConville, D. H.; Burkhardt, T. J. U.S. Patent Application 20050159299. (b) Devore, D. D.; Timmers, F. J.; Neithamer, D. R. U.S. Patent **2001**, 6,255,246. (c) Devore, D. D.; Timmers, F. J.; Neithamer, D. R. U.S. Patent **2002**, 6,365,779. (d) Gruter, G. J. M.; Van Doremaele, G. H. J.; Arts, H. J. PCT Int. Appl. WO 97 42,163.

(114) Wilson, D. R.; Hahn, S. F. PCT Int. Appl. WO 96 18,660.

(115) (a) Gavenonis, J.; Tilley, T. D. *J. Am. Chem. Soc.* **2002**, *124*, 8536. (b) Gavenonis, J.; Tilley, T. D. *Organometallics* **2002**, *21*, 5549.

(116) Lentz, M. R.; Fanwick, P. E.; Rothwell, I. P. *Organometallics* **2003**, *22*, 2259.

(117) Gutierrez, A.; Wilkinson, G.; Hussain-Bates, B.; Hursthouse, M. B. *Polyhedron* **1990**, *9*, 2081.

Mixed Metal Acetylide Complexes

PRADEEP MATHUR,* SAURAV CHATTERJEE and VIDYA D. AVASARE

Chemistry Department, Indian Institute of Technology-Bombay, Powai, Bombay 400 076, India

I
INTRODUCTION

Complexes containing C_2 hydrocarbyl ligands occupy a very important position in the development of di-, tri- and polynuclear organometallic chemistry. Recognition that monovalent anions of alk-1-ynes, $[RC{\equiv}C]^-$ are isoelectronic to cyanide ion and CO prompted the first preparation of metal alkynyl complexes in 1953.[1,2] Since then, a number of synthetic strategies have been developed and chemistry of metal acetylide complexes has grown tremendously.[3–12] Ability of the alkynyl part ($-C{\equiv}CR$) of the metal acetylides to bind to metal centres in a variety of bonding modes enables a large number of acetylide-bridged polynuclear complexes to be synthesised.[13–26] The presence of metal and electron-rich $C{\equiv}C$ moiety in acetylide complexes facilitates cluster growth reactions, and frequently, coupling of acetylide moieties to form polycarbon chains on cluster frameworks is observed.[27–30]

In this review, focus is placed on mixed-metal acetylide complexes which have been prepared from metal–acetylide precursor complexes. Particular emphasis is given to aspects of variation of acetylide bonding on mixed-metal polynuclear frameworks and to reactivity of the polycoordinated acetylide ligand. Since we recently reviewed oxo-incorporated metal acetylide complexes,[31] these are excluded from the present review. A large majority of the acetylide-bridged mixed metal complexes contains metals of the same group of the Periodic Table, and therefore, we have subdivided the review based on different groups. Although several examples exist of mixed-metal acetylide complexes containing metals from groups 6, 7, 9 and 12, quite surprisingly, we could not find any example of mixed-metal acetylide complexes containing metals only from one of these groups.

*Corresponding author.
E-mail: mathur@iitb.ac.in (P. Mathur).

ADVANCES IN ORGANOMETALLIC CHEMISTRY
VOLUME 55 ISSN 0065-3055/DOI 10.1016/S0065-3055(07)55004-5

II
MIXED GROUP 8 METAL ACETYLIDE COMPLEXES

Most of the mixed metal acetylide complexes of this particular group are those associated with Cp, Cp* and carbonyl ligands. A number of synthetic studies have been carried out for polynuclear C_2 complexes derived from ethynyl and diethynediyl iron complexes. Reaction of the ethynyliron complexes $[\{(\eta^5\text{-}C_5R_5)Fe(CO)_2\}\text{–}C{\equiv}C\text{–}H]$ (R = H, **1**; R = Me, **2**) with $[Ru_3(CO)_{12}]$ in refluxing benzene affords triruthenium μ-η^1,η^2,η^2-acetylide cluster compounds $[Ru_3(CO)_9(\mu\text{-}H)(\mu_3\text{-}\eta^1{:}\eta^2{:}\eta^2\text{-}C{\equiv}C\text{–}\{(\eta^5\text{-}C_5R_5)Fe(CO)_2\})]$ (R = H, **4**; R = Me, **5**) (Scheme 1).[32] On the other hand, the reaction of the ethynediyldiiron complex, $[\{(\eta^5\text{-}C_5Me_5)Fe(CO)_2\}\text{–}C{\equiv}C\text{–}\{(\eta^5\text{-}C_5Me_5)Fe(CO)_2\}]$ (**3**) with $[Ru_3(CO)_{12}]$, gives a complicated mixture of products, from which $[Cp^*{}_2Fe_2Ru_2(\mu_4\text{-}C_2)(CO)_{10}]$ (**6**) and $[Cp^*{}_2Fe_2Ru_6(\mu_6\text{-}C_2)(CO)_{17}]$ (**7**) have been isolated and characterised by X-ray crystallography as permetalated ethene and permetalated ethane complexes, respectively (Scheme 2).[32] The four metal atoms in **6** form an open-rectangular array embedded in which is a C_2 ligand. Compound **7** is an octanuclear compound with two C_2 units above and below a Ru_4 plane arranged perpendicular to each other. Two of the edges of the central Ru_4 plane are bridged by $Ru(CO)_3$ groups and the other two by FeCp* moieties.

On reflux of a THF solution of a mixture of $[Fe_2(CO)_9]$ and **6**, iron-substituted derivative of **5**, complex $[Ru_2Fe(\mu\text{-}H)(CO)_9(\mu_3\text{-}C{\equiv}C\text{–}\{(\eta^5\text{-}C_5Me_5)Fe(CO)_2\})]$ (**8**) is formed. Treatment of **6** with CF_3SO_3H, $HBF_4.OEt_2$ or CF_3COOH forms an unstable cationic species **9**, which has been characterised solely by spectroscopy, as it readily converts to $[\{(\eta^5\text{-}C_5Me_5)Fe(CO)]^+\ X^-$ (Scheme 3).[32]

Thermolysis of the mixed-metal tetranuclear cluster $[(\mu_3\text{-}C{\equiv}C\text{–}\{(\eta^5\text{-}C_5H_5)Fe(CO)_2\})Ru_3(\mu\text{-}H)(CO)_9]$ (**4**) affords two products: $[CpFeRu_6(\mu_5\text{-}C_2)(\mu_5\text{-}C_2H)(CO)_{16}]$ (**10**) and $[Cp_2Fe_2Ru_6(\mu_6\text{-}C_2)_2(CO)_{17}]$ (**11**) (Scheme 4).[33] The anion $[(\mu_3\text{-}C{\equiv}C\text{–}\{(\eta^5\text{-}C_5H_5)Fe(CO)_2\})Ru_3(CO)_9]^-$ (**12**) derived from **4** undergoes coupling of the $\{(C_2)FeRu_3\}$ core to give compound **11** and a heptanuclear cluster $[Cp_2Fe_2Ru_5(\mu_5\text{-}C_2)_2(CO)_{17}]$ (**14**), a coordinatively saturated compound with 112 cluster valence electrons bearing two six-electron donating C_2 ligands.

Treatment of **6** with *R*-alpha-chloropropionic acid results in an addition reaction yielding a μ-hydrido μ-carboxylato derivative $[(\mu_4\text{-}C{=}C)(\mu\text{-}H)(\mu\text{-}\kappa^1,\kappa^1\text{-}CH_3CHClCOO)Fe_2Ru_2Cp^*{}_2(CO)_8]$ (**15**). On irradiation of a THF solution of **6**,

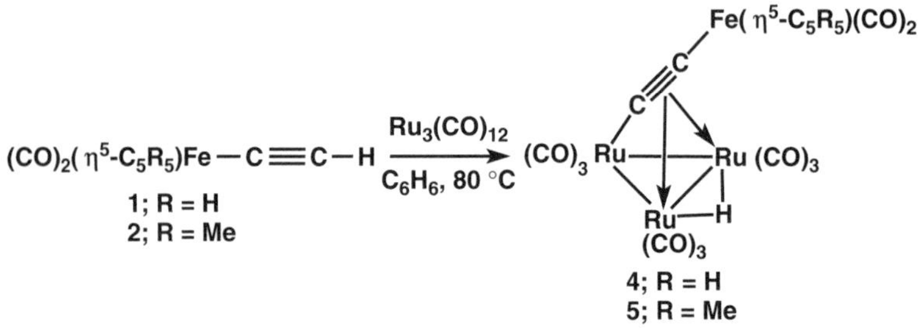

SCHEME 1.

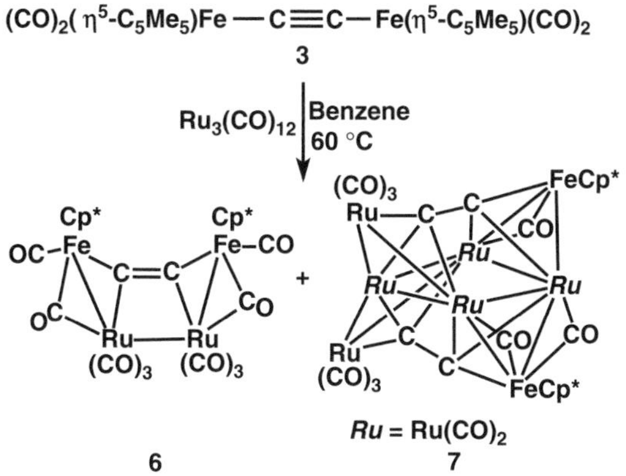

$(CO)_2(\eta^5\text{-}C_5Me_5)Fe—C{\equiv}C—Fe(\eta^5\text{-}C_5Me_5)(CO)_2$
3
$Ru_3(CO)_{12}$ Benzene 60 °C

Cp* Cp* OC–Fe Fe–CO C=C CO OC Ru—Ru $(CO)_3$ $(CO)_3$
+
$(CO)_3$ Ru—C—C FeCp* CO *Ru* Ru *Ru* *Ru* Ru Ru C C CO CO $(CO)_3$ FeCp*
Ru = $Ru(CO)_2$
6 7

SCHEME 2.

Cp* Cp* OC—Fe Fe—CO C=C CO OC Ru—Ru $(CO)_3$ $(CO)_3$
6
$Fe_2(CO)_9$
Fe (η^5-C_5Me_5) $(CO)_2$ C C $(CO)_3$ Fe Ru $(CO)_3$ Ru—H $(CO)_3$
8

HX

[Cp* Cp* OC–Fe Fe–CO C=C CO OC $(CO)_2$Ru—Ru $(CO)_2$ H $]^+X^-$ + Cp* Cp* OC–Fe Fe–CO C=C CO OC Ru—Ru $(CO)_3$ $(CO)_3$ $]^+X^-$]

9a; X = CF_3SO_3
9b; X = BF_4
9c; X = CF_3COO

$[Cp^*(CO)Fe]^+X^-$

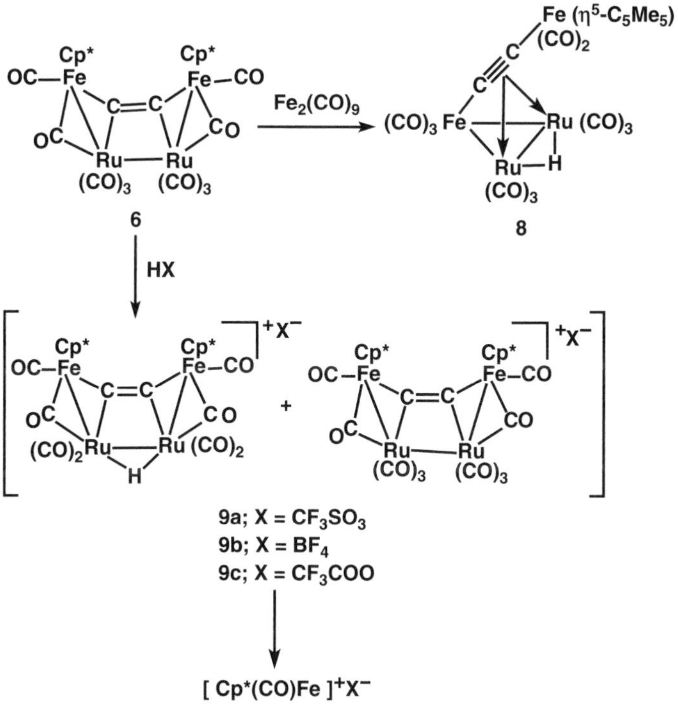

SCHEME 3.

in the presence of diphosphines, $[Ph_2P(CH_2)_nPPh_2]$ ($n = 1, 2$), the substituted products $[(\mu_4\text{-}C{=}C)Fe_2Ru_2Cp^*_2(CO)_8\{\mu\text{-}Ph_2P(CH_2)_nPPh_2\}]$ ($n = 1$, **16a**; $n = 2$, **16b**) are formed. Treatment of **16** with $HBF_4.OEt_2$ results in protonation at the C=C unit to give the cationic $\mu_4\text{-}C_2H$ complex, $[(\mu_4\text{-}C{=}CH)Fe_2Ru_2Cp^*_2(CO)_{12}(dppm)]BF_4$

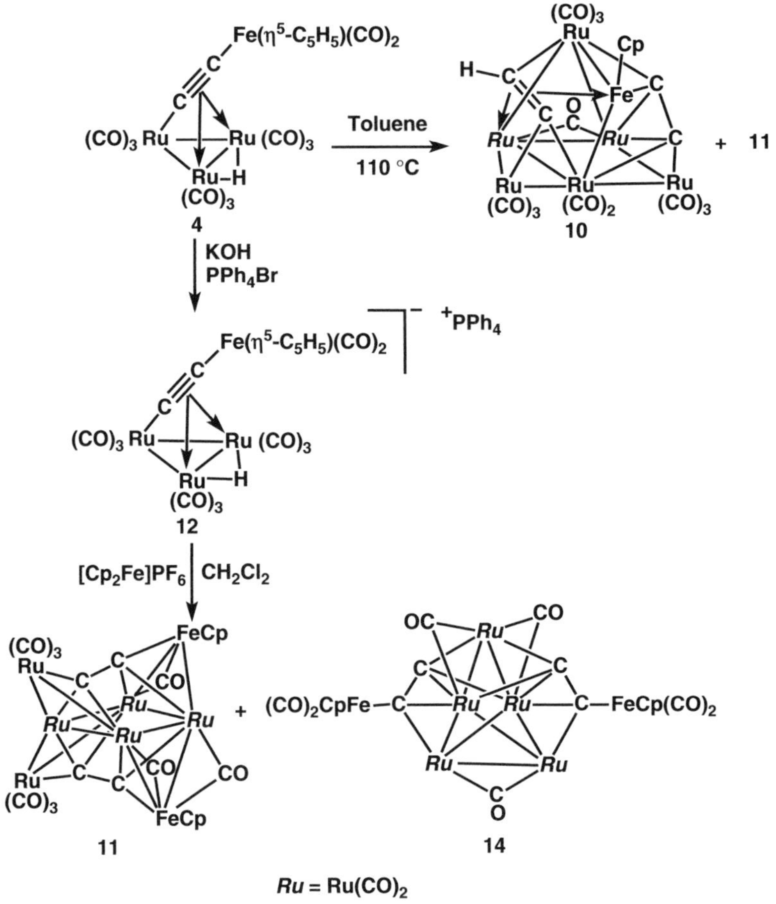

SCHEME 4.

(**17**) which on reduction with $[NEt_4]\ BH_4$ results in elimination of one of the iron fragments to afford a trinuclear acetylide cluster, $[(\mu_3\text{-C}{\equiv}\text{CH})\text{FeRu}_2\text{Cp}^*(\text{CO})_5(\text{dppm})]$ (**18a**). On the other hand, treatment of **17** with H_2SiPh_2 leads to hydrogen addition to give the trinuclear μ_3-HC=CH complex $[(\mu_3\text{-HC}{=}\text{CH})(\mu\text{-H})\text{FeRu}_2\text{Cp}^*(\text{CO})_5(\text{dppm})]$ (**18b**). Thus, the μ_3-HC=CH ligand in **18b** arises from formal hydrogenation of the μ_4-C=C ligand in **16a** and the ligand transformations are realised by the flexible coordination capability of the C_2 ligand. Cluster expansion of **16a** with $[Co_2(CO)_8]$ gives $[(\mu_3\text{-C}{\equiv}\text{CFe(CO)}_2\text{Cp}^*)\text{FeRu}_2\text{CoCp}^*(\text{CO})_{12}(\text{dppm})]$ (**19**) (Scheme 5).[34a]

Thermal reaction of a THF solution of ethynediyl complex $[\{\text{Cp}_2\text{Ru(CO)}_2\}_2(\mu\text{-C}{\equiv}\text{C})]$ (**20**) with diiron nonacarbonyl gives $[\text{Cp}_2\text{Fe}_2\text{Ru}_2(\mu_4\text{-C}{\equiv}\text{C})(\mu\text{-CO})(\text{CO})_8]$ (**21**) (Fig. 1).[34b]

Highly conjugated polycarbon–metal systems, the zwitterionic μ-but-2-yn-1-ylidene-4-ylidyne complex $[\text{Ru}_3(\text{CO})_{10}(\mu_3\text{-C–C}{\equiv}\text{C–}\mu\text{-C})\text{Fe}_2\text{Cp}^*{}_2(\text{CO})_3]$ (**23**) and

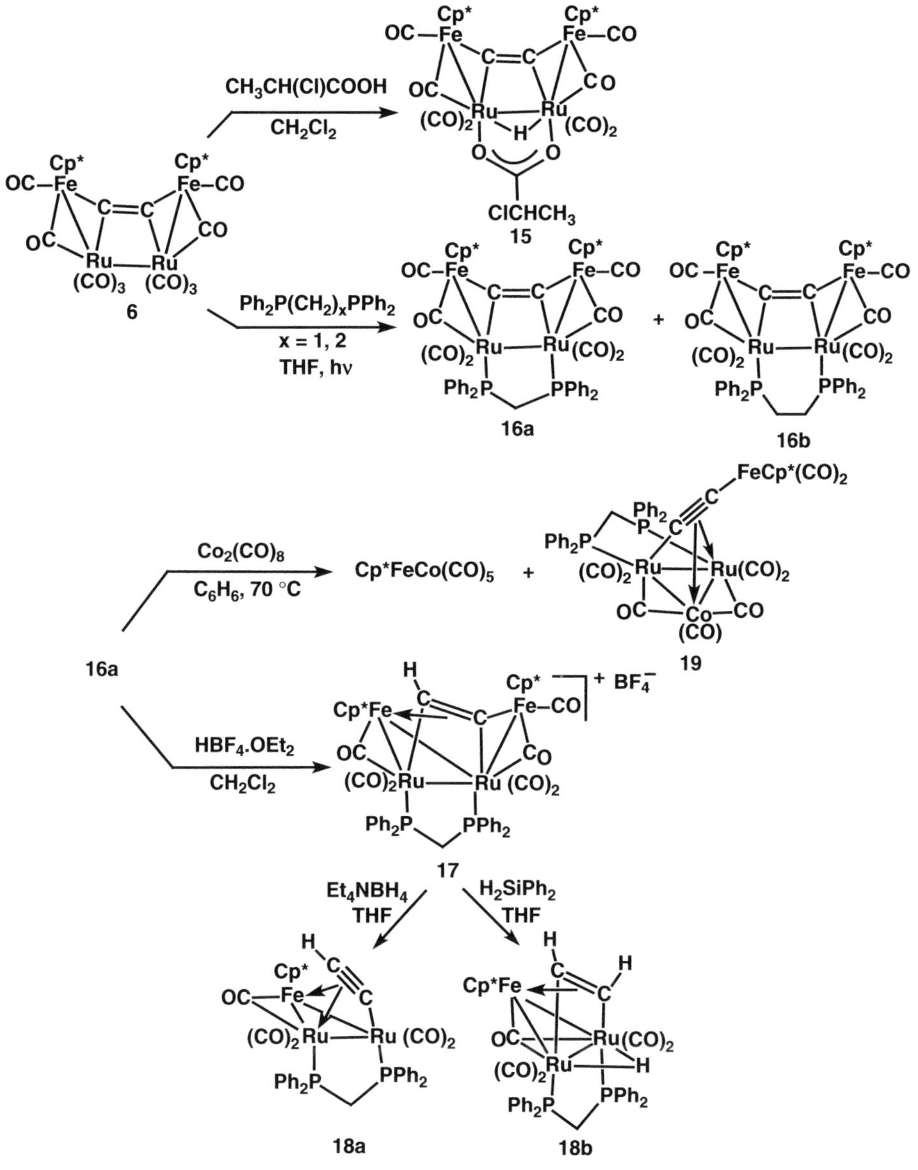

SCHEME 5.

the dimerised product with a cumulenic μ-C_8 ligand, $[(Cp^*Fe)_4Ru_2(CO)_{13}[\mu_6\text{-}C_8\text{–}C(=O)]$ (**24a**) have been isolated from the reactions of butadiynediyl-dimetal complex $[\{(\eta^5\text{-}C_5Me_5)Fe(CO)_2\}\text{–}C{\equiv}C\text{–}C{\equiv}C\text{–}Fe(\eta^5\text{-}C_5Me_5)(CO)_2]$ (**22a**) with $[Ru_3(CO)_{12}]$ (Scheme 6).

An isostructural C_8 complex **24b** is obtained on reaction of the Fe–Ru mixed metal complex $[\{(\eta^5\text{-}C_5Me_5)Fe(CO)_2\}\text{–}C{\equiv}C\text{–}C{\equiv}C\text{–}Ru(\eta^5\text{-}C_5Me_5)(CO)_2]$ (**22b**)

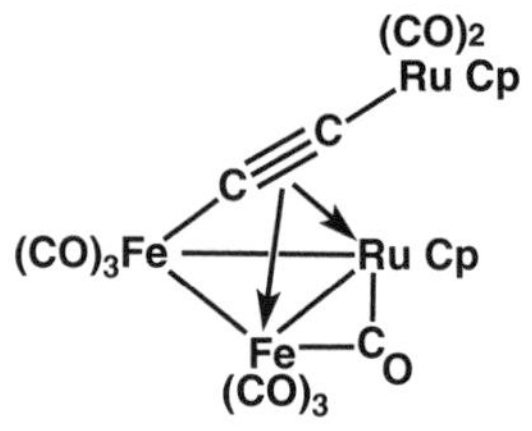

FIG. 1.

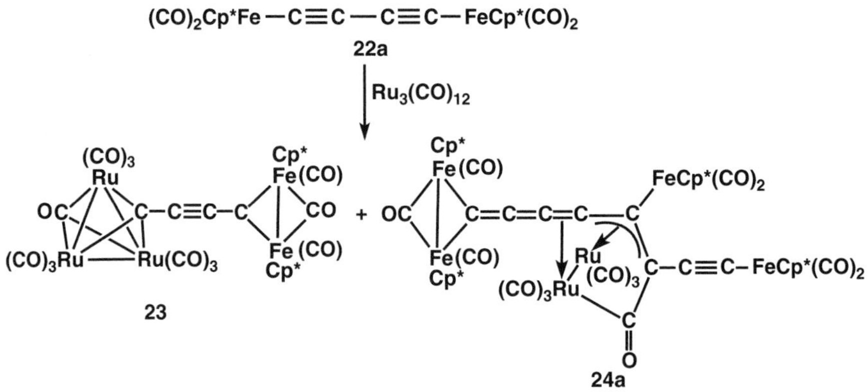

SCHEME 6.

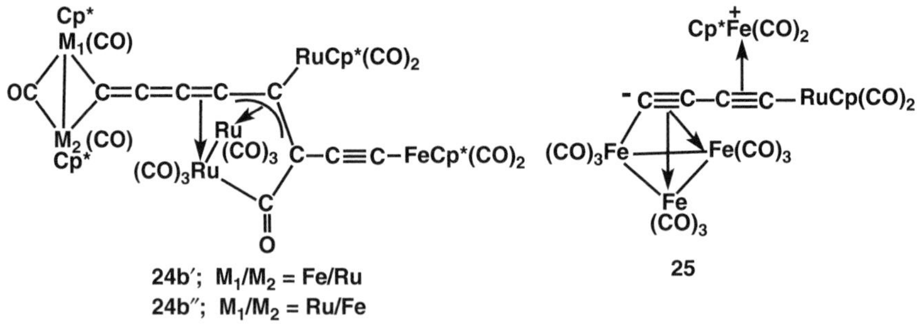

FIG. 2.

with $[Ru_3(CO)_{12}]$. A zwitterionic acetylide cluster-type product $\{(\eta^5\text{-}C_5Me_5)Fe(CO)_2\}^+[Cp(CO)_2Ru(\eta^2\text{-}C{\equiv}C)\text{–}(\mu_3\text{-}C{\equiv}C)Fe_3(CO)_9]^-$ (**25**) is formed by the reaction of butadiynediyldimetal complex $[\{(\eta^5\text{-}C_5Me_5)Fe(CO)_2\}\text{–}C{\equiv}C\text{–}C{\equiv}C\text{–}Ru(\eta^5\text{-}C_5H_5)(CO)_2]$ (**22c**) with $[Fe_2(CO)_9]$ (Fig. 2).[35a] X-ray crystallography reveals a pentanuclear structure containing a dinuclear $\mu\text{-}\eta^1,\eta^2$-acetylide cationic species and a trinuclear μ_3-acetylide anionic species.

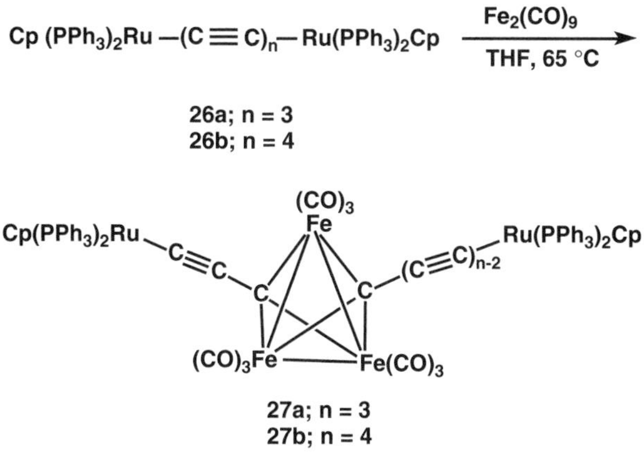

$Cp(PPh_3)_2Ru-(C\equiv C)_n-Ru(PPh_3)_2Cp \xrightarrow[\text{THF, 65 °C}]{Fe_2(CO)_9}$

26a; n = 3
26b; n = 4

$(CO)_3$ Fe
$Cp(PPh_3)_2Ru$ C≡C C C $(C\equiv C)_{n-2}$ $Ru(PPh_3)_2Cp$
$(CO)_3Fe$ $Fe(CO)_3$

27a; n = 3
27b; n = 4

SCHEME 7.

Thermolytic reactions of THF solution of $[\{CpRu(PPh_3)_2\}_2\{\mu\text{-}C\equiv C)_n\}]$ ($n = 3$, **26a**; $n = 4$, **26b**) with $[Fe_2(CO)_9]$ afford the corresponding heterometallic complexes $[Fe_3\{\mu_3\text{-}CC\equiv C\{Ru(PPh_3)Cp\}_2(CO)_9]$ (**27a**) and $[Fe_3\{\mu_3\text{-}CC\equiv C\{Ru(PPh_3)_2Cp\}$ $\{\mu_3\text{-}CC\equiv CC\equiv C\{Ru(PPh_3)_2Cp\}(CO)_9]$ (**27b**), respectively (Scheme 7).[35b]

The triosmium acetylide cluster, $[Os_3(\mu\text{-}H)(\mu_3\text{-}C\equiv CMe)(CO)_9]$ (**28**) reacts with $[Ru_3(CO)_{12}]$ in refluxing hexane to give the tetranuclear $[RuOs_3(\mu_4\text{-}HC_2Me)(CO)_{12}]$ (**29**) (Scheme 8).[36a]

Reaction of a dichloromethane solution of *cis*-$[(dppm)_2RuCl_2]$ with *trans*-$[(dppm)_2(Cl)Os(C\equiv C\text{-}p\text{-}C_6H_4\text{-}C\equiv CH)]$ and $NaPF_6$ gives a vinylidene complex, which converts to *trans*-$[(dppm)_2(Cl)Os(C\equiv C\text{-}p\text{-}C_6H_4\text{-}C\equiv C\text{-}Ru(Cl)(dppm)_2]$ (**30**) on addition of DBU in the reaction mixture (Scheme 9).[36b]

On reflux of a benzene solution containing $[Fe_2M(CO)_9(\mu_3\text{-}E)_2]$ (M = Ru; E = S, Se) (**31a,b**) and $[Fe(\eta^5\text{-}C_5H_4E'CH_2C\equiv CH)_2]$ (E′ = S, **32a**; Se, **32b**) unusual ferrocenyl containing heterometal clusters $[Fe(\eta^5\text{-}C_5H_4E'CH_2C\equiv CH)$ $(\eta^5\text{-}C_5H_4\{Fe_2M(CO)_8(\mu\text{-}E)(\mu_3\text{-}E)(E'CHCCH_2)\})]$ (M = Ru, E = Se and E′ = Se, **33a**; M = Ru, E = S and E′ = Se, **33b**) are formed (Scheme 10).[37] The molecular structure of **33b** consists of a butterfly $FeRuS_2$ unit attached to a $\{FeSeC(H) = CCH_2\}$ five-membered ring which is bonded *via* a Se atom to the C_5H_4 ring of a $\{(\eta^5\text{-}C_5H_4)(\eta^5\text{-}C_5H_4SeCH_2C\equiv CH)Fe\}$ unit.

III

MIXED GROUP 10 METAL ACETYLIDE COMPLEXES

Reaction of *trans*-bis(tri-*n*-butylphosphine)diethynylnickel (**34**) with l-alkynes in the presence of an amine complex of a copper(I) salt as a catalyst gives quantitatively alkynyl ligand exchange products. In this synthetic method, complex **34** reacts with appropriate α,ω-diethynyl compounds to afford high molecular weight

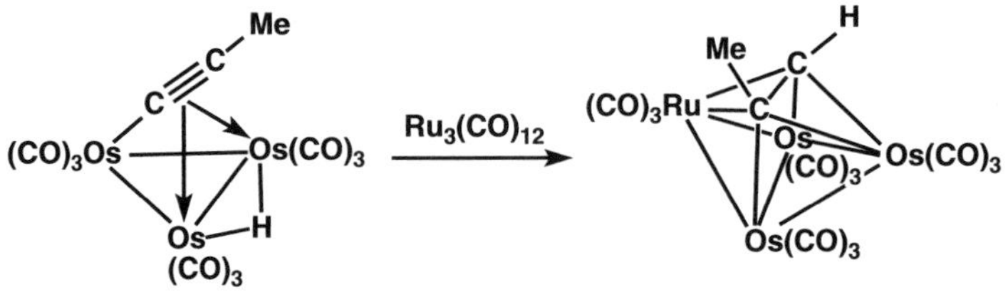

SCHEME 8.

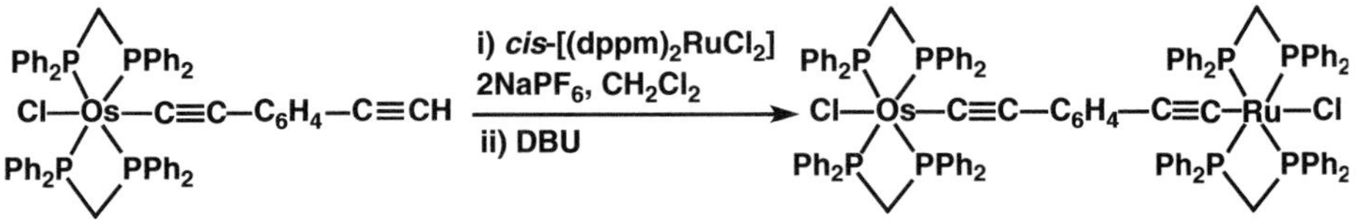

SCHEME 9.

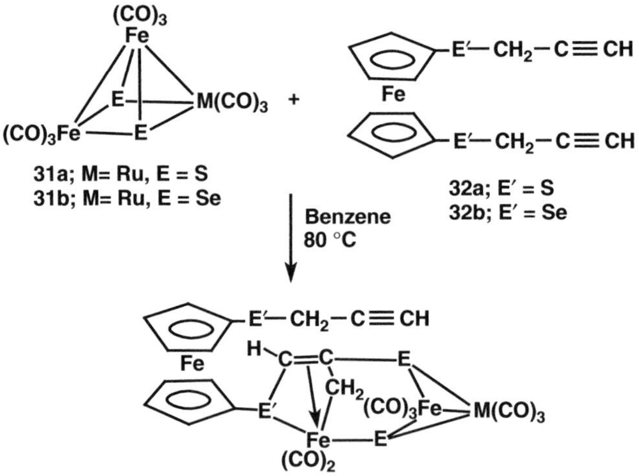

SCHEME 10.

linear polymers (**35–38**) in good yields.[38] This procedure provides a convenient route to heterometallic nickel–platinum-poly-yne polymers (Scheme 11).

Formation of heterodinuclear Pt–Pd anionic complexes $[(C_6F_5)_2Pt(CCR)_2Pd(\eta^3\text{-}C_3H_5)]^-$ (R = Ph, ^{t}Bu, $SiMe_3$) (**42a–c**) and neutral trinuclear complexes $[\{(\eta^3\text{-}C_3H_5)Pd\}_2\{Pt(CCR)_4\}]$ (R = Ph, ^{t}Bu, $SiMe_3$) (**41a–c**) have been reported from the reactions of anionic substrates **39**, $[Pt(C{\equiv}CR)_4]^{2-}$or **40**, *cis*-$[Pt(C_6F_5)_2(C{\equiv}CR)_2]^{2-}$with $[\{Pd\ (\eta^3\text{-}C_3H_5)Cl\}_2]$ complex (Scheme 12).[39]

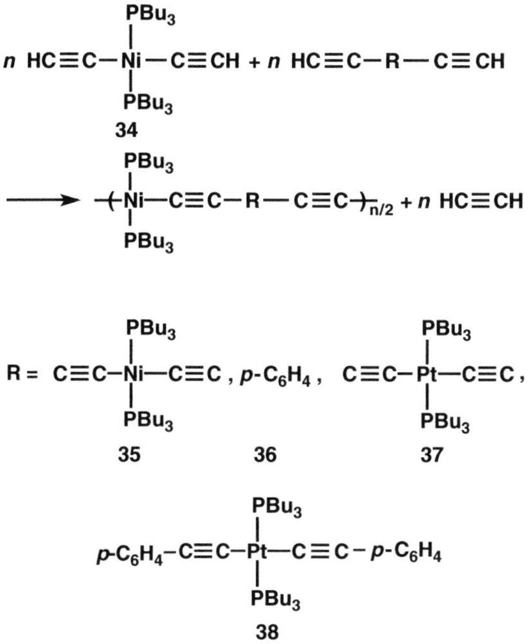

$$n\ \mathrm{HC{\equiv}C{-}Ni(PBu_3)_2{-}C{\equiv}CH} + n\ \mathrm{HC{\equiv}C{-}R{-}C{\equiv}CH}$$

34

$$\longrightarrow \mathrm{{-}\!\!(Ni(PBu_3)_2{-}C{\equiv}C{-}R{-}C{\equiv}C)\!\!{-}_{n/2}} + n\ \mathrm{HC{\equiv}CH}$$

$$\mathrm{R} = \mathrm{C{\equiv}C{-}Ni(PBu_3)_2{-}C{\equiv}C}\ (\mathbf{35}),\quad p\text{-}\mathrm{C_6H_4}\ (\mathbf{36}),\quad \mathrm{C{\equiv}C{-}Pt(PBu_3)_2{-}C{\equiv}C}\ (\mathbf{37}),$$

$$p\text{-}\mathrm{C_6H_4{-}C{\equiv}C{-}Pt(PBu_3)_2{-}C{\equiv}C{-}}p\text{-}\mathrm{C_6H_4}\ (\mathbf{38})$$

SCHEME 11.

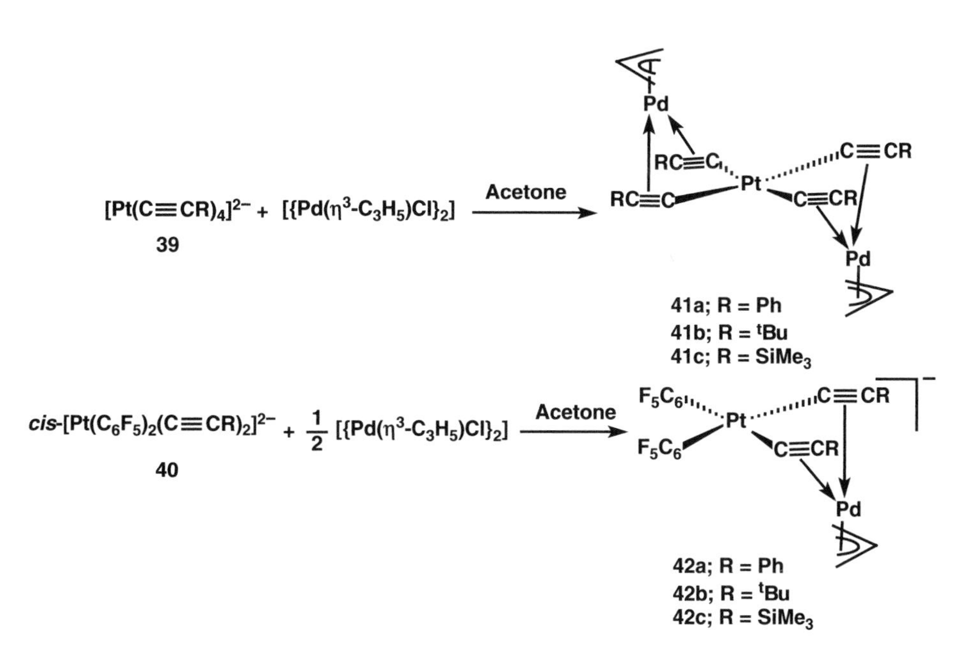

SCHEME 12.

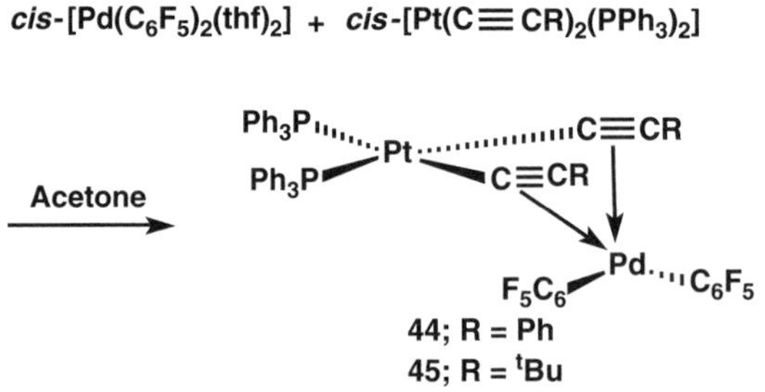

cis-$[Pd(C_6F_5)_2(thf)_2]$ + *cis*-$[Pt(C{\equiv}CR)_2(PPh_3)_2]$

Acetone

Ph_3P, Ph_3P, Pt, $C{\equiv}CR$, $C{\equiv}CR$, Pd, F_5C_6, C_6F_5

44; R = Ph
45; R = ^{t}Bu

SCHEME 13.

The reaction of mononuclear platinum acetylide complexes [*cis*-$Pt(CCR)_2L_2$] (R = Ph, ^{t}Bu; L = PPh_3) towards *cis*-$[Pd(C_6F_5)(thf)_2]$ (**43**) results in the formation of dimetallic Pt–Pd (R = Ph, **44**; R = ^{t}Bu, **45**) complexes (Scheme 13).[40]

IV

MIXED GROUP 11 METAL ACETYLIDE COMPLEXES

New aspects of chemistry and bonding have emerged from reactions of the anionic linear metal acetylide complexes $[M(C_2Ph)_2]^-$ with metal phenylacetylide polymers $[M(C_2Ph)]_n$ (M = Cu, Ag, Au). The reactions involve depolymerisation, ethynylation and condensation to form novel clusters. Thus, the homonuclear $[Ag_5(C_2Ph)_6]^-$ and heteronuclear $[Au_2Cu(C_2Ph)_4]^-$, $[Au_3M_2(C_2Ph)_6]^-$ (M = Cu, Ag), $[Ag_6Cu_7(C_2Ph)_{14}]^-$, and $[AuAg_6Cu_6(C_2Ph)_{14}]^-$ complexes have been obtained. Extension of ethynylation reactions to neutral complexes results in formation of $[Au_2Ag_2(C_2Ph)_4(PPh_3)_2]$, $[Ag_2Cu_2(C_2Ph)_4(PPh_3)_4]$, $[AuAg(C_2Ph)_2]_n$, $[AuCu(C_2Ph)_2]_n$ and $[AgCu(C_2Ph)_2]_n$.

Mixed gold–silver and gold–copper phenylacetylide polymers $[\{AuM(C_2Ph)_2\}_n]$ (M = Ag, **46**; M = Cu, **47**) have been made by the reaction of $[Au(C_2Ph)L]$ (L = $AsPh_3$, $P(OPh)_3$) with $[\{Ag(C_2Ph)\}_n]$ (**48**) and $[\{Cu(C_2Ph)\}_n]$ (**49**), respectively. The gold–silver polymer **46** has also been prepared by the reaction of $[AuClPPh_3]$ with $[\{Ag(C_2Ph)\}_n]$.[41]

Another heterometallic gold–silver cluster, $[Au_2Ag_2(C_2Ph)_4(PPh_3)_2]$ (**50**) has been prepared by the reaction of $[(PPh_3)AuC_2Ph]$ and $[AgC_2Ph]_n$ and also by the reaction of PPh_3 with the polymer $[AuAg(C_2Ph)_2]_n$ (**46**) (Scheme 14).[42] The structure of **50** consists of a linear arrangement of two phenylacetylide groups about each gold atom, with each silver atom asymmetrically π-bonded to two triple bonds and one phosphine ligand (Fig. 3).

Reaction of $[Au_2(C_2Ph)_3]^-$ and $[AgC_2Ph]$ gives a gold–silver pentanuclear cluster $[Au_3Ag_2(C_2Ph)_6]^-$ (**51**) (Scheme 15).[43]

The pentanuclear cluster $[Au_3Ag_2(C_2Ph)_6]^-$ (**51**) has also been obtained from the reaction between $[Ag(C_2Ph)_2]^-$ and the polymer complex $[\{AuAg(C_2Ph)_2\}_n]$ (**46**) (Scheme 16).[44]

Homonuclear silver complex $[Ag_5(C_2Ph)_6]^-$ (**52**) and heteronuclear silver–copper complex $[Ag_4Cu(C_2Ph)_6]^-$ (**53**) have been obtained by the reactions of silver

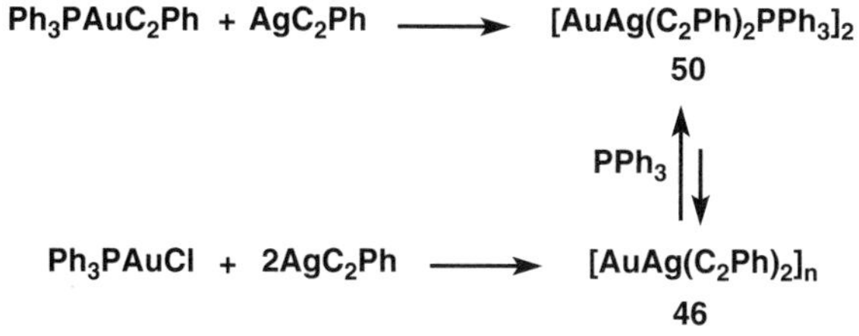

$$Ph_3PAuC_2Ph + AgC_2Ph \longrightarrow [AuAg(C_2Ph)_2PPh_3]_2 \; (\mathbf{50})$$

$$\mathbf{50} \xrightleftharpoons{PPh_3} \mathbf{46}$$

$$Ph_3PAuCl + 2AgC_2Ph \longrightarrow [AuAg(C_2Ph)_2]_n \; (\mathbf{46})$$

SCHEME 14.

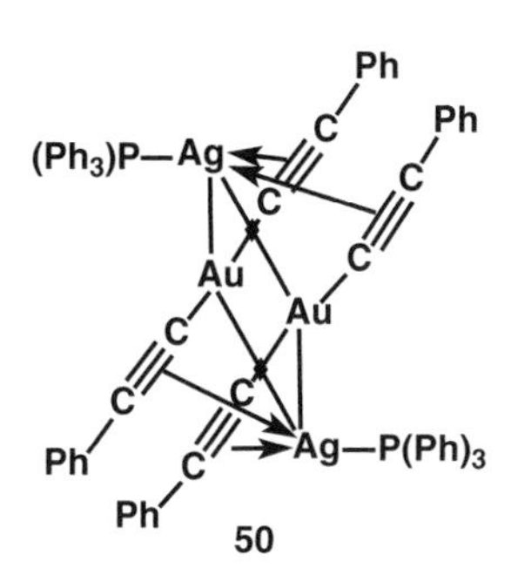

FIG. 3.

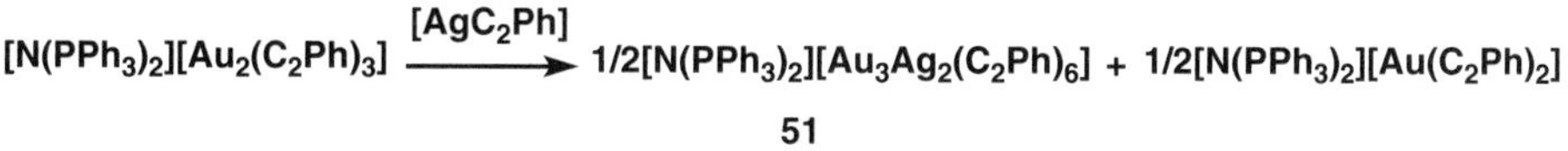

$$[N(PPh_3)_2][Au_2(C_2Ph)_3] \xrightarrow{[AgC_2Ph]} 1/2[N(PPh_3)_2][Au_3Ag_2(C_2Ph)_6] \; (\mathbf{51}) + 1/2[N(PPh_3)_2][Au(C_2Ph)_2]$$

SCHEME 15.

$$[Ag(C_2Ph)_2]^- \xrightarrow{[AuAg(C_2Ph)_2]} [Au(C_2Ph)_2]^- + 2[Ag(C_2Ph)_2]$$

$$[Au(C_2Ph)_2]^- \xrightarrow{[AuAg(C_2Ph)_2]} [Au_2Ag(C_2Ph)_4] \; \text{unstable}$$

$$2[Au_2Ag(C_2Ph)_4]^- \longrightarrow [Au_3Ag_2(C_2Ph)_6]^- \; (\mathbf{51}) + [Au(C_2Ph)_2]^-$$

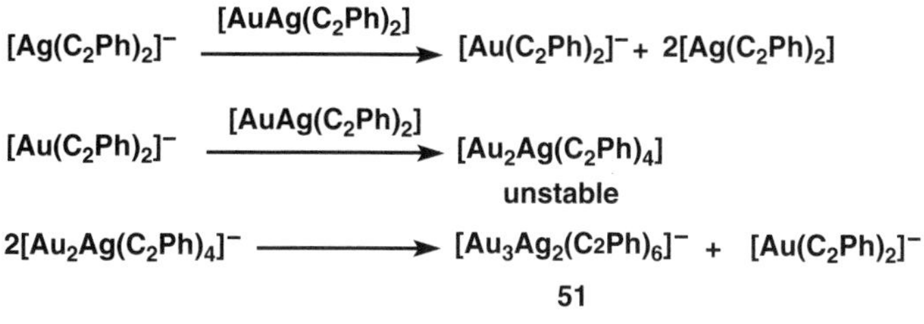

SCHEME 16.

phenylacetylide with the linear complex anions $[Ag(C_2Ph)_2]^-$ and $[Cu(C_2Ph)_2]^-$, respectively (Fig. 4).[43]

The reaction of $[Cu(C_2Ph)_2]^-$ and $[\{AuCu(C_2Ph)_2\}_n]$ (**47**) affords the cluster $[Au_2Cu(C_2Ph)_4]^-$ (**54**) or $[Au_3Cu_2(C_2Ph)_6]^-$ (**55**) depending on the molar ratio of the reactants. The pentanuclear cluster $[N^nBu_4][Au_3Cu_2(C_2Ph)_6]$ (**55**) is also obtained by the reaction of $[N^nBu_4][Au(C_2Ph)_2]$ with a mixture of $[AuC_2Ph]_n$ and $[CuC_2Ph]_n$.[45] X-ray structural data for **55** reveals a trigonal bipyramidal arrangement of metal atoms with two copper atoms in the apical and three gold atoms in

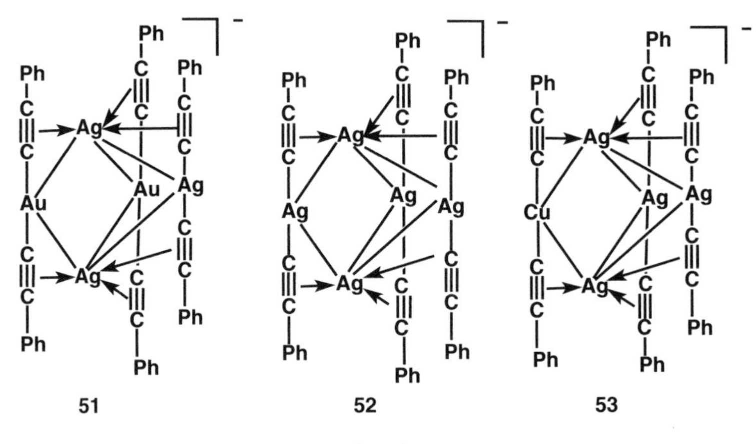

FIG. 4.

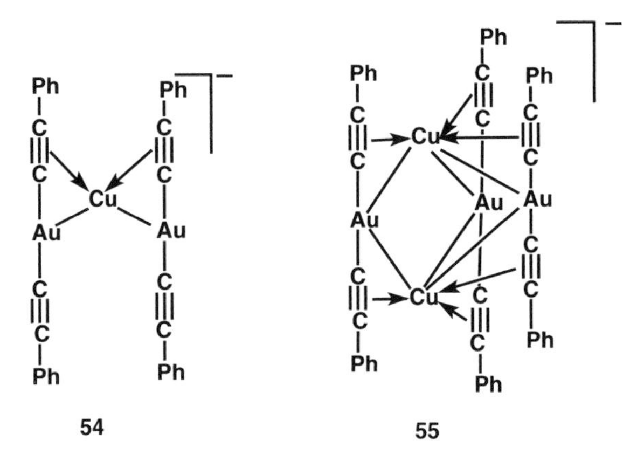

FIG. 5.

the equilateral positions. Each gold atom is σ-bonded to two acetylide groups in almost linear co-ordination while each copper atom is asymmetrically π-bonded to three alkyne groups (Fig. 5). In contrast, the reaction of $[Cu(C_2Ph)_2]^-$ and $[\{AuAg(C_2Ph)_2\}_n]$ (**46**) gives the expected complex $[Au_3Ag_2(C_2Ph)_6]^-$ (**51**) and the trimetallic $[AuAg_6Cu_6(C_2Ph)_{14}]^-$ (**56**) (Scheme 17).[46]

A trimetallic cluster $[Au_3AgCu(C_2Ph)_6]^-$ (**57**) has been prepared by the reaction of $[Au_3Cu_2(C_2Ph)_6]^-$ (**55**) with a mixture of gold phenylacetylide and silver phenylacetylide. Complex **57** is also obtained by the addition of $[Au_3Ag_2(C_2Ph)_6]^-$ (**51**) to $[\{AuCu(C_2Ph)_2\}_n]$ (**47**) or by the reaction of $[Au_2Cu(C_2Ph)_4]^-$ (**54**) with $[\{AuAg(C_2Ph)_2\}_n]$ (**46**). The high nuclearity heterometallic cluster $[AuAg_6Cu_6(C_2Ph)_{14}]^-$ (**56**) has been synthesised by the reaction of a dichloromethane solution of polymeric $[\{Ag(C_2Ph)\}_n]$, $[\{Cu(C_2Ph)\}_n]$ and $[Ag(C_2Ph)_2]^-$. Other synthetic methods to prepare **56** include reaction of $[Au_3Cu_2(C_2Ph)_6]^-$ (**55**) or $[Au_3Ag_2(C_2Ph)_6]^-$ (**51**) with a mixture of $[\{Ag(C_2Ph)\}_n]$

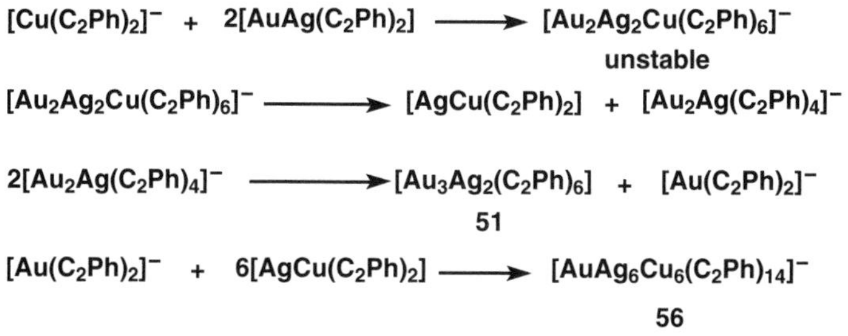

$$[Cu(C_2Ph)_2]^- + 2[AuAg(C_2Ph)_2] \longrightarrow \underset{\text{unstable}}{[Au_2Ag_2Cu(C_2Ph)_6]^-}$$

$$[Au_2Ag_2Cu(C_2Ph)_6]^- \longrightarrow [AgCu(C_2Ph)_2] + [Au_2Ag(C_2Ph)_4]^-$$

$$2[Au_2Ag(C_2Ph)_4]^- \longrightarrow \underset{\mathbf{51}}{[Au_3Ag_2(C_2Ph)_6]} + [Au(C_2Ph)_2]^-$$

$$[Au(C_2Ph)_2]^- + 6[AgCu(C_2Ph)_2] \longrightarrow \underset{\mathbf{56}}{[AuAg_6Cu_6(C_2Ph)_{14}]^-}$$

SCHEME 17.

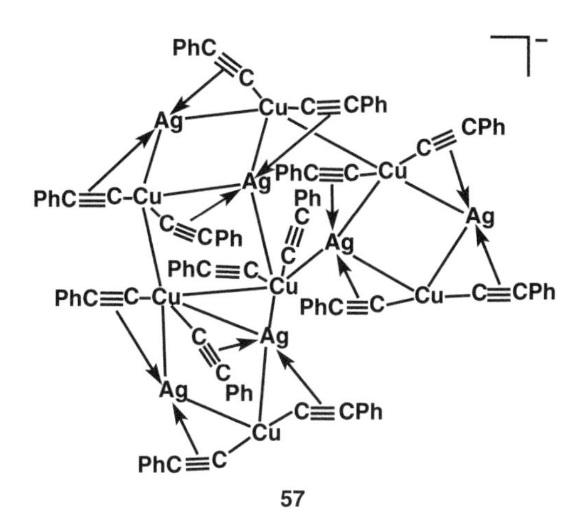

FIG. 6.

and $[\{Cu(C_2Ph)\}_n]$.[48] Room temperature reaction of the linear complex $[Ag(C_2Ph)_2]^-$ and a mixture of $[\{Cu(C_2Ph)\}_n]$ and $[\{Ag(C_2Ph)\}_n]$ in dichloromethane affords the silver–copper heteronuclear cluster $[Ag_6Cu_7(C_2Ph)_{14}]^-$ (**57**). When $[Cu(C_2Ph)_2]^-$ or halide ions, X^- ($X = Cl, Br, I$) are treated with $[\{Ag(C_2Ph)\}_n]$ and $[\{Cu(C_2Ph)\}_n]$, cluster **57** is obtained.[47]

An X-ray diffraction study of cluster **57** reveals a 13-atom bimetallic anionic cluster with two types of Cu atoms, one inside the body of the cluster and six on its surface. The surface atoms form a distorted trigonal bipyramidal geometry whereas the internal Cu atom linearly coordinates two C_2Ph groups in addition to forming six Cu–Ag bonding interactions with three tetranuclear $[Ag_2Cu_2(C_2Ph)_4]$ subclusters (Fig. 6).[49]

V
MIXED GROUP 6–9 METAL ACETYLIDE COMPLEXES

The metal acetylides $[Cp(CO)_2MC{\equiv}CR]$ (**58a**, M = Fe, R = Ph and **58b-c**, M = Ru, R = Ph, ^{t}Bu, Me) react with $[Co_2(CO)_8]$ to form the acetylide-bridged trinuclear mixed metal complexes $[Co_2(CO)_6\text{-}\mu\text{-}(Cp(CO)_2MC{\equiv}CR)]$ (M = Fe and R = Ph, **59**; M = Ru and R = Ph, ^{t}Bu, Me, **60a–c**) (Fig. 7).[50]

Compounds with a tetrahedral C_2Co_2 core, $[Co_2(CO)_6(\mu\text{-}\eta^2,\eta^2\text{-MCCH})]$ (M = $CpFe(CO)_2$, **62a**; M = $Cp^*Fe(CO)_2$, **62b**) have been obtained by the interaction of THF solution of ethynyliron complexes $[\{CpFe(CO)_2\}\text{–}C{\equiv}CH]$ (**61a**) and $[\{Cp^*Fe(CO)_2\}\text{–}C{\equiv}CH]$ (**61b**) with $[Co_2(CO)_8]$, respectively. The reaction of the ethynyldiiron complex $[\{Cp^*Fe(CO)_2\}\text{–}C{\equiv}C\text{–}\{Cp^*Fe(CO)_2\}]$ (**63**) with $[Co_2(CO)_8]$ similarly affords the deep green adduct $[Cp^*Fe[Co(CO)_2]_2(\mu\text{-}CO)_2(\mu\text{-}\eta^1,\eta^2,\eta^2\text{-CC-Fp}^*)]$ (**64**), which shows fluxional properties at ambient temperature (Scheme 18).[51]

Photolytic reaction of $[(\mu\text{-}\eta^2,\eta^2\text{-M–C}{\equiv}\text{C–H})Co_2(CO)_6]$ (M = $CpFe(CO)_2$, **62a**; M = $Cp^*Fe(CO)_2$, **62b**) produces pentanuclear clusters $[(\eta^5\text{-}C_5R_5)_2Fe_2Co_3(\mu_5\text{-CCH})(CO)_{10}]$ (R = H, **65a**; R = Me, **65b**), respectively, *via* an apparent addition reaction of a $(\eta^5\text{-}C_5R_5)FeCo(CO)_n$ fragment to **62**. On the other hand, thermolysis of **62a** gives the Fe-free hexacobalt cluster compound $[(\mu\text{-CH}{=}\text{CH})\{(\mu_3\text{-C})Co_3(CO)_9\}_2]$ (**66**), whereas when **62b** is thermolysed, the Fe–Co dimer without the C_2H ligand, $[Cp^*Fe(CO)(\mu\text{-}CO)_2Co(CO)_3]$ (**67**) is obtained, in addition to the photolysis product **65b**. Reduction of **62** with hydrosilanes gives 1,2 disilylethylene and the tetranuclear μ-vinylidene cluster $[(C_5R_5)FeCo_3(\mu_4\text{-C}{=}CH_2)(CO)_9]$

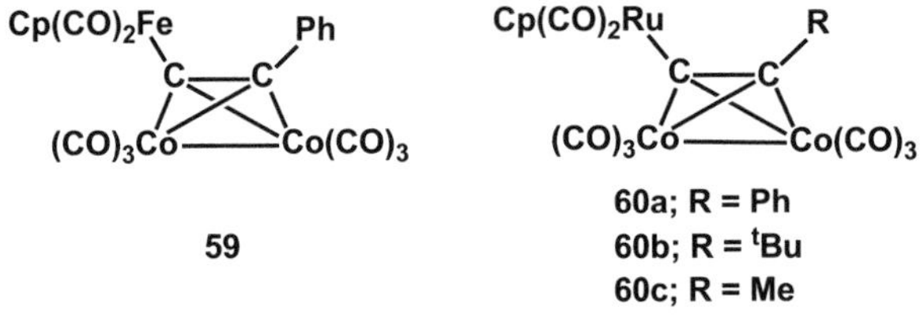

FIG. 7.

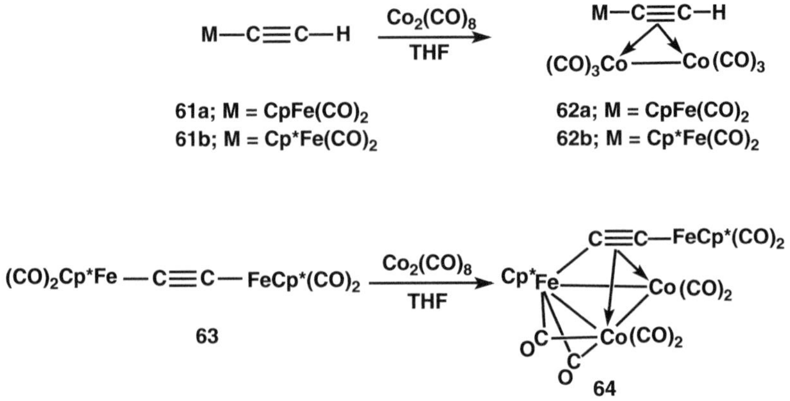

SCHEME 18.

(R = H, **68a**; R = Me, **68b**) by hydrosilylation and hydrometallation of the C_2H ligand, respectively (Scheme 19).[52]

Interaction of the trinuclear ethynyl complexes $[Co_2(CO)_6[(\mu\text{-}(\eta^5\text{-}C_5R_5)Fe(CO)_2C{\equiv}CH]$ (R=H, **62a**; R = Me, **62b**) with $[Fe_2(CO)_9]$ results in the formation of a tetranuclear vinylidene intermediate $[Co_2Fe(CO)_9[\mu_3\text{-}C{=}C(H)Fe(\eta^5\text{-}C_5R_5)(CO)_2]$ (**69**) which undergo thermal decarbonylation to produce the tetranuclear acetylide clusters $[(\mu_4\text{-}C_2H)[(\eta^5\text{-}C_5R_5)Fe]FeCo_2(CO)_{10}]$ (**70a,b**) (Scheme 20). Similarly, the addition reaction of $[(\eta^5\text{-}C_5R_5)Fe(CO)_2C{\equiv}CH]$ (**61a,b**) to the trimetallic species $[RuCo_2(CO)_{11}]$ affords the ruthenium analogue of the vinylidene complex $[Co_2Ru(CO)_9[\mu_3\text{-}C{=}C(H)Fe(\eta^5\text{-}C_5R_5)(CO)_2]$ (**71a,b**), which decarbonylates to $[(\mu_4\text{-}C_2H)[(\eta^5\text{-}C_5R_5)Fe]RuCo_2(CO)_{10}]$ (**72a,b**) (Scheme 21).[53]

A formal replacement of $(\eta^5\text{-}C_5Me_4Et)Fe(CO)$ moiety of a triiron $\mu_3,\eta^1,\eta^2,\eta^2$-acetylide cluster compound **73a** by an isolobal $Co(CO)_3$ fragment occurs when cluster **73a** reacts with $[Co_2(CO)_8]$ (Scheme 22).[54a]

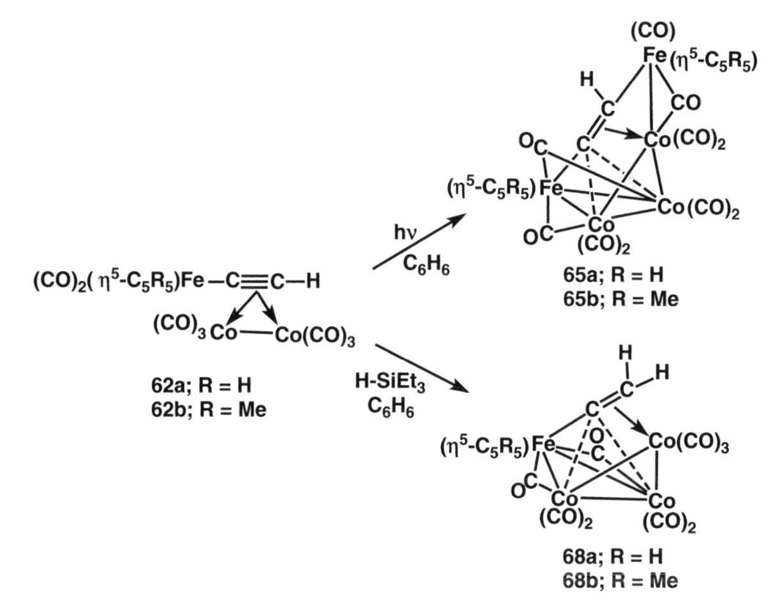

SCHEME 19.

SCHEME 20.

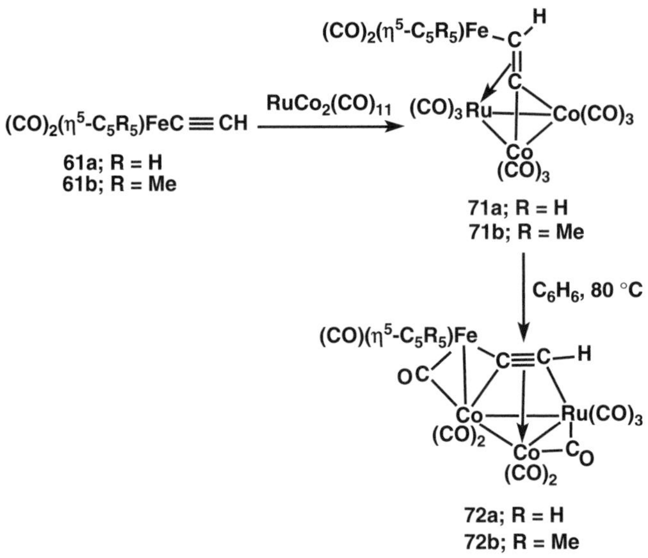

$(CO)_2(\eta^5\text{-}C_5R_5)FeC{\equiv}CH$ $\xrightarrow{RuCo_2(CO)_{11}}$ $(CO)_2(\eta^5\text{-}C_5R_5)Fe$–C(H)=C–$(CO)_3Ru$–$Co(CO)_3$–Co$(CO)_3$

61a; R = H
61b; R = Me

71a; R = H
71b; R = Me

C_6H_6, 80 °C

$(CO)(\eta^5\text{-}C_5R_5)Fe$, OC, C≡C–H, Co $(CO)_2$, $Ru(CO)_3$, Co–CO $(CO)_2$

72a; R = H
72b; R = Me

SCHEME 21.

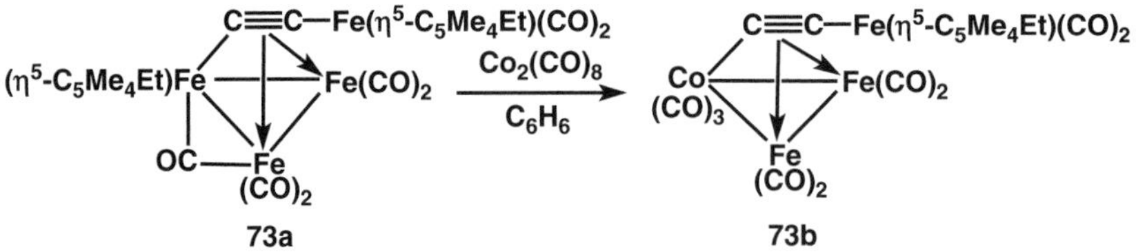

C≡C–$Fe(\eta^5\text{-}C_5Me_4Et)(CO)_2$, $(\eta^5\text{-}C_5Me_4Et)Fe$, $Fe(CO)_2$, OC–Fe $(CO)_2$ $\xrightarrow[C_6H_6]{Co_2(CO)_8}$ C≡C–$Fe(\eta^5\text{-}C_5Me_4Et)(CO)_2$, Co $(CO)_3$, $Fe(CO)_2$, Fe $(CO)_2$

73a 73b

SCHEME 22.

The triiron acetylide cluster compound [PPN][$Fe_3(CO)_9C{\equiv}C(OAc)$] (**74a**) [PPN = $(Ph_3P)_2N^+$] reacts with $[Fe(CO)_4]^{2-}$ at room temperature to produce the metalated acetylide cluster $[PPN]_2[Fe_3(CO)_9C{\equiv}CFe(CO)_4]$ (**74b**). Further metalation of **74b** with excess [$Co_2(CO)_8$] produces a hexametallic cluster [PPN] [$Fe_3Co_3(C_2)(CO)_{18}$] (**76**). The reaction proceeds by formation of two intermediate compounds, the dicarbide-containing cluster compound $[PPN]_2[Fe_4Co_2(C_2)(CO)_{18}]$ (**75a**) and the acetylide [PPN] [$Fe_2Co(CO)_9CCFe(CO)_4$] (**75b**) (Scheme 23).[54b]

A number of iron–cobalt mixed metal cluster compounds have been obtained by sequential addition of [$Co_2(CO)_8$] and [$Fe_2(CO)_9$] to [Cp*Fe $(CO)_2$–C≡C–C≡C–H] (**77**) (Scheme 24).[55] Their formation involves addition of a metal fragment to the C≡C bonds, reorganisation of the cluster framework, transfer of metal fragments, valence isomerisation of the C_4(H) linkage and 1,2-H shift of the C_4H ligand.

Sequential addition of [$Co_2(CO)_8$] and [$Fe_2(CO)_9$] to the butadiynediyl iron complex (**78**) results in formation of the mixed Fe/Co complexes **86** and **87** and the C_4-bridged Co_6 cluster **88** (Scheme 25).[55]

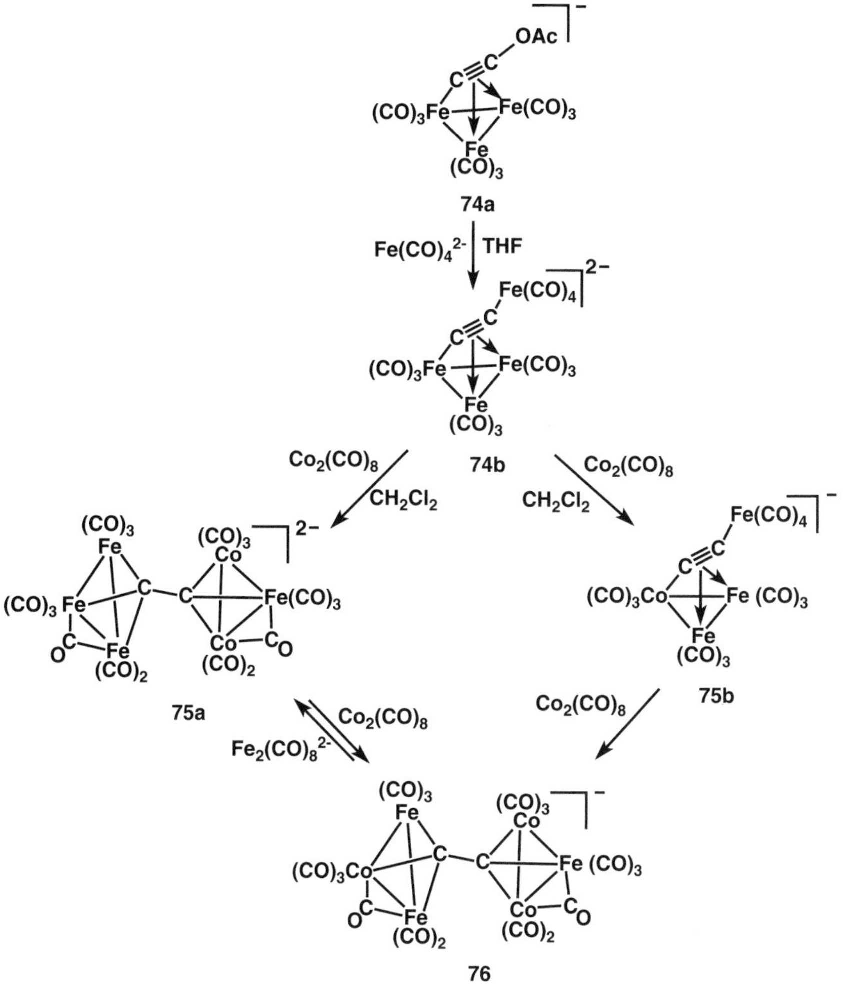

SCHEME 23.

Compound **77** also reacts with $[Mo_2Cp_2(CO)_4]$ at room temperature or with $[CpMoCo(CO)_7]$ at 50 °C to yield mixed metal acetylide complexes **89**, $[Cp^*Fe(CO)_2C_4H\{Cp_2Mo_2(CO)_4\}]$ and **90**, $[Cp^*Fe(CO)_2C_4H\{CpMoCo(CO)_5\}]$. On further reaction with $[Co_2(CO)_8]$, **89** gives **91** (Scheme 26).[55]

Reaction of $[Co_2(CO)_8]$ with the acetylide complexes, $[M(C{\equiv}CC{\equiv}CR)(CO)_3Cp]$ (M = Mo, W; R=H, $Fe(CO)_2Cp$) (**92a–c**) afford simple adducts (**93a–c**) containing a $Co_2(CO)_6$ group attached to the least sterically hindered C≡C triple bond (Scheme 27). In contrast, bis-cluster complexes $[\{Cp(OC)_8Co_2M(\mu_3\text{-}C)\}C{\equiv}C\{(\mu_3\text{-}C)Co_2M'(CO)_8Cp\}]$ (M = M′ = Mo, W; M = Mo, M′ = W) (**95a–c**) have been obtained when $[M(C{\equiv}CC{\equiv}CR)(CO)_3Cp]$ (M = Mo, W; R = $M(CO)_3Cp$ (M = Mo, W), $Fe(CO)_2Cp$) (**94a–c**) are used to react with $[Co_2(CO)_8]$ (Scheme 28). Reaction between

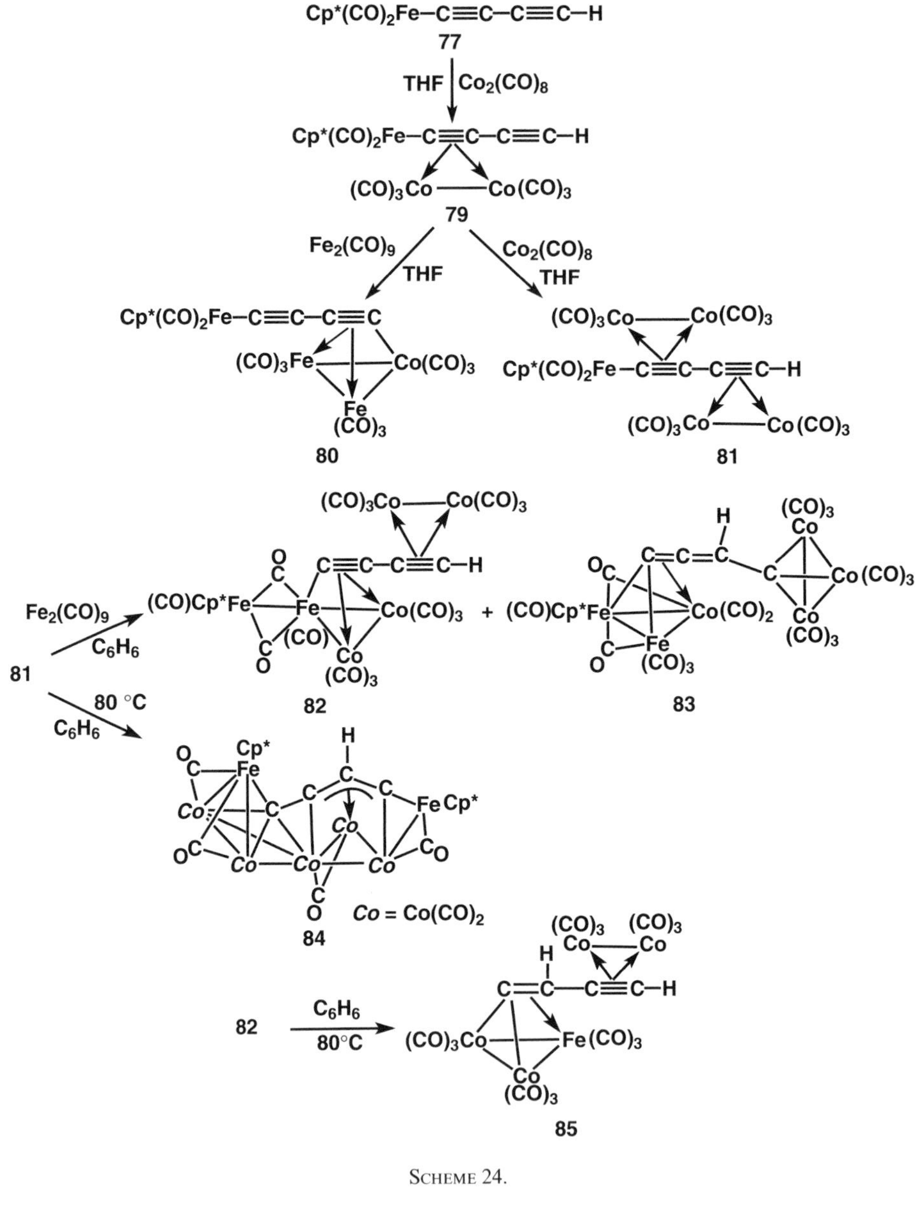

Cp*(CO)2Fe–C≡C–C≡C–H
77
THF | Co2(CO)8
Cp*(CO)2Fe–C≡C–C≡C–H
(CO)3Co——Co(CO)3
79
Fe2(CO)9 THF
Co2(CO)8 THF
Cp*(CO)2Fe–C≡C–C≡C
(CO)3Fe
Co(CO)3
Fe (CO)3
80
(CO)3Co——Co(CO)3
Cp*(CO)2Fe–C≡C–C≡C–H
(CO)3Co——Co(CO)3
81
(CO)3Co——Co(CO)3
O C
C≡C–C≡C–H
Fe2(CO)9 C6H6
(CO)Cp*Fe
Fe
Co(CO)3
(CO)
C O
Co (CO)3
82
+
H
(CO)3 Co
O C
C=C=C
C
Co(CO)3
(CO)Cp*Fe
Co(CO)2
Co (CO)3
C O
Fe (CO)3
83
81
80 °C C6H6
O C
Cp* Fe
H
C
C
C
C
Fe Cp*
Co
Co
CO
OC
Co
Co
Co
C O
Co = Co(CO)2
84
(CO)3 (CO)3
Co——Co
H
C=C–C≡C–H
82
C6H6 80°C
(CO)3Co
Fe(CO)3
Co (CO)3
85

SCHEME 24.

$[Co_2(\mu\text{-dppm})(CO)_6]$ and $[\{W(CO)_3Cp\}_2(\mu\text{-}C_4)]$ affords $[Co_2(\mu\text{-dppm})\{\mu\text{-}[Cp(OC)_3W]$ $C_2C{\equiv}C[W(CO)_3Cp](CO)_4]$ (**97**).[56a]

Reaction of a THF solution of $[\{W(CO)_3Cp\}_2(\mu\text{-}C_8)]$ and $[Co_2CO_8]$ gives mono and di-adducts, $[\{W(CO)_3Cp\}_2\{\mu\text{-}C_8[Co_2(CO)_6]\}]$ (**98a**) and $[\{W(CO)_3Cp\}_2$ $\{\mu\text{-}C_8\text{-}[Co_2(CO)_6]_2\}]$ (**98b**) (Fig. 8). Thermolytic reaction between $[\{W(CO)_3Cp\}_2$ $(\mu\text{-}C_8)]$ and $[Co_2(\mu\text{-dppm})(CO)_6]$ under benzene reflux affords $[\{W(CO)_3Cp\}_2$

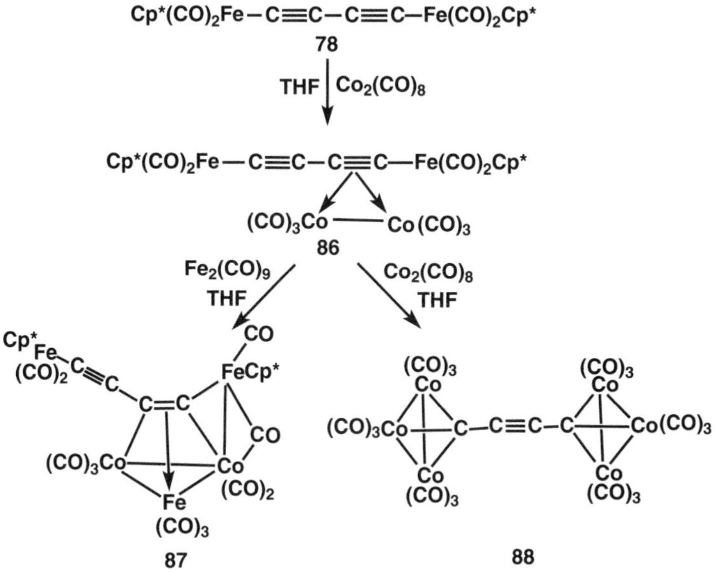

SCHEME 25.

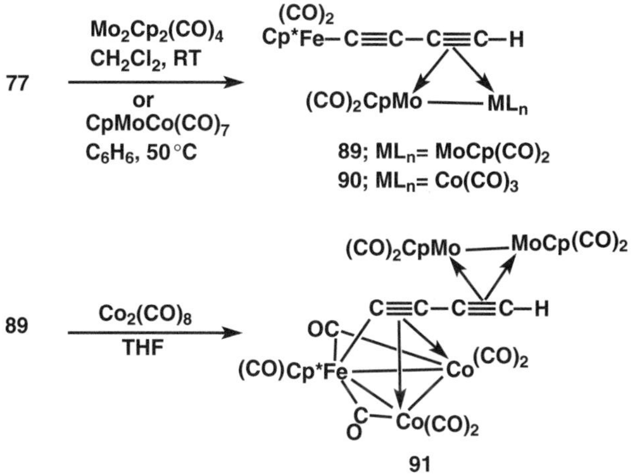

SCHEME 26.

$\{\mu\text{-}C_8[Co_2(\mu\text{-dppm})(CO)_4]\}]$ (**99a**) and two isomeric products, $[\{W(CO)_3Cp\}_2\{\mu\text{-}C_8[Co_2(\mu\text{-dppm})(CO)_4]_2\}]$ (**99b**) and $[\{W(CO)_3Cp\}_2\{\mu\text{-}C_8[Co_2(\mu\text{-dppm})(CO)_4]_2\}]$ (**99c**) (Fig. 9).[56b]

Deprotonation of the alkyne-bridged clusters $[RuCo_2(CO)_9(\mu_3\text{-}RC{\equiv}CH)]$ (**100a–d**) with triethylamine and further reaction with the organometallic halides $[Cp(CO)_2FeCl]$ (**101**), $[Cp(CO)_2RuCl]$ (**102**), $[Cp(PPh_3)NiCl]$ (**103**) and $[Cp(CO)_3MoCl]$ (**104**) in presence of catalytic amounts of copper(I) iodide results in

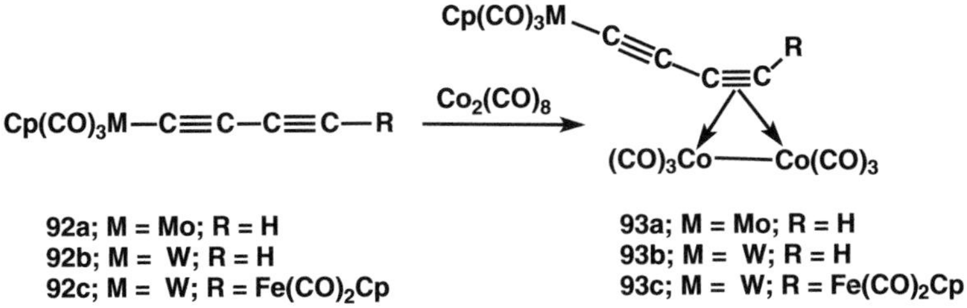
Cp(CO)3M—C≡C—C≡C—R
Co2(CO)8
Cp(CO)3M
C≡C
C≡C
R
(CO)3Co
Co(CO)3
92a; M = Mo; R = H
92b; M = W; R = H
92c; M = W; R = Fe(CO)2Cp
93a; M = Mo; R = H
93b; M = W; R = H
93c; M = W; R = Fe(CO)2Cp

SCHEME 27.

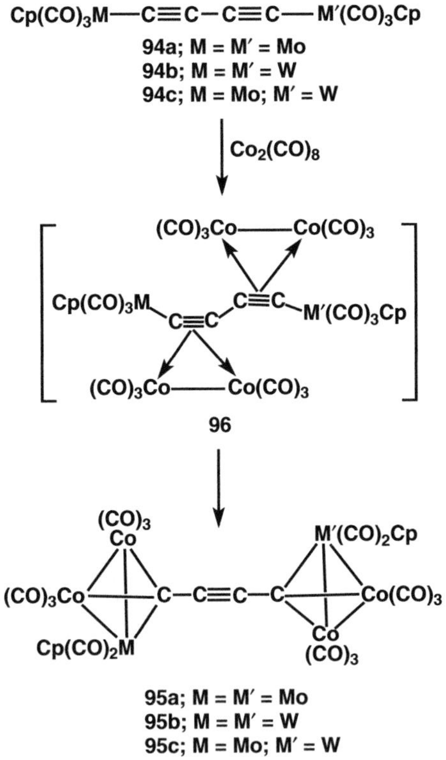
Cp(CO)3M—C≡C—C≡C—M′(CO)3Cp
94a; M = M′ = Mo
94b; M = M′ = W
94c; M = Mo; M′ = W
Co2(CO)8
(CO)3Co
Co(CO)3
Cp(CO)3M
C≡C
C≡C
M′(CO)3Cp
(CO)3Co
Co(CO)3
96
(CO)3
Co
(CO)3Co
C—C≡C—C
M′(CO)2Cp
Co(CO)3
Cp(CO)2M
Co
(CO)3
95a; M = M′ = Mo
95b; M = M′ = W
95c; M = Mo; M′ = W

SCHEME 28.

incorporation of the metal species as organometallic acetylides MC≡CR into the $RuCo_2M$ metal frameworks (**105-111**) (Scheme 29). With acids, the resulting acetylide-bridged clusters are reconverted to the starting materials.[57]

A pentanuclear heterometallic acetylide complex $[Cp_2Mo_2Ru_3(CO)_{10}(C{\equiv}CPh)_2]$ (**113**) has been prepared by the thermolysis reaction of a toluene solution of metal acetylide $[CpMo(CO)_3(C{\equiv}CPh)]$ with $[Ru_3(CO)_{12}]$ (Fig. 10).[58]

Room temperature reaction of $[W(C{\equiv}CC{\equiv}CH)(CO)_3Cp]$ with the reactive $[Ru_3(CO)_{10}(NCMe)_2]$ forms initially $[Ru_3\{\mu_3\text{-}HC_2C{\equiv}C[W(CO)_3Cp]\}(\mu\text{-}CO)(CO)_9]$ (**114**), which readily transforms to $[Ru_3(\mu\text{-}H)\{\mu_3\text{-}C_2C{\equiv}C[W(CO)_3Cp]\}(CO)_9]$ (**115**)

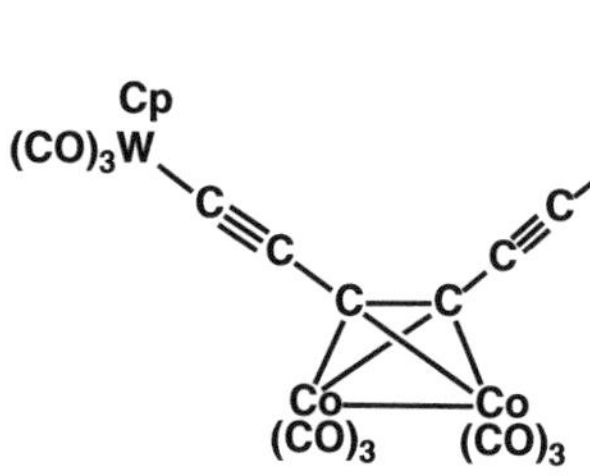

98a

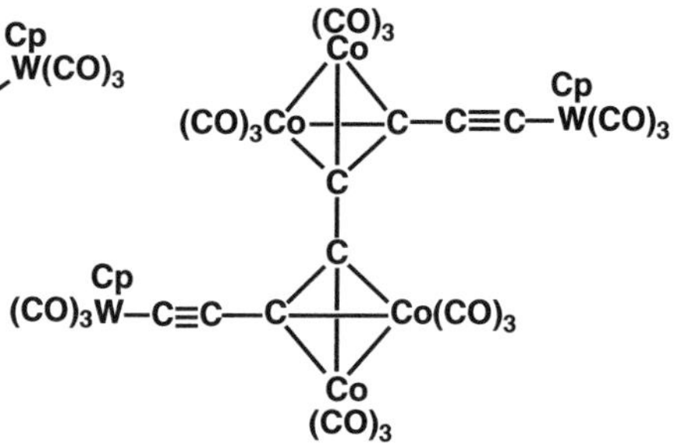

98b

FIG. 8.

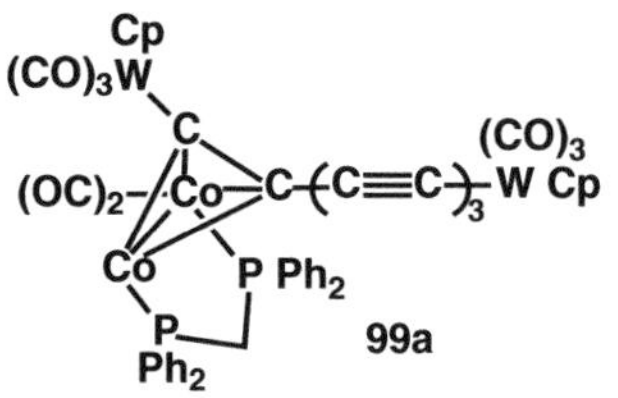

99a

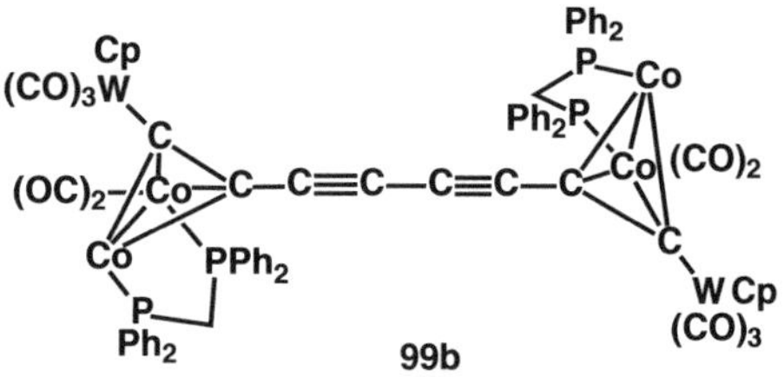

99b

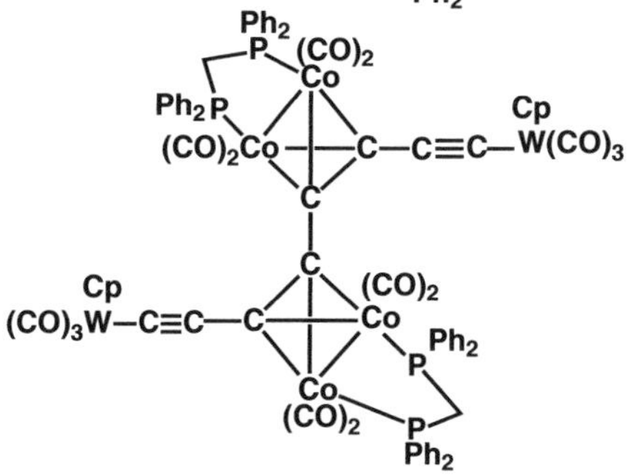

99c

FIG. 9.

on benzene reflux and to **116** on reaction with excess of $[W(C{\equiv}CC{\equiv}CH)(CO)_3Cp]$ (Scheme 30).[59] Similarly, reaction with $[Ru_3(\mu\text{-dppm})(CO)_{10}]$ forms three interconverting isomers of $[Ru_3(\mu\text{-H})\{\mu_3\text{-}C_2C{\equiv}C[W(CO)_3Cp]\}(\mu\text{-dppm})(CO)_{12}]$ (**117–119**) (Scheme 31).

Further reaction of $[Ru_3(\mu\text{-H})\{\mu_3\text{-}C_2C{\equiv}C[W(CO)_3Cp]\}(CO)_9]$ (**115**) with $[Ru_3(CO)_{12}]$ affords $[\{Ru_3(\mu\text{-H})(CO)_9\}(\mu_3\text{-}\eta^2,\mu_3\text{-}\eta^2\text{-}C_2C_2)\{Ru_2W(CO)_8Cp\}]$ (**120a**), while reaction with $[Fe_2(CO)_9]$ gives an analogous product **120b**, in which three of the ruthenium sites are partially occupied by a total of one or two iron atoms (Scheme 32). Treatment of $[Ru_3(\mu\text{-H})\{\mu_3\text{-}C_2C{\equiv}C[W(CO)_3Cp]\}(CO)_9]$ (**115**) with $[Co_2(CO)_8]$ forms a vinylidene cluster $[\{CoRu_2(CO)_9\}(\mu_3\text{-}\eta^2,\mu_3\text{-}\eta^2\text{-}CCHC_2)\{CoRuW(CO)_8Cp\}]$ (**121**) by a process involving the transfer of cluster-bound hydride to the C_4 ligand.[59]

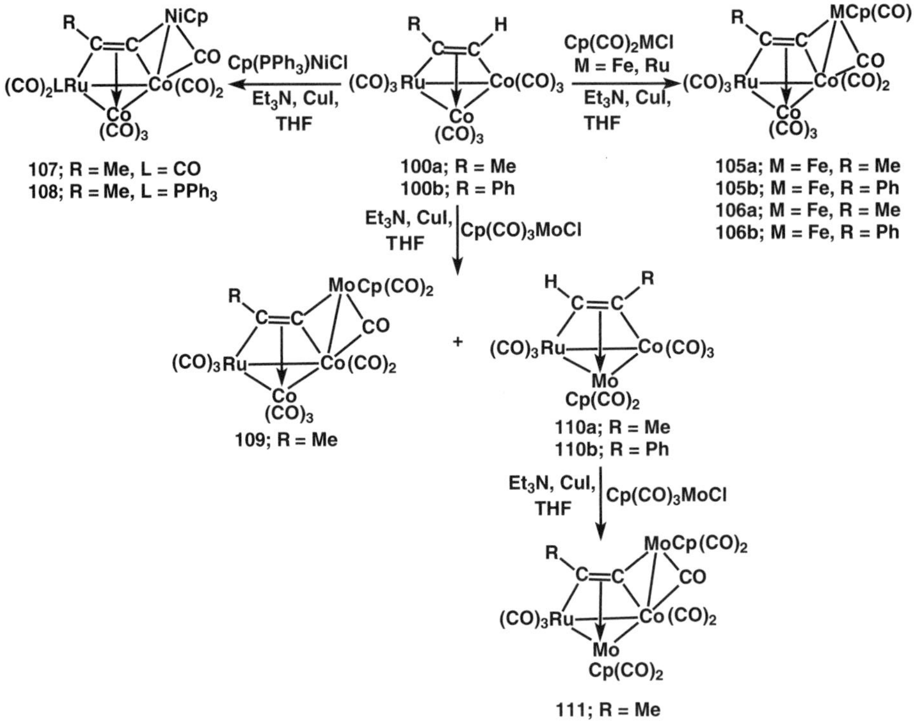

107; R = Me, L = CO
108; R = Me, L = PPh_3

100a; R = Me
100b; R = Ph

105a; M = Fe, R = Me
105b; M = Fe, R = Ph
106a; M = Fe, R = Me
106b; M = Fe, R = Ph

109; R = Me

110a; R = Me
110b; R = Ph

111; R = Me

SCHEME 29.

Ru = $Ru(CO)_2$

113

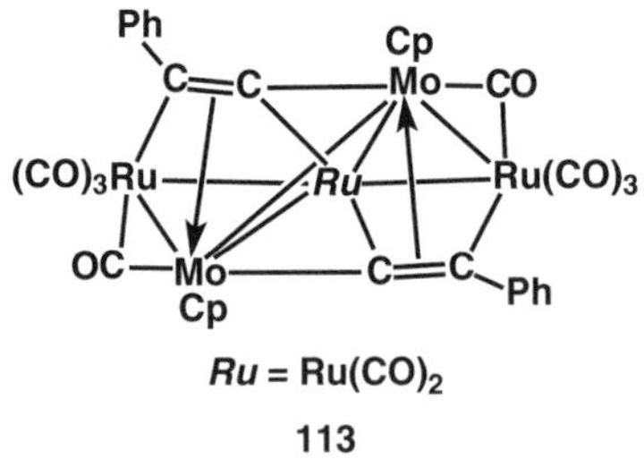

FIG. 10.

Reaction of a dichloromethane solution of 1,6-bis(trimethylsilyl)hexa-1,3,5-triyne (**122**) with $[Os_3(CO)_{10}(NCMe)_2]$ yields $[Os_3(CO)_9(\mu\text{-}CO)(\mu_3\text{-}\eta^1,\eta^1,\eta^2\text{-}Me_3SiC{\equiv}CC_2C{\equiv}CSiMe_3)]$ (**123**), which on reflux in heptane with $[Ru_3(CO)_{12}]$ gives $[Os_3Ru(CO)_{12}(\mu_4\text{-}\eta^1,\eta^2,\eta^1,\eta^2\text{-}Me_3SiC{\equiv}CC_2C{\equiv}CSiMe_3)]$ (**124**). However, thermal reaction between **122** and $[Ru_3(CO)_{12}]$ forms a butterfly cluster $[Ru_4(CO)_{12}(\mu_4\text{-}\eta^1,\eta^2,\eta^1,\eta^2\text{-}Me_3SiC{\equiv}CC_2C{\equiv}CSiMe_3)]$ (**125a**) and the ruthenole complex $[Ru_2(CO)_6\{\mu\text{-}\eta^2,\eta^5\text{-}C(C{\equiv}CSiMe_3)C(C{\equiv}CSiMe_3)C(C{\equiv}CSiMe_3)C(C{\equiv}CSiMe_3)\}]$ (**125b**). In the room temperature reaction between cluster **125a** and $[Co_2(CO)_8]$, a slippage of the butterfly cluster core along the hexatriyne chain occurs and

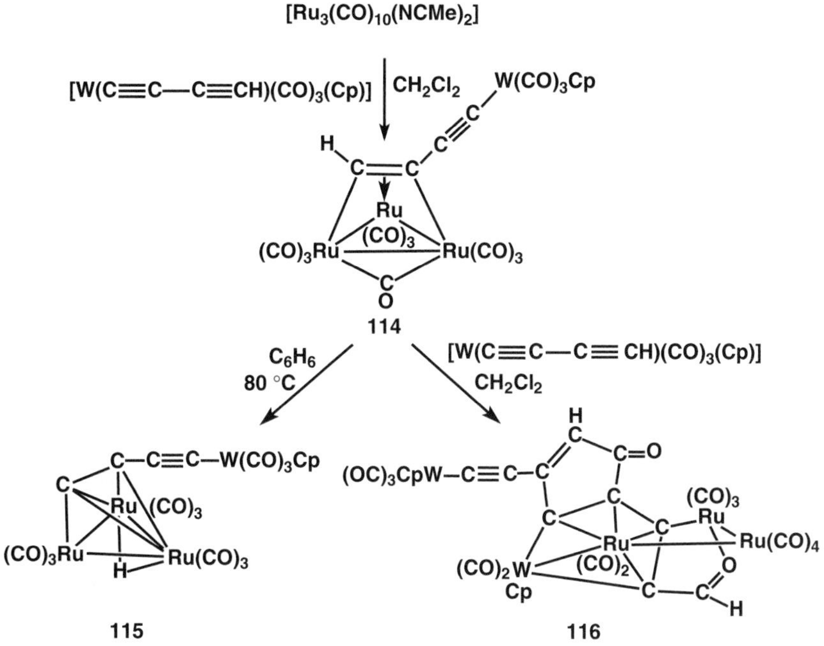

$[Ru_3(CO)_{10}(NCMe)_2]$

$[W(C{\equiv}C{-}C{\equiv}CH)(CO)_3(Cp)]$ | CH_2Cl_2

W(CO)$_3$Cp
C
C
H
C
C
Ru
(CO)$_3$
(CO)$_3$Ru
Ru(CO)$_3$
C
O

114

C_6H_6
80 °C

$[W(C{\equiv}C{-}C{\equiv}CH)(CO)_3(Cp)]$
CH_2Cl_2

C—C≡C—W(CO)$_3$Cp
C
Ru (CO)$_3$
(CO)$_3$Ru
H
Ru(CO)$_3$

115

H
C
C=O
(OC)$_3$CpW—C≡C—C
C
(CO)$_3$
C
Ru
C
Ru
Ru(CO)$_4$
(CO)$_2$W
(CO)$_2$
O
Cp
C
C
H

116

SCHEME 30.

$[Ru_3(\mu\text{-dppm})(CO)_{10}]$

THF, 65 °C | HC≡C-C≡CW(CO)$_3$Cp

(CO)$_3$Cp W—C≡C—C
Ru (CO)$_3$
H
C
(CO)$_3$Ru
Ru(CO)$_3$
Ph$_2$ P
PPh$_2$

117

(CO)$_3$Cp W – C≡C—C
Ru (CO)$_3$
H
C
(CO)$_3$Ru
Ru(CO)$_3$
Ph$_2$ P
P Ph$_2$

118

(CO)$_3$Cp W—C≡C—C
Ru(CO)$_3$
C
(CO)$_3$Ru
Ru(CO)$_3$
H
Ph$_2$ P
PPh$_2$

119

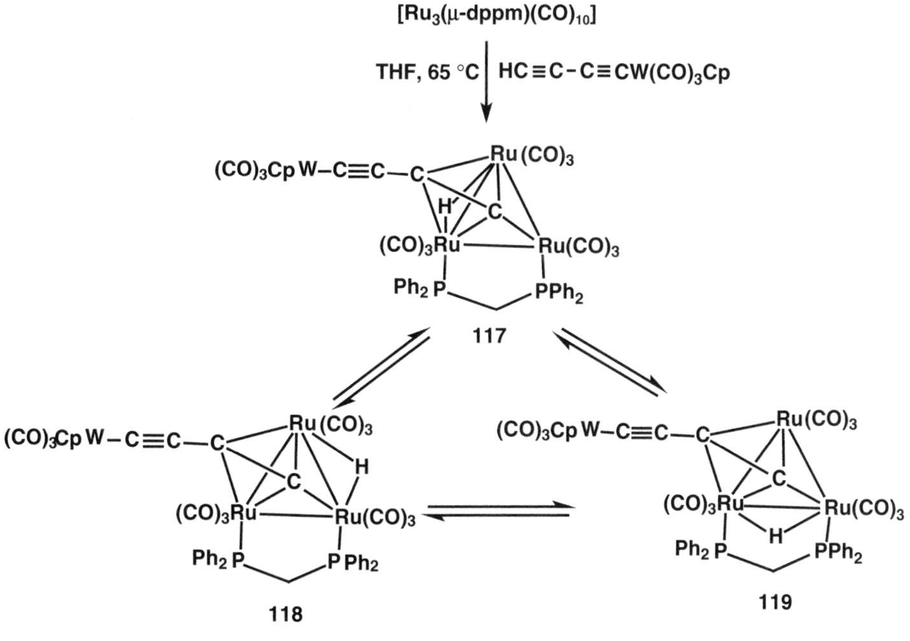

SCHEME 31.

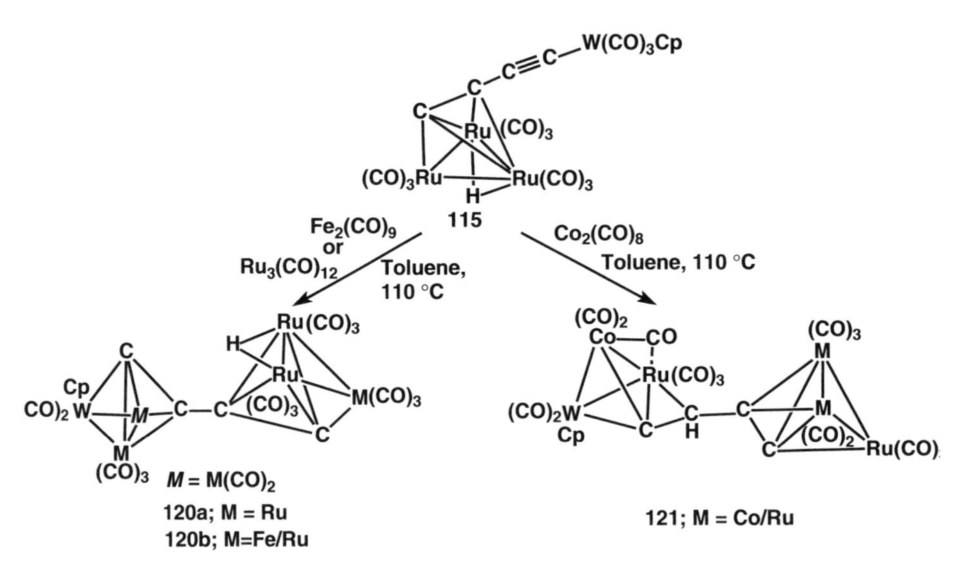

SCHEME 32.

$[\{Ru_4(CO)_{12}\}\{Co_2(CO)_6\}(\mu_4\text{-}\eta^1,\eta^2,\eta^1,\eta^2\text{:}\mu\text{-}\eta^2,\eta^2\text{-}Me_3SiC_2C{\equiv}CC_2SiMe_3)]$ (**126**) is obtained (Scheme 33).[60]

Reaction of a benzene solution of $[\{Cp(PPh_3)_2Ru\}_2(\mu\text{-}C{\equiv}C)_3\}]$ with $[Co_2(CO)_8]$ or $[Co_2(\mu\text{-dppm})(CO)_6]$ gives $[\{Ru(PPh_3)_2Cp\}_2\{\text{-}C{\equiv}CC_2\{Co_2(CO)_6\}(C{\equiv}C)\}]$ (**127a**) and $[\{Ru(PPh_3)_2Cp\}_2\{\mu\text{-}C{\equiv}CC_2\{Co_2(\mu\text{-dppm})(CO)_4\}C{\equiv}CC{\equiv}C\}]$ (**127b**), while $[\{Ru(PPh_3)_2Cp\}_2\{\mu\text{-}C{\equiv}CC_2\{Co_2(\mu\text{-dppm})(CO)_4\}C{\equiv}CC{\equiv}C\}]$ (**127c**) has been obtained from a thermal reaction of $[\{Cp(PPh_3)_2Ru\}_2(\mu\text{-}C{\equiv}C)_4\}]$ with $[Co_2(\mu\text{-dppm})(CO)_6]$ (Scheme 34).[35b]

A heterometallic $CoRu_5$ cluster, $[CoRu_5(\mu_4\text{-}PPh)(\mu_4\text{-}C_2Ph)(\mu\text{-}PPh_2)(CO)_{12}(\eta^5\text{-}C_5H_5)]$ (**129**) has been isolated from the reaction between $[Ru_5(\mu_5\text{-}C_2PPh_2)(\mu\text{-}PPh_2)(CO)_{13}]$ (**128**) and $[Co(CO)_2(\eta\text{-}C_5H_5)]$ (Fig. 11).[61] Its structure consists of a square pyramidal Ru_5Co unit, and a $Ru(Ph)C_2$-group attached to it through two ruthenium and one cobalt atoms.

The reaction between $[Ru_3(\mu_3\text{-}\eta^2\text{-}PhC_2C_2Ph)(\mu\text{-}CO)(CO)_9]$ and $[Co_2(CO)_8]$ affords a pentametallic cluster $[Co_2Ru_3(\mu_5\text{-}\eta^2,\eta^2\text{-}PhC_2C_2Ph)(CO)_{14}]$ (**130**) in quantitative yield.[62a] The X-ray determined molecular structure consists of a Co_2Ru_3 bow-tie cluster straddled by the PhC_2C_2Ph ligand, the two $C{\equiv}C$ triple bonds are attached to the five metal atoms by $2\sigma(2Ru)$ and $\pi(Co)$ bonding modes (Fig. 12).

A dihydride complex $[Cp^*Rh(\mu_2\text{-}1,2\text{-}S_2C_6H_4)(\mu_2\text{-}H)Ru(H)(PPh_3)_2]$ (**131**) reacts with excess of *p*-tolylacetylene at room temperature to afford a mixture of *cis* and *trans* isomers of alkynyl hydride complex $[Cp^*Rh(\mu_2\text{-}1,2\text{-}S_2C_6H_4)(\mu_2\text{-}H)Ru(C{\equiv}CTol\text{-}p)(PPh_3)_2]$ (**132a**). On the other hand reaction of **131** with trimethylsilylacetylene gives, exclusively the *cis* isomer of $[Cp^*Rh(\mu_2\text{-}1,2\text{-}S_2C_6H_4)(\mu_2\text{-}H)Ru(C{\equiv}CSiMe_3)(PPh_3)_2]$ (**132b**) (Scheme 35).[62b] The mechanism involves hydrogen transfer from **131** to the alkyne, followed by oxidative addition of a second equivalent of the alkyne to the dinuclear core of $[Cp^*Rh(\mu_2\text{-}1,2\text{-}S_2C_6H_4)Ru(PPh_3)_2]$ unit.

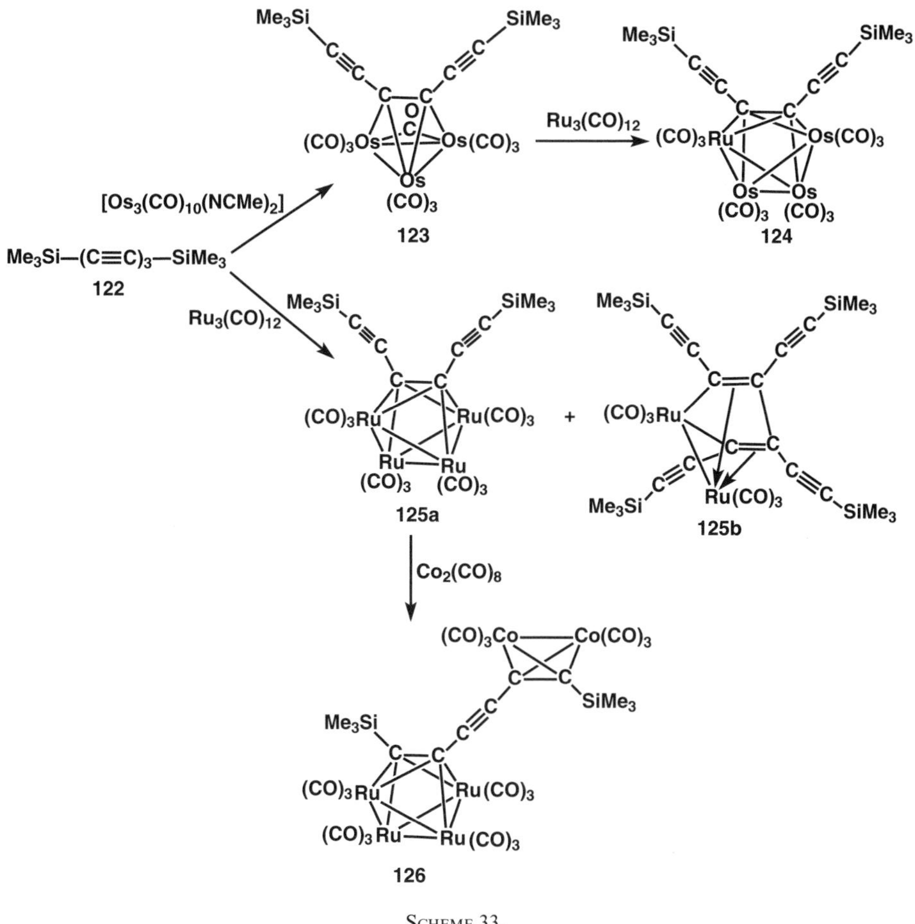

SCHEME 33.

$Cp(PPh_3)_2Ru$–$(C{\equiv}C)_n$–$Ru(PPh_3)_2Cp$

n = 3, 4

$[Co_2(CO)_6(dppm)_2]$ or $Co_2(CO)_8$

127a; n = 3, L = CO

127b; n = 3, L_2 = dppm

127c; n = 4, L_2 = dppm

SCHEME 34.

Thermal reaction of $[Fe_3(CO)_9(\mu_3\text{-}E)_2]$ (E = Se, **133a**; Te, **133b**) with $[(\eta^5\text{-}C_5H_5)M(CO)_3(C{\equiv}CPh)]$ (M = Mo, W) (**134a**, **b**) in the presence of trimethylamine-N-oxide (TMNO), in acetonitrile solvent yield the mixed-metal clusters $[(\eta^5\text{-}C_5H_5)MFe_2(\mu_3\text{-}E)_2(CO)_7(\eta^1\text{-}CCPh)]$ (E = Se and M = Mo, **135**; E = Se and

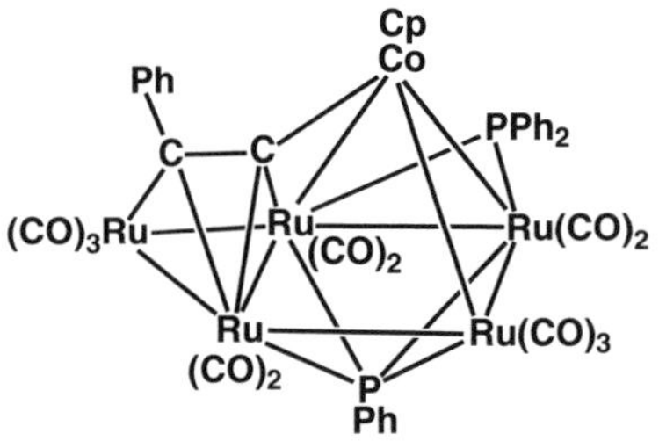

129

FIG. 11.

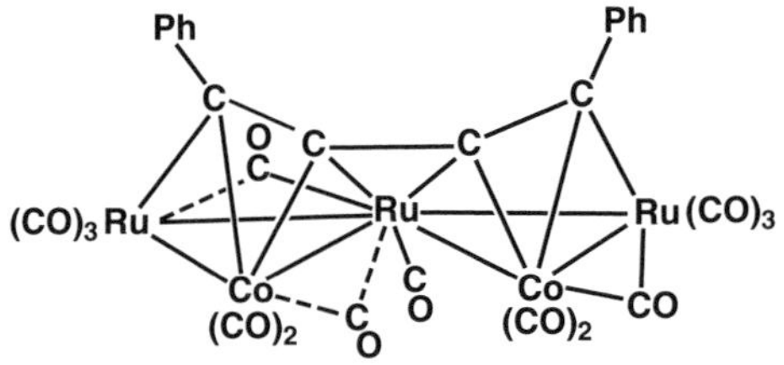

130

FIG. 12.

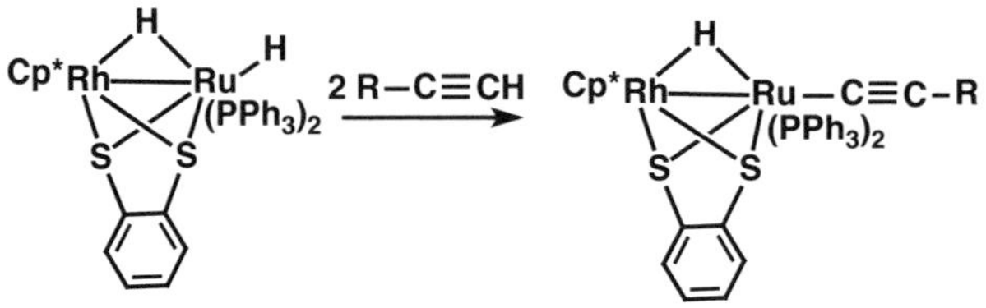

131

132a; R = *p*-tolyl (*cis:trans* = 8:2)
132b; R = SiMe$_3$ (*cis*)

SCHEME 35.

M = W, **136**; E = Te and M = Mo, **137**; E = Te and M = W, **138**) bearing η^1-acetylide groups. Further reaction of **135** or **137** with dicobaltoctacarbonyl at room temperature gives the mixed-metal cluster compound [(η^5-C_5H_5)$MoFe_2Co_2$ (μ_3-E)$_2$(CO)$_9$(η-CCPh)] (**139**, **140**) (Scheme 36).[63]

When a toluene solution containing [Fe_3(CO)$_9$(μ_3-E)$_2$] (E = S, Se or Te) and [W(η^5-C_5Me_5)(CO)$_3$(CCPh)] is subjected to reflux, mixed metal clusters [W_2Fe_3 (η^5-C_5Me_5)$_2$(μ_3-E)$_2${μ_4-CC(Ph)C(Ph)C}] (E = S, **141**; Se, **142**) are formed.[64] The molecule is made up of a Fe_3W_2 metal core which is enveloped by terminal and bridging carbonyl ligands and a μ_4-{CC(Ph)C(Ph)C} unit. The five metal atoms are arranged in the form of an open triangular bipyramidal polyhedron wherein the three Fe atoms occupy the basal plane while the two W centres are at the axial positions (Scheme 37).

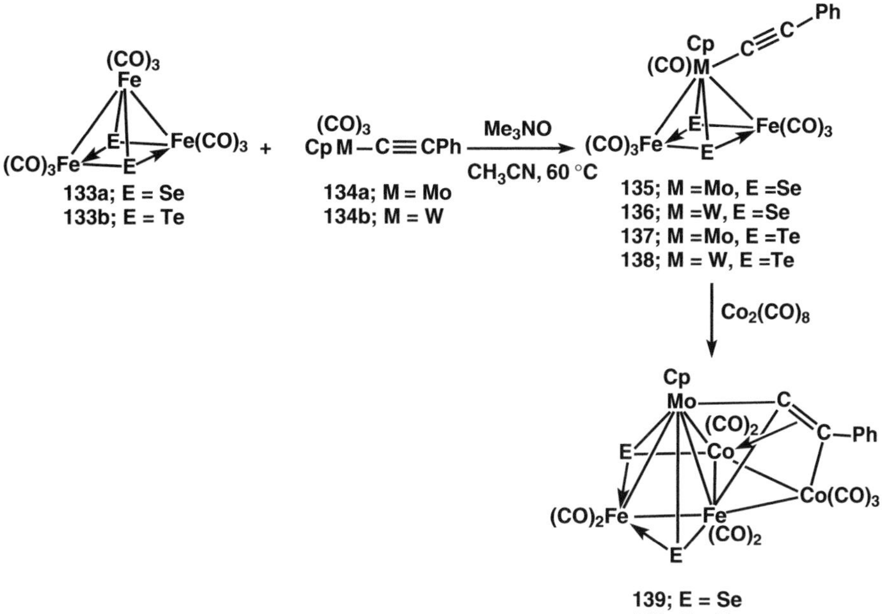

(CO)3 Fe (CO)3Fe E E Fe(CO)3 + (CO)3 CpM—C≡CPh —Me3NO / CH3CN, 60 °C→ (CO)3Fe E E Fe(CO)3 (CO)M Cp C≡C Ph

133a; E = Se
133b; E = Te

134a; M = Mo
134b; M = W

135; M = Mo, E = Se
136; M = W, E = Se
137; M = Mo, E = Te
138; M = W, E = Te

Co2(CO)8

Cp Mo C C—Ph (CO)2 E Co Co(CO)3 (CO)2Fe Fe (CO)2 E

139; E = Se
140; E = Te

SCHEME 36.

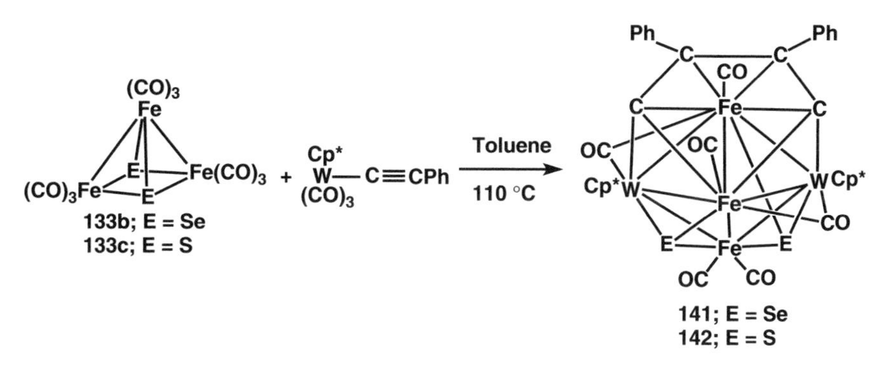

SCHEME 37.

When a toluene solution of the acetylide complex, $[Mo(\eta^5\text{-}C_5H_5)(CO)_3(C{\equiv}CPh)]$ is thermally reacted with $[Fe_3(CO)_9(\mu_3\text{-}E)_2]$ (E = S, Se), formation of mixed metal clusters, $[(\eta^5\text{-}C_5H_5)_2Mo_2Fe_3(CO)_8(\mu_3\text{-}E)_2\{\mu_5\text{-}CC(Ph)CC(Ph)\}]$ (E = S, **143**; Se, **144**), $[(\eta^5\text{-}C_5H_5)_2Mo_2Fe_4(CO)_9(\mu_3\text{-}E)_2(\mu_4\text{-}CCPh)_2]$ (E = S, **145**; Se, **146**) and $[(\eta^5\text{-}C_5H_5)_2Mo_2Fe_3(CO)_7(\mu_3\text{-}E)_2\{\mu_5\text{-}CC(Ph)C(Ph)C\}]$ (E = S, **147**; Se, **148**) are observed featuring head to tail coupling of two acetylide groups, two uncoupled acetylide groups and a tail to tail coupling of two acetylide groups, respectively, on the chalcogen bridged Fe/Mo cluster framework (Scheme 38).[65] The head-to-tail acetylide-coupled μ_5-{CC(Ph)CC(Ph)} unit of **143**/**144** acts as an eight-electron

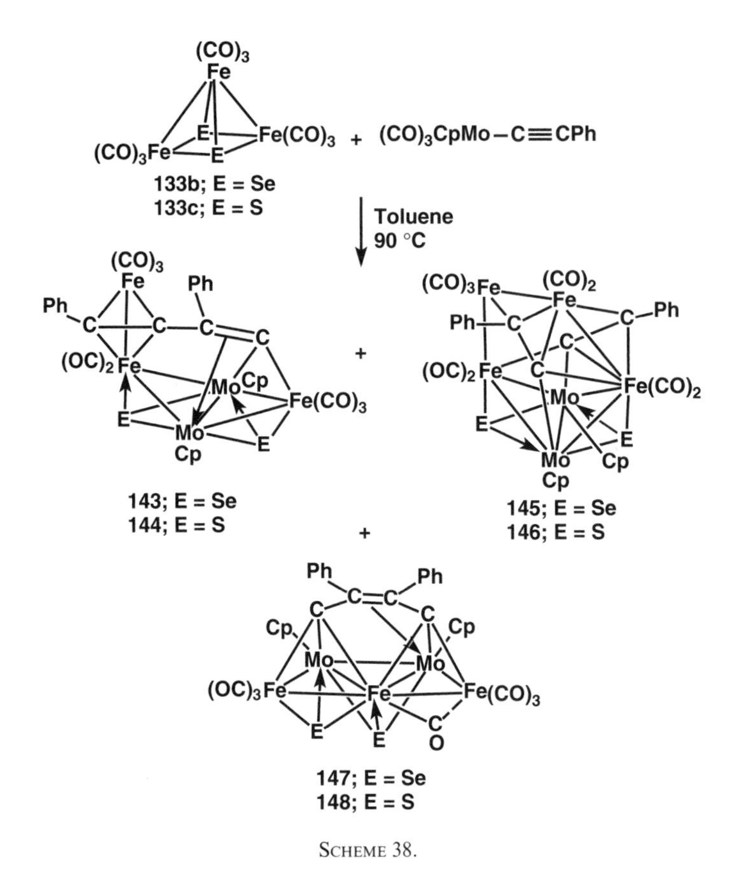

SCHEME 38.

donor to the cluster core. The basic cluster geometry of **147**/**148** consists of a twisted bow-tie type Fe_3MoE unit. One of the faces is capped by a second Mo atom, and the second face is capped by one terminal carbon atom of a {(CC(Ph)C(Ph)C} unit.

Under similar thermolytic conditions, formation of clusters $[(\eta^5\text{-}C_5H_5)_2W_2Fe_3(CO)_7(\mu_3\text{-}E)_2(\mu_3\text{-}\eta^2\text{-}CCPh)(\mu_3\text{-}\eta^1\text{-}CCH_2Ph)]$ (E = S, **149**; Se, **150**) and $[(\eta^5\text{-}C_5H_5)WFe_2(CO)_8(\mu\text{-}CCPh)]$ (**151**) were observed in reactions of $[W(\eta^5\text{-}C_5H_5)(CO)_3(C{\equiv}CPh)]$ with $[Fe_3(CO)_9(\mu_3\text{-}E)_2]$ (E = S, Se) (Scheme 39).[65]

When a benzene solution containing $[Fe_3(CO)_9(\mu_3\text{-}S)_2]$, $[(\eta^5\text{-}C_5R_5)Mo(CO)_3(C{\equiv}CPh)]$ (R = H, Me), H_2O and Et_3N are photolysed with continuous bubbling of argon a rapid formation of cluster $[(\eta^5\text{-}C_5R_5)MoFe_2(CO)_6(\mu_3\text{-}S)(\mu\text{-}SCCH_2Ph)]$ (R = H, **152**; R = Me, **153**) is observed (Scheme 40).[66] The H_2O molecule acts as a source of protons as confirmed by a labelling experiment, and the Et_3N works as a phase-transfer catalyst.

Photolytic reaction of metal acetylide complexes with simple metal carbonyls has been used to form heterometallic acetylide bridged complexes. For instance, photolytic reaction of a benzene solution of $[Fe(CO)_5]$ and $[(\eta^5\text{-}C_5R_5)Mo(CO)_3(C{\equiv}CPh)]$ (R = H, Me) under continuous bubbling of argon results in a rapid

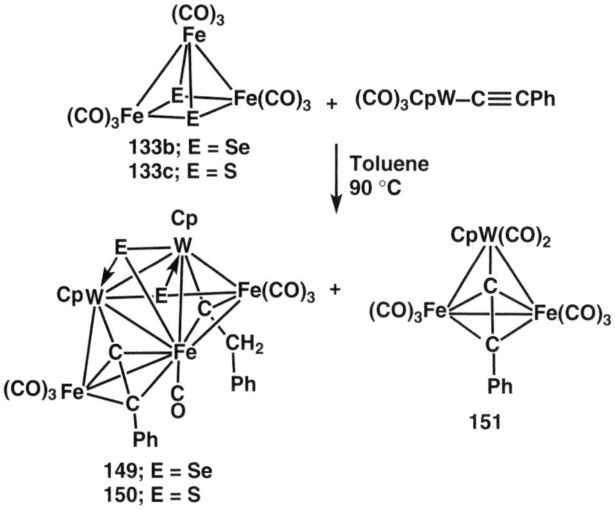

Scheme 39.

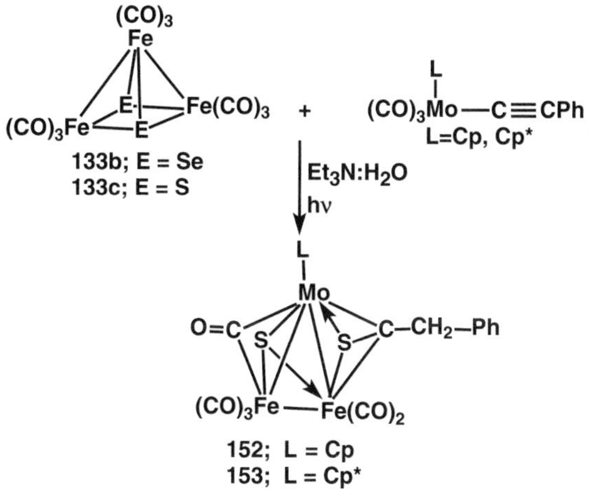

Scheme 40.

formation of $[(\eta^5\text{-}C_5R_5)Fe_2Mo(CO)_8(\mu_3\text{-}\eta^1{:}\eta^2{:}\eta^2\text{-}CCPh)]$ (R = H, **154**; Me, **155**) and $[(\eta^5\text{-}C_5H_5)Fe_3Mo(CO)_{11}(\mu_4\text{-}\eta^1{:}\eta^1{:}\ \eta^2{:}\eta^1\text{-}CCPh)]$ (**156**) (Scheme 41).[67] The structure of **154** consists of a MoFe triangle and a $\mu_3\text{-}\eta^1,\eta^2,\eta^2$ acetylide ligand. The molecular structure of **156** comprises of an open Fe_3Mo butterfly arrangement with an acetylide group bonded in $\mu_3\text{-}\eta^1$ mode.

However, when a benzene solution containing a mixture of $[(\eta^5\text{-}C_5Me_5)Mo(CO)_3(C{\equiv}CPh)]$, $[Fe(CO)_5]$ and $[PhC{\equiv}CH]$ are photolysed, formation of alkyne–acetylide coupled mixed metal cluster $[(\eta^5\text{-}C_5Me_5)Fe_2Mo(CO)_7(\mu_3\text{-}C(H)C(Ph)C(Ph)C)]$ (**157**) is obtained in moderate yield (Scheme 42).[67] The molecular

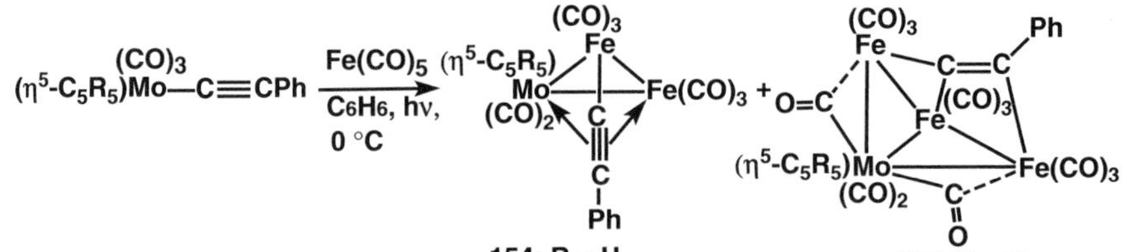

SCHEME 41.

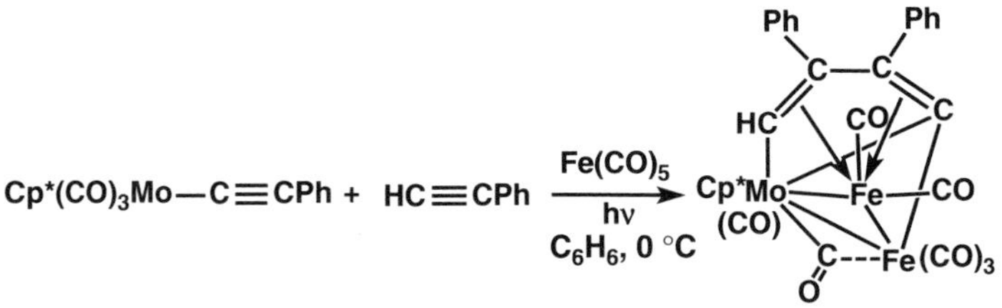

SCHEME 42.

structure of **157** consists of a Fe_2Mo triangle with a {C(H)$=$C(Ph)C(Ph)$=$C} ligand triply bridging the Fe_2Mo face.

When a mixed metal cluster [$Fe_2Ru(CO)_9(\mu_3\text{-}E)_2$] (E = S or Se) was reacted with [(η^5-C_5H_5)M$(CO)_3$(C$\equiv$CPh)] (M = Mo or W) in toluene reflux condition, heterometallic clusters [(η^5-$C_5H_5)_2Fe_2RuM_2(CO)_6(\mu_3\text{-}E)_2\{\mu_4\text{-}CC(Ph)C(Ph)C\}$] (M = Mo and E = S, **158**; M = Mo and E = Se, **159**; M = W and E = S, **160**; M = W and E = Se, **161**) and [(η^5-$C_5H_5)_2Fe_2Ru_2M_2(CO)_9(\mu_3\text{-}E)_2\{\mu_3\text{-}CCPh\}_2$] (M = W and E = S, **162**; M = W and E = Se, **163**), was isolated from the reaction mixtures (Scheme 43).[68] The molecular structure of **158** consists of an open trigonal bipyramidal Fe_2RuMo_2 metal core enveloped by terminal and bridging carbonyl ligands and a μ_4-{CC(Ph)C(Ph)C} unit. Structure of **162** can be described as a $FeRuW_2S$ distorted square pyramid core, in which the WRu edge is bridged by a $Fe(CO)_3S$ and the RuFe edge by a $Ru(CO)_3$ unit. One acetylide group caps the W_2Ru face in a η^1, η^2, η^2 fashion and another caps the open FeRuW face in a similar bonding mode.

When a toluene solution of a Fe/W mixed metal cluster, [$Fe_2W(CO)_{10}(\mu_3\text{-}S)_2$] is heated at 80 °C with the tungsten acetylide complex [(η^5-C_5H_5)W$(CO)_3$(CCPh)], the Fe_2W_3 mixed metal [(η^5-$C_5H_5)_2W_3Fe_2S_2(CO)_{12}(CCPh)_2$] (**164**) is isolated (Scheme 44).[69] Its structure consists of a Fe_2WS_2 trigonal bipyramidal unit in which the W atom is attached to two separate molecules of [(η^5-C_5H_5)W$(CO)_3$(CCPh)] through the acetylide group.

When [$Fe_2Mo(CO)_{10}(\mu_3\text{-}S)_2$] is made to interact with [(η^5-C_5Me_5)W $(CO)_3$(C$\equiv$CPh)] under mild thermolytic condition in argon atmosphere, a mixed metal acetylide cluster [(η^5-C_5Me_5)MoW$Fe_4(\mu_3\text{-}S)_3(\mu_4\text{-}S)(CO)_{14}$(CCPh)] (**165**) is isolated (Scheme 45).[70] Molecular structure of **165** consists of a {Fe_2MoS_2}

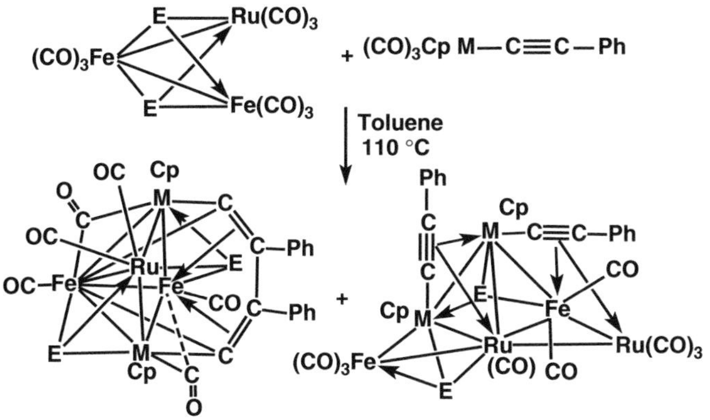

158; M = Mo, E = S
159; M = Mo, E = Se
160; M = W, E = S
161; M = W, E = Se
162; M = W, E = S
163; M = W, E = Se

SCHEME 43.

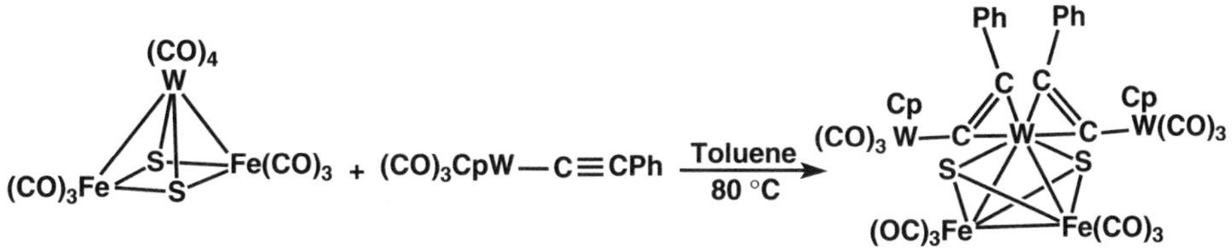

164

SCHEME 44.

distorted square pyramid unit in which the basal Mo atom is bonded to the two sulphur atoms of an open $\{Fe_2S_2\}$ butterfly unit and a $\{(\eta^5\text{-}C_5H_5)W(CO)_2\}$ group. A single phenylacetylide ligand bridges the Mo–W bond in $\mu\text{-}\eta^1,\eta^2$ fashion.

Photolysis of a benzene solution of a mixture of $[Fe_3(CO)_9(\mu_3\text{-}S)_2]$ $[(\eta^5\text{-}C_5Me_5)M(CO)_3(C{\equiv}CPh)]$ (M = Mo, W) and $HC{\equiv}CR$ (R = Ph, *n*-Bu and Fc) (Fc = ferrocene) forms two types of clusters depending on the nature of acetylene used. When phenylacetylene or *n*-butyl acetylene are used, clusters $[(\eta^5\text{-}C_5Me_5)MFe_3(CO)_6(\mu_3\text{-}S)\{(\mu_3\text{-}C(H) = C(R)S\}(\mu_3\text{-}CCPh)]$ (**166–169**) are formed (Scheme 46).[71] In these compounds a new C–S bond is formed yielding a $\{HC{=}C(R)S\}$ ligand which bridges a $\{MFe_2\}$ face of the cluster framework. Use of ferrocenylacetylene in the reaction medium results in the formation of $[(\eta^5\text{-}C_5Me_5)MFe_3(CO)_7(\mu_3\text{-}S)\{\mu_3\text{-}C(Fc){=}C(H)S\}(\mu_3\text{-}CCPh)]$ (**170**, **171**) in which a formal switching of the α and β carbons of the coordinated acetylide moiety is observed (Scheme 47). The nature of the sulfido-acetylene coupling may be an important factor in determining which cluster type is formed, with the bulky ferrocene group disfavouring the expected $S{-}C_{subs}$ coupling. The structure of **166** can be described as consisting of a $\{MoFe_3\}$ butterfly core in which the hinge is composed of the Mo and a Fe atom. One $\{MoFe_2\}$ face is capped by μ_3-S ligand as well as an unusual

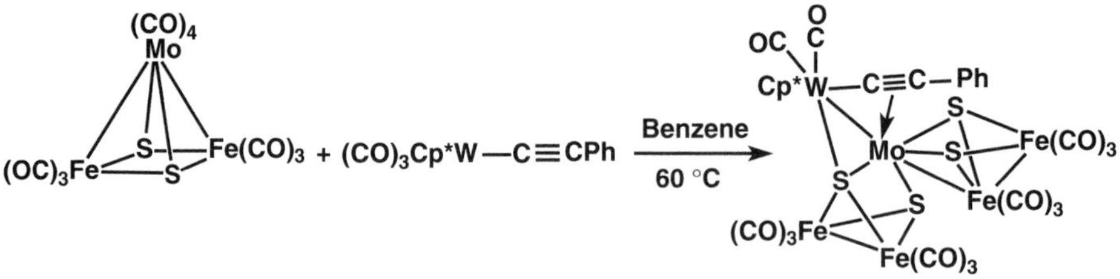

$(CO)_4$
Mo
$(OC)_3Fe$ S S $Fe(CO)_3$ + $(CO)_3Cp^*W—C\equiv CPh$ $\xrightarrow[60\ °C]{Benzene}$
OC C O
$Cp^*W—C\equiv C—Ph$
Mo S $Fe(CO)_3$
S S $Fe(CO)_3$
$(CO)_3Fe$ $Fe(CO)_3$
165

SCHEME 45.

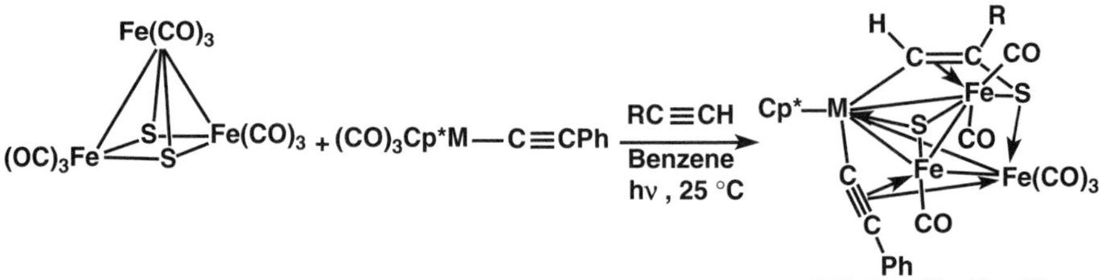

$Fe(CO)_3$
$(OC)_3Fe$ S S $Fe(CO)_3$ + $(CO)_3Cp^*M—C\equiv CPh$ $\xrightarrow[Benzene\ h\nu,\ 25\ °C]{RC\equiv CH}$
H R
C=C CO
$Cp^*—M$ Fe—S
S CO
C Fe $Fe(CO)_3$
C CO
Ph
166; M = Mo, R = Ph
167; M = Mo, R = ^{n}Bu
168; M = W, R = Ph
169; M = W, R = ^{n}Bu

SCHEME 46.

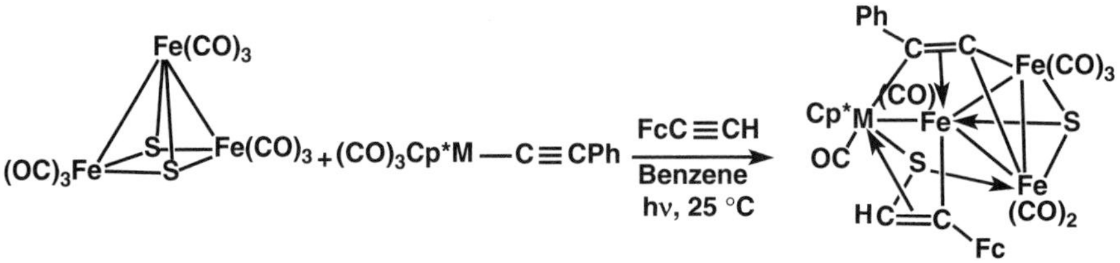

$Fe(CO)_3$
$(OC)_3Fe$ S S $Fe(CO)_3$ + $(CO)_3Cp^*M—C\equiv CPh$ $\xrightarrow[Benzene\ h\nu,\ 25\ °C]{FcC\equiv CH}$
Ph
C=C $Fe(CO)_3$
(CO)
Cp^*M—Fe S
OC S Fe
HC=C $(CO)_2$
Fc
170 M = Mo
171 M = W

SCHEME 47.

{HC = C(Ph)S} ligand. The second butterfly face is capped by a {η^1:η^2:η^2-CCPh} group. The wing-tip Fe atoms bear two and three carbonyl groups each and the hinge Fe atom has one terminal carbonyl bonded to it.

Compound **170** consists of a Fe_3 triangle, which is capped by a sulfido ligand. One of the iron atoms is bonded to a {(η^5-C_5Me_5)W} group, and the Fe–W bond is bridged by a ferrocenylacetylene group. A second sulfido ligand triply bridges the terminal carbon atom of this acetylene with the tungsten and one of the iron atoms. A {CCPh} group forms an unusual quadruple bridge in which the β-carbon of the acetylide is now η^1-bonded to the tungsten atom. Seven terminal carbonyls distributed on the three iron atoms and on the tungsten atom complete the ligand environment of the cluster.[72]

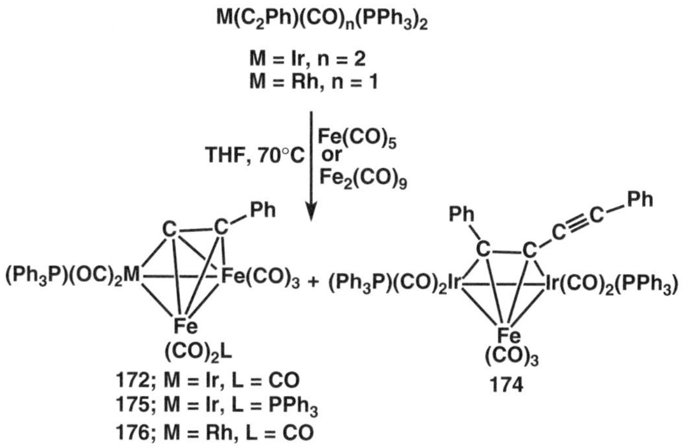

$M(C_2Ph)(CO)_n(PPh_3)_2$

M = Ir, n = 2
M = Rh, n = 1

THF, 70°C; $Fe(CO)_5$ or $Fe_2(CO)_9$

$(Ph_3P)(OC)_2M$ – $Fe(CO)_3$ – Fe $(CO)_2L$ + $(Ph_3P)(CO)_2Ir$ – $Ir(CO)_2(PPh_3)$ – Fe $(CO)_3$

172; M = Ir, L = CO
175; M = Ir, L = PPh_3
176; M = Rh, L = CO

174

SCHEME 48.

Reactions between $[Ir(C_2Ph)(CO)_2(PPh_3)_2]$ and iron carbonyls ($Fe(CO)_5$ or $Fe_2(CO)_9$) result in the formation of iron–iridium clusters $[Fe_2Ir(\mu_3\text{-}C_2Ph)(CO)_{9-n}(PPh_3)_n]$ ($n = 1$, **172**; 2, **173**) and $[FeIr_2(\mu_3\text{-}PhC_2C_2Ph)(CO)_{12}(PPh_3)_2]$ (**174**). The Rh analogue of **172** has been obtained by a similar reaction (Scheme 48). Substitution of the iridium bound PPh_3 by PEt_3 and addition of PEt_3 to C_α of the acetylide ligand in **172** gives zwitterionic $[Fe_2Ir(\mu_3\text{-}PhC_2PEt_3)(CO)_8(PEt_3)]$ (**177**), which on heating converts to $[Fe_2Ir(\mu_3\text{-}C_2Ph)(CO)_7(PEt_3)_2]$ (**179**) by migration of PEt_3 from carbon to iridium (Scheme 49).[73] Analogous complexes containing cluster-bound PMe_2Ph and $P(OMe)_3$ have also been obtained by such type of reactions.

Reaction of Fe_2Ir cluster (**172**) with dihydrogen gives complex **184** whereas stepwise addition of H^- and H^+ results in the formation of **184** and the isomeric hydrido alkyne derivative **185** (Scheme 50).[74]

Treatment of the triruthenium imido complex $[Ru_3(CO)_{10}(\mu_3\text{-}NPh)]$ (**186**) with metal hydride complex $[LW(CO)_3H]$ in refluxing toluene produces a trinuclear heterometallic imido cluster $[LWRu_2(CO)_8(\mu\text{-}H)(\mu_3\text{-}NPh)]$ (L = Cp, **187a**; Cp*, **187b**), which on reaction with tungsten acetylide $[CpW(CO)_3CCPh]$ produces a tetranuclear imidoalkyne complex $[LCpW_2Ru_2(CO)_6(\mu\text{-}NPh)(\mu\text{-}\eta^2\text{-}CH{=}CPh)]$ (L=Cp, Cp*) (**188a**,**b**) (Scheme 51).[75]

Convenient and widely applicable synthetic routes to trinuclear heterometallic acetylide complexes $[LMM'_2(CO)_8(CCR)]$ have been developed which involve the reaction of metal acetylide $[LM(CO)_3(CCR)]$ (L=Cp or Cp*; M = W or Mo; R = Ph, C_5H_4F, C_5H_4OMe, ^{t}Bu and ^{n}Pr) with $[Os_3(CO)_{10}(NCMe)_2]$ or $[Ru_3(CO)_{12}]$. Thus, reaction of $[CpW(CO)_3C{\equiv}CR]$ (R = Ph, ^{t}Bu) with $[Os_3(CO)_{10}(CH_3CN)_2]$ in refluxing toluene produces tetranuclear mixed-metal acetylide complexes $[CpWOs_3(CO)_{11}(C{\equiv}CR)]$ (R = Ph, **189a**; ^{t}Bu, **189b**).[76] The molybdenum analogue $[MoOs_3(CO)_{11}(C{\equiv}CPh)(\eta\text{-}C_5H_5)]$ (**190**) has also been prepared by a similar procedure. Carbonylation of **189a,b** at 120 °C under pressurised CO induces cluster fragmentation, giving $[CpWOs_2(CO)_8(C{\equiv}CR)]$ (**191a,b**) in 80–85% yield.

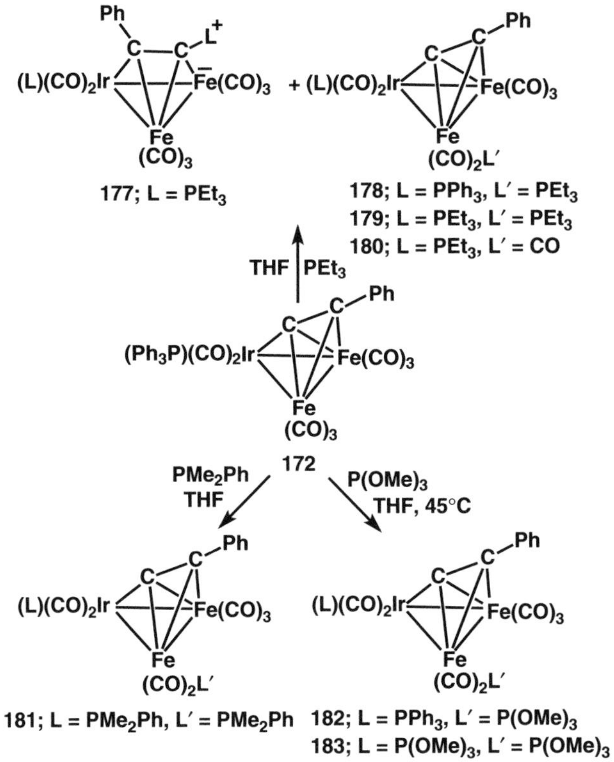

SCHEME 49.

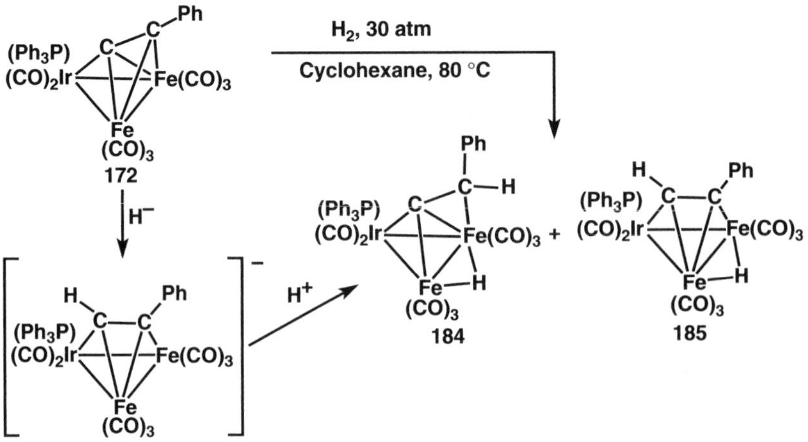

SCHEME 50.

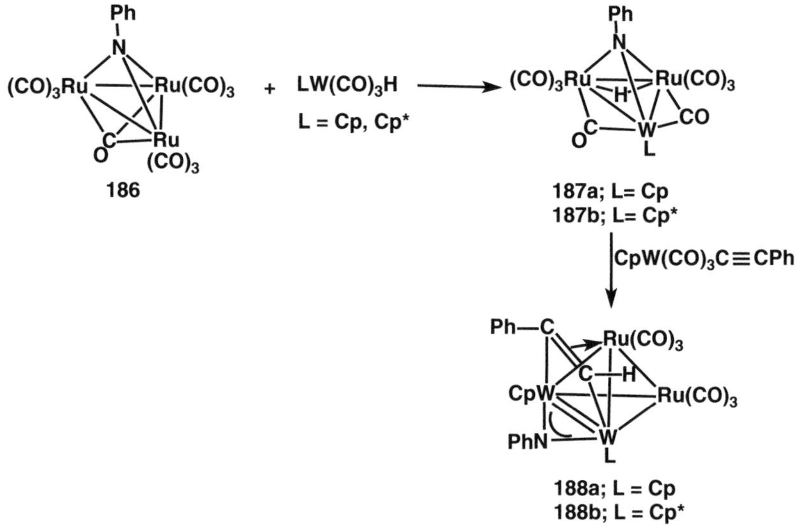

Ph
N
$(CO)_3Ru$ $Ru(CO)_3$
C Ru
O $(CO)_3$
186

+ $LW(CO)_3H$
L = Cp, Cp*

Ph
N
$(CO)_3Ru$ H $Ru(CO)_3$
C W CO
O L

187a; L= Cp
187b; L= Cp*

$CpW(CO)_3C{\equiv}CPh$

Ph—C
$Ru(CO)_3$
C—H
CpW $Ru(CO)_3$
PhN W
L

188a; L = Cp
188b; L = Cp*

SCHEME 51.

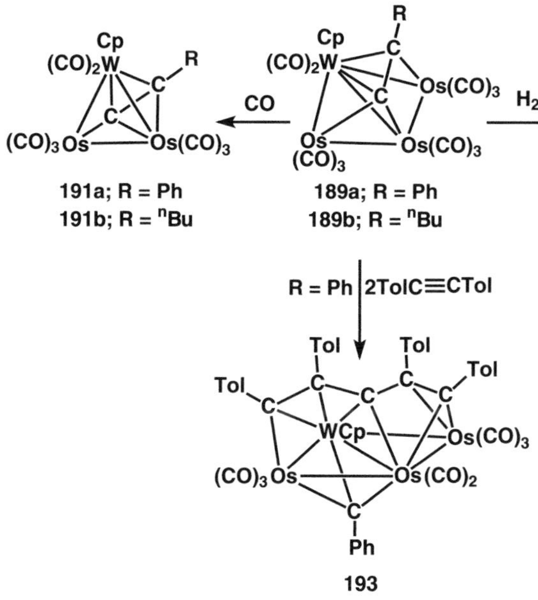

Cp
$(CO)_2W$ R
C
C
$(CO)_3Os$ $Os(CO)_3$

191a; R = Ph
191b; R = ^{n}Bu

CO

R
Cp C
$(CO)_2W$
$Os(CO)_3$
C
Os $Os(CO)_3$
$(CO)_3$

189a; R = Ph
189b; R = ^{n}Bu

H_2

$(CO)_2$ CH_2R
Cp W C
$(CO)_3Os$ $Os(CO)_3$
$Os(CO)_3$

192 R = ^{n}Bu

R = Ph | $2TolC{\equiv}CTol$

Tol Tol
Tol C C Tol
C C C
WCp $Os(CO)_3$
$(CO)_3Os$ $Os(CO)_2$
C
Ph

193

SCHEME 52.

Hydrogenation of **189b** produces an alkylidyne complex $[CpWOs_3(CO)_{11}(\mu_3\text{-}CC_5H_{11})]$ (**192**), while treatment of **189a** with excess of ditolylacetylene affects the scission of the acetylide ligand to give $[CpWOs_3(CO)_8(\mu_3\text{-}CPh)\{\mu_4\text{-}\eta^5\text{-}C(C_2Tol_2)_2\}]$ (**193**) (Scheme 52).[77]

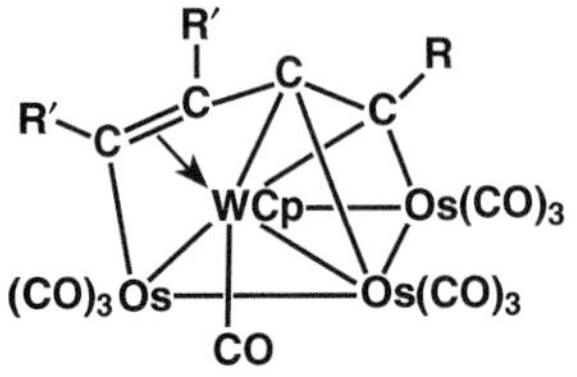

194a; R = Ph, R′ = CO_2Me
194b; R = Ph, R′ = CO_2Et
194c; R = ^nBu, R′ = CO_2Me

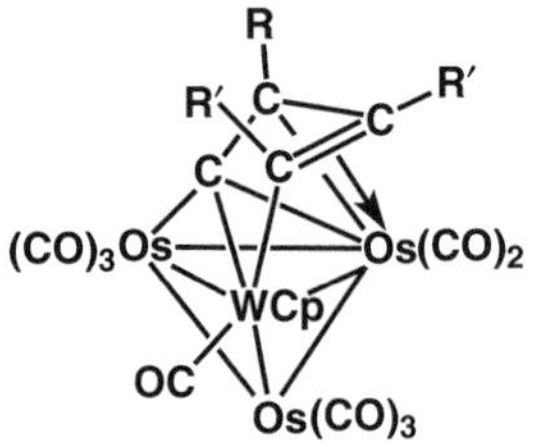

195a; R = Ph, R′ = CO_2Me
195b; R = Ph, R′ = CO_2Et
195c; R = ^nBu, R′ = CO_2Me

FIG. 13.

In contrast, treatment of complexes **189a** and **b** with alkynes containing electron-withdrawing groups produces the tetranuclear alkyne–acetylide coupling products $[CpWOs_3(CO)_{10}\{CR'CR'CCR\}]$ (R = Ph and R′ = CO_2Me, **194a**; R = Ph and R′ = CO_2Et, **194b**; R = ^nBu and R′ = CO_2Et, **194c**) and $[CpWOs_3(CO)_9\{CCRCR'CR'\}]$ (R = Ph and R′ = CO_2Me, **195a**; R = Ph and R′ = CO_2Et, **195b**; R = ^nBu and R′ = CO_2Et, **195c**) (Fig. 13).[76]

The cluster acetylide complexes $[MOs_3(CO)_{11}(C{\equiv}CPh)(\eta\text{-}C_5H_5)]$ (M = Mo, **190**; M = W, **189a**) react with $[Mo(CO)_3(C{\equiv}CPh)(\eta\text{-}C_5H_5)]$ to give planar pentanuclear complexes $[MMoOs_3(CO)_{11}(CCPhCCPh)(\eta\text{-}C_5H_5)_2]$ (M = Mo, **196**; M = W, **197**) which contain a C_4 hydrocarbon fragment derived from head-to-tail coupling between the two acetylide fragments. Reaction of complex **190** with $[W(CO)_3(C{\equiv}CPh)(\eta\text{-}C_5H_5)]$ does not produce the coupling product but induces C–C bond scission of $[W(CO)_3(C{\equiv}CPh)(\eta\text{-}C_5H_5)]$ giving a novel carbide–alkylidyne complex $[MoWOs_3(CO)_8(\mu_4\text{-}C)(\mu_3\text{-}CPh)(CCPh)(\eta\text{-}C_5H_5)_2]$ (**198**) (Fig. 14).[78]

Treatment of the acetylide complexes $[CpWOs_2(CO)_8(C{\equiv}CR)]$ (R = Ph, **191a**; R = ^nBu, **191b**) with Me_3NO in acetonitrile followed by reaction with various disubstituted alkynes, $C_2R'_2$ in refluxing toluene facilitates acetylide–alkyne coupling and formation of two isomeric trinuclear complexes, $[CpWOs_2(CO)_{12}\{C(R')C(R')CCR\}]$ (R = Ph and R′ = Tol, **199**; R = ^nBu and R′ = Tol, **200**; R = Ph and R′ = CO_2Et, **201**; R = ^nBu and R′ = CO_2Et, **202**; R = Ph and R′ = CF_3, **203**) (Scheme 53).[79]

When $[Ru_3(CO)_{12}]$ has been used to react with $[CpW(CO)_3C{\equiv}CPh]$ in refluxing toluene, $[CpWRu_2(CO)_8(C{\equiv}CPh)]$ (**204**) is isolated. In solution the variable temperature NMR suggests that the acetylide ligand of **204** undergoes a 360° rotation on the face of the WRu_2 triangle.[80] With excess of $[CpW(CO)_3C{\equiv}CR]$, **204** gives two heterometallic acetylide clusters $[Cp_2W_2Ru_2(CO)_9(CCRCCR)]$ (R = Ph, **205a**; R = p-C_6H_4F, **205b**), with C–C bond coupling and $[Cp_2W_2Ru_2(CO)_6(C{\equiv}CR)_2]$ (R = Ph, **206a**; R = p-C_6H_4F, **206b**), without C–C bond coupling (Scheme 54).[81]

In a similar reaction, thermolysis of a mixture of $[Ru_3(CO)_{12}]$ and $[CpMo(CO)_3C{\equiv}CPh]$ in a molar ratio 2:3 in refluxing toluene gives the trinuclear acetylide derivative $[CpMoRu_2(CO)_8C{\equiv}CPh]$ (**207**) and a pentanuclear heterometallic acetylide complex $[Cp_2Mo_2Ru_3(CO)_{10}(C{\equiv}CPh)_2]$ (**208**) (Fig. 15).[82]

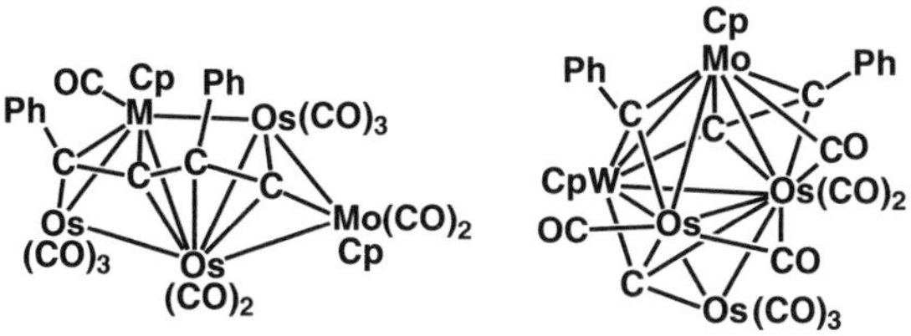

196; M = Mo
197; M = W

198

FIG. 14.

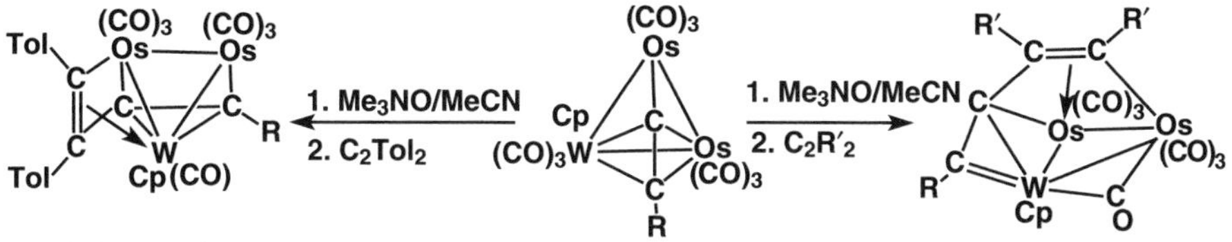

1. $Me_3NO/MeCN$
2. C_2Tol_2

1. $Me_3NO/MeCN$
2. $C_2R'_2$

199; R = Ph
200; R = ^{n}Bu

191a; R = Ph
191a; R = ^{n}Bu

201; R = Ph, R′ = Co_2Et
202; R = ^{n}Bu, R = Co_2Et
203; R = Ph, R = CF_3

SCHEME 53.

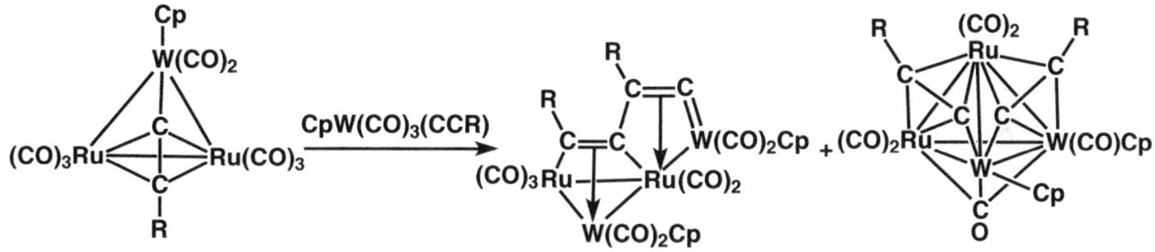

$CpW(CO)_3(CCR)$

204a;R = Ph
204b;R = C_6H_4F-*p*

205a;R = Ph
205b;R = C_6H_4F-*p*

206a;R = Ph
206b;R = C_6H_4F-*p*

SCHEME 54.

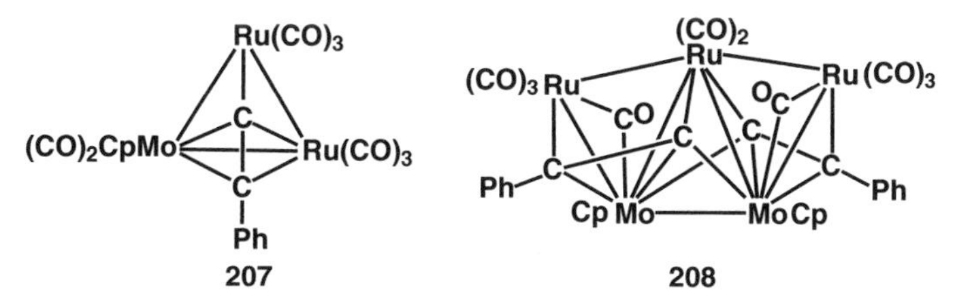

FIG. 15.

Condensation of triosmium alkyne complexes $[Os_3(CO)_{10}(C_2R_2)]$ (R = Tol, **209a**; R = Me, **209b**) with mononuclear tungsten acetylide complexes $[LW(CO)_3C{\equiv}CR']$ (L = Cp, Cp*, R′ = Ph, ^{t}Bu) generates six WOs_3 cluster complexes *via* 1:1 combination of the starting materials. Treatment of $[Os_3(CO)_{10}(C_2Me_2)]$ with

[$CpW(CO)_3C{\equiv}CPh$] under refluxing toluene forms two heterometallic complexes [$CpWOs_3(CO)_9\{CC(Ph)C(Me)C(Me)\}$] (**210a**) and [$CpWOs_3(CO)_{10}\{C(Me)C(Me)CC(Ph)\}$] (**211**). On the other hand, thermal reaction of [$Os_3(CO)_{10}(C_2Tol_2)$] with [$CpW(CO)_3C{\equiv}CPh$] gives [$CpWOs_3(CO)_9\{CC(Ph)C(Tol)C(Tol)\}$ (**210b**) and [$CpWOs_3(CO)_9(CCTolCTol)(\mu_3\text{-}CPh)$] (**212b**), whereas only one heterometallic cluster, [$LWOs_3(CO)_9\{CC(R')C(Tol)C(Tol)\}$] (L = Cp* and R′ = Ph, **210c**; L = Cp and R′ = ^{t}Bu, **210d**) is obtained on thermolysis of [$Os_3(CO)_{10}(C_2Tol_2)$] with [$LW(CO)_3C{\equiv}CR'$] (L = Cp*, R′ = Ph or L = Cp, R′ = ^{t}Bu). Addition of 1:1 molar equivalent of [Me_3NO] to compound **211**, followed by heating under refluxing toluene affords [$CpWOs_3(CO)_9(CCMeCMe)(\mu_3\text{-}CPh)$] (**212a**). When a toluene solution of [$Os_3(CO)_{10}(C_2Me_2)$] is thermally reacted with [$Cp^*W(CO)_3C{\equiv}CPh$], five heterometallic clusters are formed (**210e**, **212e**, **213**, **214** and **215**) (Fig. 16).[83]

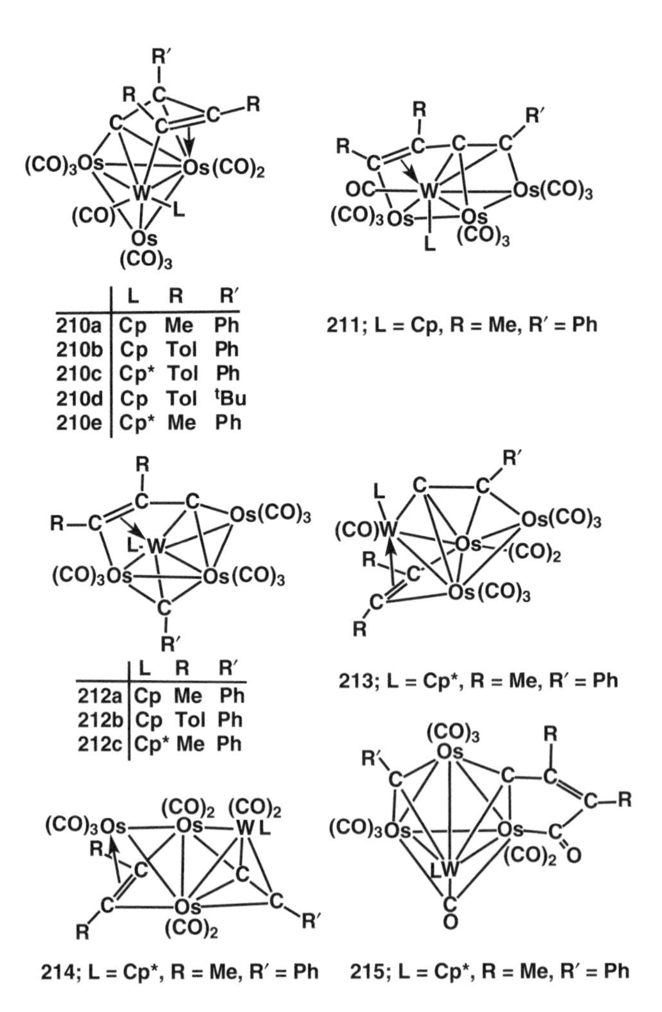

	L	R	R′
210a	Cp	Me	Ph
210b	Cp	Tol	Ph
210c	Cp*	Tol	Ph
210d	Cp	Tol	^{t}Bu
210e	Cp*	Me	Ph

	L	R	R′
212a	Cp	Me	Ph
212b	Cp	Tol	Ph
212c	Cp*	Me	Ph

FIG. 16.

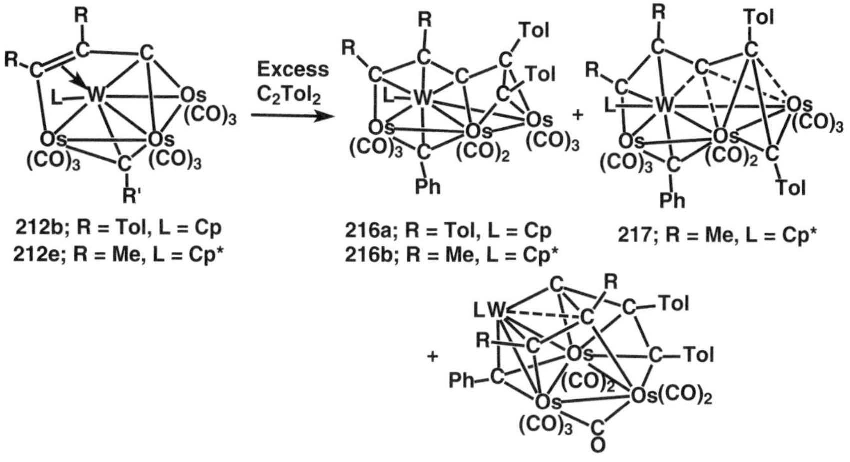

212b; R = Tol, L = Cp
212e; R = Me, L = Cp*

Excess C_2Tol_2

216a; R = Tol, L = Cp
216b; R = Me, L = Cp*

217; R = Me, L = Cp*

218; R = Me, L = Cp*

SCHEME 55.

Thermolysis of a toluene solution of complex **212b** with ditolylacetylene yields the planar clusters [CpWOs$_3$(CO)$_8$(μ_3-CPh){C(Tol)C(Tol)CC(Tol)C(Tol)}] (**216a**). On the other hand, reaction of **212e** in refluxing xylene solvent gives [Cp*WOs$_3$(CO)$_8$(μ_3-CPh){C(Tol)C(Tol)CC(Me)C(Me)}] (**216b**), [WOs$_3$(C$_5$Me$_5$)(CO)$_{12}$(μ_3-CPh){CMeCMeCC(Tol)C(Tol)}] (**217**) and **218** (Scheme 55).[84]

Treatment of **211** with Me$_3$NO followed by thermolysis in refluxing toluene yields the spiked triangular cluster [WOs$_3$Cp(CO)$_9$(μ-H){CMeCMeCC(μ_2-η^2-C$_6$H$_4$)}] (**219**) as a major product, which on further thermolysis converts to a butterfly cluster [WOs$_3$Cp(CO)$_9${CMeCMeCHC(μ_2-η^2-C$_6$H$_4$)}] (**220**) and then to a tetrahedral cluster [WOs$_3$Cp(CO)$_8${CMeCMeCHC(μ_2-η^2-C$_6$H$_4$)}] (**221**) *via* hydride migration followed by loss of CO (Scheme 56).[85]

The tetranuclear clusters **222**–**225** shown in Scheme 57 are formed in the reaction of [Cp*W(CO)$_3$C$\equiv$CR] (R = Ph, ^{n}Bu, CH$_2$OMe, CH$_2$Ph) with [Os$_3$(CO)$_{10}$(NCMe)$_2$]. While **222** is stable in a single isomeric form, clusters **223**–**225** exist in two interconverting isomeric forms on heating. When **222** is treated with Me$_3$NO, followed by heating to reflux in toluene, **226** is obtained. On similar treatment, complex **223** affords two isomeric vinylidene clusters [Cp*WOs$_3$(μ_4-C)(μ-H)(μ-CCH^{n}Pr)(CO)$_9$] (**227a,b**) as an inseparable mixture, while thermal reaction of **224a,b** with Me$_3$NO forms two methoxy derivatives [Cp*WOs$_3$(μ_4-C)(μ-H)(μ-CCHOMe)(CO)$_9$] (**228a,b**) (Scheme 58).[86]

Thermal reaction of **227** with CO in toluene leads to regeneration of **223a,b**, while **228** gives **224a,b** along with an alkenyl cluster [Cp*WOs$_3$(μ_4-C)(μ-CHCHOMe)(CO)$_{10}$] (**229**) (Scheme 59).[86]

A mixture of two isomers (**230** and **231**) of a benzofuryl complex [Cp*WOs$_3$(μ_4-C)(μ-H)$_2$(μ-C$_8$H$_5$O)(CO)$_9$] is formed on thermolysis of **225**. Clusters **230** and **231** have also been isolated on thermolysis of a 1:1 mixture of [Cp*W(CO)$_3$C$\equiv$CCH$_2$OPh] with [Os$_3$(CO)$_{10}$(NCMe)$_2$] in refluxing toluene (Scheme 60).[86]

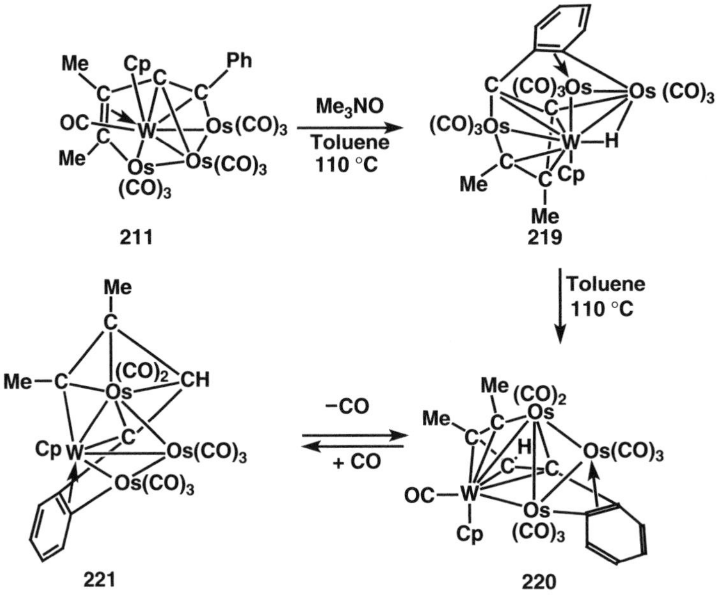

SCHEME 56.

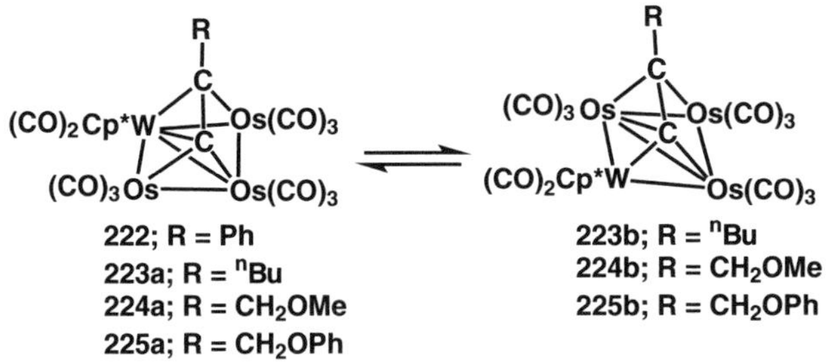

SCHEME 57.

Hydrogenation of the CH_2OMe derivative, **224** affords four compounds (**232**–**235**) (Scheme 61).[86]

When the carbide cluster complex $[Cp^*WOs_3(\mu_4\text{-}C)(\mu\text{-}H)(CO)_{11}]$ (**236**) is treated with excess of Me_3NO and 4-ethynyltoluene, the formation of an alkylidyne compound $[(C_5Me_5)WOs_3(\mu_3\text{-}CCHCHTol)(CO)_{11}]$ (**237**) is observed, while the corresponding reaction of **236** with 3-phenyl-1-propyne affords alkenyl carbido cluster $[(C_5Me_5)WOs_3(\mu_4\text{-}C)(CHCHCH_2Ph)(CO)_{10}]$ (**238**) and an alkylidyne compound $[(C_5Me_5)WOs_3[\mu_3\text{-}CC(CH_2Ph)(CH_2)](CO)_{10}]$ (**239**) (Scheme 62).[87]

On the other hand, coupling of **236** with an electron deficient alkyne, diisopropyl acetylenedicarboxylate (DPAD) results in the formation of a dimetallaallyl cluster $[Cp^*WOs_3(CO)_{10}\{C_3H(CO_2^iPr)_2\}]$ (**240**), which undergoes C–C metathesis to afford a second dimetallaallyl cluster (**241**) (Scheme 63).[88] Trace amounts of a related

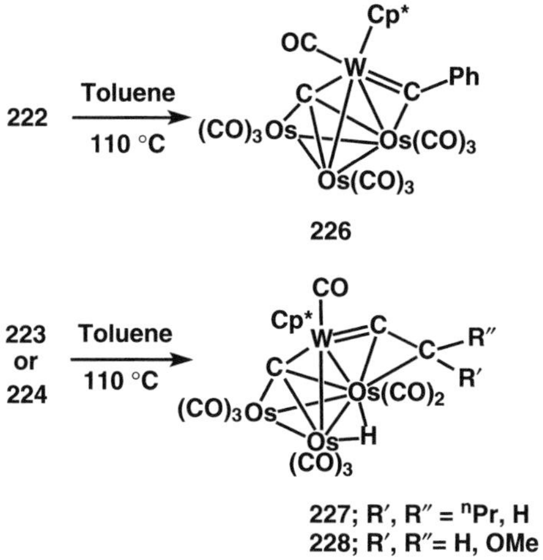

SCHEME 58.

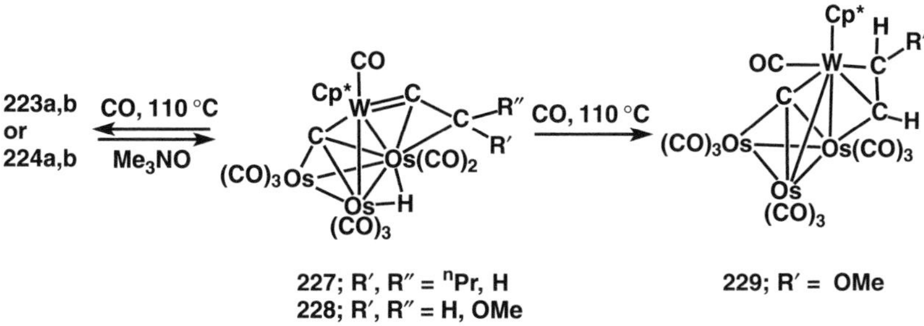

SCHEME 59.

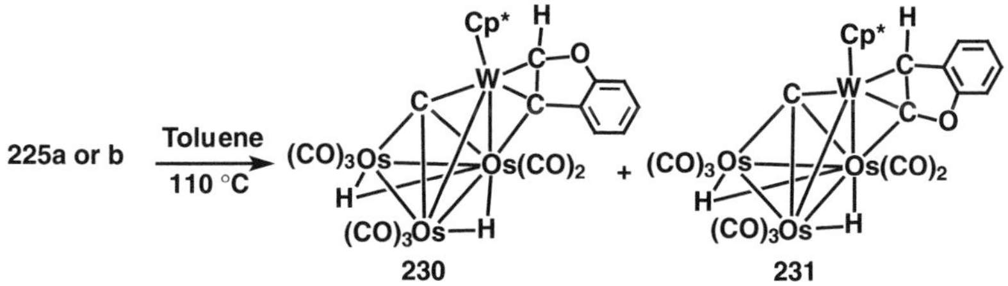

SCHEME 60.

vinyl-alkylidene cluster **242** were observed in the reaction of **236** with DPAD in refluxing toluene solution.

Reaction of [$CpWOs_2(CO)_8(C{\equiv}CPh)$] with 1.2 equiv. of Me_3NO in a mixture of dichloromethane–acetonitrile solvent at room temperature followed by *in situ* reaction with hydride complexes [$LW(CO)_3H$] (L = Cp and Cp*) in refluxing toluene produces the acetylide cluster complexes [$CpLW_2Os_2(CO)_9(CCPh)(\mu\text{-}H)$]

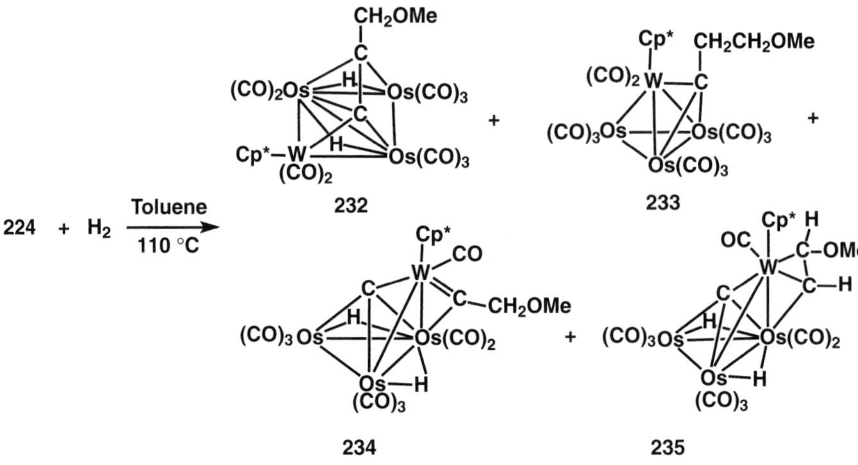

224 + H_2 —Toluene, 110 °C→

CH_2OMe | C | $(CO)_2Os$ H $Os(CO)_3$ C H Cp^*—W $(CO)_2$ $Os(CO)_3$

232

+

Cp^* CH_2CH_2OMe $(CO)_2$W—C $(CO)_3Os$ $Os(CO)_3$ $Os(CO)_3$

233

+

Cp^* CO W C C—CH_2OMe H $(CO)_3Os$ $Os(CO)_2$ Os—H $(CO)_3$

234

+

Cp^* H OC W C—OMe C—H C H $(CO)_3Os$ $Os(CO)_2$ Os—H $(CO)_3$

235

SCHEME 61.

$(CO)_2$ WCp^* C $(CO)_3Os$ $Os(CO)_3$ H Os $(CO)_3$

236

—TolC≡CH→

(CO) Cp^* OC W C R $(CO)_3Os$ $Os(CO)_3$ Os $(CO)_3$

237; R = CH=CHTol

↓ $PhCH_2C≡CH$

Ph CH_2 (CO) WCp^* H C C C H $(CO)_3Os$ Os $(CO)_3$ Os $(CO)_3$

238

+

Cp^* CH_2Ph $(CO)_2$W—C C C H H $(CO)_3Os$ $Os(CO)_2$ Os $(CO)_3$

239

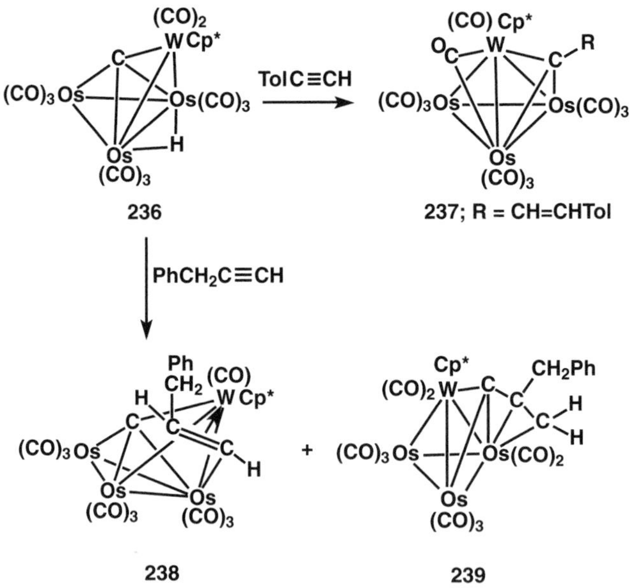

SCHEME 62.

(L = Cp, **243a**; L = Cp*, **243b**) and vinylidene cluster complexes [$CpLW_2Os_2(CO)_9$(CCHPh)] (L = Cp, **244a**; L = Cp*, **244b**). Heating of the acetylide or vinylidene complex in refluxing toluene induces a reversible rearrangement giving a mixture of two isomeric complexes (Scheme 64).[78]

The molybdenum analogue [$Cp_2Mo_2Os_2(CO)_9(CCPh)(\mu\text{-}H)$] (**243c**) has been obtained by the addition of Me_3NO to dichloromethane–acetonitrile solution of [$CpMoOs_2(CO)_8(CCPh)$], followed by a thermolytic reaction with [$CpW(CO)_3H$].[89]

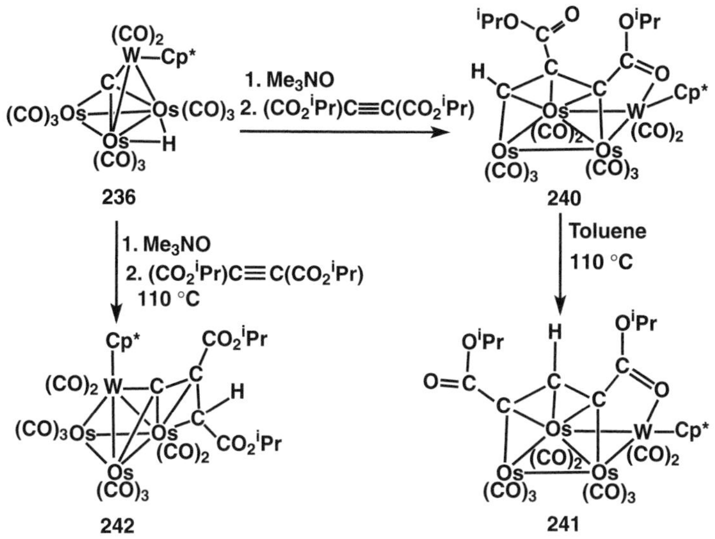

236

1. Me_3NO
2. $(CO_2{}^iPr)C{\equiv}C(CO_2{}^iPr)$

240

Toluene
110 °C

241

1. Me_3NO
2. $(CO_2{}^iPr)C{\equiv}C(CO_2{}^iPr)$
110 °C

242

Scheme 63.

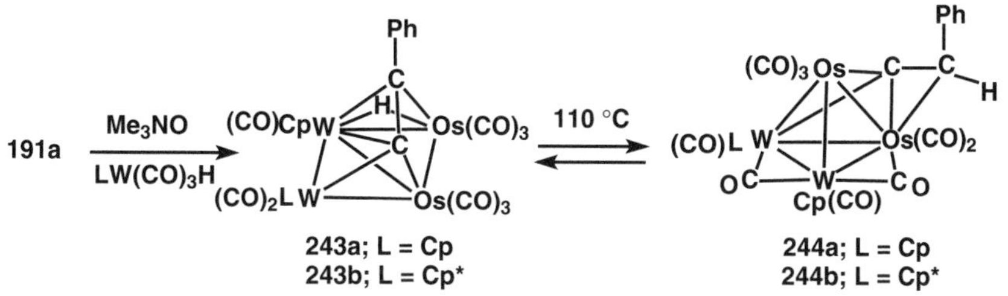

191a

Me_3NO
$LW(CO)_3H$

110 °C

243a; L = Cp
243b; L = Cp*

244a; L = Cp
244b; L = Cp*

Scheme 64.

Treatment of the carbido cluster $[Ru_5(\mu_5\text{-}C)(CO)_{15}]$ with Me_3NO followed by addition of the tungsten acetylide complexes $[LW(CO)_3(CCPh)]$ (L = Cp, Cp*) affords the heterometallic cluster complexes $[LWRu_5(\mu_5\text{-}C)(CCPh)(CO)_{15}]$ (L = Cp, **245a**; L = Cp*, **245b**) and $[LWRu_5(\mu_5\text{-}C)(CCPh)(CO)_{13}]$ (L = Cp, **246a**; L = Cp*, **246b**). Thermolysis of **245** results in an irreversible formation of **246**. Hydrogenation of **246b** gives two cluster compounds $[(C_5Me_5)WRu_5(\mu_6\text{-}C)(\mu\text{-}CCH_2Ph)(\mu\text{-}H)_2(CO)_{13}]$ (**247**) and $[(C_5Me_5)WRu_5(\mu_4\text{-}C)(\mu_3\text{-}CCH_2Ph)(\mu\text{-}H)_4(CO)_{12}]$ (**248**), *via* 1,1-addition of H_2 to the ligated acetylide and concurrent formation of two or four bridging hydrides (Scheme 65).[90]

Heterometallic vinylacetylide clusters $[Cp^*WRe_2(CO)_9(CCR)]$(R = –C(Me)= CH_2, **249**; R = C_6H_9, **250**) have been obtained from the condensation of mononuclear tungsten acetylide complexes $[Cp^*W(CO)_3(C{\equiv}CR)]$ (R = –C(Me)= CH_2, C_6H_9) and rhenium carbonyl complex $[Re_2(CO)_8(NCMe)_2]$. Treatment of the vinylacetylide complex $[Cp^*WRe_2(CO)_9\{C{=}CC(Me){=}CH_2\}]$ (**249**) with alcohols, ROH (R = Me, Et, Ph) in refluxing toluene solution afford complexes $[Cp^*WRe_2$

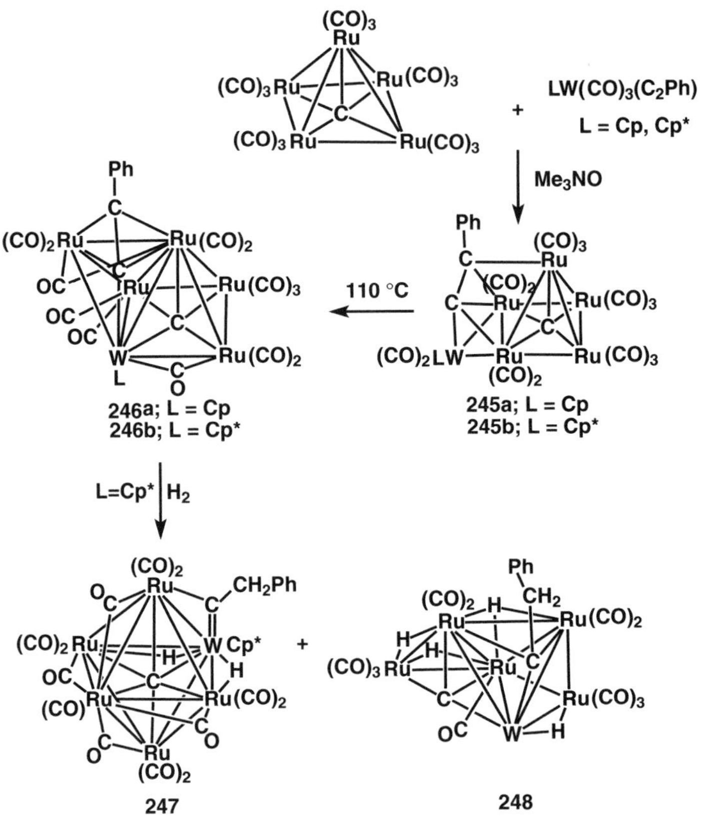

SCHEME 65.

$(CO)_8(\mu\text{-}OR)(C{=}C{=}CMe_2)]$ (R = Me, **251a**; R = Et, **251b**; R = Ph, **251c**), which contain an unusual μ_3-η^3-allylidene ligand and a bridging alkoxide ligand, and show fluxional behaviour in solution. When **249** is heated in the presence of hydrogen, a metallacyclopentadienyl complex [$Cp^*WRe_2(CO)_7(\mu\text{-}H)\{CHCHC(Me)CH\}$] (**252a,b**) is formed as a mixture of two non-interconvertible isomers (Scheme 66).[91] On the other hand, hydrogenation of **250** in toluene under refluxing condition initially produces a dihydride cluster [$Cp^*WRe_2(CO)_8(\mu\text{-}H)_2\{C{=}C(C_6H_9)\}$] (**253**) which on heating in toluene converts to **254** and **255** (Scheme 67).[91]

The acetylide complex [$Co_2(CO)_6\{\mu\text{-}PhC{=}CRe(CO)_5\}$] (**256**) (obtained by the treatment of the binuclear acetylide–hydride carbonyl complex of rhenium [$Re_2(CO)_8(\mu\text{-}H)(\mu\text{-}C{\equiv}CPh)$] with [$Co_2(CO)_8$] undergoes nondestructive reaction with oxygen resulting in the net loss of one acetylide carbon atom together with a CO ligand and formation of the carbyne cluster complex [$Co_2Re(CO)_{10}(\mu_3\text{-}CPh)$] (**257**) (Scheme 68).[92]

A dichloromethane solution of rhenium ethynyl complex [$(\eta^5\text{-}C_5Me_5)Re(NO)(PPh_3)\{(C{\equiv}C)_nH\}$] ($n$ = 1–3) (**258a–c**) and triosmium complex [$Os_3(CO)_{10}$

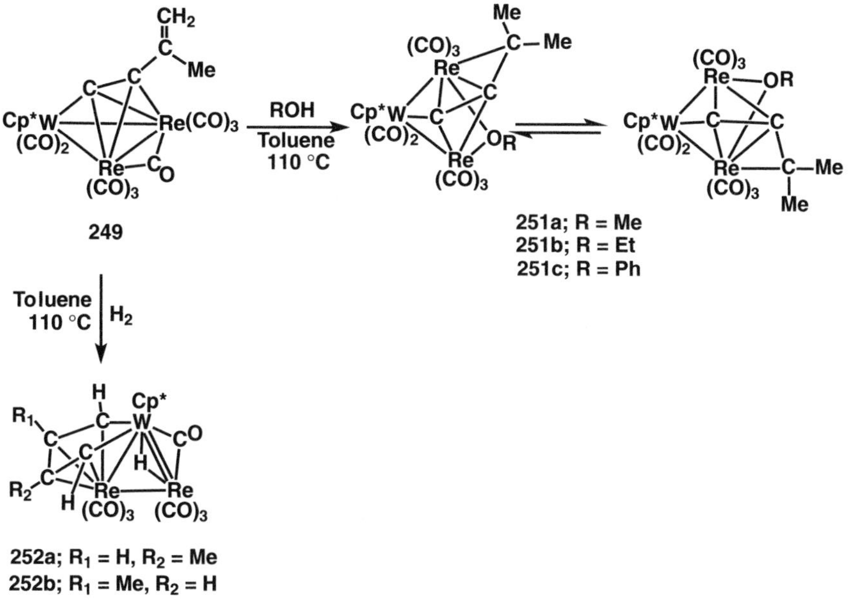

SCHEME 66.

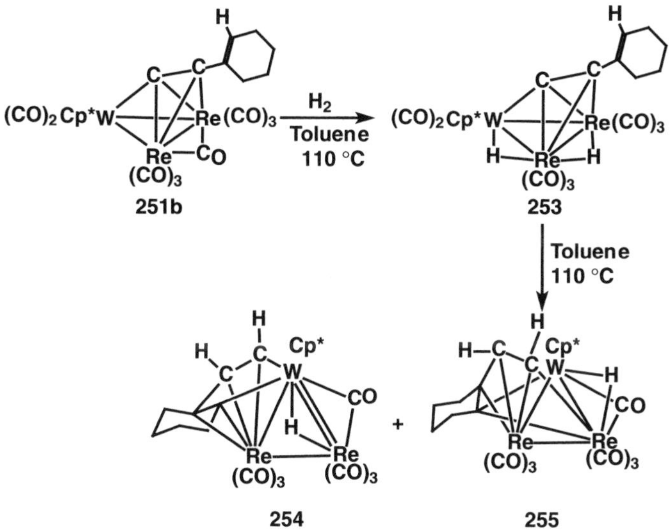

SCHEME 67.

$(NCMe)_2]$ react at room temperature to give $[(\eta^5\text{-}C_5Me_5)Re(NO)(PPh_3)\ (CC)_n\ Os_3(H)(CO)_{10}]$ ($n = 1$–3) (**259a–c**) (Scheme 69). Thermal reaction in hexane converts **259b** to the nonacarbonyl complex $[(\eta^5\text{-}C_5Me_5)Re(NO)(PPh_3)(CCCC)\ Os_3(CO)_9(H)]$ (**260**) (Scheme 70).[93,94]

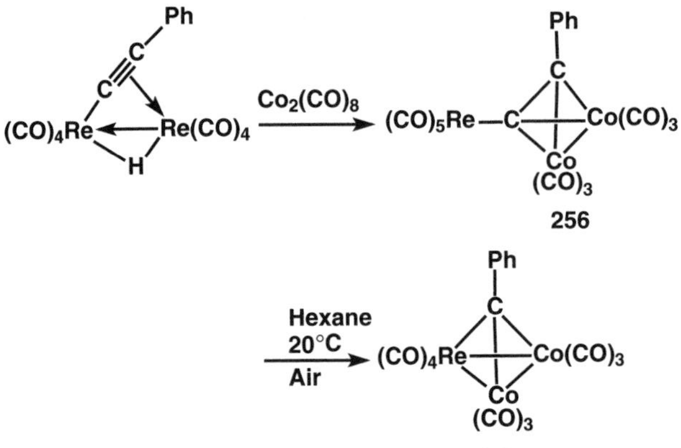

Ph
C
C
$(CO)_4Re$ $Re(CO)_4$
H
$Co_2(CO)_8$
Ph
C
$(CO)_5Re$—C $Co(CO)_3$
Co
$(CO)_3$
256

Ph
C
Hexane
20°C
Air
$(CO)_4Re$ $Co(CO)_3$
Co
$(CO)_3$
257

SCHEME 68.

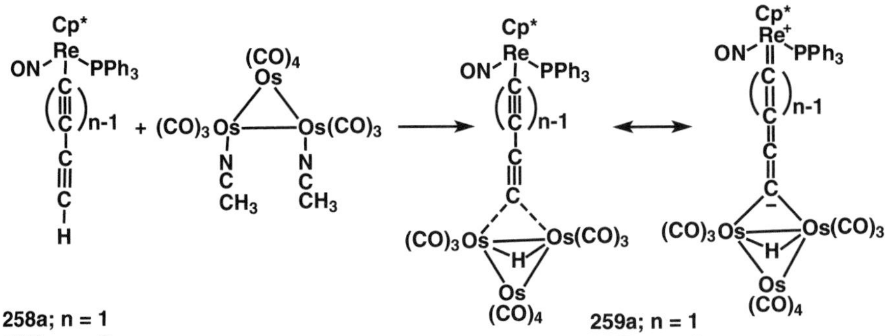

Cp*
Re
ON PPh3
C
C
n-1
C
C
H
+ $(CO)_3Os$ $Os(CO)_3$
$(CO)_4$
Os
N
C
CH_3
N
C
CH_3

Cp*
Re
ON PPh3
C
C
n-1
C
C
C
$(CO)_3Os$ $Os(CO)_3$
H
Os
$(CO)_4$

Cp*
Re^+
ON PPh3
C
C
n-1
C
C
$(CO)_3Os$ $Os(CO)_3$
H
Os
$(CO)_4$

258a; n = 1
258b; n = 2
258c; n = 3

259a; n = 1
259b; n = 2
259c; n = 3

SCHEME 69.

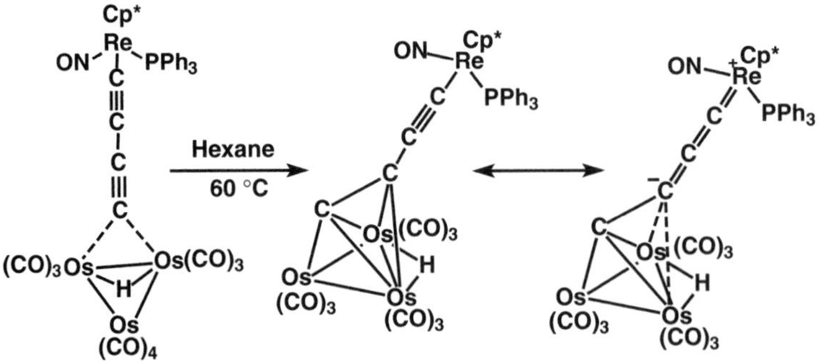

Cp*
Re
ON PPh3
C
C
C
C
$(CO)_3Os$ $Os(CO)_3$
H
Os
$(CO)_4$
259b

Hexane
60 °C

ON—Re Cp*
PPh3
C
C
C
C
Os $(CO)_3$
H
Os
$(CO)_3$
Os
$(CO)_3$

ON—Re Cp*
PPh3
C
C
C
C
Os $(CO)_3$
H
Os
$(CO)_3$
Os
$(CO)_3$
260

SCHEME 70.

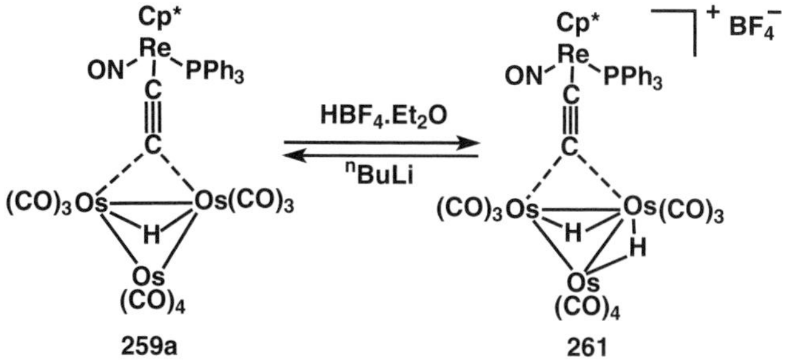

Scheme 71.

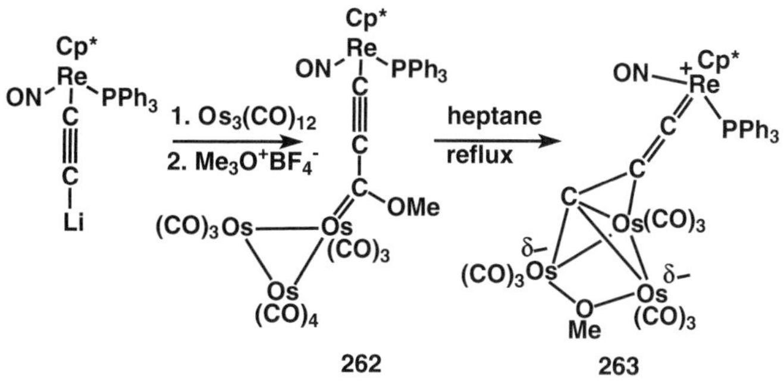

Scheme 72.

Addition of HBF_4. Et_2O (1 equiv.) to compound **259a** gives the cationic dihydride complex $[(\eta^5\text{-}C_5Me_5)Re(NO)(PPh_3)(CC)Os_3(CO)_{10}(H)_2]^+$ BF_4^- (**261**) (Scheme 71).[93]

Synthesis of a C_3OMe complex $[(\eta^5\text{-}C_5Me_5)Re(NO)(PPh_3)(C{\equiv}CC(OMe))Os_3(CO)_{11}]$ (**262**) from the reaction of $[(\eta^5\text{-}C_5Me_5)Re(NO)(PPh_3)(C{\equiv}CLi)]$ with $[Os_3(CO)_{12}]$ and $[Me_3O^+BF_4^-]$ has been reported.[93] On refluxing in heptane it forms **263** (Scheme 72).

Products obtained from the reaction between molybdenum dimer $[Mo_2(CO)_4(\eta\text{-}C_5H_5)_2]$ and $[M(CCR)(CO)_2(\eta\text{-}C_5H_5)]$ (M = Ru or Fe, R = Me or Ph) depends on the alkynyl metal used. The ruthenium containing reactant gives the dimolybdenum alkynyl adducts $[Mo_2\{Ru(\mu\text{-}CCR)(CO)_2(\eta\text{-}C_5H_5)\}(CO)_4(\eta\text{-}C_5H_5)_2]$ (R = Me (**264a**), Ph (**264b**)) as the only isolable product. In contrast, the iron alkynyls undergo Fe–C bond cleavage to give the alkyne adducts $[Mo_2(\mu\text{-}\eta^2\text{-}HC_2R)(CO)_4(\eta\text{-}C_5H_5)_2]$ (R = Me, **265a**; R = Ph, **265b**) (Scheme 73).[95]

On the other hand, reaction of $[\{Ru(CO)_2(\eta\text{-}C_5H_5)\}_2(\mu\text{-}C{\equiv}C)]$ with $[Mo_2(CO)_4(\eta\text{-}C_5H_5)_2]$ gives $[MoRu_2(\mu_2\text{-}CO)_3[\mu_3\text{-}C{\equiv}C\{Ru(CO)_2(\eta\text{-}C_5H_5)\}](\eta\text{-}C_5H_5)_3]$ (**266**), but not the dimolybdenum 'alkynyl' adduct, **264a,b** (Scheme 74).[96]

Reaction of $[(OC)_4Fe(\eta^1\text{-}PPh_2C{\equiv}CPh)]$ (**267**) with $[Co_2(CO)_8]$ at room temperature affords the heterotrimetallic complex $[(OC)_4Fe(\mu\text{-}\eta^1{:}\eta^2\text{-}PPh_2CCPh)Co_2(CO)_6]$ (**268**), in which the alkynic moiety is bound to a $Co_2(CO)_6$ unit. Both the mono- and di-substituted complexes, $[(OC)_4Fe(\mu\text{-}\eta^1{:}\eta^2\text{-}PPh_2C{\equiv}CPh)Co_2(CO)_5\{P(OMe)_3\}]$

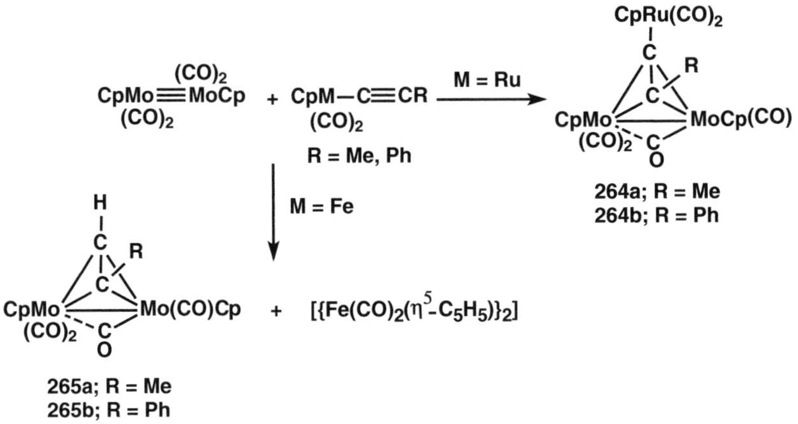

SCHEME 73.

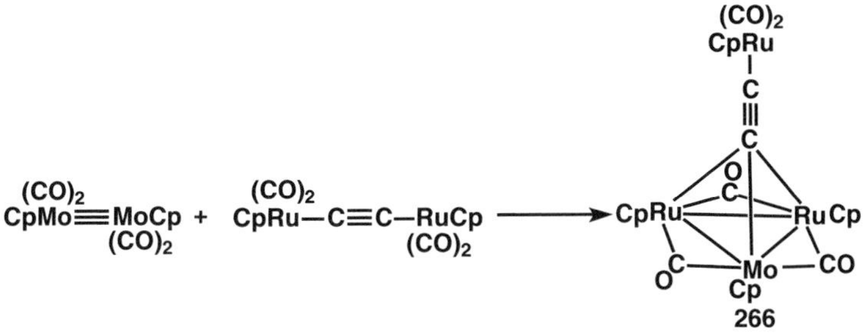

SCHEME 74.

(**269a**) and $[(OC)_4Fe(\mu\text{-}\eta^1\text{:}\eta^2\text{-}PPh_2CCPh)Co_2(CO)_4\{P(OMe)_3\}_2]$ (**269b**), have been obtained on reaction of **268** with an excess of trimethylphosphite at elevated temperature. Thermolysis of **269a** results in phosphorus–carbon bond cleavage and iron–cobalt bond formation to yield the acetylide bound mixed-metal triangular cluster $[FeCo_2(CO)_6\{\mu_3\text{-}\eta^2\text{-}CCPh\}\{P(OMe)_3\}(\mu\text{-}PPh_2)]$ (**270**). Substitution of a Co-bound carbonyl ligand of **270** with triphenylphosphine gives $[FeCo_2(CO)_5\{\mu_3\text{-}\eta^2\text{-}CCPh\}\{P(OMe)_3\}(PPh_3)(\mu\text{-}PPh_2)]$ (**271**) (Scheme 75).[97]

Reaction between $[CpFe(CO)_2(C_2Me)]$ and $[Co_2(CO_8)]$ affords a μ-alkyne complex $[CpFe(C_2Me)Co_2(CO)_8]$ (**272a**).[98] In contrast the reaction of phenylacetylide complex $[CpFe(CO)_2C_2Ph]$ and $[Co_2(CO_8)]$ forms $[(\eta\text{-}C_5H_5)Fe(C_2Ph)Co_2(CO)_6]$ (**273**) and $[(\eta\text{-}C_5H_5)Fe(C_2Ph)Co_2(CO)_8]$ (**272b**) (Fig. 17).[99]

Tungsten–iridium tetrahedral cluster $[WIr_3(\mu\text{-}CO)_3(CO)_8(\eta\text{-}C_5R_5)]$ ($R_5 = H_5$, **274a**; $R_5 = Me_5$, **274b**; $R_5 = H_4Me$, **274c**), prepared from the reaction between $[WH(CO)_3(\eta\text{-}C_5R_5)]$ and $[IrCl(CO)_2(p\text{-toluidine})]$ under CO atmosphere, has been used extensively to synthesise several mixed metal acetylide complexes. Thermal reaction of $[WIr_3(\mu\text{-}CO)_3(CO)_8(\eta\text{-}C_5Me_5)]$ (**274b**) with $[W(C{\equiv}CPh)(CO)_3(\eta\text{-}C_5H_5)]$ results in the isolation of an edge-bridged tetrahedral cluster $[W_2Ir_3(\mu_4\text{-}\eta^2\text{-}C_2Ph)(\mu\text{-}CO)(CO)_9(\eta\text{-}C_5H_5)(\eta\text{-}C_5Me_5)]$ (**275a**) and an edge-bridged

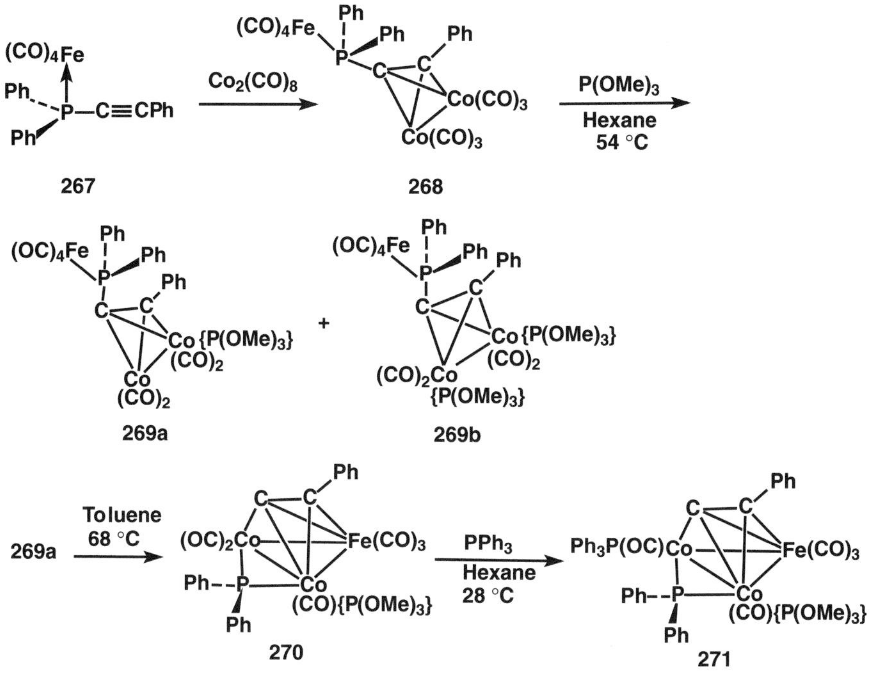

SCHEME 75.

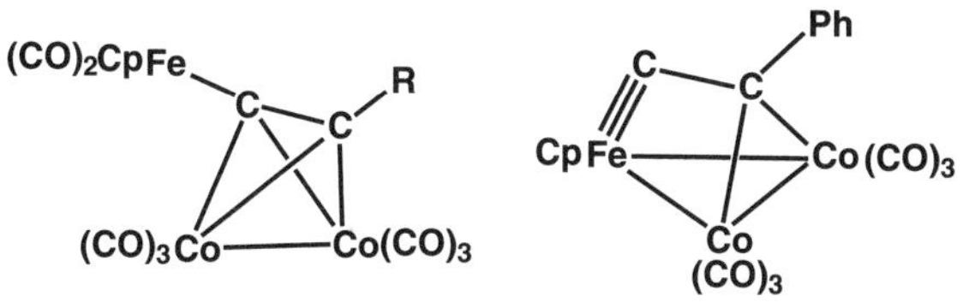

FIG. 17.

trigonal-bipyramidal cluster $[W_3Ir_3(\mu_4\text{-}\eta^2\text{-}C_2Ph)(\mu\text{-}\eta^2\text{-}C{=}CHPh)(Cl)(CO)_8(\eta\text{-}C_5Me_5)(\eta\text{-}C_5H_5)_2]$ (**275b**) (Scheme 76).[100] Cluster **275a** is formed by the insertion of $[W(C{\equiv}CPh)(CO)_3(\eta\text{-}C_5H_5)]$ into Ir–Ir and W–Ir bonds, and contains a $\mu_4\text{-}\eta^2$ alkynyl ligand. Cluster **275b** also contains an alkynyl ligand bonded to two iridium atoms and two tungsten atoms in a $\mu_4\text{-}\eta^2$ fashion and a vinylidene ligand bridges the W–W bond.

A similar reaction of a THF solution of W–Ir cluster $[WIr_3(CO)_{11}(\eta\text{-}C_5R_5)]$ (R = H, **274a**; R = Me, **274b**) with $[(\eta\text{-}C_5H_5)(CO)_2Ru(C{\equiv}C)Ru(CO)_2(\eta\text{-}C_5H_5)]$ leads to cluster compound $[Ru_2WIr_3(\mu_5\text{-}\eta^2\text{-}C_2)(\mu\text{-}CO)_3(CO)_7(\eta\text{-}C_5H_5)_2(\eta\text{-}C_5R_5)]$ [R = H, **276a**; R = Me, **276b**] containing a WIr_3 butterfly core capped by Ru atoms. The reaction involves an insertion of a C_2 unit into a W–Ir bond and scission of Ru–C bond of the diruthenium ethyndiyl precursor (Scheme 77). Another mixed

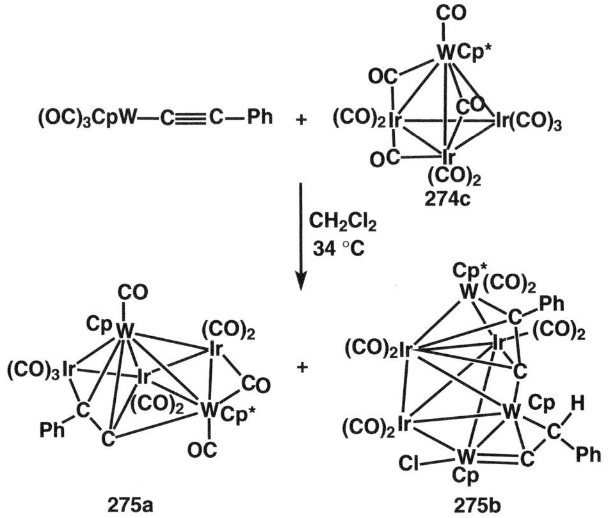

SCHEME 76.

metal cluster with a butterfly W_2Ir_2 unit capped by a Ru(η-C_5H_5) group, [$RuW_2Ir_2\{\mu_4$-η^2-($C_2C{\equiv}C$)Ru(CO)$_2$(η-C_5H_5)$\}$(μ-CO)$_2$(CO)$_6$(η-C_5H_5)$_3$] (**277**) has been isolated by the reaction of [$W_2Ir_2(CO)_{10}$(η-C_5H_5)$_2$] (**274d**) with the diruthenium ethyndiyl reagent (Scheme 78). When the reaction between [$WIr_3(CO)_{11}$ (η-C_5R_5)] ($R_5 = H_5$, **274a**; $R_5 = Me_5$, **274b**; $R_5 = H_4Me$, **274c**) and [(η-C_5H_5) $(CO)_3W(C{\equiv}CC{\equiv}C)W(CO)_3$($\eta$-$C_5H_5$)] is performed in refluxing toluene solution, metallaethynyl type of clusters [$W_2Ir_3\{\mu_4$-η^2-($C_2C{\equiv}C$)W(CO)$_3$(η-C_5H_5)$\}$(μ-CO)$_2$ (CO)$_2$(η-C_5H_5)(η-C_5R_5)] (R = H, **278a**; R = Me, **278b**; $R_5 = H_4Me$, **278c**) are obtained (Scheme 79).[100a]

Mixed group 6–9 transition metal complexes, [$\{Cp(OC)_3W\}C{\equiv}CC{\equiv}C\{M$ $(CO)(PPh_3)_2\}$] (M = Ir, **279a**; M = Rh, **279b**) have been obtained from the reaction between [$W(C{\equiv}CC{\equiv}CH)(CO)_3Cp$] and [$M(OTf)(CO)(PPh_3)_2$] (M = Ir, Rh) in presence of diethylamine. Thermal reaction of a THF solution of **279a** or **279b** with [$Fe_2(CO)_9$] gives [$\{Cp(OC)_3W\}C{\equiv}CC_2\{Fe_2M(CO)_8(PPh_3)\}$] (M = Ir, **280a**; M = Rh, **280b**) and [$\{Cp(OC)_8Fe_2W\}C_2C_2\{Fe_2M(CO)_8(PPh_3)\}$] (M = Ir, **281a**; M = Rh, **281b**) (Scheme 80).[100b]

VI

MIXED GROUP 10–12 METAL ACETYLIDE COMPLEXES

Platinum–acetylide complexes have been used in reactions with copper, silver or gold salts to form heteropolynuclear Pt–Ag, Pt–Cu or Pt–Au complexes bridged by alkynyl ligands. A synthetic strategy for heterometallic complexes with bridging alkynyl ligands have been developed by Fornies and co-workers,[102,110] in which alkynyl substrates containing linear acetylide units are allowed to react with electrophilic metal centres containing potential leaving groups. Reaction of alkynyl platinate(II)

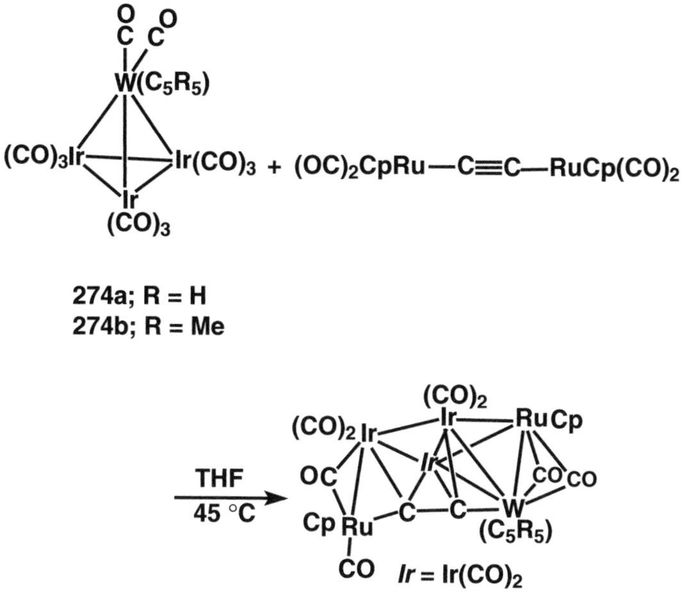

(CO)3Ir
Ir(CO)3
(OC)2CpRu—C≡C—RuCp(CO)2
W(C5R5)
THF
45 °C
Ir = Ir(CO)2

274a; R = H
274b; R = Me

276a; R = H
276b; R = Me

SCHEME 77.

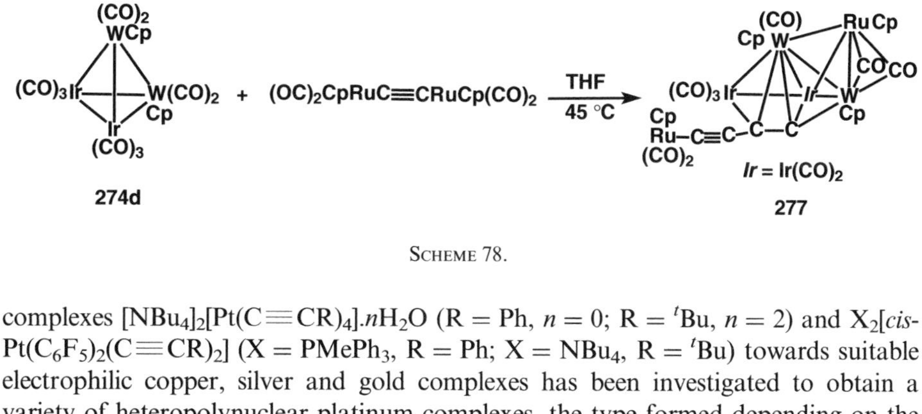

(OC)2CpRuC≡CRuCp(CO)2
THF
45 °C
Ir = Ir(CO)2

274d

277

SCHEME 78.

complexes $[NBu_4]_2[Pt(C{\equiv}CR)_4].nH_2O$ (R = Ph, $n = 0$; R = ^{t}Bu, $n = 2$) and $X_2[cis\text{-}Pt(C_6F_5)_2(C{\equiv}CR)_2]$ (X = $PMePh_3$, R = Ph; X = NBu_4, R = ^{t}Bu) towards suitable electrophilic copper, silver and gold complexes has been investigated to obtain a variety of heteropolynuclear platinum complexes, the type formed depending on the nature of the platinum starting material and on the molar ratio of the reactants. Treatment of $[Pt(C{\equiv}CR)_4]^{2-}$ (**39**) (R = Ph (**a**), ^{t}Bu (**b**)) with $AgClO_4$, or with CuCl or [AuCl(tht)] (tht = tetrahydrothiophene) in the presence of $NaClO_4$ in 1:2 molar ratio forms a hexanuclear complex $[Pt_2M_4(C{\equiv}CR)_8]$ (M = Ag, Cu, Au) (R = Ph, ^{n}Bu) (**282a,b–284a,b**).[102] The Cu analogue, $[Pt_2Cu_4(C{\equiv}CPh)_8]$ has been prepared by the reaction of $[Pt\{C_5H_4Fe(C_5H_5)\}Cl(cod)]$ with an excess of phenylacetylene and CuI.[101] An anionic tetranuclear complex $[Pt_2M_2(C_6F_5)_4(C{\equiv}CR)_4]^{2-}$ (**285a,b–286a,b**) has been obtained from the reaction of $[cis\text{-}Pt(C_6F_5)_2(C{\equiv}CR)_2]^{2-}$ (**40**) with $AgClO_4$,

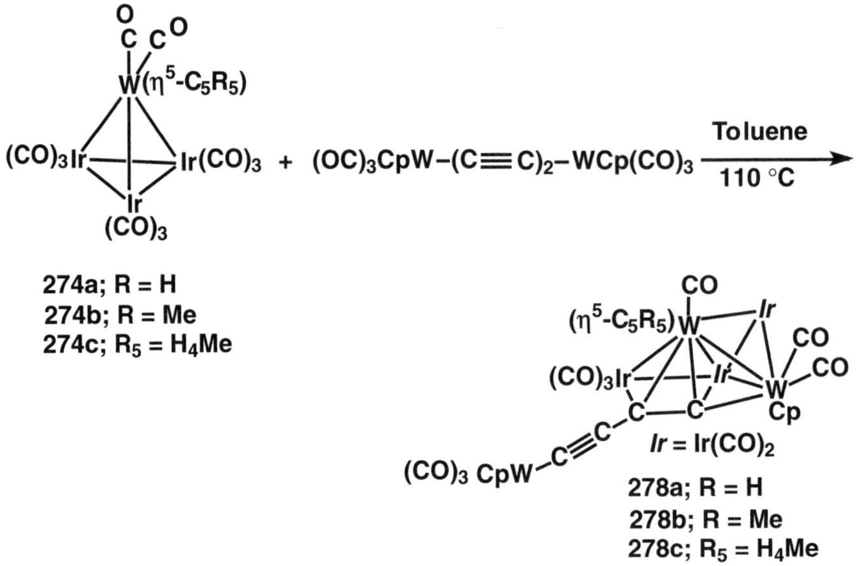

SCHEME 79.

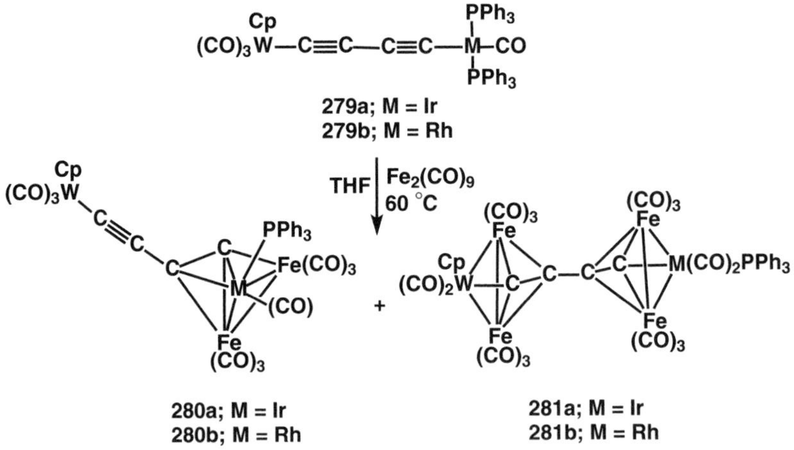

SCHEME 80.

AgCl or CuCl in a 1:1 molar ratio.[103] On treatment of complex **285a,b** with $AgClO_4$ (1:2), the Pt–Ag mixed-metal acetylide complex $[\{PtAg_2(C_6F_5)_2(C{\equiv}CR)_2\}_n]$ (**287a,b**) is obtained.[104] An alternative reaction of $[Pt(C_6F_5)_2(C{\equiv}CR)_2]^{2-}$ (**40**) with two equivalents of $AgClO_4$ also gives complex **287a,b**. X-ray diffraction study of **301** (an acetone solvated derivative of **287**) reveals a polymeric nature of the complex (Scheme 81). Trimetallic Pt–Ag complexes $[Pt_2Ag(C{\equiv}CR)_4L_4]ClO_4$ (**291**–**293**) have been synthesised by the reaction of *cis*-$[Pt(C{\equiv}CR)_2L_2]$ (**288**–**290**) with $AgClO_4$ (2:1 molar ratio) (Scheme 82).[102]

Treatment of complex **285a,b** with PPh_3 or PEt_3 (1:2 or 1:4 ratio) or with dppe results in the formation of anionic complexes **294**–**295** and dianionic complexes **296**

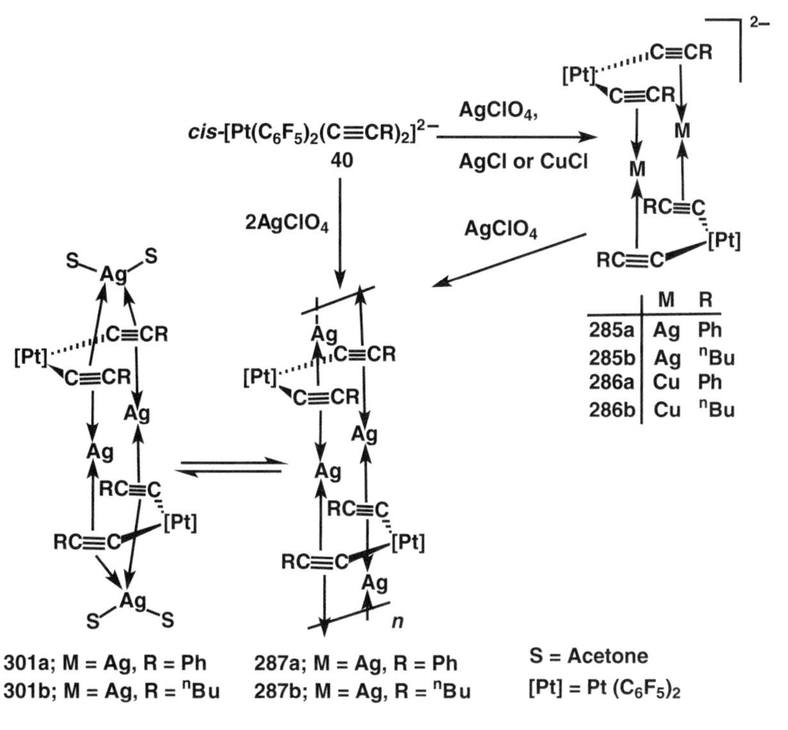

SCHEME 81.

respectively.[105] On the other hand, reaction of polymeric complex **287** with phosphines, tert-butylisocyanide or pyridine in a Ag:L molar ratio of 2:1 produces hexanuclear complexes $[Pt_2Ag_4(C_6F_5)_4(C{\equiv}CR)_4L_2]$ (**297**–**300**) which are structurally related to complex **301** (Scheme 76). The X-ray structure of **297a** reveals a tridentate behaviour of the two *cis*-$[Pt(C_6F_5)_2(C{\equiv}CR)_2]$ fragments towards silver atoms. The alkynyl ligands exhibit an unsymmetrical μ_3-η^2 bonding mode. In contrast, trinuclear complexes $[PtAg_2(C_6F_5)_2(C{\equiv}CR)_2L_2]$ (**302**–**305**) are isolated when a higher proportion of the ligand L is used with complex **287**. The tetranuclear **285** is obtained when **287** is reacted with NBu_4Br (Scheme 83).[106,110]

Two different Pt–M (M = Ag, Cu) complexes (**306**–**313**) are isolated when the hexanuclear complexes **282a,b** and **283a,b** are reacted with anionic ligands X^- (X = Cl, Br) or neutral ligands L (L = ^{t}CNBu, pyridine). For instance, the reaction of $[Pt_2M_4(C{\equiv}CPh)_8]$ (M = Ag, **282a**; M = Cu, **283a**) with four equivalents of anionic (X = Cl, Br) or neutral ligands (L = CNBu^t, pyridine) gives the corresponding trinuclear anionic or neutral complexes (**306**–**310**), while reaction of $[Pt_2M_4(C{\equiv}C^tBu)_8]$ (M = Ag, **282b**; M = Cu, **283b**) with the same ligands gives hexanuclear complexes (**311**–**313**). Addition of eight equivalents of the ligands to **282b** or **283b** afford the trinuclear complexes (**306**–**310**) (Scheme 84).[107]

The platinum mononuclear complexes $[Pt(C{\equiv}CR)_4]^{2-}$ (**39**) and $[Pt(C_6F_5)_2(C{\equiv}CR)_2]^{2-}$ (**40**) have been used to synthesise di- and trinuclear complexes containing doubly bridged alkynyl systems. Their reactions with mercury halides afford 1:2 adducts (**314**–**316**) or 1:1 adducts (**317**–**319**) (Scheme 85).[108]

The hexanuclear platinum–copper complex $[Pt_2Cu_4(C_6F_5)_4(C{\equiv}C^tBu)_4(acetone)_2]$ (**320**) and the polynuclear derivative $[PtCu_2(C_6F_5)_2(C{\equiv}CPh)_2]_x$ (**321**), which

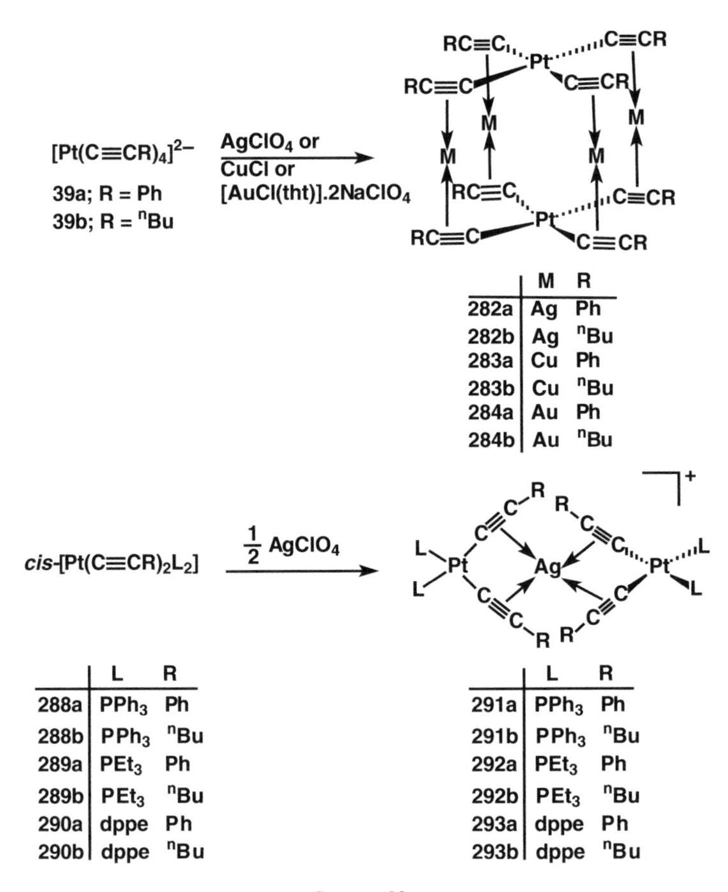

SCHEME 82.

crystallises in acetone as $[Pt_2Cu_4(C_6F_5)_4(C{\equiv}CPh)_4(acetone)_4]$ (**322**), have been prepared from the reaction of $[cis\text{-}Pt(C_6F_5)_2(THF)_2]$ with the corresponding copper-acetylide, $[Cu(C{\equiv}CR)]_x$ (R = Ph, ^{t}Bu) (molar ratio 1:2) as starting materials. The Ag-analogues (**323**–**325**) have been synthesised similarly (Scheme 86).[104,109]

Addition of four equivalents of 2,2′-bipyridine to a solution of **320** or **321** yields neutral trinuclear alkynyl bridged complexes $[\{cis\text{-}Pt(C_6F_5)_2(\mu\text{-}C{\equiv}CR)_2\}\{Cu(bipy)\}_2]$ (R = ^{t}Bu, **326**; R = Ph, **327**). An analogous trinuclear complex $[\{cis\text{-}Pt(C_6F_5)_2(\mu\text{-}C{\equiv}C^tBu)_2\}\{Cu(dppe)_2]$ (**328**) was obtained by the reaction of **320** with four equivalents of dppe. In contrast, the reaction of **321** with dppe produces mixtures of mononuclear platinum or copper complexes (Scheme 87). A comparison of the photoluminescent spectra of **320** and **321** with those of the related platinum–silver species $[PtAg_2(C_6F_5)_2(C{\equiv}CR)_2]_x$ and the monomeric $[cis\text{-}Pt(C_6F_5)_2(C{\equiv}CR)_2]_2^-$ suggest the presence of emitting states bearing a large cluster $[PtM_2]_x$ -to-ligand (alkynide) charge transfer (CLCT).[109]

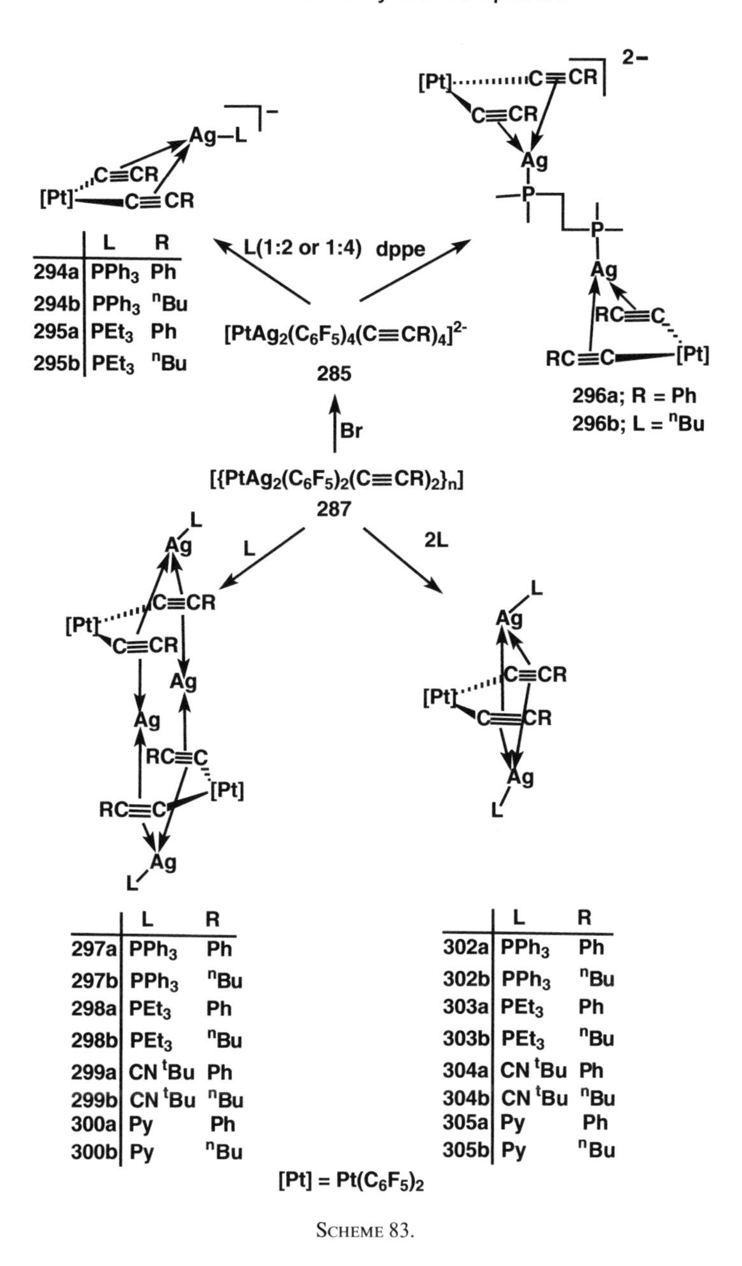

	L	R
294a	PPh_3	Ph
294b	PPh_3	^{n}Bu
295a	PEt_3	Ph
295b	PEt_3	^{n}Bu

	L	R
297a	PPh_3	Ph
297b	PPh_3	^{n}Bu
298a	PEt_3	Ph
298b	PEt_3	^{n}Bu
299a	CN ^{t}Bu	Ph
299b	CN ^{t}Bu	^{n}Bu
300a	Py	Ph
300b	Py	^{n}Bu

	L	R
302a	PPh_3	Ph
302b	PPh_3	^{n}Bu
303a	PEt_3	Ph
303b	PEt_3	^{n}Bu
304a	CN ^{t}Bu	Ph
304b	CN ^{t}Bu	^{n}Bu
305a	Py	Ph
305b	Py	^{n}Bu

SCHEME 83.

VII
OTHER MIXED METAL ACETYLIDE COMPLEXES

A variety of acetylide complexes have been reported with mixed early–late transition metal or mid-late transition metal complexes. The synthetic procedures either

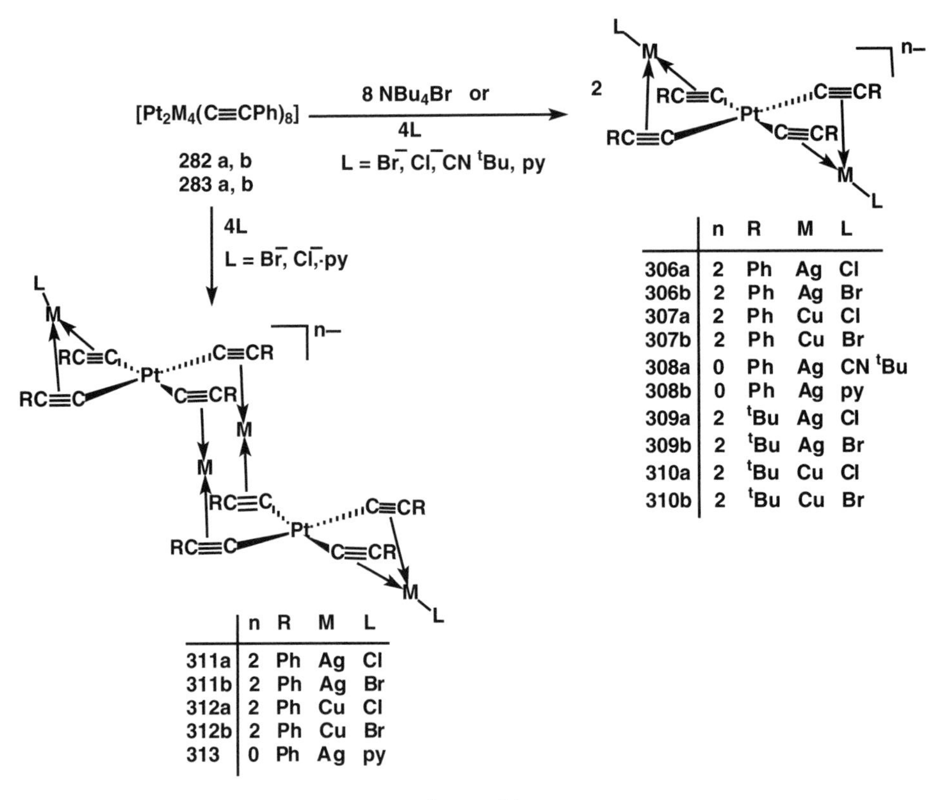

SCHEME 84.

involve reactions of early transition metal acetylide complexes with late transition metal complexes containing labile ligands or vice versa. Mixed early-mid transition metal acetylide complexes are rather rare.

The reaction of mononuclear $[Cp_2Ti(C{\equiv}C^tBu)_2]$ with *cis*-$[M(C_6F_5)(thf)_2]$ (M = Pt) results in the formation of dimetallic Pt–Ti acetylide complex $[Cp_2Ti(\mu\text{-}\eta^1\text{-}C{\equiv}CBu^t)_2Pt(C_6F_5)_2]$ (**329**) (Scheme 88).[111]

When $[Pt(C_2H_4)(PPh_3)_2]$ is added to $[Ti(\eta\text{-}C_5H_5)_2(C{\equiv}CR)_2]$ (R = Bu^t, Ph), products $[Cp_2Ti(\mu\text{-}\eta^1,\eta^1\text{-}C{\equiv}CBu^t)(\mu\text{-}\eta^2,\eta^1\text{-}C{\equiv}CBu^t)Pt(PPh_3)]$ (**330**) and $[Cp_2Ti(\mu\text{-}\eta^2,\eta^1\text{-}C{\equiv}CPh)_2Pt(PPh_3)]$ (**331**) are obtained (Scheme 89).[112]

A solution of tetranuclear heteroleptic arylcopper aggregate $[Cu_4R_2Br_2]$ (R = $C_6H_3(CH_2NMe_2)_2$-2,6) in diethylether reacts at room temperature with the titanium complex $[(\eta^5\text{-}C_5H_4SiMe_3)_2Ti(C{\equiv}CSiMe_3)_2]$ (**332**), in 1:4 molar ratio, to give a 1,1-bis-metallaalkenyl complex $[(\eta^5\text{-}C_5H_4SiMe_3)_2Ti(C{\equiv}CSiMe_3)\{\mu\text{-}C{=}C(SiMe_3)(R)\}Cu]$ (**334**) resulting from an intramolecular addition of a Cu–C bond across the alkyne triple bond (Scheme 90).[113] A probable formation of complex **333** as an intermediate has been proposed.

Dinuclear Pt–Rh (**336a-c**, **339**) and Pt–Ir (**337a-c**, **338**) complexes with doubly alkynyl bridging systems have been isolated from the reactions between platinum alkynyl complex, **40** and from cyclooctadiene complexes of rhodium and iridium (Scheme 91).[114]

$[Pt(C{\equiv}CR)_4]^{2-}$ (39) $\xrightarrow{2HgX_2}$

	X	R
314a	Cl	${}^tBu$
314b	Cl	$SiMe_3$
315a	Br	${}^tBu$
315b	Br	$SiMe_3$
316a	I	${}^tBu$
316b	I	$SiMe_3$

cis-$[Pt(C_6F_5)_2(C{\equiv}CR)_2]^{2-}$ (40) $\xrightarrow{HgX_2}$

	X	R
317a	Cl	${}^tBu$
317b	Cl	$SiMe_3$
318a	Br	${}^tBu$
318b	Br	$SiMe_3$
319a	I	${}^tBu$
319b	I	$SiMe_3$

SCHEME 85.

The tetranuclear mixed metal cluster $[(\eta^5\text{-}C_5H_5)_2Ni_2Fe_2(CO)_5(\mu\text{-}PPh_2)(\mu_4\text{-}\eta^2\text{-}C{\equiv}CPh)]$ (**340**) has been synthesised in high yield *via* condensation of $[Fe_2(CO)_6(\mu\text{-}PPh_2)(\mu_2\text{-}\eta^2\text{-}C{\equiv}CPh)]$ and $[(\eta^5\text{-}C_5H_5)_2Ni_2(CO)_2]$ in refluxing benzene.[115] X-ray analysis has revealed a μ_4-acetylide complex coordinated on a spiked triangular metal skeleton (Fig. 18).

A triangular Ni–Fe acetylide complex $[Cp_2NiFe_2(CO)_6(C_2^tBu)]$ (**341**) is obtained when $[Cp_2Ni_2(HC_2^tBu)]$ is reacted with $[Fe_3(CO)_{12}]$ in refluxing heptane (Fig. 19).[116]

A pentanuclear Ni–Ru cluster $[NiRu_4(CO)_9(\mu\text{-}PPh_2)_2(\mu_4\text{-}C{\equiv}C^iPr)_2]$ (**342**) was synthesised by the addition of electron rich $[(\eta\text{-}Cp)Ni(CO)]_2$ to a carbocationic μ_3-acetylide group in $[Ru_3(CO)_9(\mu_3\text{-}C{\equiv}C^iPr)(PPh_2)]$.[117] Its structure consists of a Ru_4 butterfly, zipped up by a Ni(CO) group, with the two acetylide groups in $\mu_4\text{-}\eta^2$-bonding mode (Fig. 20).

The reaction of the terminal alkyne, $[HC{\equiv}C\text{-}^iPr]$ with $[Ru_3(CO)_{12}]$ in boiling heptane gives $[(\mu\text{-}H)Ru_3(CO)_9(\mu_4\text{-}\eta^2\text{-}C{=}CH^iPr)]$ (**343**) which on further reaction

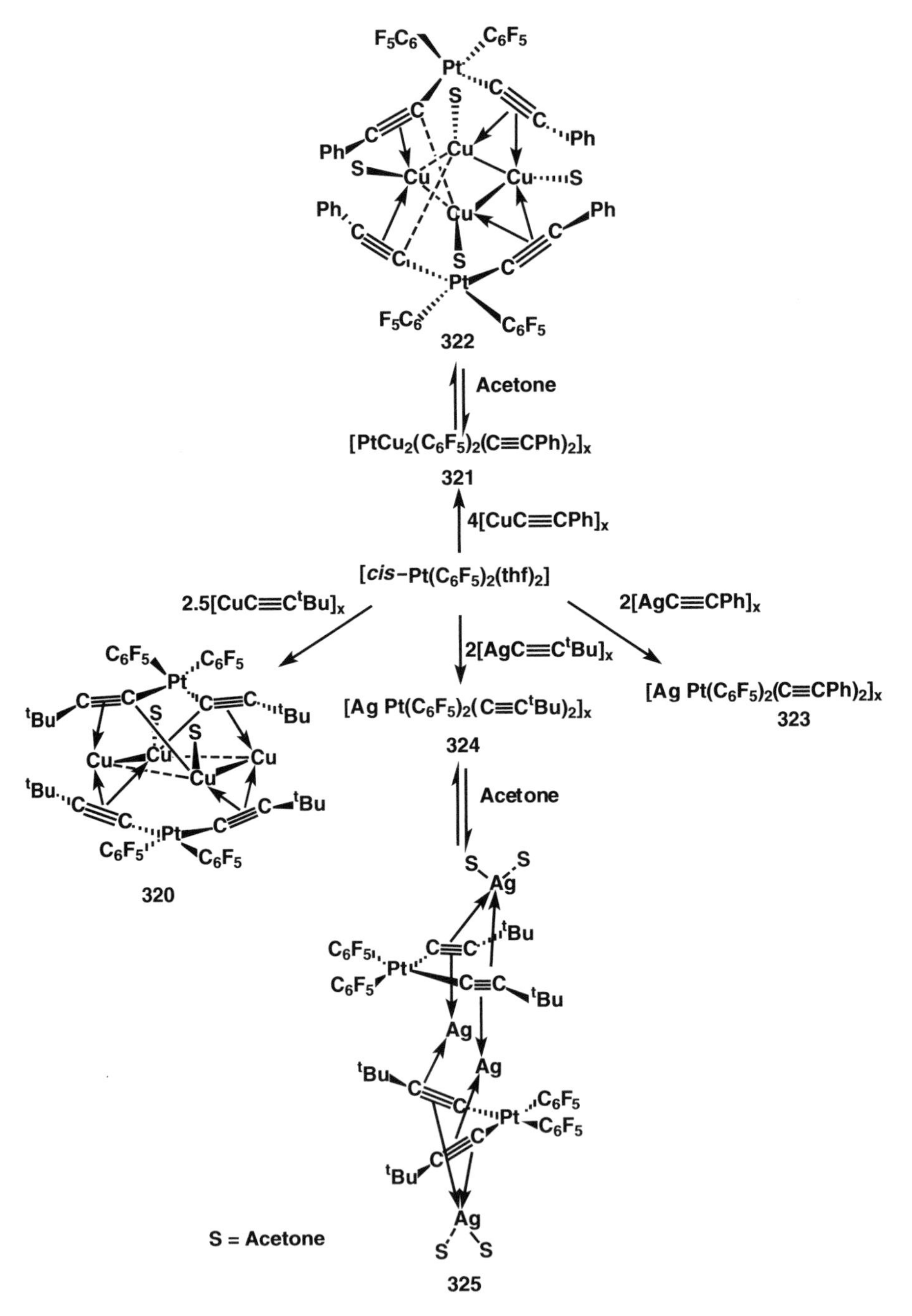

322
Acetone
[PtCu2(C6F5)2(C≡CPh)2]x
321
4[CuC≡CPh]x
[cis-Pt(C6F5)2(thf)2]
2.5[CuC≡CtBu]x
2[AgC≡CtBu]x
2[AgC≡CPh]x
[Ag Pt(C6F5)2(C≡CtBu)2]x
324
[Ag Pt(C6F5)2(C≡CPh)2]x
323
320
Acetone
325
S = Acetone

SCHEME 86.

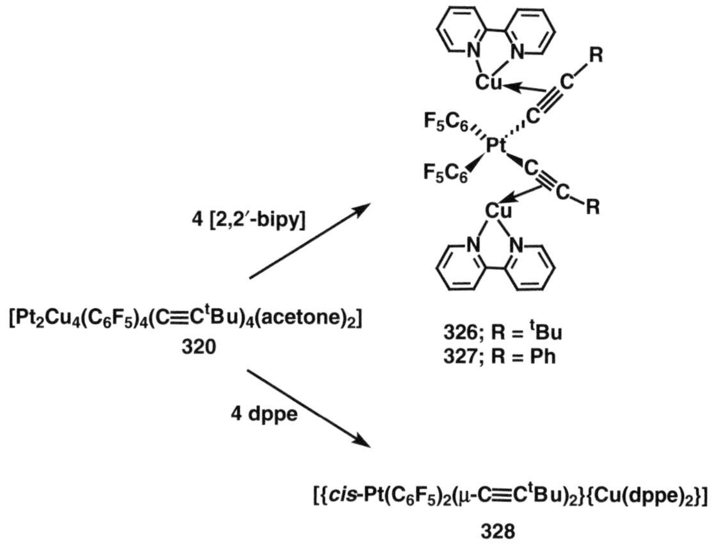

$[Pt_2Cu_4(C_6F_5)_4(C{\equiv}C^tBu)_4(acetone)_2]$
320

4 [2,2′-bipy]

326; R = ^{t}Bu
327; R = Ph

4 dppe

$[\{cis\text{-}Pt(C_6F_5)_2(\mu\text{-}C{\equiv}C^tBu)_2\}\{Cu(dppe)_2\}]$
328

SCHEME 87.

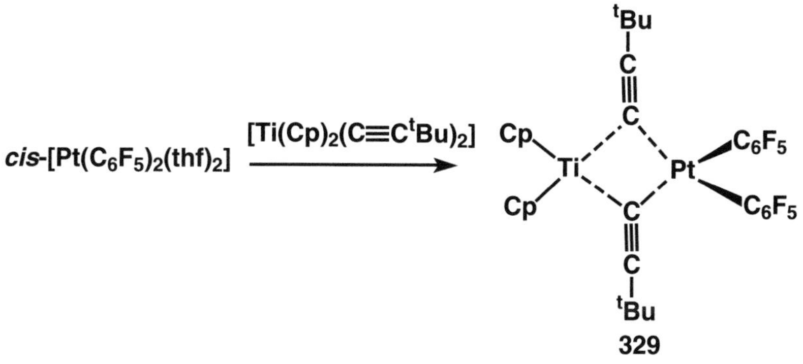

$cis\text{-}[Pt(C_6F_5)_2(thf)_2]$ $\xrightarrow{[Ti(Cp)_2(C\equiv C^tBu)_2]}$

329

SCHEME 88.

with an octane solution of $[(\eta^5\text{-}C_5H_5)Ni(CO)]_2$ under reflux condition forms the vinylidene cluster $[(\mu\text{-}H)(\eta^5\text{-}C_5H_5Ni)Ru_3(CO)_9(\mu_4\text{-}\eta^2\text{-}C{=}CH^iPr)]$ (**344**) (Fig. 21).[118] Structure of **344** consists of a butterfly arrangement of Ru_3Ni moiety, with the vinylidene ligand σ-bonded to two ruthenium and one nickel atoms and η^2-coordinated to the third ruthenium atom.

Reaction of a THF solution of $[Ru_2(CO)_6(\mu\text{-}PPh_2)(\mu\text{-}\eta^1,\eta^2\text{-}C{\equiv}C\text{–}C{\equiv}CR)]$ (R = ^{t}Bu, **345a**;R = Ph, **345b**) with $[Pt(PPh_3)_2(\eta\text{-}C_2H_4)]$ leads to a trimetallic heteronuclear complex $[Ru_2Pt(CO)_7(PPh_3)(\mu_3\text{-}\eta^1,\eta^1,\eta^1\text{-}C{=}C\text{–}C{\equiv}CR)(\mu\text{-}PPh_2)]$ (**346a,b**) (Scheme 92).[119] A THF solution of **345a** also reacts under reflux condition with $[Ni(cod)_2]$ or $[Ni(CO)_4]$ to give a pentanuclear cluster $[Ru_4Ni(CO)_{12}(\mu\text{-}PPh_2)_2(\mu_4\text{-}\eta^1,\eta^1,\eta^2,\eta^4\text{-}{}^tBuC{\equiv}CC_4C{\equiv}C^tBu)]$ (**347**), resulting from stoichiometric coupling of two moieties of **345a** and the incorporation of a nickel atom bonded

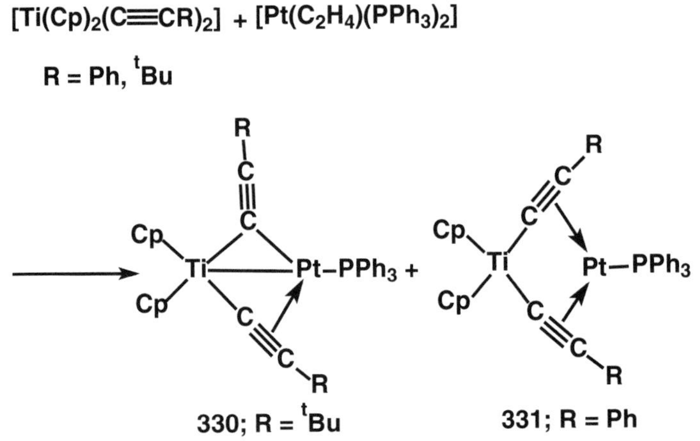

$[Ti(Cp)_2(C{\equiv}CR)_2] + [Pt(C_2H_4)(PPh_3)_2]$

R = Ph, ^{t}Bu

330; R = ^{t}Bu 331; R = Ph

SCHEME 89.

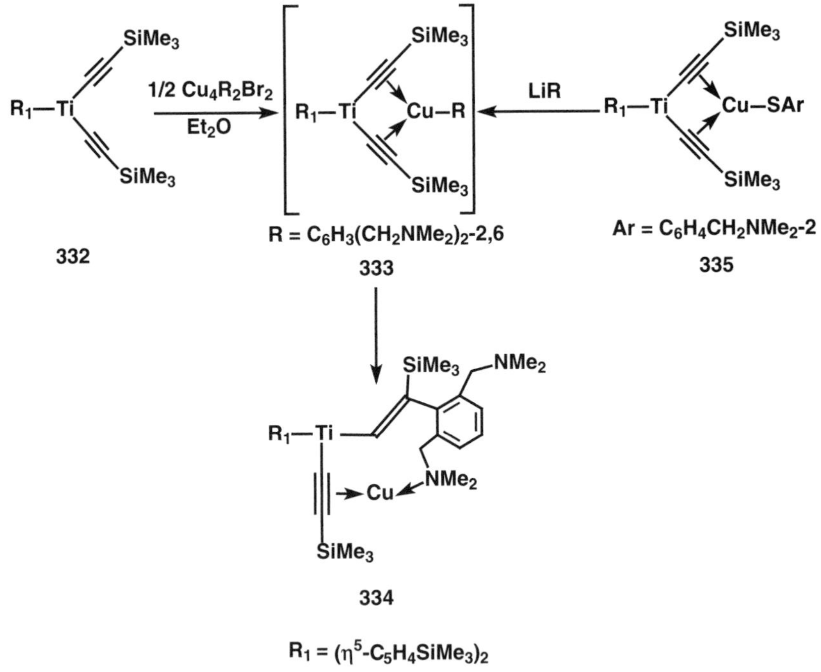

1/2 $Cu_4R_2Br_2$, Et_2O; LiR

332

R = $C_6H_3(CH_2NMe_2)_2$-2,6

333

Ar = $C_6H_4CH_2NMe_2$-2

335

334

$R_1 = (\eta^5\text{-}C_5H_4SiMe_3)_2$

SCHEME 90.

to two Ru_2 units. The structure of **347** also shows a head-to-head coupling of two butadienyl ligands to form a C_8 chain.

A triangular triosmium–platinum cluster complex $[Os_3Pt(\mu\text{-}H)(\mu_4\text{-}\eta^2\text{-}C{\equiv}CPh)(CO)_{10}(PCy_3)]$ (**349**) has been synthesised by the treatment of the unsaturated $[Os_3Pt(\mu\text{-}H)_2(CO)_{10}(PCy_3)]$ (**348**) with $[LiC{\equiv}CPh]$ followed by protonation.[120]

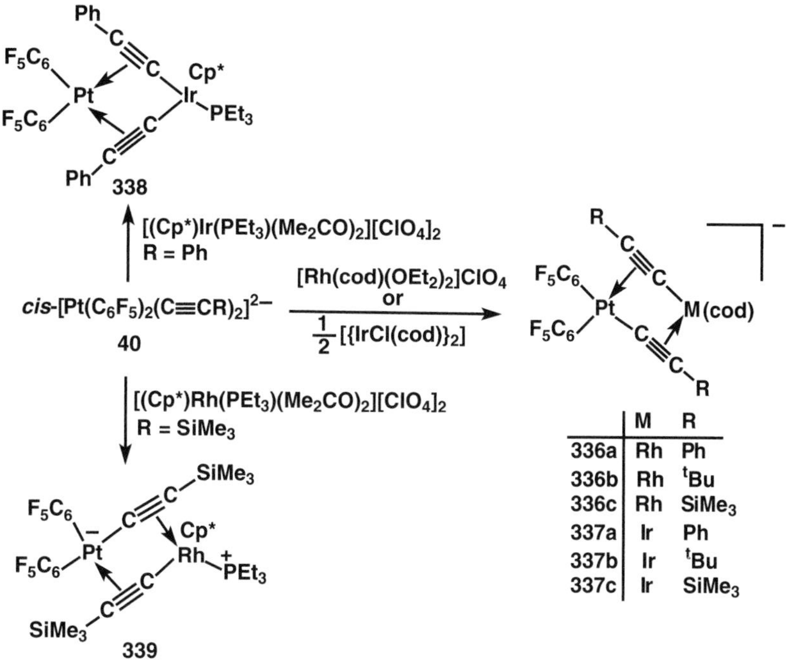

Ph
C
C
F_5C_6
Pt
F_5C_6
Cp*
Ir
PEt_3
C
C
Ph
338

$[(Cp^*)Ir(PEt_3)(Me_2CO)_2][ClO_4]_2$
R = Ph

cis-$[Pt(C_6F_5)_2(C{\equiv}CR)_2]^{2-}$
40

$[Rh(cod)(OEt_2)_2]ClO_4$
or
$\frac{1}{2}[\{IrCl(cod)\}_2]$

R
C
C
F_5C_6
Pt
F_5C_6
M(cod)
C
C
R
−

	M	R
336a	Rh	Ph
336b	Rh	^{t}Bu
336c	Rh	$SiMe_3$
337a	Ir	Ph
337b	Ir	^{t}Bu
337c	Ir	$SiMe_3$

$[(Cp^*)Rh(PEt_3)(Me_2CO)_2][ClO_4]_2$
R = $SiMe_3$

$SiMe_3$
C
C
F_5C_6
−
Pt
F_5C_6
Cp*
Rh
+
PEt_3
C
C
$SiMe_3$
339

SCHEME 91.

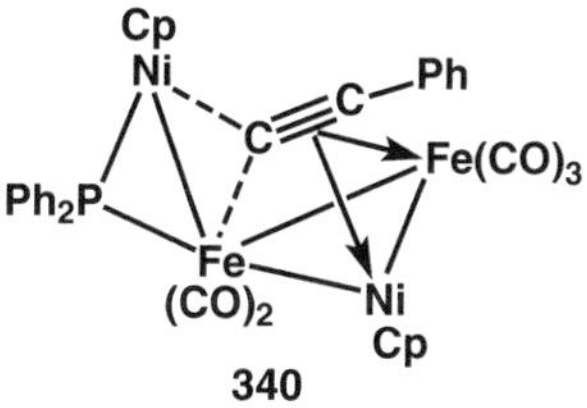

Cp
Ni
C
C
Ph
$Fe(CO)_3$
Ph_2P
Fe
$(CO)_2$
Ni
Cp
340

FIG. 18.

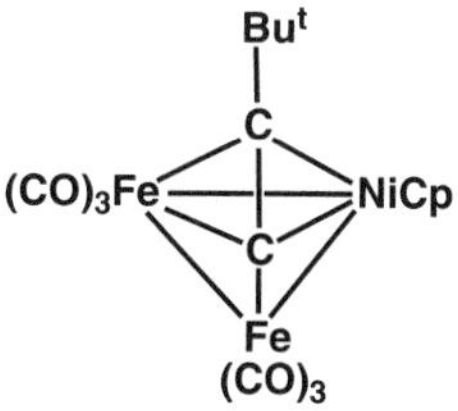

Bu^t
C
$(CO)_3Fe$
NiCp
C
Fe
$(CO)_3$
341

FIG. 19.

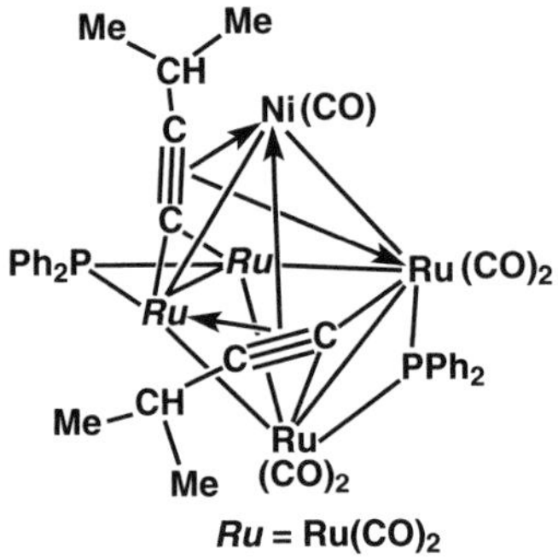

Ru = $Ru(CO)_2$

342

FIG. 20.

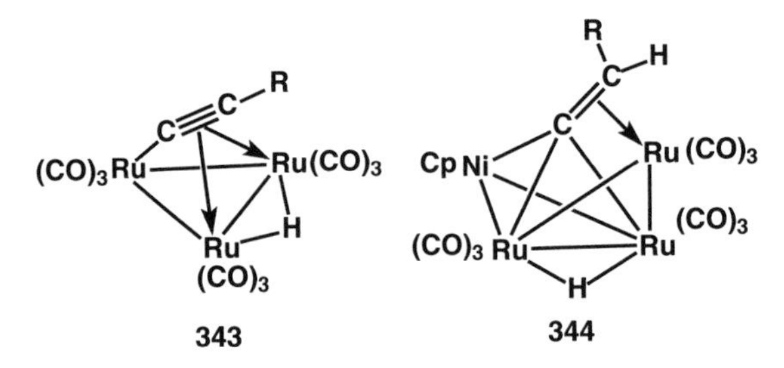

FIG. 21.

X-ray structural analysis reveals a μ_4-η^2-C$\equiv$CPh ligand about the triosmium framework with a platinum atom coordinated to one of the three osmium atoms in a spiked triangular arrangement (Fig. 22).

To achieve alkyne oligomerisation, bis(acetylide) Pt(II) complexes have been used to facilitate acyclic dimerisation of alkynyls through head-to-head C–C bond coupling.[121,122] Thermal reaction of a toluene solution of [*cis*-Pt(C$\equiv$CPh)$_2$(dppe)] (**350a**) with [$Mn_2(CO)_9(CH_3CN)$] gives a mixed Mn–Pt compound, [Mn_2Pt (PhCCCCPh)$(CO)_6$(dppe)] (**351**),[121] whereas reaction of **350a** with [$Ru_3(CO)_{12}$] in refluxing toluene forms two isomeric Ru/Pt compounds (**352a,b**) (Scheme 93).[122]

In refluxing toluene, compound **350b** reacts with [$Ru_3(CO)_9(PPh_3)_3$] to form the μ-1,3-diyne compound [Pt(η^2-PhCCCCPh)$(PPh_3)_2$] (**350c**). Successive reaction of **350c** with [$Fe(CO)_5$] and [$Ru_3(CO)_{12}$] under benzene reflux lead to the isolation of a dimetallacyclic compound, [FePt(μ_2-η^1:η^1:η^2-C(O)PhC$=$C$=$C$\equiv$CPh) $(CO)_3(PPh_3)_2$] (**353**) and a triangular cluster [MPt_2(μ_3-η^1:η^1:η^2-Ph-C$=$CC$\equiv$CPh) $(CO)_5(PPh_3)_2$] (M = Fe, **354a**; M = Ru, **354b**) (Scheme 94).[121]

Addition of [$(dppm)_2RuCl_2$] to *trans*-[$(PEt_3)_2$Pt(Ph)(C$\equiv$C–*p*-C_6H_4–C$\equiv$CH)] followed by treatment with DBU forms a heterometallic acetylide complex [$(PEt_3)_2$Pt(C$\equiv$C–*p*-C_6H_4–C$\equiv$C)Ru(Cl)$(dppm)_2$)] (**355**). Thermal reaction of a methanol solution of [Cp$(PPh_3)_2$RuCl] with *trans*-[$(PEt_3)_2$Pt(Ph)(C$\equiv$C–*p*-C_6H_4–C$\equiv$CH)] results in the formation of *trans*-[Cp$(PPh_3)_2$Ru(C$\equiv$C–*p*-C_6H_4–C$\equiv$C) Pt$(Et_3P)_2$(Ph)] (**356**) (Scheme 95).[36b]

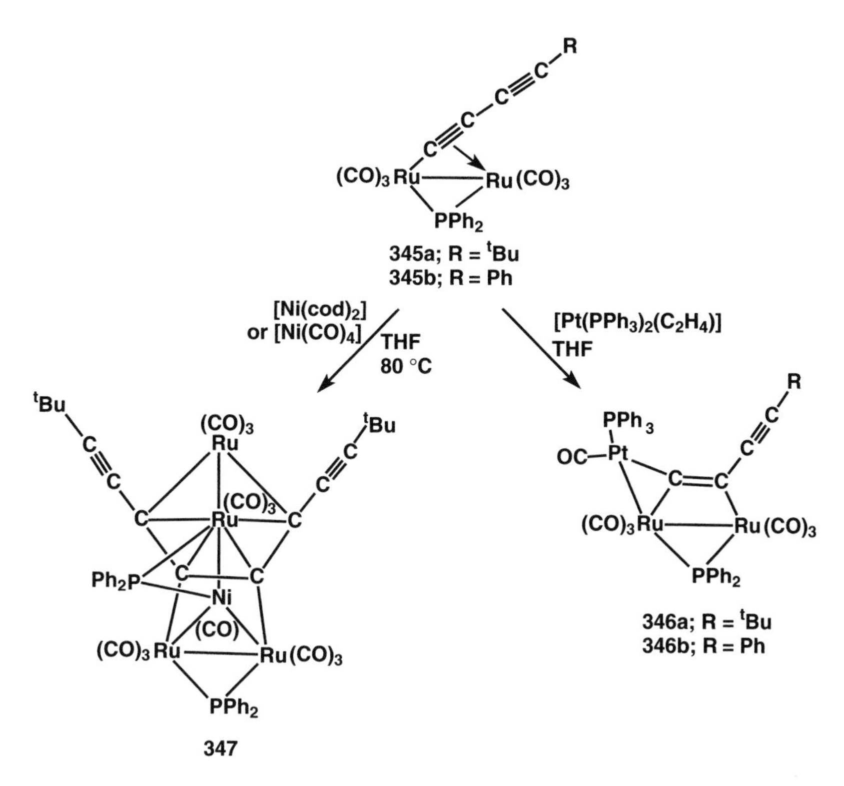

SCHEME 92.

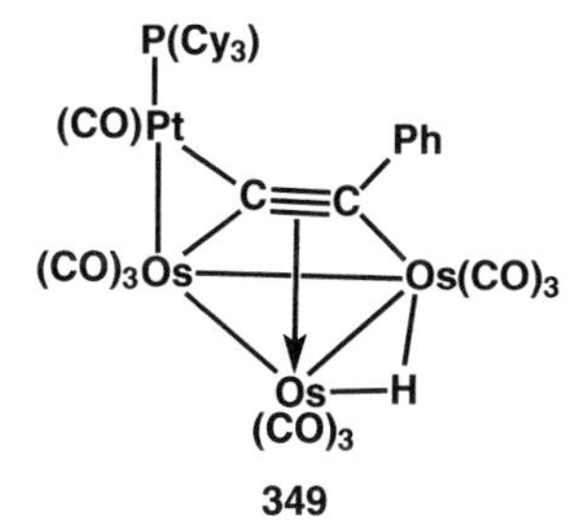

FIG. 22.

Reduction of **172** with Na/Hg followed by addition of $[O\{Au(PPh_3)\}_3][BF_4]$ or $[AuCl(PPh_3)]$ gives complexes $[AuFe_2Ir(\mu_3\text{-}C_2HPh)(CO)_8(PPh_3)_2]$ (**357a**) and $[Au_2Fe_2Ir(\mu_4\text{-}C_2Ph)(CO)_7(PPh_3)_2]$ (**357b**) (Scheme 96). Reaction of complex **172** with $K[BH(CHMeEt)_3]$ followed by auration gives a Au_3FeIr cluster $[Au_3Fe_2Ir(C_2HPh)(CO)_{12}(PPh_3)_4]$ (**358**). The Rh analogue of **357** has also been prepared by a similar procedure.[123]

Mono-, di- and tri-gold containing ruthenium clusters $[Ru_3Au(\mu_3\text{-}C_2^tBu)(CO)_9(PPh_3)]$ (**360**), $[Ru_3Au_2(\mu_3\text{-}C{=}CH^tBu)(CO)_9(PPh_3)_2]$ (**361**) and $[Ru_3Au_3(C_2^tBu)(CO)_8(PPh_3)_3]$ (**362**) were isolated by a deprotonation reaction of $[HRu_3(CO)_9C_2^tBu]$

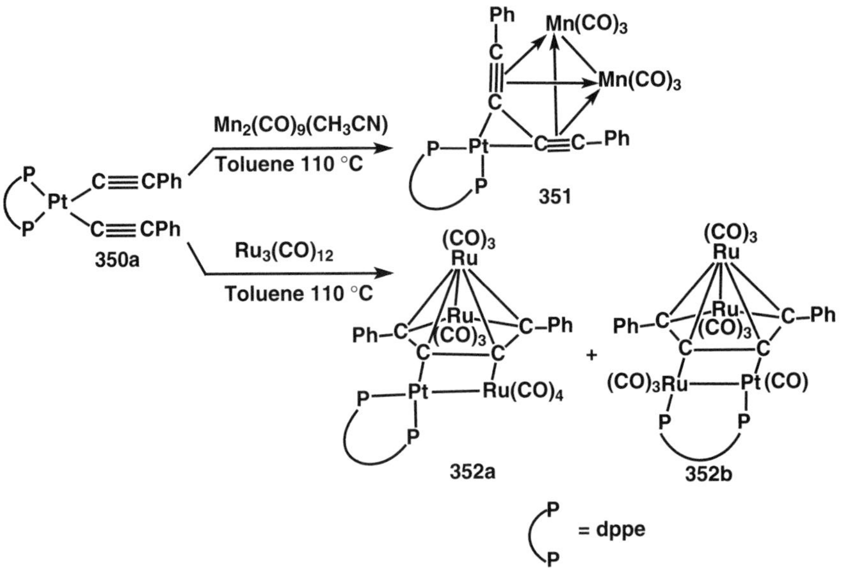

SCHEME 93.

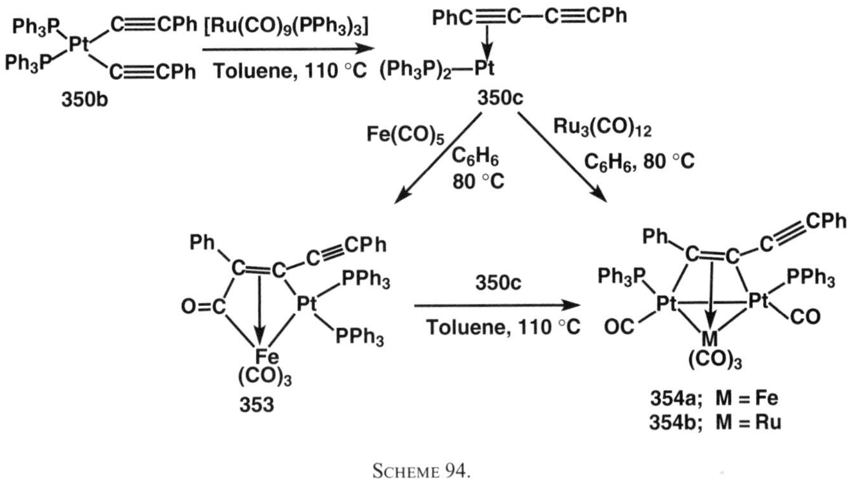

SCHEME 94.

(**359**) with K[HB(CHMeEt)$_3$], followed by auration with [O{Au(PPh$_3$)}$_3$][BF$_4$] (Scheme 97).[124] The molecular structure of **361** contains a trigonal pyramidal Ru$_3$Au$_2$ core with the Ru$_3$ face bridged by a vinylidene ligand, σ-bonded to two ruthenium atoms and η^2-coordinated to the third ruthenium atom.

Addition of a THF solution of K[BH^{s}Bu$_3$] to a solution of [Ru$_3$(μ-H)(μ_3-C$_2$H)(CO)$_9$] (**363**) in THF followed by auration with [AuCl(PPh$_3$)] results in the isolation of [AuRu$_3$(μ-H)(μ_3-C$_2$H$_2$)(CO)$_9$(PPh$_3$)] (**364**) and [Au$_2$Ru$_3$(μ_3-C=CH$_2$)(CO)$_9$(PPh$_3$)$_2$] (**365**) (Scheme 98).[125] Complex **364** crystallises in a dark red

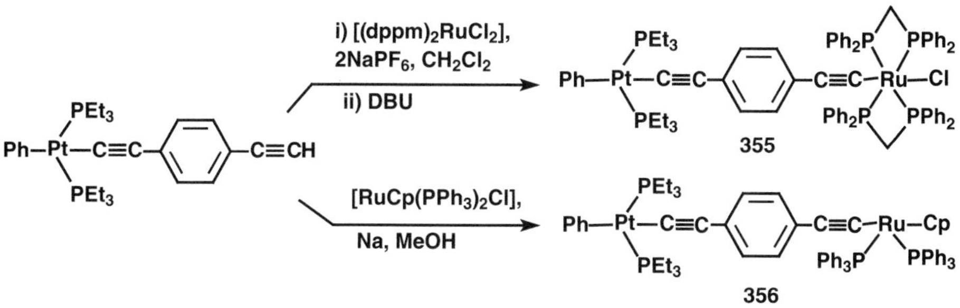

SCHEME 95.

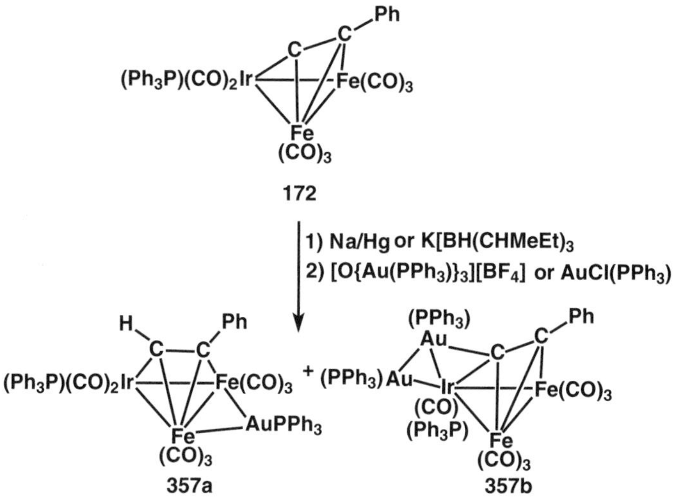

SCHEME 96.

form, in which the $AuRu_3$ core forms a tetrahedron (**364a**), and a yellow form, in which the $AuRu_3$ core has a butterfly structure (**364b**). The C_2H_2 ligand is $2\eta^1$, η^2 coordinated to the Ru_3 face in both the isomeric forms. In cluster **365** the Au_2Ru_3 core has a distorted square pyramidal conformation with the CCH_2 ligand attached to the Ru_3 face.

Thermal reactions of a benzene solution of $[RuCl(PPh_3)_2(C_5H_5)]$ and $[Cu(C_2Ph)]$ afford $[RuCuCl(C_2Ph)(PPh_3)_2(C_5H_5)]_2$ (**366**) and $[RuCuCl(C_2Ph)(PPh_3)_2(C_5H_5)]$ (**367a**), whereas similar reaction with $[Cu(C_2C_6H_4Me\text{-}p)]$ or $[Cu(C_2C_6H_4F\text{-}p)]$ results in the formation of a halogen free complex $[RuCu(C_2R)_2(PPh_3)(C_5H_5)]$ (**368a,b**) (R = p-MeC_6H_4 or p-FC_6H_4) and $[RuCuCl(C_2C_6H_4Me\text{-}p)(PPh_3)_2(C_5H_5)]$ (**367b**) (Fig. 23).[126] The Me-analogue, $[RuCuCl(C_2Me)(PPh_3)_2(C_5H_5)]$ (**367c**) has been obtained by heating reaction of $[Cu(C_2Me)]$ and ruthenium chloride in benzene. On reflux, a benzene solution of $[RuCl(PPh_3)_2(C_5H_5)]$ and $[Cu(C_2C_6F_5)]$ gives a tetranuclear complex $[\{RuCu(C_2C_6F_5)_2(PPh_3)(C_5H_5)\}_2]$ (**369**). Room temperature reaction of a benzene solution of $[Fe_2(CO)_9]$ and **366** gives the trinuclear cluster

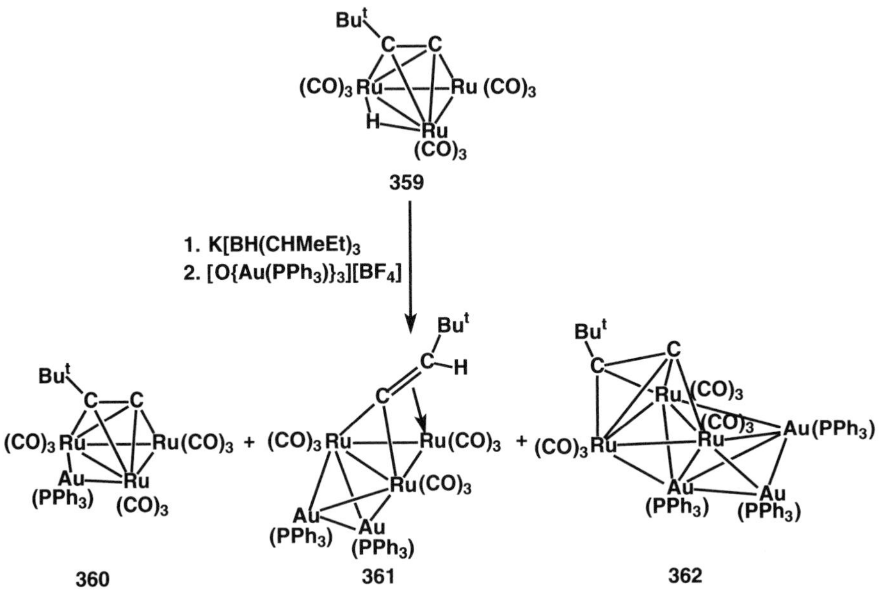

SCHEME 97.

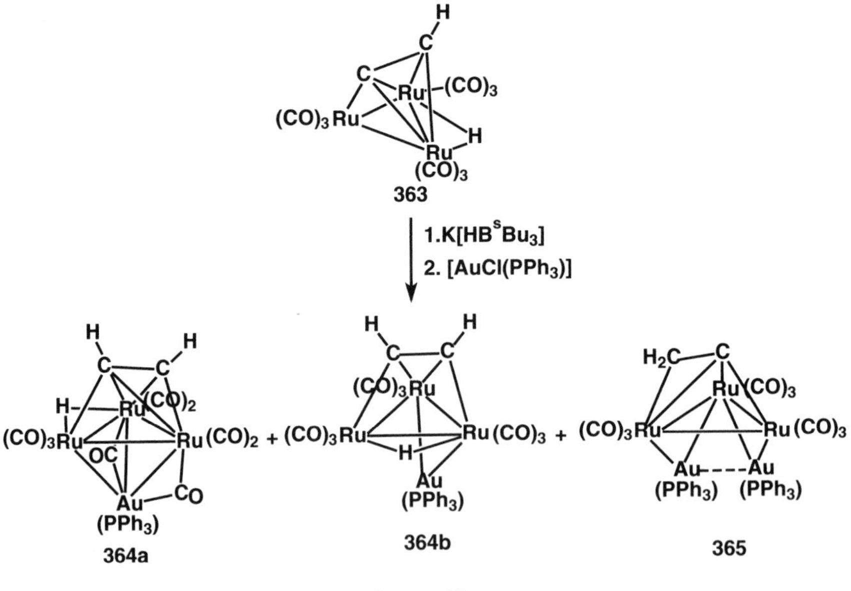

SCHEME 98.

$[Fe_2Ru(C_2Ph)(CO)_6(PPh_3)(C_5H_5)]$ (**370**). Thermal reaction of a mixture of *cis*-$[ReCl(CO)_3(PPh_3)_2]$ and $Cu(C_2Ph)$ in THF affords $[ReCuCl(C_2Ph)(CO)_3(PPh_3)_2]$ (**371**), while, thermolysis of $[ReCl(CO)_3(PPh_3)_2]$ and $Cu(C_2C_6F_5)$ in benzene gives $[ReCu(C_2C_6F_5)_2(CO)_3(PPh_3)_2]$ (**372**) (Fig. 24).[126]

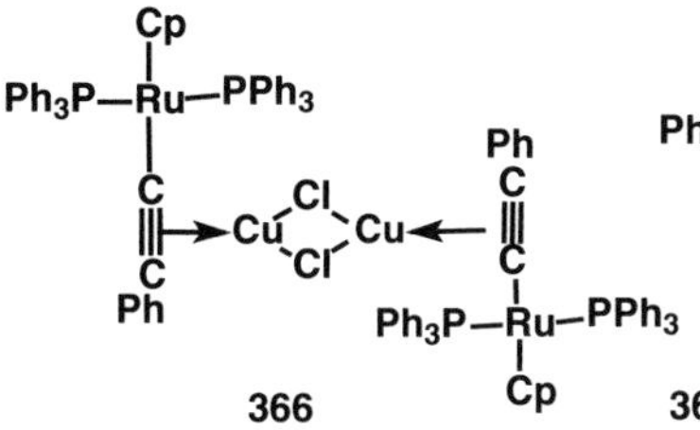

366

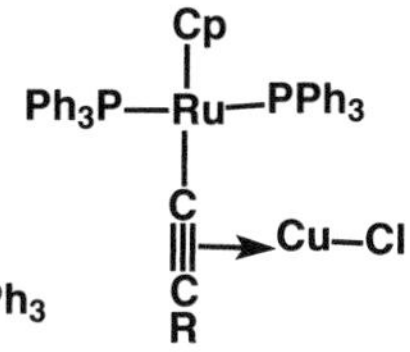

367a; R = Ph
367b; R= p-MeC$_6$H$_4$
367c; R = Me

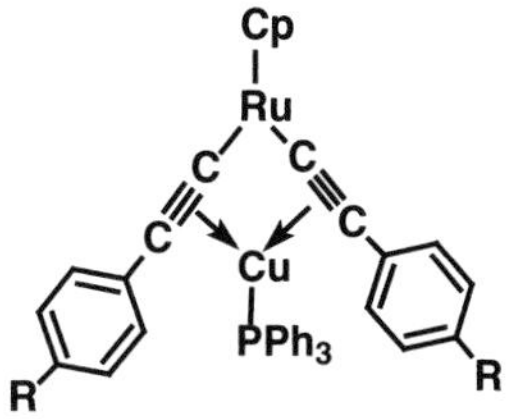

368a; R= p-MeC$_6$H$_4$
368b; R= p-FC$_6$H$_4$

FIG. 23.

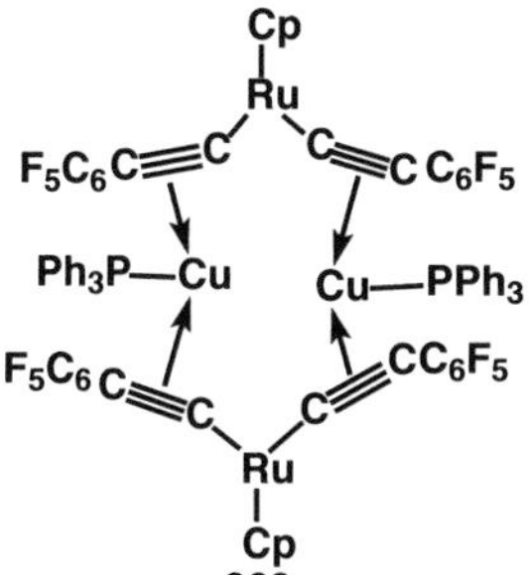

369

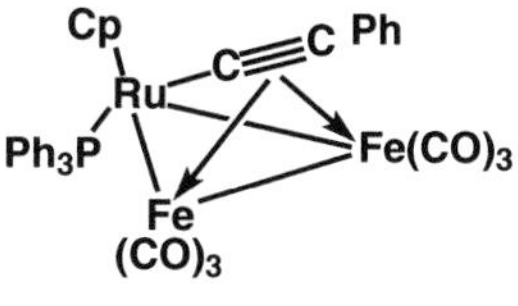

370

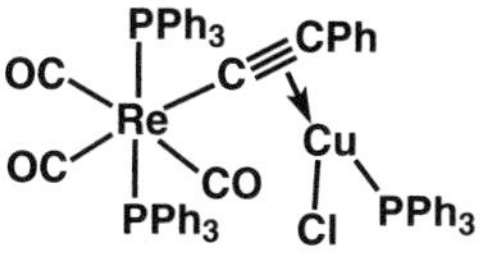

371

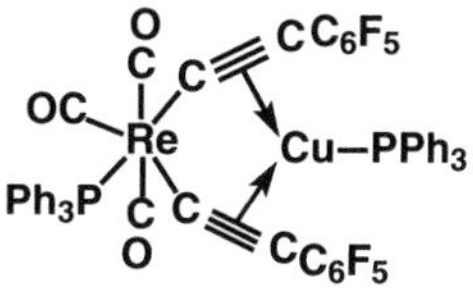

372

FIG. 24.

The compounds [Ag$_3$({Ru(CO)$_2$(η-C$_5$H$_4$R)}$_2$(μ-C$\equiv$C))$_3$](BF$_4$)$_3$ (R = H, **374a**; R = Me, **374b**), have been prepared from [{Ru(CO)$_2$(η-C$_5$H$_4$R)}$_2$(μ_2-C$\equiv$C)] (R = H, **373a**; Me, **373b**) and AgBF$_4$ (Fig. 25), and used in the reaction with [CpRuCl(CO)$_2$] to give [{Ru(CO)$_2$(η-C$_5$H$_4$R)}$_3$(η^1,η^1-C$\equiv$C)][BF$_4$] (R = H, Me).[127]

An addition reaction of dichloromethane solution of [Au(C$\equiv$CPh)L] (L = PPh$_3$ or PMe$_2$Ph) and [Os$_3$(CO)$_{10}$(MeCN)$_2$] gives [Os$_3$(C$\equiv$CPh)(AuL)(CO)$_{10}$] (**375a,b**) which contains a butterfly Os$_3$Au metal core and a μ-η^2-phenylethynyl ligand bridging two osmium atoms. Decarbonylation of **375a,b** in refluxing heptane gives [Os$_3$(μ_3,η^2-C$\equiv$CPh)(μ-AuL)(CO)$_9$] (**376a,b**) (Scheme 99).[128]

Numerous examples exist of reactions between [M(C$_2$R)(PR'$_3$)] (M = Au, Ag, Cu; R = Ph, C$_6$F$_5$; R' = Me, Ph) and [Os$_3$(μ-H)$_2$(CO)$_{10}$], which might be expected to proceed by oxidative addition of [RC$_2$M(PR'$_3$)] and addition of the cluster-bound hydrogen to the acetylide moiety. In toluene at –11 °C, a rapid reaction occurs between [Os$_3$(μ-H)$_2$(CO)$_{10}$] and [M(C$_2$C$_6$F$_5$)(PPh$_3$)] (M = Au, Ag, Cu) to give [Os$_3$M(μ-CH$=$CHC$_6$F$_5$)(CO)$_{10}$(PPh$_3$)] (**377a–c**) in quantitative yield,[129] while

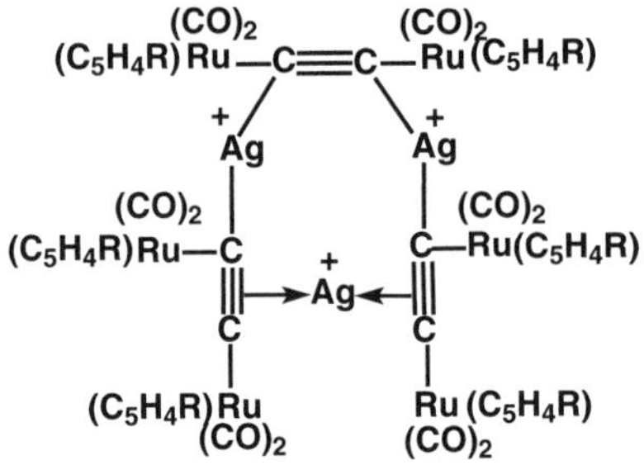

FIG. 25.

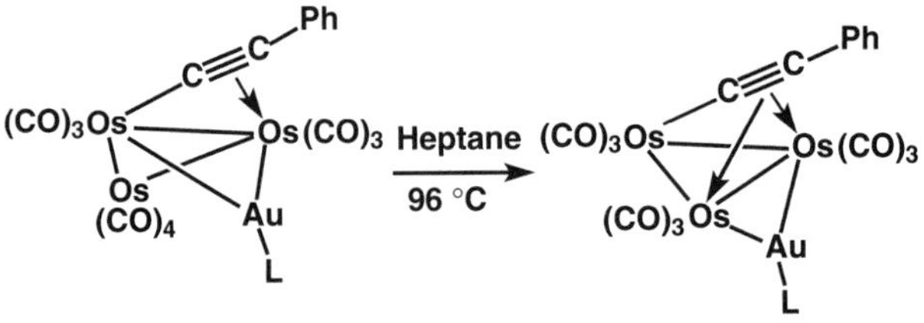

SCHEME 99.

the reaction between $[H_2Os_3(CO)_{10}]$ and $[M(C_2Ph)(PR_3)]$ (M = Cu, Ag, Au;R = Me or Ph) affords $[HOs_3M(CO)_{10}(PR_3)]$ (M = Au, R = Me, Ph; M = Ag, Cu, R = Ph), $[Os_3M(\mu\text{-}CH{=}CHPh)(CO)_{10}(PR_3)]$ (M = Au and R = Me, **378a**; M = Au and R = Ph, **378b**; M = Ag and R = Ph, **378c**; M = Cu and R = Ph, **378d**), $[Os_3M(\mu\text{-}CH{=}CHPh)(CO)_9(PR_3)]$ (M = Au and R = Me, **379a**; M = Au and R = Ph, **379b**; M = Ag, R = Ph, **379c**), $[HOs_3M(\mu_3\text{-}HCCPh)(CO)_8]$ (M = Au, **380a**; M = Ag, **380b**; M = Cu, **380c**) and $[Os_3M(\eta\text{-}CH{=}CHPh)(CO)_9(PR_3)_2]$ (M = Au and R = Me, **381a**; M = Au and R = Ph, **381b**; M = Ag and R = Ph, **381c**; M = Cu and R = Ph, **381d**) (Fig. 26).[129] X-ray crystallography of compounds **377a** and **378b** reveals the presence of a butterfly metal core and a $\mu\text{-}\eta^1,\eta^2$-vinyl ligand which is bonded to two osmium atoms of the Os_3Au core.

Thermal reaction of a THF solution of $[AgC_2Ph]$ and $[RhCl(PPh_3)_3]$ results in the formation of a hexanuclear cluster compound $[Rh_2Ag_4(C_2Ph)_8(PPh_3)_2]$ (**382a**).[130] The analogous Ir_2Ag_4 compound $[Ir_2Ag_4(C_2Ph)_8(PPh_3)_2]$ (**383a**) has been obtained from AgC_2Ph and $[trans\text{-}IrCl(CO)(PPh_3)_2]$ on toluene reflux. Thermolysis of a solution of $[RhCl(PPh_3)_3]$ and $[AgC_2C_6F_5]$ in 1,2-dimethoxyethane yields three compounds: $[Rh_2Ag_4(C_2C_6F_5)_8(PPh_3)_2]$ (**382b**), $[Ag(PPh_3)]^+[Rh(C_2C_6F_5)_4(PPh_3)_2]^-$ (**384a**) and $\{[Ag(PPh_3)]_2^+[Rh(C_2C_6F_5)_5(PPh_3)]^{2-}\}$ (**385**). The analogous iridium compounds $[Ir_2Ag_4(C_2C_6F_5)_8(PPh_3)_2]$ (**383b**) and $[Ag(PPh_3)]^+[Ir(C_2C_6F_5)_4(PPh_3)_2]^-$ (**384b**) are

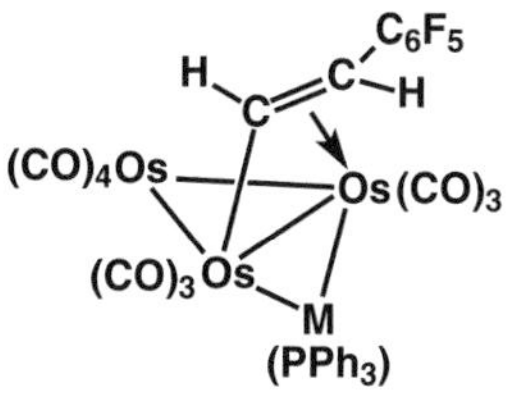

377a; M = Au
377b; M = Ag
377c; M = Cu

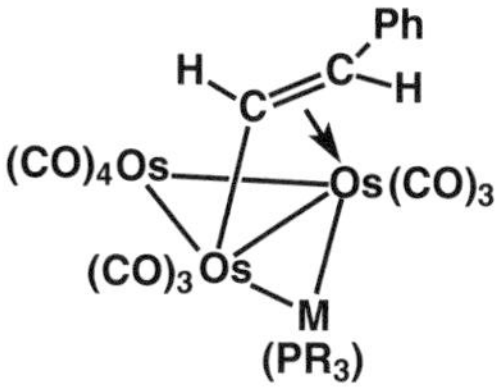

	M	R
378a	Au	Me
378b	Au	Ph
378c	Ag	Ph
378d	Cu	Ph

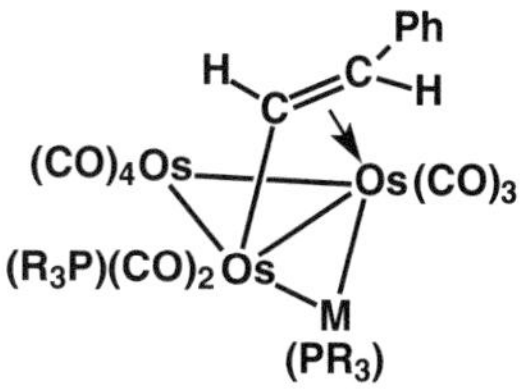

	M	R
381a	Au	Me
381b	Au	Ph
381c	Ag	Ph
381d	Cu	Ph

FIG. 26.

obtained when a toluene solution of [*trans*-$IrCl(CO)(PPh_3)_2$] and $AgC_2C_6F_5$ are heated to reflux (Fig. 27).

Thermolytic reaction between [(η-C_5H_5)$Fe(CO)_2Cl$] and [CuC_2Ph] affords a mixed metal complex [(η-C_5H_5)$Fe(CO)_2(C_2Ph)CuCl]_2$ (**386**) (Fig. 28).[131] Complex **386** contains a planar Cu_2Cl_2 ring, and each copper atom is symmetrically bonded to a C_2 unit of the phenylethynyl group.

In toluene solution [*trans*-Pt(C$\equiv$CH)$_2$(PMe_2Ph)$_2$] and [$W_2(O^tBu)_6$] react at 30 °C to give [*trans*-Pt(C$\equiv$CH)[$C_2W_2(O^tBu)_5$](PMe_2Ph)$_2$] (**387**) and [*trans*-Pt {$C_2W_2(O^tBu)_5$}$_2$(PMe_2Ph)$_2$] (**388**) (Fig. 29).[132,133a]

Reaction of a THF solution of [Cp(CO)(NO)W(C$\equiv$CR)]$^-$ (R = C_6H_5, **390a**; R = $C_6H_4CH_3$, **390b**; R = $C(CH_3)_3$, **390c**) with [$(C_6H_5)_3PAuCl$] results in the formation of a tungsten–gold acetylide complex [CpW(NO)(μ-CO)(μ-C$\equiv$CR) Au($P(C_6H_5)_3$)] (**391**) (Scheme 100).[133b] Reaction of a THF solution of [Cp(CO) (NO)W(C$\equiv$CR)]$^-$ (R = C_6H_5, **390a**; R = $C_6H_4CH_3$, **390b**; R = $C(CH_3)_3$, **390c**; R = $SiMe_3$, **390d**) with [$Cp(CO)_2Fe(THF)]BF_4$ as electrophile gives a bimetallic complex [{Cp(CO)(NO)W}η^2-{$Cp(CO)_2$FeC$\equiv$CR}] (**392a-d**). Photolytic decarbonylation of complexes **392a–d** gives diastereomeric mixtures of heterometallic acetylides [{Cp(CO)(NO)W}η^2-{Cp(CO)FeC$\equiv$CR}] (**393a-d/394a-d**) (Scheme 100).[133c]

Reaction of *fac*-[Mn(CCR)$(CO)_3$(dppe)] (R = CH_2OMe, **395a**; R = ^{t}Bu, **395b**; R = Ph, **395c**) with a suspension of CuCl in dichloromethane affords [MnCuCl (μ-CCR)$(CO)_3$(dppe)] (**396a–c**). On the other hand, the reaction of **395a** and **395b** with [Au(C_6F_5)(tht)] (tht = tetrahydrothiophene) gives [MnAu(C_6F_5)(μ-CCR$(CO)_3$ (dppe)] (R = CH_2OMe, **397a**; R = Bu^t, **397b**). Cationic mixed metal complexes [Mn_2Cu(μ-CCR)$_2(CO)_6$(dppe)$_2$]PF_6 (R = CH_2OMe, **398a**; R = Bu^t, **398b**) have been isolated from a dichloromethane solution of compound **396a** or **396b** on addition of $TlPF_6$ in presence of **395a** or **395b**, respectively. The silver and gold complexes, [Mn_2M(μ-CC^{t}Bu)$_2(CO)_6$(dppe)$_2$]PF_6 (M = Ag, **399**; Au, **400**) are formed when [$AgBF_4$] or [AuCl(tht)] are reacted with $TlPF_6$ and two fold excess of **395b**. Addition of $TlPF_6$ to complex **396b** in presence of [P(C_6H_4Me-2)$_3$] gives a cationic mixed metal complex [MnCu(μ-CC^{t}Bu)$(CO)_3$(dppe){P(C_6H_4Me-2)$_3$}]PF_6 (**401**) (Fig. 30).[133d]

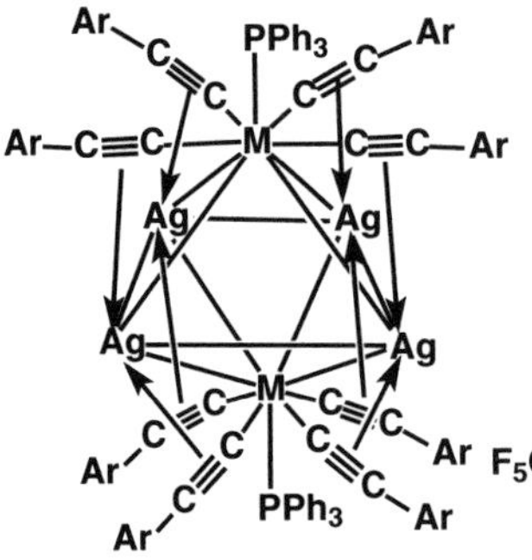

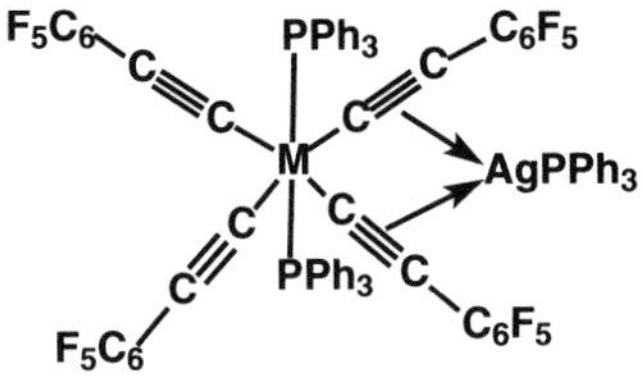

382a; M = Rh, Ar = Ph
382b; M = Rh, Ar = C_6F_5
383a; M = Ir, Ar = Ph
383b; M = Ir, Ar = C_6F_5

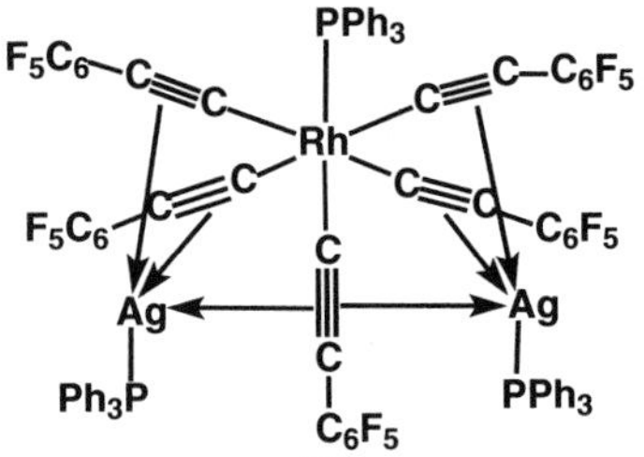

384a; M = Rh
384b; M = Ir

385

Fig. 27.

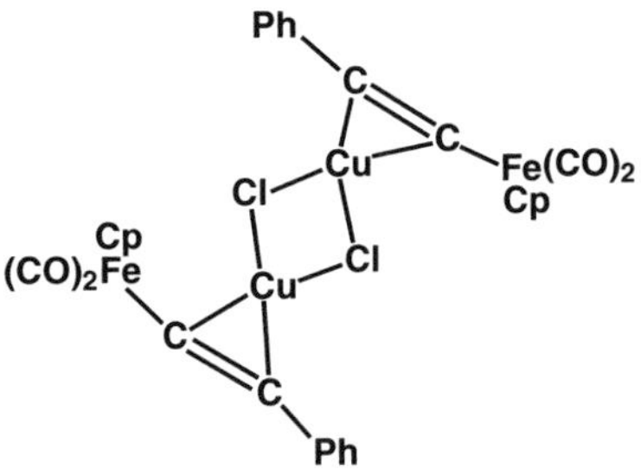

386

Fig. 28.

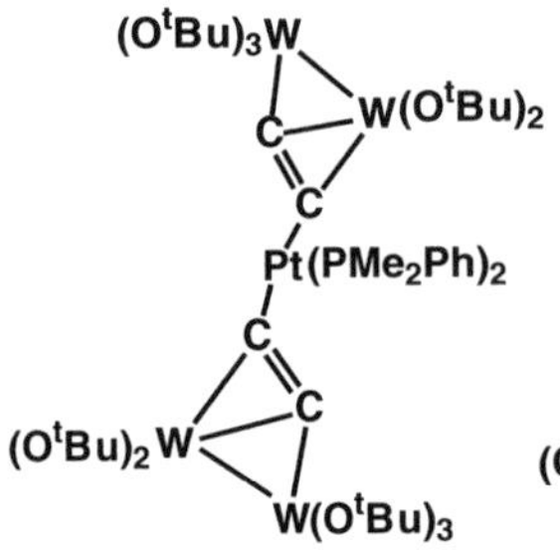

388

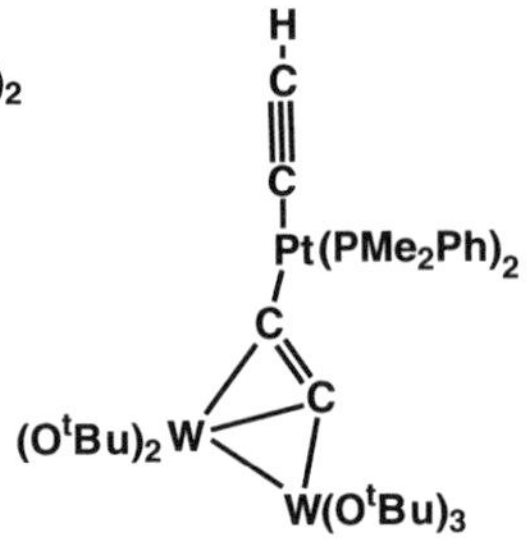

387

Fig. 29.

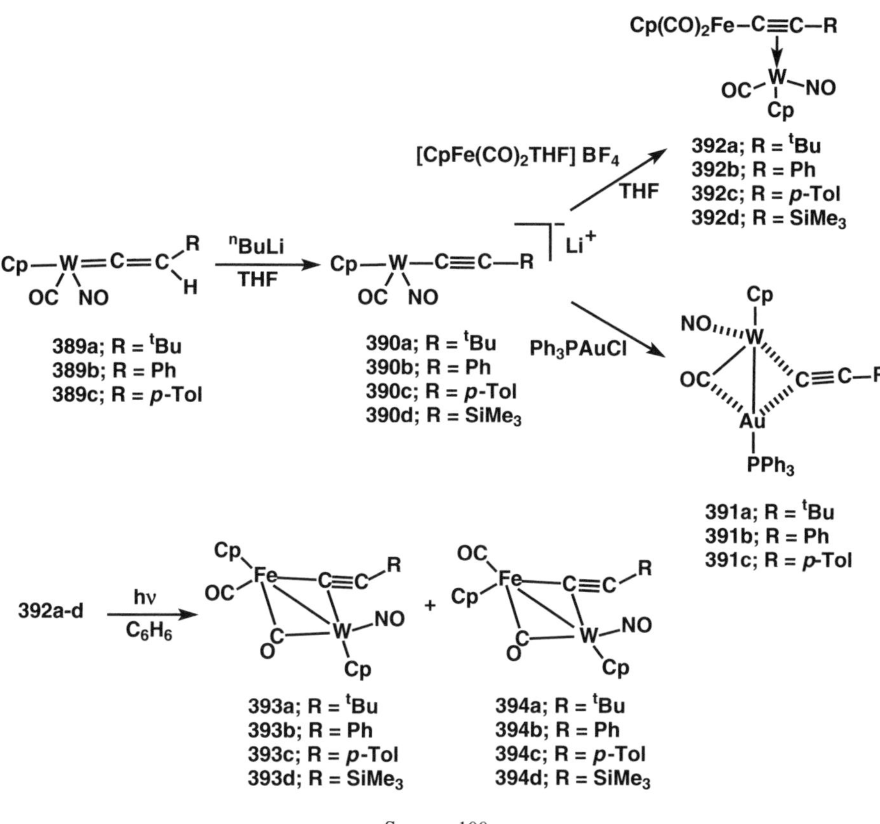

Scheme 100.

(dppe)(CO)$_3$Mn–C≡C–R → Cu–Cl

396a; R = CH_2OMe
396b; R = ^{t}Bu
396c; R = Ph

(dppe)(CO)$_3$Mn–C≡C–R → Au(C_6F_5)

397a; R = CH_2OMe
397b; R = ^{t}Bu

[(dppe)(CO)$_3$Mn–C≡C–R → M ← (dppe)(CO)$_3$Mn–C≡C–R]$^+$

	M	R
398a	Cu	CH_2OMe
398b	Cu	^{t}Bu
399	Ag	^{t}Bu
400	Au	^{t}Bu

[(dppe)(CO)$_3$Mn–C≡C–R → M–P(C_6H_4Me-2)$_3$]$^+$

401; M = Cu, ^{t}Bu

Fig. 30.

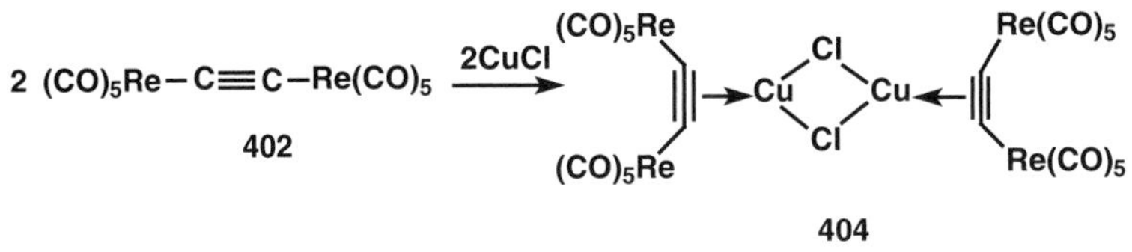

SCHEME 101.

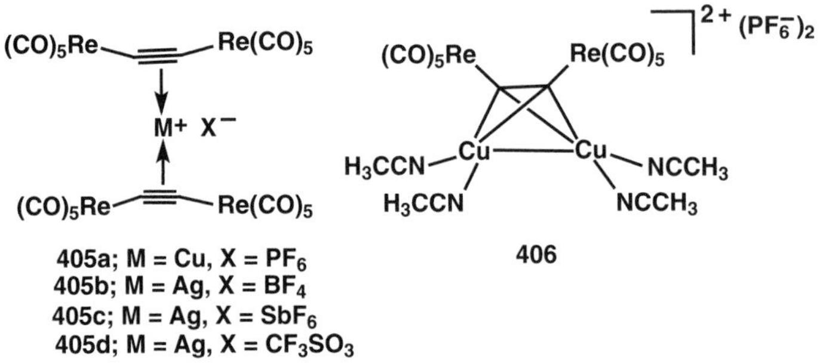

FIG. 31.

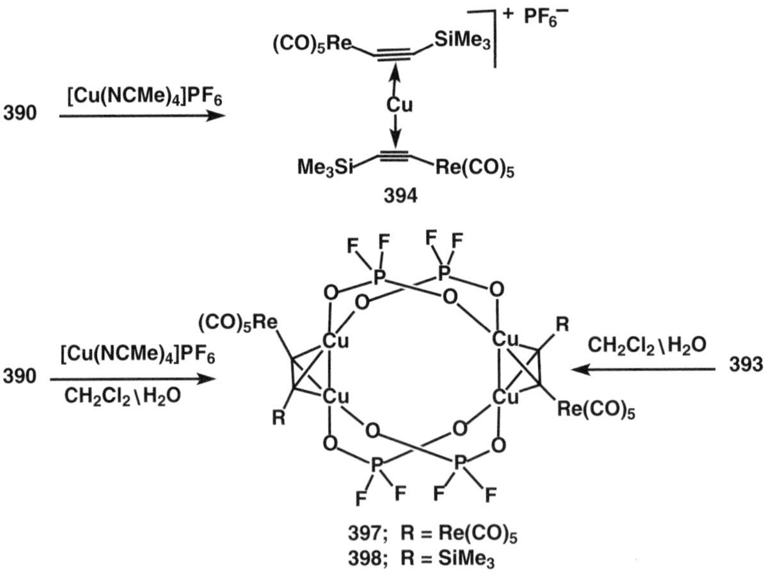

SCHEME 102.

Like organic alkynes or some monometalated acetylides ($L_nMC\equiv CR$), the dirhenioethyne $[(OC)_5ReC\equiv CRe(CO)_5]$ (**402**) has been found to behave as an η^2-ligand towards Cu^I, Ag^I and Au^I. Thus, when a THF solution of **402** reacts with CuCl, complex $[\{(\eta^2\text{-}C\equiv C\{Re(CO)_5\}_2)Cu(\mu\text{-}Cl)\}_2]$ (**404**) is isolated (Scheme 101).[134]

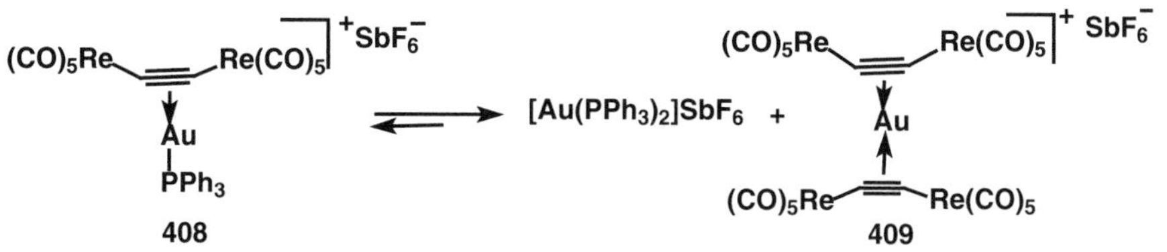

Scheme 103.

Reaction of **402** with $[Cu(NCMe)_4]PF_6$, $[Ag(NCMe)_4]BF_4$, $[AgSbF_6]$ or $[AgO_2SOCF_3]$ gives cationic bis-(alkyne) complexes $[(\eta^2\text{-}C{\equiv}C\{Re(CO)_5\}_2)_2M]^+X^-$ (M = Cu, Ag; X = PF_6, BF_4, SbF_6 or CF_3SO_2) (**405a–d**) and a cationic dimetallic tetrahedron $[(\mu\text{-}\eta^2,\eta^2\text{-}C{\equiv}C\{Re(CO)_5\}_2)Cu_2(NCMe)_4]^{2+}$ (**406**) (Fig. 31).[134]

On the other hand, a dichloromethane solution of $[(OC)_5ReC{\equiv}CSiMe_3]$ (**403**) on reaction with $[Cu(NCMe)_4]PF_6$ yields the cationic complex $[(\eta^2\text{-}C{\equiv}CRe(CO)_5SiMe_3)_2Cu]^+$ (**407**). Hydrolysis of $[(\mu\text{-}\eta^2{:}\eta^2\text{-}C{\equiv}C\{Re(CO)_5\}_2)Cu_2(NCMe)_4](PF_6)_2$ (**406**) or treatment of **403** with $[Cu(MeCN)_4]PF_6$ in moist CH_2Cl_2 affords the difluorphosphate-bridged complexes $[\{\mu\text{-}\eta^2{:}\eta^2\text{-}[(OC)_5ReC{\equiv}CR]\}_2Cu_4(\mu_2\text{-}O_2PF_2)_4]$ (R = $Re(CO)_5$, **410**; R = $SiMe_3$, **411**) (Scheme 102). The gold complex $[(\eta^2\text{-}C{\equiv}C\{Re(CO)_5\}_2)AuPPh_3]SbF_6$ (**408**) obtained from the reaction of $[Au(PPh_3)SbF_6]$ with **402**, exists in equilibrium in solution with $[Au(PPh_3)_2SbF_6]$ and $[(\eta^2\text{-}C{\equiv}C\{Re(CO)_5\}_2)_2Au]SbF_6$ (**409**) (Scheme 103).[134]

VIII
ABBREVIATIONS

Cp = $(\eta^5\text{-}C_5H_5)$
Cp* = $(\eta^5\text{-}C_5Me_5)$
dppe = Diphenylphosphinoethane
dppm = Diphenylphosphinomethane
DPAD = Diisopropylacetylenedicarboxylate
DBU = 1,5-Diazabicyclo(5.4.0)undec-5-en
Fc = $(C_5H_5)(C_5H_4)Fe$
PPN = Bis(triphenylphosphine)nitrogen(1+), $[(Ph_3P)_2N^+]$
tht = Tetrahydrothiophene

References

(1) Nast, R. Z *Naturforsch. Teil B* **1953**, *8*, 381.
(2) Nast, R. *Angew. Chem.* **1960**, *72*, 26.
(3) Manna, J.; John, K. D.; Hopkins, M. D. *Adv. Organomet. Chem.* **1995**, *38*, 79.
(4) Nast, R. *Coord. Chem. Rev.* **1982**, *47*, 89.
(5) Rourke, J. P.; Bruce, D. W.; Marder, T. B. *J. Chem. Soc., Dalton Trans.* **1995**, 317.
(6) Takahashi, S.; Morimoto, H.; Murata, E.; Kataoka, S.; Sonogashira, K.; Hagihara, N. *J. Polym. Sci., Polym. Chem. Ed.* **1982**, *20*, 565.
(7) Hagihara, N.; Sonogashira, K.; Takahashi, S. *Adv. Polym. Sci.* **1981**, *41*, 149.

(8) Stang, P. J. *Chem. Eur. J.* **1998**, *4*, 19.
(9) Stang, P. J.; Olenyuk, B. *Acc. Chem. Res.* **1997**, *30*, 502.
(10) Akita, M.; Terada, M.; Moro-oka, Y. *Organometallics* **1992**, *11*, 1825.
(11) Bruce, M. I.; Humphrey, M. G.; Matisons, J. G.; Roy, S. K.; Swincer, A. G. *Aust. J. Chem.* **1984**, *37*, 1955.
(12) Bruce, M. I.; Humphrey, M. G.; Snow, M. R.; Tiekink, E. R. T. *J. Organomet. Chem.* **1986**, *314*, 213.
(13) Chiang, S.-J.; Chi, Y.; Su, P.-C.; Peng, S.-M.; Lee, G.-H. *J. Am. Chem. Soc.* **1994**, *116*, 11181.
(14) Lotz, S.; van Rooyen, P. H.; Meyer, R. *Adv. Organomet. Chem.* **1995**, *37*, 219.
(15) Weidmann, T.; Weinrich, V.; Wagner, B.; Robl, C.; Beck, W. *Chem. Ber.* **1991**, *124*, 1363.
(16) Adams, R. D.; Bunz, U. H. F.; Fu, W.; Roidl, G. J. *J. Organomet. Chem.* **1999**, *578*, 55.
(17) Low, P. J.; Udachin, K. A.; Enright, G. D.; Carty, A. J. *J. Organomet. Chem.* **1999**, *578*, 103.
(18) Low, P. J.; Enright, G. D.; Carty, A. J. *J. Organomet. Chem.* **1998**, *565*, 279.
(19) Blenkiron, P.; Enright, G. D.; Carty, A. J. *Chem. Commun.* **1997**, 483.
(20) Ward, M. D. *Chem. Ind.* **1996**, 568.
(21) Ward, M. D. *Chem. Soc. Rev.* **1995**, *24*, 121.
(22) Bunz, U. H. F. *Angew. Chem.* **1994**, *106*, 1127.
(23) Schumm, J. S.; Pearson, D. L.; Tour, J. M. *Angew. Chem.* **1994**, *106*, 1445.
(24) Metzger, R. M. *Acc. Chem. Res.* **1999**, *32*, 950.
(25) Dembinski, R.; Bartik, T.; Bartik, B.; Jaeger, M.; Gladysz, J. A. *J. Am. Chem. Soc.* **2000**, *122*, 810.
(26) Bruce, M. I.; Low, P. J.; Costuas, K.; Halet, J.-F.; Best, S. P.; Heath, G. A. *J. Am. Chem. Soc.* **2000**, *122*, 1949.
(27) Wu, C. H.; Chi, Y.; Peng, S. M.; Lee, G. H. *J. Chem. Soc., Dalton Trans.* **1990**, 3025.
(28) Akita, M.; Sugimoto, S.; Terada, M.; Moro-oka, Y. *J. Organomet. Chem.* **1993**, *447*, 103.
(29) Ustynyuk, N. A.; Vinogradova, V. N.; Korneva, V. N.; Kravtsov, D. N.; Andrianov, V. G.; Struchkov, Y. T. *J. Organomet. Chem.* **1984**, *277*, 285.
(30) Mathur, P.; Ahmed, M. O.; Dash, A. K.; Walawalkar, M. G.; Puranik, V. G. *J. Chem. Soc., Dalton Trans.* **2000**, 2916.
(31) Mathur, P.; Chatterjee, S. *Comments Inorg. Chem.* **2006**, *26*, 255.
(32) Akita, M.; Sugimoto, S.; Hirakawa, H.; Kato, S.-I.; Terada, M.; Tanaka, M.; Morooka, Y. *Organometallics* **2001**, *20*, 1555.
(33) Akita, M.; Hirakawa, H.; Tanaka, M.; Morooka, Y. *J. Organomet. Chem.* **1995**, *485*, C14.
(34) (a) Terada, M.; Akita, M. *Organometallics* **2003**, *22*, 355. (b) Koutsantonis, G. A.; Selegue, J. P.; Wang, J.-G. *Organometallics* **1992**, *11*, 2704.
(35) (a) Akita, M.; Chung, M.-C.; Sakurai, A.; Moro-oka, Y. *Chem. Commun.* **2000**, 1285. (b) Bruce, M. I.; Kelly, B. D.; Skelton, B. W.; White, A. H. *J. Organomet. Chem.* **2000**, *604*, 150.
(36) (a) Koridze, A. A.; Sheloumov, A. M.; Dolgushin, F. M.; Yanovsky, A. I.; Struchkov, Y. T.; Petrovskii, P. V. *J. Organomet. Chem.* **1997**, *536–537*, 381. (b) Younus, M.; Long, N. J.; Raithby, P. R.; Lewis, J. *J. Organomet. Chem.* **1998**, *570*, 55.
(37) Mathur, P.; Singh, V. K.; Mobin, S. M.; Srinivasu, C.; Trivedi, R.; Bhunia, A. K.; Puranik, V. G. *Organometallics* **2005**, *24*, 367.
(38) Sonogashira, K.; Ohga, K.; Takahashi, S.; Hagihara, N. *J. Organomet. Chem.* **1980**, *188*, 237.
(39) (a) Berenguer, J. R.; Forniés, J.; Lalinde, E.; Martínez, F. *J. Organomet. Chem.* **1994**, *470*, C15. (b) Forniés, J.; Lalinde, E. *J. Chem. Soc., Dalton Trans.* **1996**, 2587.
(40) Forniés, J.; Gomez-Saso, M. A.; Lalinde, E.; Martínez, F.; Moreno, M. T. *Organometallics* **1992**, *11*, 2873.
(41) Abu-Salah, O. M. *J. Organomet. Chem.* **1990**, *387*, 123.
(42) Abu-Salah, O. M.; Knobler, C. B. *J. Organomet. Chem.* **1986**, *302*, C10.
(43) Abu-Salah, O. M.; Al-Ohaly, A. R. A.; Mutter, Z. F. *J. Organomet. Chem.* **1990**, *389*, 427.
(44) Abu-Salah, O. M.; Al-Ohaly, A. R. A. *J. Chem. Soc., Dalton Trans.* **1988**, 2297.
(45) Abu-Salah, O. M.; Al-Ohaly, A. R. A.; Knobler, C. B. *Chem Commun.* **1985**, 1502.
(46) Hussain, M. S.; Abu-Salah, O. M. *J. Organomet. Chem.* **1993**, *445*, 295.
(47) Abu-Salah, O. M.; Al-Ohaly, A. R. A.; Mutter, Z. F. *J. Organomet. Chem.* **1990**, *391*, 267.
(48) Abu-Salah, O. M. *Polyhedron* **1992**, *11*, 951.

(49) Abu-Salah, O. M.; Hussain, M. S.; Schlemper, E. O. *Chem Commun.* **1988**, 212.
(50) Bernhardt, W.; Vahrenkamp, H. *J. Organomet. Chem.* **1990**, *383*, 357.
(51) Akita, M.; Terada, M.; Moro-oka, Y. *Organometallics* **1992**, *11*, 1825.
(52) Akita, M.; Terada, M.; Ishii, N.; Hirakawa, H.; Moro-oka, Y. *J. Organomet. Chem.* **1994**, *473*, 175.
(53) Akita, M.; Hirakawa, H.; Sakaki, K.; Moro-oka, Y. *Organometallics* **1995**, *14*, 2775.
(54) (a) Akita, M.; Chung, M.-C.; Terada, M.; Miyauti, M.; Tanaka, M.; Moro-oka, Y. *J. Organomet. Chem.* **1998**, *565*, 49. (b) Jensen, M. P.; Phillips, D. A.; Sabat, M.; Shriver, D. F. *Organometallics* **1992**, *11*, 1859.
(55) Chung, M.-C.; Sakurai, A.; Akita, M.; Moro-oka, Y. *Organometallics* **1999**, *18*, 4684.
(56) (a) Bruce, M. I.; Halet, J.-F.; Kahlal, S.; Low, P. J.; Skelton, B. W.; White, A. H. *J. Organomet. Chem.* **1999**, *578*, 155. (b) Bruce, M. I.; Kelly, B. D.; Skelton, B. W.; White, A. H. *J. Chem. Soc., Dalton Trans.* **1999**, 847.
(57) Bernhardt, W.; Vahrenkamp, H. *J. Organomet. Chem.* **1988**, *355*, 427.
(58) Hwaug, D.-K.; Chi, Y.; Peng, S.-M.; Lee, G.-H. *J. Organomet. Chem.* **1990**, *389*, C7.
(59) Bruce, M. I.; Low, P. J.; Zaitseva, N. N.; Kahlal, S.; Halet, J.-F.; Skelton, B. W.; White, A. H. *J. Chem. Soc., Dalton Trans.* **2000**, 2939.
(60) Low, P. J.; Udachin, K. A.; Enright, G. D.; Carty, A. J. *J. Organomet. Chem.* **1999**, *578*, 103.
(61) Adams, C. J.; Bruce, M. I.; Skelton, B. W.; White, A. H. *J. Organomet. Chem.* **1991**, *420*, 95.
(62) (a) Bruce, M. I.; Zaitseva, N. N.; Skelton, B. W.; White, A. H. *Polyhedron* **1995**, *14*, 2647. (b) Takemoto, S.; Shimadzu, D.; Kamikawa, K.; Matsuzaka, H.; Nomura, R. *Organometallics* **2006**, *25*, 982.
(63) Mathur, P.; Bhunia, A. K.; Kumar, A.; Chatterjee, S.; Mobin, S. M. *Organometallics* **2002**, *21*, 2215.
(64) Mathur, P.; Ahmed, M. O.; Dash, A. K.; Walawalkar, M. G. *J. Chem. Soc., Dalton Trans.* **1999**, 1795.
(65) Mathur, P.; Ahmed, M. O.; Kaldis, J. H.; McGlinchey, M. J. *J. Chem. Soc., Dalton Trans.* **2002**, 619.
(66) Mathur, P.; Srinivasu, C.; Mobin, S. M. *J. Organomet. Chem.* **2003**, *665*, 226.
(67) Mathur, P.; Bhunia, A. K.; Mobin, S. M.; Singh, V. K.; Srinivasu, C. *Organometallics* **2004**, *23*, 3694.
(68) Mathur, P.; Srinivasu, C.; Ahmed, M. O.; Puranik, V. G.; Umbarkar, S. B. *J. Organomet. Chem.* **2002**, *669*, 196.
(69) Mathur, P.; Mukhopadhyay, S.; Ahmed, M. O.; Lahiri, G. K.; Chakraborty, S.; Puranik, V. G.; Bhadbhade, M. M.; Umbarkar, S. B. *J. Organomet. Chem.* **2001**, *629*, 160.
(70) Mathur, P.; Mukhopadhyay, S.; Ahmed, M. O.; Lahiri, G. K.; Chakraborty, S.; Walawalkar, M. G. *Organometallics* **2000**, *19*, 5787.
(71) Mathur, P.; Bhunia, A. K.; Srinivasu, C.; Mobin, S. M. *J. Organomet. Chem.* **2003**, *670*, 144.
(72) Mathur, P.; Bhunia, A. K.; Srinivasu, C.; Mobin, S. M. *Phosphorus, Sulfur, and Silicon* **2004**, *179*, 899.
(73) Bruce, M. I.; Koutsantonis, G. A.; Tiekink, E. R. T. *J. Organomet. Chem.* **1991**, *407*, 391.
(74) Bruce, M. I.; Koutsantonis, G. A.; Tiekink, E. R. T. *J. Organomet. Chem.* **1991**, *408*, 77.
(75) Adams, C. J.; Bruce, M. I.; Skelton, B. W.; White, A. H. *Polyhedron* **1998**, *17*, 2795.
(76) Su, P.-C.; Chiang, S.-J.; Chang, L.-L.; Chi, Y.; Peng, S.-M.; Lee, G.-H. *Organometallics* **1995**, *14*, 4844.
(77) Chi, Y.; Lee, G.-H.; Peng, S.-M.; Wu, C.-H. *Organometallics* **1989**, *8*, 1574.
(78) Wu, C.-H.; Chi, Y.; Peng, S.-W.; Lee, G.-H. *J. Chem. Soc., Dalton Trans.* **1990**, 3025.
(79) Chi, Y.; Hüttner, G.; Imhof, W. *J. Organomet. Chem.* **1990**, *384*, 93.
(80) Hwang, D.-K.; Chi, Y.; Peng, S.-M.; Lee, G.-H. *Organometallics* **1990**, *9*, 2709.
(81) Hwang, D.-K.; Lin, P.-J.; Chi, Y.; Peng, S.-M.; Lee, G.-H. *J. Chem. Soc., Dalton Trans.* **1991**, 2161.
(82) Hwang, D.-K.; Chi, Y.; Peng, S.-M.; Lee, G.-H. *J. Organomet. Chem.* **1990**, *389*, C7.
(83) Chi, Y.; Lin, R.-C.; Chen, C.-C.; Peng, S.-M.; Lee, G.-H. *J. Organomet. Chem.* **1992**, *439*, 347.
(84) Chi, Y.; Hsu, H.-F.; Peng, S.-M.; Lee, G.-H. *Chem. Commun.* **1991**, 1019.

(85) Chi, Y.; Hsu, H.-Y.; Peng, S.-M.; Lee, G.-H. *Chem. Commun.* **1991**, 1023.
(86) Chi, Y.; Chung, C.; Chou, Y.-C.; Su, P.-C.; Chiang, S.-J.; Peng, S.-M.; Lee, G.-H. *Organometallics* **1997**, *16*, 1702.
(87) Chi, Y.; Chung, C.; Tseng, W.-C.; Chou, Y.-C.; Peng, S.-M.; Lee, G.-H. *J. Organomet. Chem.* **1999**, *574*, 294.
(88) Chung, C.; Tseng, W.-C.; Chi, Y.; Peng, S.-M.; Lee, G.-H. *Organometallics* **1998**, *17*, 2207.
(89) Chi, Y.; Wu, C.-H.; Peng, S.-M.; Lee, G.-H. *Organometallics* **1991**, *10*, 1676.
(90) Chao, W.-J.; Chi, Y.; Way, C.-J.; Mavunkal, I. J.; Wang, S.-L.; Liao, F.-L.; Farrugia, L. J. *Organometallics* **1997**, *16*, 3523.
(91) Peng, J.-J.; Horng, K.-M.; Cheng, P.-S.; Chi, Y.; Peng, S.-M.; Lee, G.-H. *Organometallics* **1994**, *13*, 2365.
(92) Shaposhnikova, A. D.; Stadnichenko, R. A.; Kamalov, G. L.; Pasynskii, A. A.; Eremenko, I. L.; Nefedov, S. E.; Struchkov, Y. T.; Yanovsky, A. I. *J. Organomet. Chem.* **1993**, *453*, 279.
(93) Falloon, S. B.; Szafert, S.; Arif, A. M.; Gladysz, J. A. *Chem. Eur. J.* **1998**, *4*, 1033.
(94) Falloon, S. B.; Arif, A. M.; Gladysz, J. A. *Chem. Commun.* **1997**, 629.
(95) Griffith, C. S.; Koutsantonis, G. A.; Skelton, B. W.; White, A. H. *Chem. Commun.* **1998**, 1805.
(96) Byrne, L. T.; Griffith, C. S.; Koutsantonis, G. A.; Skelton, B. W.; White, A. H. *J. Chem. Soc., Dalton Trans.* **1998**, 1575.
(97) Mays, M. J.; Sarveswaran, K.; Solan, G. A. *Inorg. Chim. Acta* **2003**, *354*, 21.
(98) Yasufuku, K.; Yamazaki, H. *Bull. Chem. Soc. Jpn.* **1972**, *45*, 2664.
(99) Bruce, M. I.; Duffy, D. N.; Humphrey, M. G. *Aust. J. Chem.* **1986**, *39*, 159.
(100) (a) Dalton, G. T.; Viau, L.; Waterman, S. M.; Humphrey, M. G.; Bruce, M. I.; Low, P. J.; Roberts, R. L.; Willis, A. C.; Koutsantonis, G. A.; Skelton, B. W.; White, A. H. *Inorg. Chem.* **2005**, *44*, 3261. (b) Bruce, M. I.; Ellis, B. G.; Skelton, B. W.; White, A. H. *J. Organomet. Chem.* **2000**, *607*, 137.
(101) Tanaka, S.; Yoshida, T.; Adachi, T.; Yoshida, T.; Onitsuka, K.; Sonogashira, K. *Chem. Lett.* **1994**, 877.
(102) Espinet, P.; Forniés, J.; Martínez, F.; Tomás, M.; Lalinde, E.; Moreno, M. T.; Ruiz A.; Welch, A. J. *J. Chem. Soc., Dalton Trans.* **1990**, 791.
(103) Espinet, P.; Forniés, J.; Martínez, F.; Sotes, M.; Lalinde, E.; Moreno, M. T.; Ruiz, A.; Welch, A. J. *J. Organomet. Chem.* **1991**, *403*, 253.
(104) Forniés, J.; Gomez-Saso, M. A.; Martínez, F.; Lalinde, E.; Moreno, M. T.; Welch, A. J. *New J. Chem.* **1992**, *16*, 483.
(105) Forniés, J.; Lalinde, E.; Martínez, F.; Moreno, M. T.; Welch, A. J. *J. Organomet. Chem.* **1993**, *455*, 271.
(106) Ara, I.; Forniés, J.; Lalinde, E.; Moreno, M. T.; Tomás, M. *J. Chem. Soc., Dalton Trans.* **1995**, 2397.
(107) Forniés, J.; Lalinde, E.; Martin, A.; Moreno, M. T. *J. Organomet. Chem.* **1995**, *490*, 179.
(108) Berenguer, J. R.; Forniés, J.; Lalinde, E.; Martin, A.; Moreno, M. T. *J. Chem. Soc., Dalton Trans.* **1994**, 3343.
(109) Ara, I.; Berenguer, J. R.; Eguizábal, E.; Forniés, J.; Gómez, J.; Lalinde, E. *J. Organomet. Chem.* **2003**, *670*, 221.
(110) Ara, I.; Forniés, J.; Lalinde, E.; Moreno, M. T.; Tomás, M. *J. Chem. Soc., Dalton Trans.* **1994**, 2735.
(111) Berenguer, J. R.; Falvello, L. R.; Forniés, J.; Lalinde, E.; Tomás, M. *Organometallics* **1993**, *12*, 6.
(112) Berenguer, J. R.; Forniés, J.; Lalinde, E.; Martin, A. *Angew. Chem. Int. Ed. Engl.* **1994**, *33*, 2083.
(113) Janssen, M. D.; Smeets, W. J. J.; Spek, A. L.; Grove, D. M.; Lang, H.; van Koten, G. *J. Organomet.Chem.* **1995**, *505*, 123.
(114) Berenguer, J. R.; Forniés, J.; Lalinde, E.; Martínez, F. *Chem. Commun.* **1995**, 1227.
(115) Weatherell, C.; Taylor, N. J.; Carty, A. J.; Sappa, E.; Tiripicchio, A. *J. Organomet. Chem.* **1985**, *291*, C9.
(116) Marinetti, A.; Sappa, E.; Tiripicchio, A.; Camellini, M. T. *J. Organomet. Chem.* **1980**, *197*, 335.
(117) Lanfranchi, M.; Tiripicchio, A.; Sappa, E.; MacLaughlin, S. A.; Carty, A. J. *Chem. Commun.* **1982**, 538.
(118) Carty, A. J.; Taylor, N. J.; Sappa, E.; Tiripicchio, A. *Inorg. Chem.* **1983**, *22*, 1871.

(119) Blenkiron, P.; Enright, G. D.; Carty, A. J. *Chem. Commun.* **1997**, 483.
(120) Ewing, P.; Farrugia, L. J. *J. Organomet. Chem.* **1987**, *320*, C47.
(121) Yamazaki, S.; Taira, Z.; Yonemura, T.; Deeming, A. J. *Organometallics* **2005**, *24*, 20.
(122) Yamazaki, S.; Taira, Z.; Yonemura, T.; Deeming, A. J.; Nakao, A. *Chem. Lett.* **2002**, 1174.
(123) Bruce, M. I.; Koutsantonis, G. A.; Tiekink, E. R. T. *J. Organomet. Chem.* **1991**, *408*, 77.
(124) Bruce, M. I.; Horn, E.; Shawkataly, O. B.; Snow, M. R. *J. Organomet. Chem.* **1985**, *280*, 289.
(125) Bruce, M. I.; Zaitseva, N. N.; Skelton, B. W.; White, A. H. *J. Chem. Soc., Dalton Trans.* **1999**, 2777.
(126) Abu Salah, O. M.; Bruce, M. I. *J. Chem. Soc., Dalton Trans.* **1975**, 2311.
(127) Griffith, C. S.; Koutsantonis, G. A.; Skelton, B. W.; White, A. H. *Chem. Commun.* **2002**, 2174.
(128) Deeming, A. J.; Donovan-Mtunzi, S.; Hardcastle, K. *J. Chem. Soc., Dalton Trans.* **1986**, 543.
(129) Bruce, M. I.; Horn, E.; Matisons, J. G.; Snow, M. R. *J. Organomet. Chem.* **1985**, *286*, 271.
(130) Abu Salah, O. M.; Bruce, M. I. *Aust. J. Chem.* **1977**, *30*, 2639.
(131) (a) Bruce, M. I.; Clark, R.; Howard, J.; Woodward, P. *J. Organomet. Chem.* **1972**, *42*, C107. (b) Abu Salah, O. M.; Bruce, M. I. *J. Chem. Soc., Dalton Trans.* **1974**, 2302.
(132) Blau, R. J.; Chisholm, M. H.; Folting, K.; Wang, R. J. *J. Am. Chem. Soc.* **1987**, *109*, 4552.
(133) (a) Blau, R. J.; Chisholm, M. H.; Folting, K.; Wang, R. J. *Chem. Commun.* **1985**, 1582. (b) Ipaktschi, J.; Munz, F. *Organometallics* **2002**, *21*, 977. (c) Ipaktschi, J.; Mirzaei, F.; Müller, B. G.; Beck, J.; Serafin, M. *J. Organomet. Chem.* **1996**, *526*, 363. (d) Carriedo, G. A.; Miguel, D.; Riera, V.; Soláns, X. *J. Chem. Soc., Dalton Trans.* **1987**, 2867.
(134) Mihan, S.; Sünkel, K.; Beck, W. *Chem. Eur. J.* **1999**, *5*, 745.

Transition Metal Organometallic Synthesis Utilising Diorganoiodine(III) Reagents

ALLAN J. CANTY,[a,*] THOMAS RODEMANN[a] and JOHN H. RYAN[b]

[a]*School of Chemistry, University of Tasmania, Hobart, Tasmania 7001, Australia*
[b]*CSIRO Molecular and Health Technologies, Ian Wark Laboratory, Bayview Avenue, Clayton, Victoria 3169, Australia*

I
INTRODUCTION

Hypervalent iodine reagents have a prominent role in organic synthesis,[1–12] but there is only a slowly emerging chemistry employing these reagents for transition metal organometallic synthesis, despite the early promising results reported in the 1960s,[13,14] and extensive reports of the application of diorganoiodine(III) reagents in organic synthesis involving metal centres. This review summarises the fundamental aspects of mechanisms proposed for organic chemistry applications, reviews representative reports in organometallic chemistry with this perspective, and highlights the potential for further developments in the field.

Diorganoiodine(III) reagents belong to a class of hypervalent compounds that contain two carbon ligands attached to the central iodine atom. A significant number of accounts have appeared on this subject;[1] particularly noteworthy are the comprehensive reviews of Varvoglis[3–5] and Stang and Zhdankin,[6–8] as well as recent books dedicated to the topic.[9–12] While a wide range of diorganoiodine(III) compounds have been prepared and studied, this review will focus on those that have been utilised in transition metal organometallic synthesis, i.e. diaryl-, alkynyl(aryl)- and alkenyl(aryl)iodine(III) reagents **1-3** (Scheme 1). These reagents usually contain a non-nucleophilic anion such as triflate or tetrafluoroborate as the third ligand or counterion, although for diaryliodine(III) reagents the counterion can also be a

*Corresponding author.
E-mail: Allan.Canty@utas.edu.au (A.J. Canty).

ADVANCES IN ORGANOMETALLIC CHEMISTRY
VOLUME 55 ISSN 0065-3055/DOI 10.1016/S0065-3055(07)55005-7

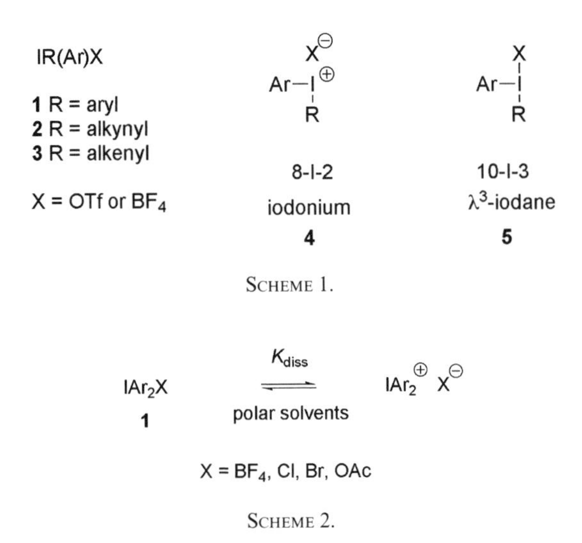

SCHEME 1.

SCHEME 2.

halide. According to the conventional nomenclature these reagents can be classified as either an 8-I-2 structure **4** ("8 e^- and 2 bonds", an iodonium salt), generally considered as an ionic structure with pseudotetrahedral geometry around the central iodine atom, or a 10-I-3 structure **5** ("10 e^- and 3 bonds", a λ^3-iodane) which is considered to have a distorted trigonal bipyramidal structure with the most electronegative ligands found in the axial position.[6]

There is a lack of clarity regarding the nomenclature associated with these species. Conventionally, they are known as iodonium salts due to their dissociation in polar solvents,[15] although their solid-state structures resemble typical λ^3-iodanes.[7] Depending on the nature of the diorganoiodine(III) species, the third ligand and the solvent, aggregation phenomena and/or dissociation of the covalent λ^3-diorganoiodane structure into iodonium ions in polar solvents (Scheme 2) have been observed.[16]

The structures of diorganoiodine(III) compounds in the solid state have been established from single crystal X-ray data, generally indicating an approximate T-shaped geometry with the least electronegative of the two carbon ligands (usually a phenyl group) occupying the equatorial position, and the most electronegative of the two organic ligands along with the heteroatom ligand occupying the axial or apical positions.[3,6,9,10] In a recent example, the single crystal X-ray structural data obtained for the *p*-iodophenyl(phenyl)iodonium salt **6** show the phenyl group occupying the equatorial position and the more electronegative *p*-iodophenyl group in the axial position, with a C–I–C bond angle of 90.6(1)° (Scheme 3).[17] As an indication of the complexity in geometry obtained for some species, the structural data obtained for the trimethylsilylalkynyl(phenyl)iodonium salt **7** reveals a dimeric structure where the shortest I···O interaction, 2.664(2) Å, is opposite the more electronegative alkynyl group (C–I–C 92.76(13))°.[18] The structural features of diorganoiodine(III) compounds have been explained using the hypervalent bonding model, which involves a linear three-centre four-electron hypervalent bond which results in a weaker, longer, and hence more reactive I–C bond than the equatorial I–C bond.[19,20] Reed and

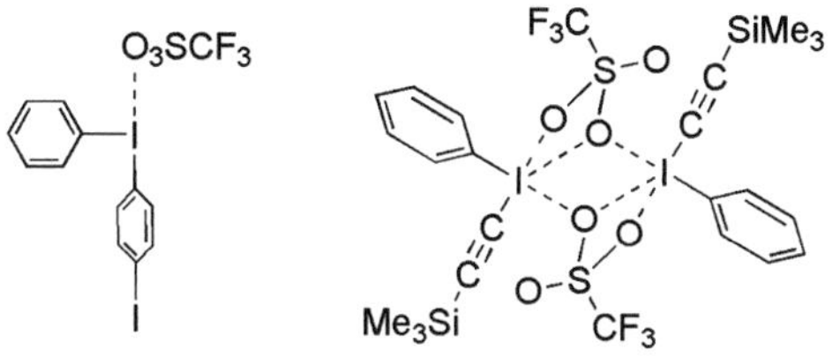

6 7

SCHEME 3.

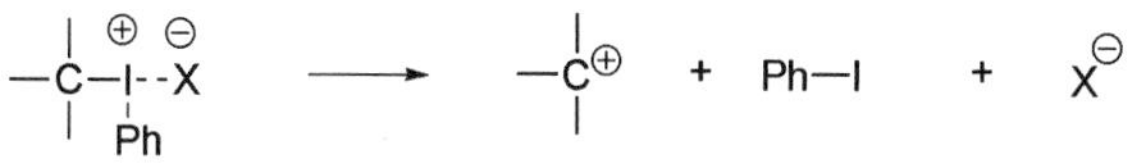

SCHEME 4.

Schleyer have described the bonding in terms of the dominance of ionic bonding and negative hyperconjugation over d-orbital participation.[21]

In view of the weak interactions with anions observed in the solid state, and dissociation in solution, diorganoiodine(III) species in this review are generally represented as ionic species or with a weak $I^{+} \cdots X^{-}$ interaction.

II
MECHANISMS OF REACTIONS OF DIORGANOIODINE(III) REAGENTS

The majority of the synthetically useful diorganoiodine(III) reagents contain at least one aryl ligand, usually a phenyl ligand. The aryliodanyl group, departing as ArI in reactions, has been termed a hyperleaving group as it shows a leaving group ability about 10^6 times greater than triflate, itself termed a superleaving group.[22] The driving force for the hyperleaving nature of the aryliodanyl group is the reductive elimination of the hypervalent iodine in the common –I oxidation state with an octet electron count (Scheme 4).[22] A wide range of reaction mechanisms have been elucidated for diorganoiodine(III) reagents, showing that polar (ligand coupling), radical, transition-metal catalysed and photochemical processes can operate.[7] In some cases these processes lead to other highly reactive intermediates, such as benzyne or alkylidene carbenes which in turn have a rich chemistry.[8] As a background to this review of transition metal organometallic synthesis using diorganoiodine(III) reagents, we restrict our overview to the polar, radical and transition metal catalysed processes involved in nucleophilic substitution reactions of the diorganoiodine(III) reagents **1**–**3**.

In the following sections, reactions involving diaryl-, alkynyl(aryl)- and alkenyl(aryl)iodine(III) reagents are addressed separately. However, reactivities in the presence or absence of transition metal species are treated together, emphasising the relationship between mechanisms for these systems.

A. *Diaryliodine(III) Reagents*

The oldest and most studied of the reagents are the diaryliodine(III) reagents **1**.[2–4,6,7,23,24] These species have been used in polar arylation reactions with charged nucleophiles (anionic species containing an alkali metal or tetralkylammoniun counterion) ranging from carbon nucleophiles, such as enolates of β-dicarbonyl compounds,[25–28] enol ethers,[29,30] nitroalkanates[31] and cyanide;[32] oxygen nucleophiles such as alkoxides[25] and phenoxides,[25,33,34] carboxylates,[25,33] hydroxamates[35] and hydroxyamines;[36] nitrogen nucleophiles such as arylamines,[25,33] imides,[33] sulphonamides,[25] *N*-hydroxycarbamates,[37] benzotriazinonates,[38] sodium nitrite[25] and azide;[32] sulphur nucleophiles such as arylthiolates,[39] arylsulphinates,[25,40] carbonodithioates,[41] dithiocarbamates,[42–44] dithioates,[45] thiocyanate[32,46] thiocarboxylates,[47] thiophosphates,[48] thiosulphonates[49] and sodium sulphite;[25] phosphorus nucleophiles such as dialkylphosphites,[50] main group nucleophiles such as telluride,[51] and arylselenides[52] and aryltellurides (Scheme 5).[53]

In addition, arylations of several neutral nucleophiles to give cationic products are known (Scheme 6). Examples include carbon monoxide in the presence of alcohols (non-catalysed methoxycarbonylation)[54] and amines;[55] tertiary amines and pyridine;[33] oxygen nucleophiles such as pyridine *N*-oxides;[56] sulphur nucleophiles such as diaryl-,[33] and alkyl(aryl)sulphides;[57] phosphorus nucleophiles such as triphenylphosphine[33,58] and triethylphosphite (undergoes an Arbuzov-type reaction);[59] diphenylselenide, triphenylarsine and triphenylstibine.[33] A mild method has been reported for formation of benzyne analogues involving the fluoride promoted elimination of benzyne from 2-trimethylsilylaryl(phenyl)iodine(III) triflates **8** (Scheme 7).[60]

A number of mechanisms have been proposed for the arylation of nucleophiles by diaryliodine(III) reagents including radical coupling processes, nucleophilic aromatic substitution (S_NAr) and ligand coupling on iodine(III). It is now thought

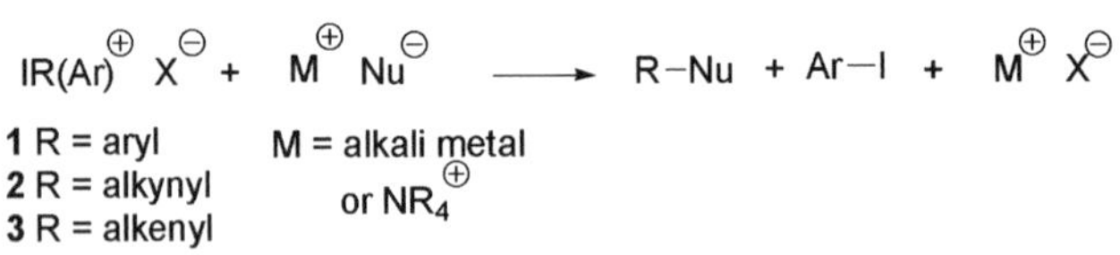

$$IR(Ar)^{\oplus} X^{\ominus} + M^{\oplus} Nu^{\ominus} \longrightarrow R{-}Nu + Ar{-}I + M^{\oplus} X^{\ominus}$$

1 R = aryl
2 R = alkynyl
3 R = alkenyl

M = alkali metal or $NR_4^{\oplus}$

SCHEME 5.

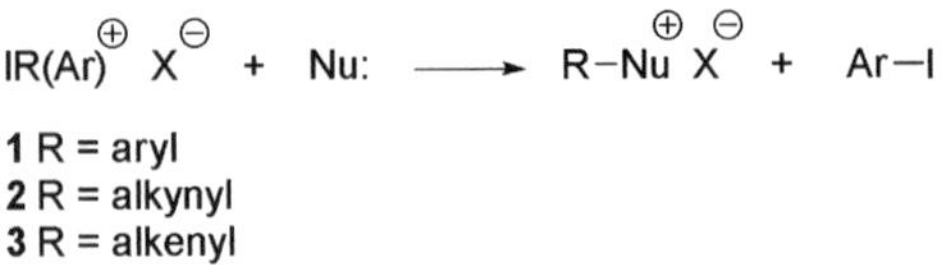

$$IR(Ar)^{\oplus} X^{\ominus} + Nu{:} \longrightarrow R{-}Nu^{\oplus} X^{\ominus} + Ar{-}I$$

1 R = aryl
2 R = alkynyl
3 R = alkenyl

SCHEME 6.

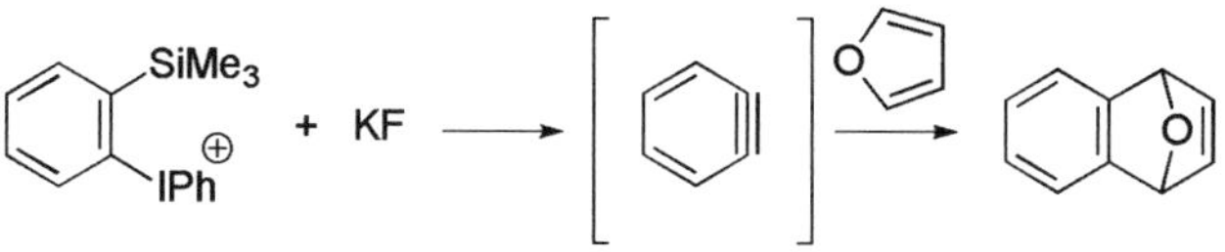

SCHEME 7.

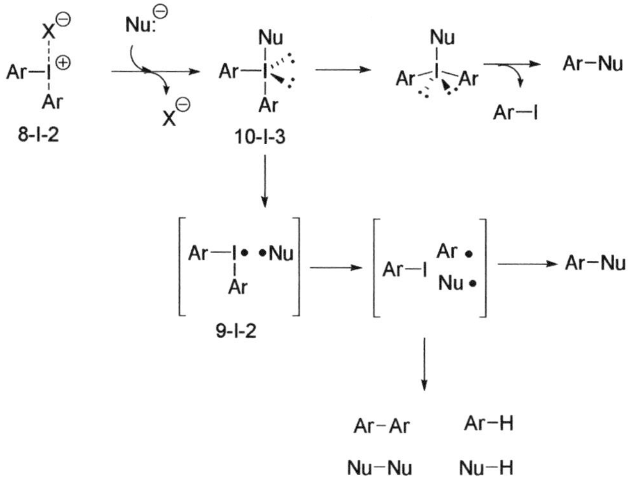

SCHEME 8.

that the reactions that proceed to give synthetically useful yields of arylated nucleophiles predominantly involve polar processes rather than free radical processes (Scheme 8).[2–4,6,7,23,24,61] Early work by Beringer and coworkers showed that radicals were generated during the arylation of nucleophiles by diaryliodine(III) reagents.[62] These observations led to suggestions that the arylation reaction proceeded *via* a radical coupling pathway.[62] Tanner elucidated the proposed radical coupling pathway in a study of the reaction of unsymmetrically substituted diaryliodine(III) reagents with sodium phenolate to give diarylethers.[63] The first step would involve electron transfer from the anionic nucleophile to the cationic iodine to form a nucleophile radical and a 9-I-2 intermediate or alternatively nucleophilic attack at iodine(III) to give a 10-I-3 intermediate that decomposes to the nucleophile radical and the 9-I-2 intermediate. The 9-I-2 radical would then decompose to an aryl radical and aryl iodide. The final step involves efficient coupling of the nucleophile radical and an aryl radical (Scheme 8).

Important studies by McEwan and coworkers[32,64] showed the existence of two processes, an induced radical-chain process involving formation of aromatic hydrocarbon (Ar• → ArH) and a non-radical process involving the formation of arylated nucleophile (Ar-Nu), at that time thought to be an aromatic S_NAr process. This work was supported by Barton and coworkers[65] who detected the presence of phenyl radicals during the reactions of diaryliodine(III) reagents with enolates. Thus, the use of a large excess of radical trapping agent diphenylethylene (DPE) resulted in an increased yield of the arylated nucleophile (Ar-Nu); however, a good recovery of the radical trap was observed. These results indicated that the arylation reaction did not involve radicals, but rather a non-radical process and that free radical chain processes were competing side reactions that led to decomposition of the iodine(III) starting materials.

For the non-radical process there are at least two possibilities, a nucleophilic aromatic *ipso* substitution (S_NAr) or a ligand coupling process on iodine(III). It is now widely understood that the non-radical process generally involves polar reactions (ligand coupling) of iodine(III) (Scheme 8).[61] In the general case of a negatively charged nucleophile ($Nu:^-$), the first step appears to involve attack of the nucleophile at iodine and displacement of the non-nucleophilic counterion (e.g. triflate or tetrafluoroborate) to form a λ^3-iodane species and salt. The λ^3-iodane species can then undergo a ligand coupling process[23,66] with concomitant reductive elimination of aryl iodide to give the arylated nucleophile and aryl iodide.

The polar (or ligand coupling) mechanism for nucleophilic substitution of diaryliodine(III) reagents has been studied in detail.[61] Grushin has proposed a unified mechanism, the first step of which is the formation of a 10-I-3 hypervalent iodine intermediate.[67] While the 10-I-3 species is present as a trigonal bipyramidal structure in the solid state, in solution the trigonal bipyramidal compounds undergo topological transformations such as pseudorotation to C_{4v} symmetric tetragonal pyramidal intermediates, which can then undergo an allowed synchronous cheletropic extrusion of the iodoarene (Scheme 8). This mechanism has been used to explain the observation that in polar reactions with diaryliodine(III) reagents the nucleophile prefers to bind to the more bulky ligand (*ortho* effect).[68–70] Binding to the more bulky ligand would result in a greater decrease in steric strain. The mechanism also explains why, when there is no significant steric difference between the aryl groups of **1**, the nucleophile will prefer to bind to the more electron-deficient aryl group. On doing so, this will induce a larger decrease of the positive charge on the iodine atom.

When at least one of the aryl groups is substituted with at least one strongly electron withdrawing group (e.g. nitro or cyano) at the *ortho* or *para* position then there is evidence that the S_NAr mechanism, involving the formation of Meisenheimer-type complexes, can become a competitive pathway (Scheme 9).[63]

The reaction of diaryliodine(III) reagents with neutral nucleophiles is less well understood but has been postulated by Finet to proceed *via* the formation of a

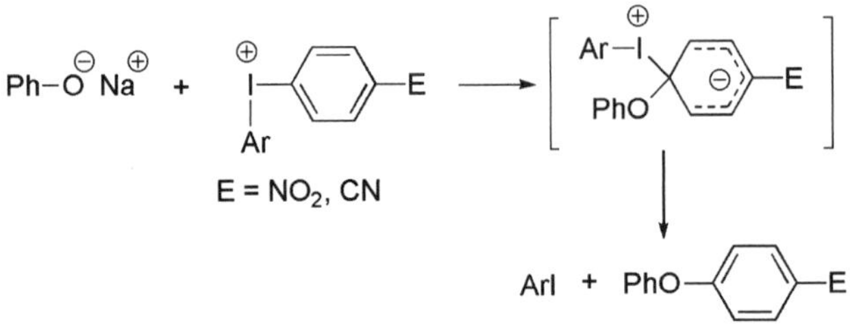

Ph–O⁻ Na⁺ + [Ar–I⁺–C₆H₄–E] → [Meisenheimer complex: Ar–I⁺, PhO, E] → ArI + PhO–C₆H₄–E

E = NO_2, CN

SCHEME 9.

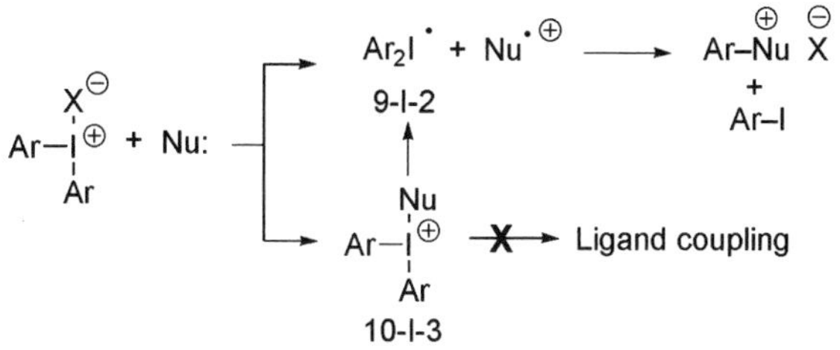

SCHEME 10.

positively charged 10-I-3 intermediate.[61] In this intermediate the positive charge is localised on the atom which is sharing its lone pair with the iodine atom. Electrostatic interactions between the unshared pair of electrons of the iodine atom and the positive charge results in an increase in the energy barrier to rotation. In such a case the intermediate then cannot reach the favourable transition state for ligand coupling and therefore follows the homolytic (radical) pathway (Scheme 10).

Diaryliodine(III) reagents **1** can effect arylation of a range of organometallic reagents and an increasing range of nucleophilic species catalysed by transition metals such as copper and palladium. Reagents **1** react with Grignard reagents,[25,71] aryl lithium reagents[25,71–73] and other organometallic reagents to give cross-coupled products in varying yields, e.g. biaryls from reactions with arylmetal reagents.

Acyltetracarbonylferrates react with diaryliodine(III) reagents **1** to afford arylketones.[74] Examples of cross couplings of **1** with organocopper reagents include arylation of alkenylcopper reagents[75] and the use of stoichiometric amount of CuCN to effect arylation of ketone lithium enolates.[76] Copper-catalysed arylations with **1** include CuI-catalysed coupling with terminal alkynes,[77] organoboranes and organostannanes and carbonylative cross-coupling with organoboranes and organostannanes.[78] Copper(I) or copper(II) salts catalyse the arylation of diarylsulphides and diphenylselenides[79,80] and the double arylation of thiophenolates in the presence of tertiary amines.[39] Recently, there has been a surge in interest in Pd(0)-catalysed cross-coupling reactions with diaryliodine(II) reagents **1** including Heck-type arylation of olefins,[81–83] arylation of allenes substituted with nucleophiles,[84] cross couplings with organoboron compounds,[85–87] organostannanes,[88,89] organofluorosilanes,[90] organolead compounds,[91] triarylbismuth(V) derivatives,[92] organozirconium compounds,[93] Grignard reagents in the presence of $ZnCl_2$,[94] alkynylcopper reagents,[95] terminal alkynes[96] and *O,O*-dialkylphosphites.[97] A range of Pd(0)-catalysed carbonylative cross coupling methods have been developed including those with alkali,[98] alcohols,[99] amidoximes,[100] benzotriazoles,[101] organoboron compounds,[102,103] organostannanes[89,104] and organofluorosilanes.[90] Whereas amine nucleophiles do not readily undergo arylation with reagents **1** under polar conditions, palladium catalysis has been shown to effectively catalyse arylation of secondary amines in the presence of strong bases.[105] An emerging strategy involves Pd(0) and Cu(I) co-catalysed cross coupling of reagents **1** with terminal alkynes,[106,107] and benzotriazole[108] as well as carbonylative cross coupling with terminal alkynes[109] and double carbonylative coupling with secondary amines.[110] Other transition metal species have been shown to catalyse cross

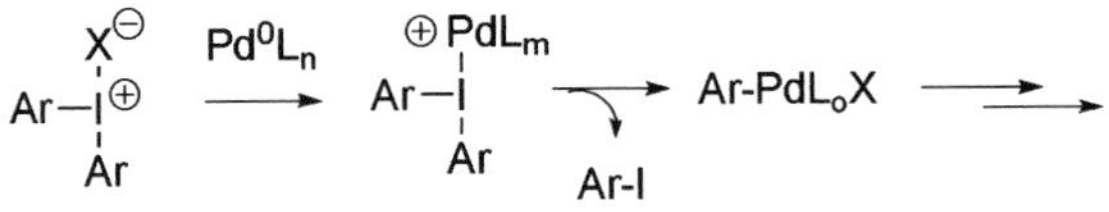

SCHEME 11.

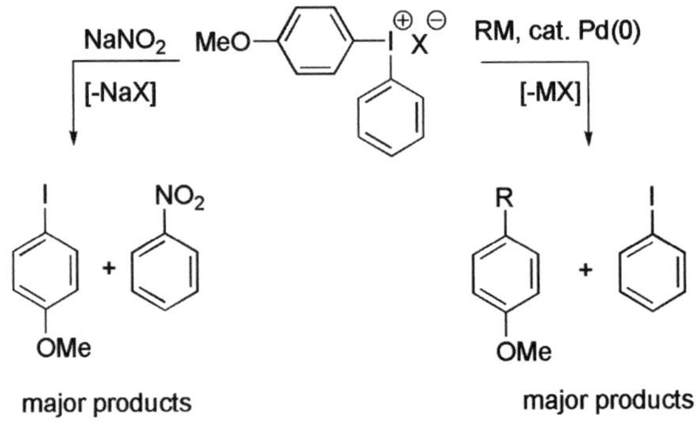

SCHEME 12.

coupling reactions with reagents **1**, for example cross coupling and carbonylative cross couplings with $MnCl_2 \cdot 4H_2O$[111] and $Ni(acac)_2$.[112]

While little mechanistic information is available it has been proposed that the first step in the palladium catalysed cross-couplings involves oxidative addition of the Pd(0) species to give a cationic palladium(II)-iodane species (Scheme 11).[107] Reductive elimination of iodoarene *via* either a polar or radical process would then give a Pd(II) species similar to that postulated for Pd(0)-catalysed cross-couplings using aryl halides, aryl triflates, etc.[113] In these processes the coupling of organic ligands occurs on palladium rather than iodine and this could account for the differing selectivity that is observed between catalysed and uncatalysed couplings with reagents **1**. For instance reaction of *p*-methoxyphenyl(phenyl)iodine(III) reagent **10** with sodium nitrite gives nitrobenzene as the major coupled product *via* selective coupling of the phenyl group on the iodane intermediate.[40] In contrast, a range of Pd(0)-catalysed couplings lead to the *p*-methoxyphenylated product *via* selective coupling on Pd(II) (Scheme 12).[86,91,102]

Very recently, diaryliodine(III) reagents have been used in palladium(II) catalysed arylation where the involvement of Pd(IV) intermediates is implicated, e.g. *ortho*-arylation proceeding *via ortho*-palladation,[114] and 2-arylation of indoles proceeding *via* direct palladation,[115] followed by oxidation to form **11** and **12** respectively, and reductive elimination (Scheme 13).

B. *Alkynyl(aryl)iodine(III) Reagents*

The chemistry of alkynyl(aryl)iodine(III) reagents **2** has been largely developed over the last 20 years and is the subject of a number of reviews.[6,7,22,23,116–118] The

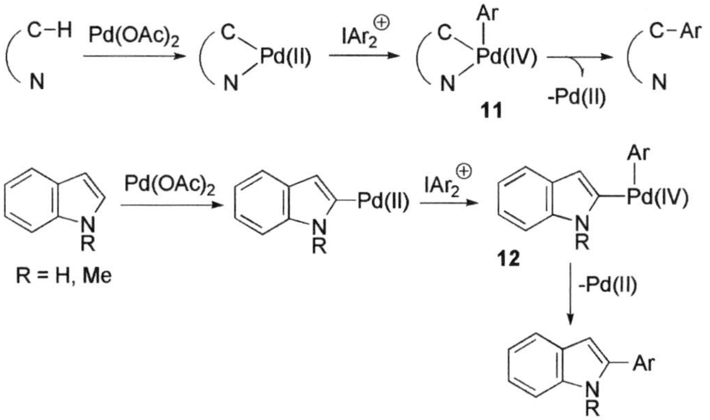

SCHEME 13.

reaction of reagents **2** with anionic nucleophiles often results in overall coupling of the alkynyl group and the nucleophile (Scheme 5). Typically, soft or stabilised nucleophiles readily undergo alkynylation reactions with **2** including carbon centred nucleophiles such as anions of *β*-dicarbonyl compounds,[119–122] rhenium diynyls;[123] nitrogen nucleophiles such as anions of diphenylamine,[124] benzotriazole,[125] *N*-alkylsulphonamides;[126] oxygen nucleophiles such as phenolates,[127] carboxylates[128,129] and diorganophosphates;[130] phosphorus nucleophiles such as dialkylphosphonates;[131] sulphur nucleophiles such as thiophenolate,[127] thiocyanate,[122,129] arylsulphinates,[132,133] arylthiosulphonates[134] and *O*,*O*-dialkylphosphorodithioates;[135] arylselenolates[136,137] and aryltellurolates.[136,138] A few examples of alkynylation of neutral nucleophiles exist (Scheme 6), for example alkynylation of triphenylphosphine,[127,139,140] tetraphosphacubane,[141] trialkylphosphites (to afford alkynylphosphonates by an Arbuzov type process)[142] and triphenylarsine.[143]

The alkynylation of anionic nucleophiles by reagents **2** has been described as an umpulong process with reagent **2** serving as electrophilic synthetic equivalent of acetylene and the mechanism of this transformation is now well understood.[118] Data accumulated for the alkynylation of nucleophiles indicate that a polar mechanism operates but that this process is generally different from that which operates in arylation reactions with diaryliodine(III) reagents **1**. Thus, through ^{13}C-labelling and other experiments, Ochiai, Stang and coworkers were able to show that the anionic nucleophile undergoes a conjugate addition to the *β*-carbon of the acetylene to give an alkylidenecarbene ("vinylidene") iodine(III) ylide (Scheme 14).[144,145] The ylide can be protonated to give an alkenyliodonium salt, as shown,[145,146] or can undergo reductive elimination of iodoarene to afford a free alkylidene carbene. The alkylidene carbene intermediate can undergo a wide range of chemistry, including cyclisation as shown, which has been reviewed.[118] A 1,2-rearrangment of the alkylidene carbene results in migration of either the nucleophilic group or the substituent (-R) to give the acetylene.

The processes involved in alkynylation of neutral nucleophiles by iodine(III) reagents to give onium salts are not fully understood and a range of mechanisms have been proposed depending on the iodine(III) counterion and the type of

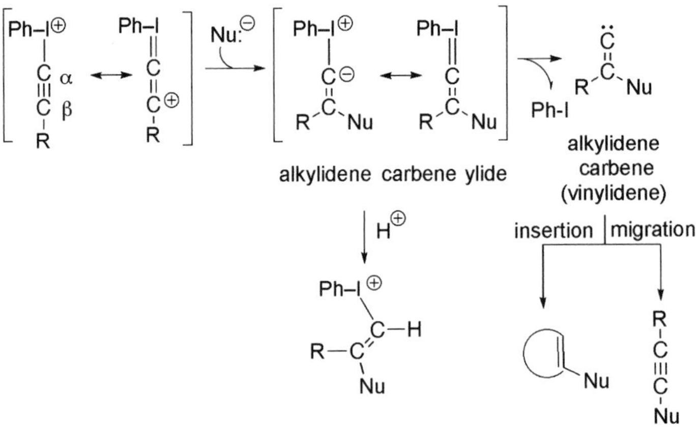

SCHEME 14.

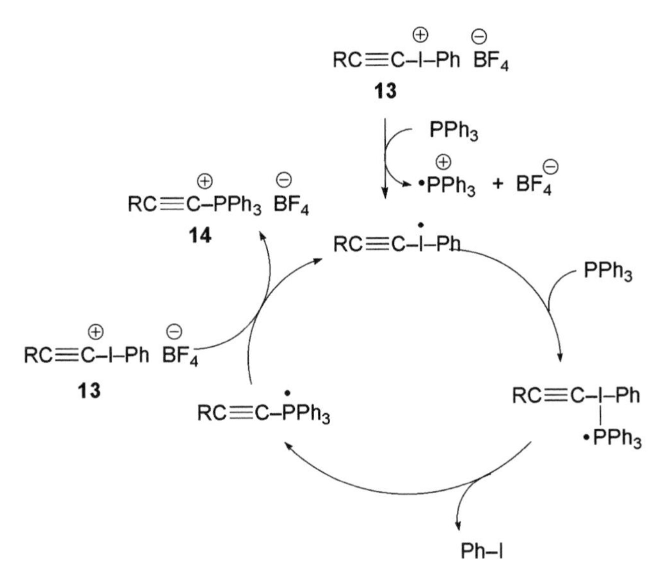

SCHEME 15.

nucleophile. For example, Ochiai and coworkers reported that a range of β-alkylalkynyl(phenyl)iodine(III) tetrafluoroborates **13** react with triphenylphosphine in sunlight at $-78°C$ to afford high yields of alkynylphosphonium salts **14** (Scheme 15).[139] The observations, that the reaction required sunlight and did not proceed in the dark, were used to support a radical chain mechanism (Scheme 15).[139] In contrast, Stang and coworkers found that alkynylation of triphenylphosphine with alkynyl(phenyl)iodonium triflates **15** also gave high yields of alkynylphosphonium salts **16**; however, the latter reactions did not require sunlight, proceeded readily in the dark and were not inhibited by radical traps such as O_2 or 2,6-di-*tert*-butyl-4-methylphenol (BHT).[140] For the triflate salts, a standard

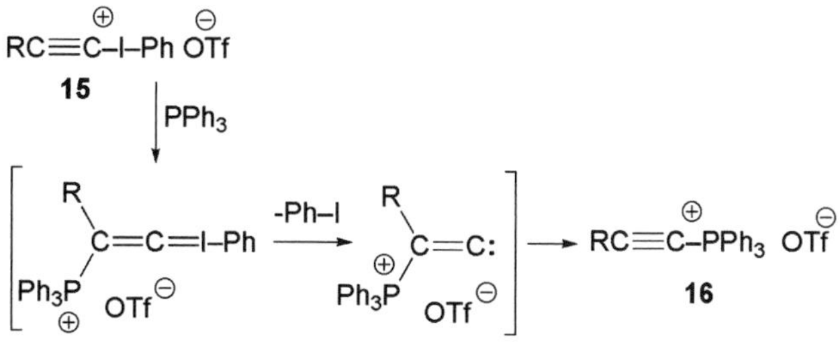

$RC{\equiv}C\text{–}I^{+}\text{–}Ph\ OTf^{-}$ **15**

PPh_3

$[\,R(Ph_3P^{+})C{=}C{=}I\text{–}Ph\ OTf^{-} \xrightarrow{-Ph\text{–}I} R(Ph_3P^{+})C{=}C{:}\ OTf^{-}\,] \rightarrow RC{\equiv}C\text{–}PPh_3^{+}\ OTf^{-}$ **16**

SCHEME 16.

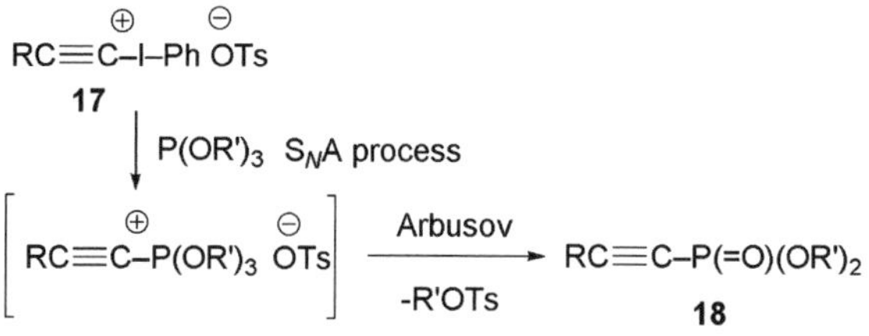

$RC{\equiv}C\text{–}I^{+}\text{–}Ph\ OTs^{-}$ **17**

$P(OR')_3$ S_NA process

$[\,RC{\equiv}C\text{–}P^{+}(OR')_3\ OTs^{-}\,] \xrightarrow[-R'OTs]{Arbusov} RC{\equiv}C\text{–}P({=}O)(OR')_2$ **18**

SCHEME 17.

$R\overset{*}{C}{\equiv}C\text{–}I^{+}\text{–}Ph\ BF_4^{-}$ **19** $*{:}\,^{13}C$ $\xrightarrow{AsPh_3}$ $[\,R\overset{*}{C}{\equiv}C\text{–}I(\text{–}Ph)\text{–}As^{+}Ph_3\ BF_4^{-}\,]$ **21** $\xrightarrow{-Ph\text{–}I}$ $R\overset{*}{C}{\equiv}C\text{–}As^{+}Ph_3\ BF_4^{-}$ **20**

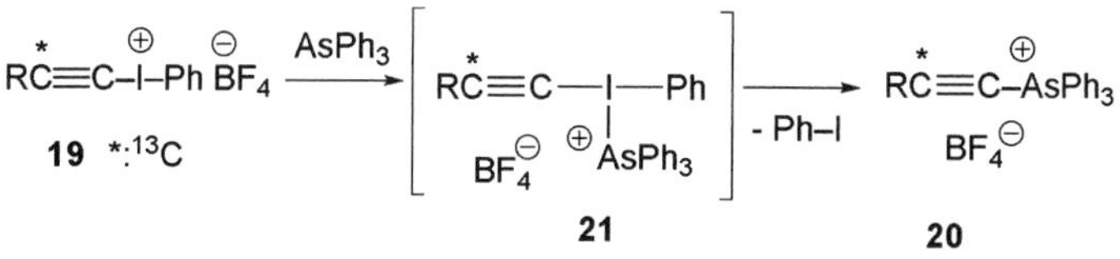

SCHEME 18.

nucleophilic acetylenic substitution (S_NA) process involving iodonium ylide and alkylidene carbene intermediates was invoked (Scheme 16).[140] In a similar manner, alkynylation of trialkylphosphites with alkynyl(phenyl)iodine(III) tosylates **17** was proposed to proceed *via* a S_NA process with a concomitant Arbusov-type reaction to give the observed alkynylphosphonates **18** (Scheme 17).[142] Ochiai and coworkers have studied the reaction of triphenylarsine with ^{13}C-labelled alkynyl(phenyl) iodine(III) tetrafluoroborate **19** and shown that it proceeds to give the corresponding alkynylarsonium salt **20** with retention of the ^{13}C label at the formerly β-acetylenic carbon (Scheme 18).[143] This result was interpreted as more likely due to a ligand coupling pathway involving formation of an arsonium salt **21** rather than an S_NAr process.

The organometallic chemistry of alkynyl(aryl)iodine(III) reagents is not nearly as well explored as that of the diaryl reagents. Notwithstanding, organocopper reagents such a vinylcopper compounds,[147,148] alkynylcopper compounds[149] and cubyl cuprate[150] undergo coupling reactions with alkynyliodine(III) reagents to give alkynylated products. Reactions of reagents (X = sulphonate) with catalytic amounts of CuOTf, AgOTf, AgOTs or $(Ph_3P)_4Pd$ leads to the formation of

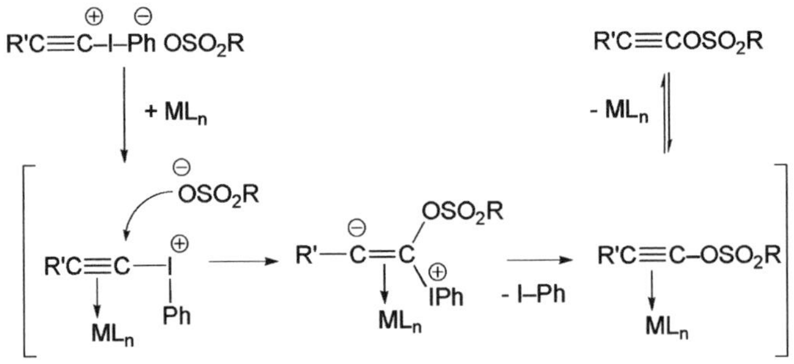

R'C≡C–I–Ph OSO$_2$R
+ ML$_n$
R'C≡COSO$_2$R
- ML$_n$
OSO$_2$R
R'C≡C—I
ML$_n$
Ph
R'—C=C
OSO$_2$R
IPh
ML$_n$
- I–Ph
R'C≡C–OSO$_2$R
ML$_n$

SCHEME 19.

alkynylsulphonate esters.[129,151] Examples of palladium-catalysed coupling reactions include Heck-type alkynylation of olefins,[82] and coupling reactions with vinylzirconium reagent,[152] organoboron compounds[87] and terminal alkynes.[96] Examples of palladium-catalysed carbonylative cross couplings include with alcohols[153] and organoboron compounds.[102] Copper-catalysed carbonylative cross coupling of reagents **2** with vinylzirconium reagents has been reported.[154]

Little is known about the mechanism of organometallic coupling processes with **2**. For reactions with organocopper reagents it has been assumed that the reactions proceed through oxidative addition of the alkynyl species **2** to give a Cu(III) intermediate, followed by reductive elimination and ligand coupling.[147] For the CuOTf-catalysed formation of alkynylsulphonates (mesylates and *p*-toluenesulphonates) the studies by Stang and coworkers favoured a "metal-assisted nucleophilic acetylenic displacement *via* an addition-elimination process" (Scheme 19).[151]

C. *Alkenyl(aryl)iodine(III) Reagents*

The chemistry of alkenyl(aryl)iodine(III) reagents is the subject of a number of reviews.[6,7,15,23,24,155] These reagents undergo reactions with a broad range of anionic nucleophiles generally resulting in coupling of the alkenyl group and the nucleophile (Scheme 5). Examples of carbon nucleophiles include enolates of β-dicarbonyl compounds[18,156,157] and cyanide;[158] limited types of nitrogen nucleophiles such as azide[159] and nitrite;[160] limited types of oxygen nucleophiles such as alkoxides[159,161] carboxylates,[162] and tosylate;[158] a broad range of sulphur nucleophiles such as thiophenolates,[156] arylsulphinates,[158,163,164] dithiocarbamates[165] and carbonotrithioates,[166] selenium nucleophiles such as sodium selenolates[167] and *O*,*O*-dialkylphosphoroselenoates;[168] sodium tellurolates[169] and halides.[158,170]

Neutral nucleophiles are also alkenylated to give alkenylonium salts (Scheme 6). Examples include secondary amines (with concomitant deprotonation to afford enamines)[158] and tertiary amines,[171,172] acetic acid (with deprotonation to give alkenylacetates);[173] formamides (that give alkenylformates),[174] sulphur-centred nucleophiles such as sulphides,[171,172] thioamides (to give thioalkenes), thioureas and heterocyclic thiols;[175] group 15 nucleophiles triphenylphosphine, -arsine

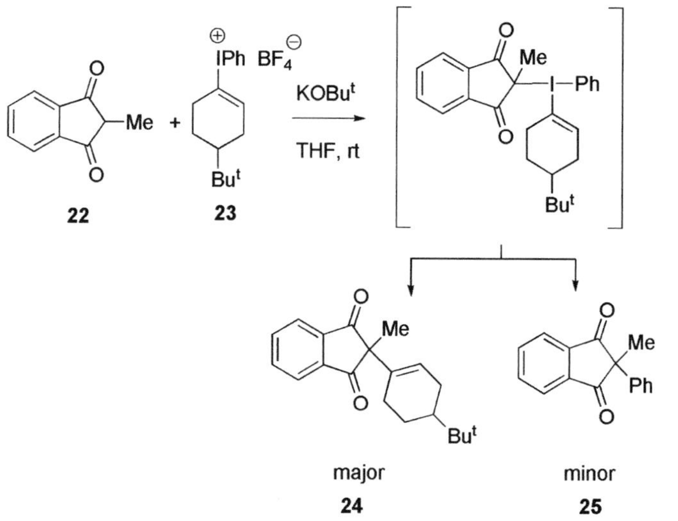

SCHEME 20.

and -stibine[159,172] as well as group 16 nucleophiles diphenylsulphide, -selenide and -telluride.[172]

A variety of mechanisms have been show to operate in alkenylation of nucleophiles including ligand coupling, addition-elimination and elimination-addition reactions for anionic nucleophiles and vinylic S_N2, vinylic S_N1 and elimination-addition reactions for neutral nucleophiles. The mechanisms so far elucidated have been extensively reviewed by Ochiai[15] and Okuyama[176] and herein we provide a summary of these processes.

The ligand coupling mechanism has been invoked for a number of nucleophilic substitution reactions of alkenyl(aryl)iodine(III) reagents **3**.[15] The reactions are characterised by complete retention of stereochemistry in the vinylic substitution and in some instances products are obtained that are derived from competitive aromatic substitution pathway. For example, the enolate ion derived from **22** reacts with alkenyliodane species **23** to give mainly the vinyl coupled product **24** with a minor amount of aryl coupled product **25** (Scheme 20).[177] As would be predicted from a ligand coupling process, the amount of aryl coupled product can be reduced by using an aryl group substituted with an electron-donating group such as a *p*-methoxy substituent. The aryl radical trap 1,1-diphenylethylene inhibits radical induced decomposition pathways and improves the yield of the alkenylation product.

Addition-elimination processes are thought to occur in the nucleophilic substitution of alkenyliodine(III) species containing extra electron-withdrawing group on the alkene.[15] For example, in the case of a (*Z*)-β-sulphonylalkenyliodine(III) reagent **26** α-vinylic substitution with tetrabutylammonium halides occurs with retention of stereochemistry to give alkenyl halides **27** (Scheme 21).[178] The proposed mechanism for this process involves β-addition of the halide to give the α-sulphonyl stabilised carbanion which undergoes a 60° rotation followed by elimination of

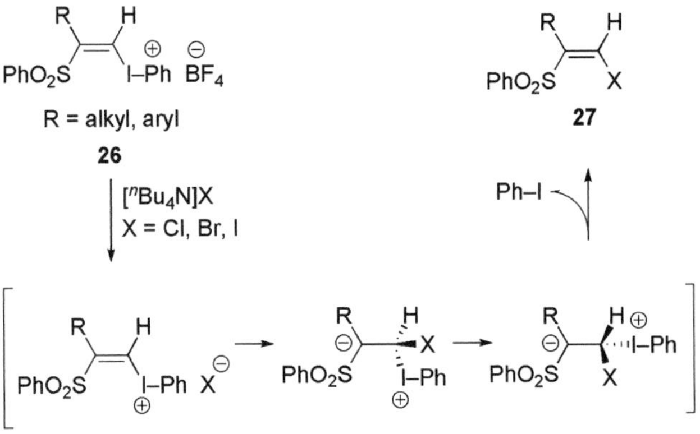

Scheme 21.

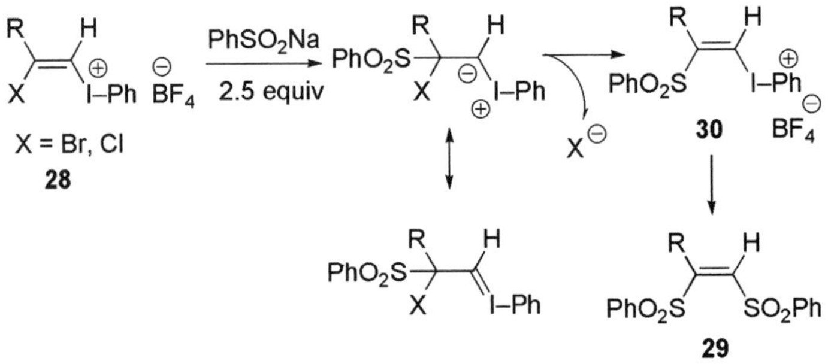

Scheme 22.

iodobenzene. A β-addition-elimination process followed by an α-addition-elimination process have been proposed for the conversion of the (Z)-β-haloalkenyliodine(III) species **28** into a (Z)-1,2-disulphone **29** *via* the isolated intermediate (Z)-β-sulphonylalkenyliodine(III) species **30** (Scheme 22).[163]

An elimination-addition mechanism has also been invoked for the nucleophilic substitution of cyclohexenyliodonium salts with acetate ion.[179] For example, the reaction of either the 4- or 5-substituted cyclohex-1-enyl(phenyl)iodonium tetrafluoroborate (**31** or **32** respectively) with tetrabutylammonium acetate in aprotic solvents gives the *ipso* and *cine* acetate substitution products (**33** and **34** and vice-versa, respectively) in almost the same ratio (Scheme 23). These results were consistent with an elimination-addition mechanism involving 4-substituted cyclohexyne **35** as a common intermediate. The presence of a cyclohexyne intermediate was confirmed by deuterium labelling and trapping studies leading to [4+2]-cycloaddition products.

The vinylic S_N2 reaction of reagents **3** has been proposed to proceed through reversible ligand exchange followed by an in-plane bimolecular substitution process (Scheme 24).[15] These reactions are characterised by exclusive inversion of configuration and can be complicated by elimination side reactions to form alkynes. The ratio of S_N2 substitution versus elimination depends on the nature of the

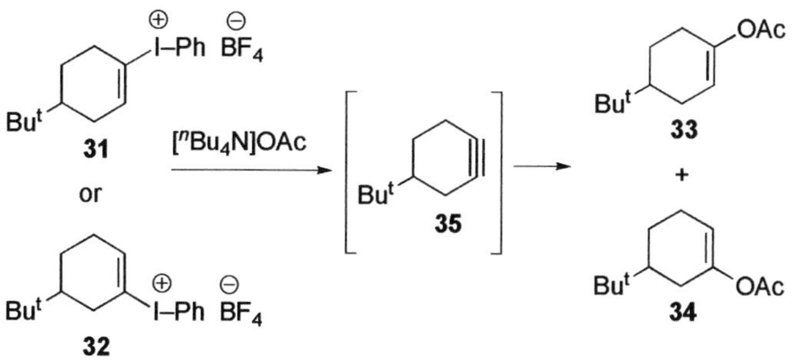

SCHEME 23.

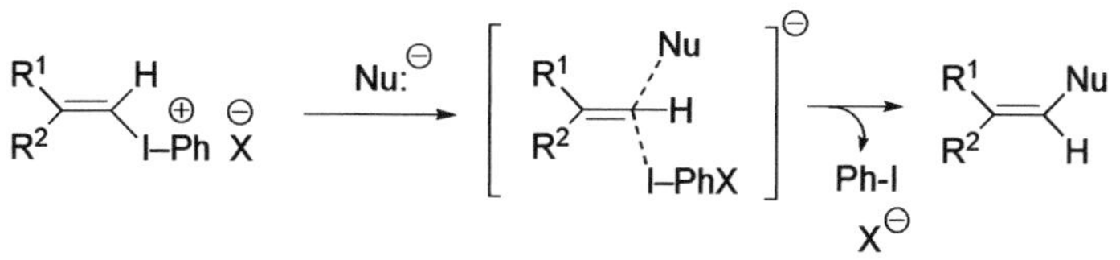

SCHEME 24.

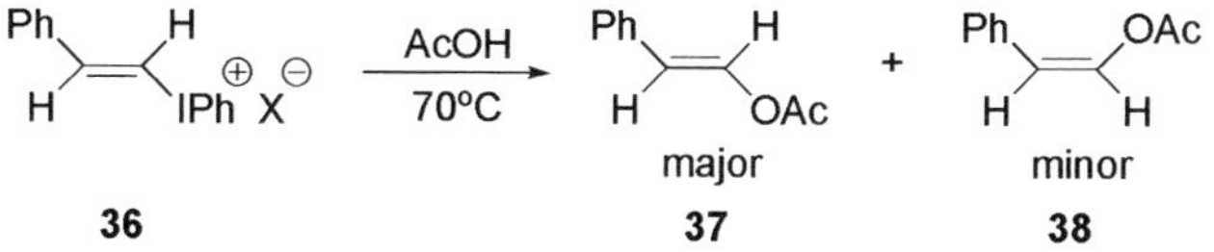

SCHEME 25.

substituents on the alkenyliodine(III) reagent, the nucleophile and the solvent. This reaction process is typically observed for a range of weakly or non-basic nucleophiles including sulphides, selenides, carboxylic acids, phosphoroselenoates, methanol and dialkylformamides (which give vinyl formates after hydrolysis of the initially formed Vilsmeier–Haack salts). More basic nucleophiles tend to result in α-elimination of the vinyliodane to generate alkylidene carbenes which can undergo a range of other reactions as discussed earlier.

In the case of (*E*)-phenylvinyl(aryl)iodine(III) tetrafluoroborates **36**, nucleophilic substitution can occur *via* a at least two pathways. The reaction of **36** with acetic acid proceeds to give a mixture of (*E*)- and (*Z*)- acetates **37** and **38** with the retention of configuration product **37** being the major product (Scheme 25).[15,173,180] For the (*Z*)- product **37**, deuterium labelling experiments showed complete scrambling between the olefinic protons whereas for the (*E*)- product **38** deuterium remained at the original position on the olefin. This was rationalised in terms of two different substitution pathways: an S_N1-type with a vinylenebenzenium ion intermediate **39** leading to retention of configuration (Scheme 26) and a vinylic S_N2-type with direct attack by the nucleophile leading to inversion (as in Scheme 24). In support of this hypothesis, exclusive retention of configuration was observed for the more polar but less nucleophilic solvent 2,2,2-trifluoroethanol.

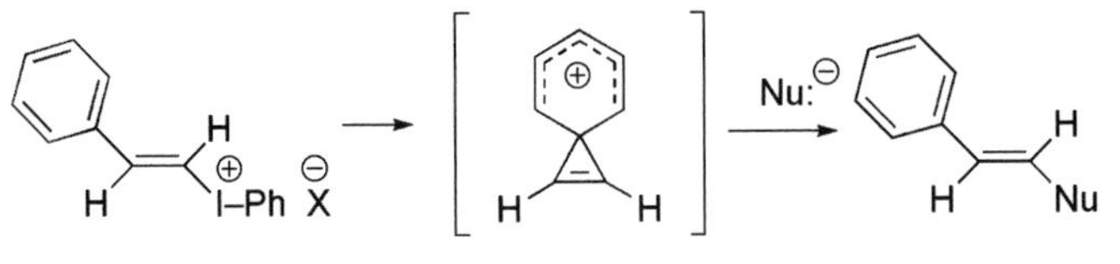

SCHEME 26.

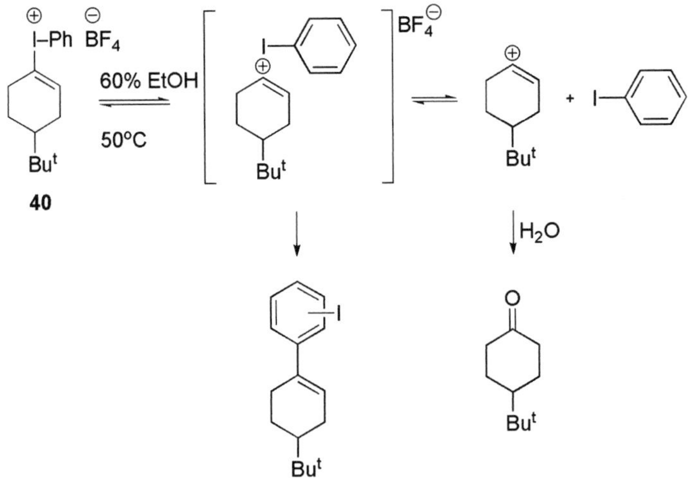

SCHEME 27.

There are a number of examples where vinylic S_N1 processes involving vinyl cation intermediates have been proposed to operate.[15,176] For example, the solvolysis of cyclohexenyl(aryl)iodine(III) species **40** in DMSO to give a mixture of cyclohexanone **41**, iodobenzene and iodophenylcyclohexenes **42** with the formation of cyclohexanone **41** used to support evidence of the cyclohexenyl cation intermediate (Scheme 27).[181]

An elimination-addition process has been invoked for the reaction of neutral Group 15 and 16 nucleophiles with the alkenyliodine(III) species to give onium salts (Scheme 28).[15,172] Thus, deprotonation of the acidic α-proton of **43** with a base gives an iodonium ylide **44** which undergoes reductive loss of iodobenzene to afford an alkylidene carbene **45** which is then trapped with the Group 15 or 16 nucleophile with concomitant protonation to give the cationic product **46**. Hindered tertiary amine bases such as diisopropylethylamine were found to be suitable catalysts as they are not strong nucleophiles. In the cases of unsymmetrical alkenes the reaction is not stereospecific and gives a mixture of stereoisomers. Interestingly, studies have shown that subjection of either the (*E*)- or (*Z*)-stereoisomer to these reaction conditions gives a similar ratio of products. The high degree of stereoconvergence suggests that the reaction proceeds by a free alkylidene carbene **45**. The reaction

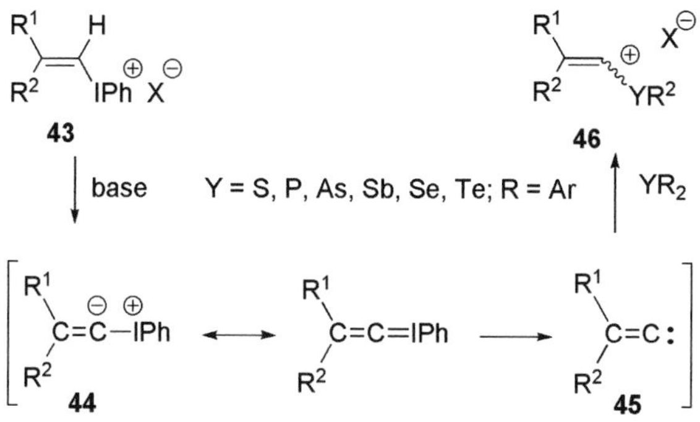

SCHEME 28.

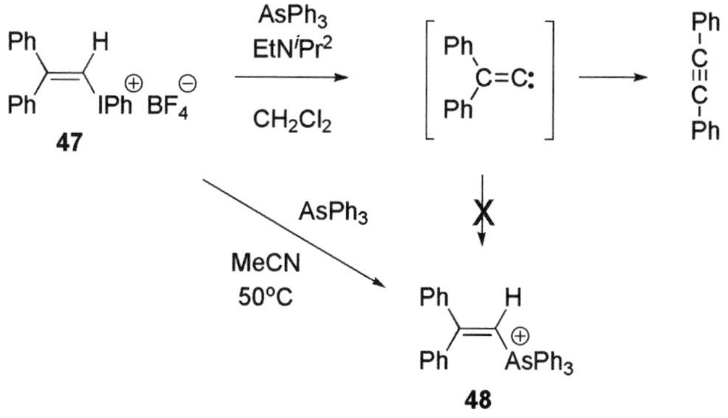

SCHEME 29.

pathway is quite dependent on the nature of the substituents on the alkene. For instance, with a β,β-diphenylsubstituted alkene **47** and triphenylarsine, the base-catalysed process only gives diphenylacetylene as a result of rearrangement of the intermediate alkylidene carbene, whereas in the absence of base the arsonium salt **48** was formed (Scheme 29).[172] These results were accounted for by either an in-plane vinylic S_N2 substitution mechanism, an ionic mechanism involving a vinylbenzenium ion intermediate or a ligand coupling mechanism on hypervalent iodine(III).

Intramolecular Friedel–Crafts vinylation reactions of aryl-tethered (*E*)-alkenyl-(aryl)iodanes **49** to afford 1,2-dihydronaphthalenes and 2*H*-chromenes **50** have also been reported to occur *via* an addition-elimination process (Scheme 30) or alternatively an intramolecular S_N2 displacement at the vinylic centre.[182] The intramolecular displacement was deemed energetically unstable on steric grounds but could not be ruled out. In these cases the presence of a vinylic cation was ruled out as the ionising power of the solvent has little effect on the cyclisation and the corresponding (*Z*)-isomers do not undergo ring closure.

Like the diaryliodine(III) reagents **1**, in recent times there has been a significant increase in the range of transition metal-assisted or -catalysed coupling of

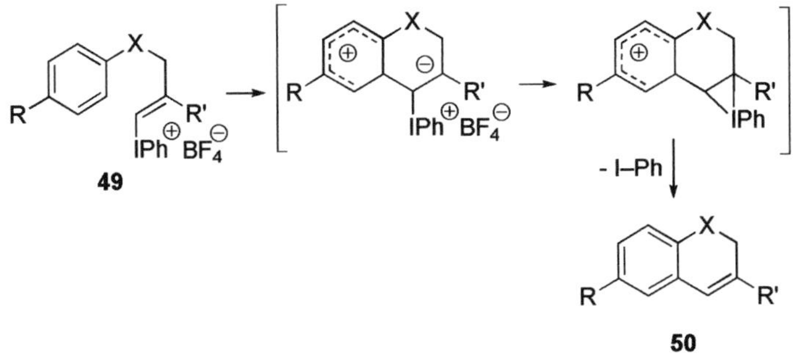

SCHEME 30.

alkenyl(aryl)iodine(III) reagents **3** with nucleophiles to afford alkenylated nucleophiles. Examples of cross coupling of **3** with organocopper reagents include lithium diorganocuprates,[156,183] and organo(cyano)cuprates.[184–186] Recently, cross coupling of reagents **3** with benzylic organozinc reagents was reported.[187] A range of copper-catalysed cross coupling reactions have been reported including cross coupling with organoboron compounds,[78] Grignard reagents,[188] terminal alkynes,[77] *O,O*-dialkylphosphorothioates[189] and *O,O*-dialkylphosphorodithioates,[190] cyanide,[162] nitrite[156,158] and halides.[156,159,162] A wide range of palladium-catalysed coupling reactions of reagents **3** have been reported including Heck-type alkenylation of olefins,[82,84,96,183,191,192] and cross coupling with organostannanes,[89,193–195] organoboron compounds,[78,87] organolead compounds,[91] organofluorosilanes[90] and terminal alkynes.[96] A range of palladium-catalysed carbonylative cross-couplings have been reported including couplings with alcohols,[99,156] amidoximes[100] and organoboron compounds.[102,103] Examples of palladium and copper co-catalysed cross couplings include couplings with terminal alkynes,[196] carbonylative coupling with terminal alkynes[109] and couplings with organostannanes[195] and cross coupling of alkenyl-1,2-diiodine(III) species with alkynylstannanes to give ene-diynes.[197] Other transition metal catalysed reactions of **3** include $MnCl_2 \cdot 4H_2O$ catalysed cross couplings and carbonylative cross couplings with organostannanes[111] and $Ni(acac)_2$ catalysed carbonylative cross coupling with organostannanes.[112]

The effect of copper on nucleophilic substitution reaction can be quite dramatic. Nucleophilic substitution of (*E*)-(*β*-alkylvinyl)phenyliodanes **51** with halide ions in the absence of copper were shown to proceed with inversion of configuration by a S_N2 process (Scheme 31).[170] In contrast, nucleophilic substitution of these species with halides in the presence of cuprous halides gave products with exclusive retention of configuration (Scheme 32).[170] The reactions of alkenyl(aryl)iodine(III) reagents with alkyllithium reagents were found to give a complex mixture of products, whereas reactions with diaryl- and dialkyl-lithium cuprates give vinylic substitution products in a stereospecific manner with exclusive retention of configuration.[156] The mechanism that has been proposed for substitutions involving copper starts with ligand exchange on iodine(III) to give a λ^3-iodane **52** with the less electronegative copper group occupying the equatorial position. Ligand coupling on iodine(III) then gives the alkenylcopper(III) species **53** which then

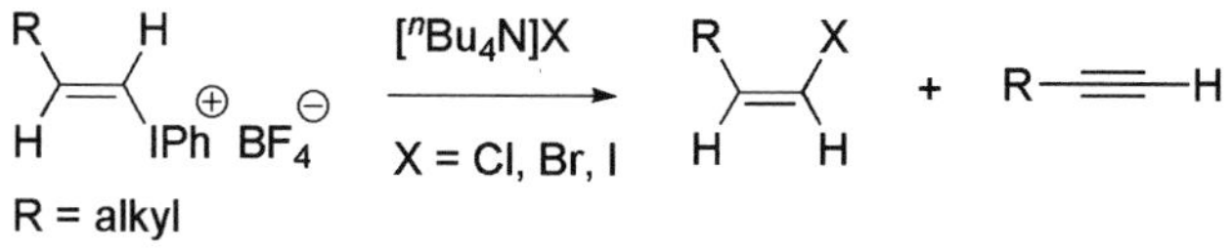

SCHEME 31.

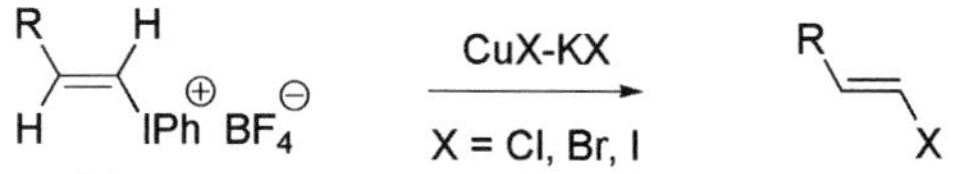

SCHEME 32.

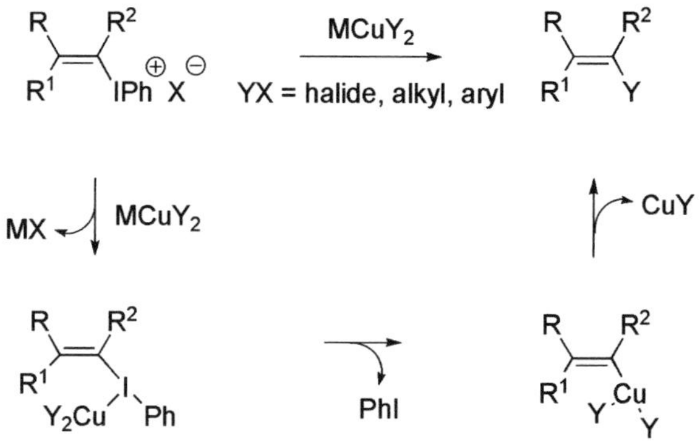

SCHEME 33.

undergoes ligand coupling on copper to give the observed substituted product (Scheme 33).[156]

Palladium-catalysed coupling of alkenyl(phenyl)iodine(III) reagents **3** with nucleophiles occur with retention of configuration. For example, the palladium-catalysed coupling of **3** with olefins gives the corresponding dienes **54** with retention of configuration of the alkenyl species.[191] The general mechanism proposed (Scheme 34) involves reaction of the highly electrophilic iodine(III) species with the catalytic species Pd(0) (formed by reduction of Pd(II) under the reaction conditions) followed immediately by reductive elimination of iodobenzene and ligand coupling to yield an alkenyl palladium intermediate **55**. Addition of the intermediate to the olefin followed by reductive elimination would then lead to the coupled product. An alternative initial step was proposed involving transmetallation between the alkenyliodine(III) species **3** and the palladium(II) catalyst to yield an alkenyl palladium intermediate similar to **55**.

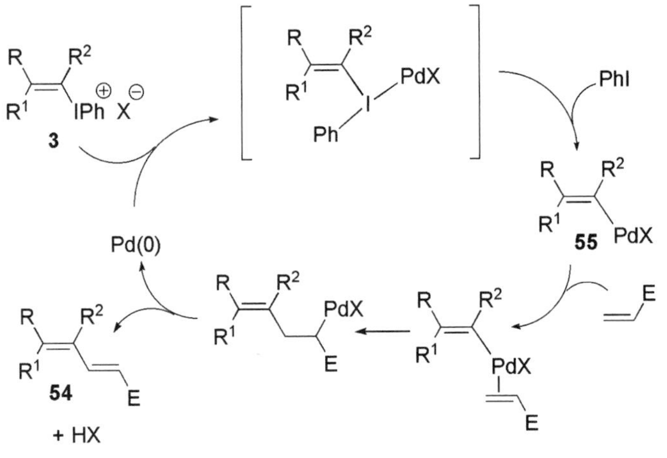

SCHEME 34.

III
ORGANOMETALLIC CHEMISTRY

A. *Diaryliodine(III) Reagents*

The reactivity of several metal carbonyl species with diaryliodonium reagents has been studied.[13,14,198–200] In an early study $[IPh_2]X$ (X = I, BF_4) was found to react in an apparently straightforward manner with the indenyl complex ion $[W(\eta^5\text{-}C_9H_7)(CO)_3]^-$ to form $WPh(\eta^5\text{-}C_9H_7)(CO)_3$, but reactions with the chromium and molybdenum analogues were unsuccessful.[199] The η^5-cyclopentadienyl complex $WPh(Cp)(CO)_3$ has been prepared from $[WCp(CO)_3]^-$,[198] and $[IPh_2]F$ reacts with $[FeCp(CO)_2]^-$ to form $FePh(Cp)(CO)_2$, but in this case some biphenyl is formed, consistent with radical formation during the reaction.[14] Reaction of the tetrafluoroborate salt $[IPh_2]BF_4$ with $[FeCp(CO)_2]^-$ gives both biphenyl and the dimer $[FeCp(CO)_2]_2$.[14] The radical process can be assumed to involve $[FeCp(CO)_2]\bullet$ and $IPh_2\bullet$, in the manner discussed above for $1e^-$ reduction of diaryliodine(III) cations in organic reactions (Scheme 8), fragmentation of $IPh_2\bullet$ to form PhI and $Ph\bullet$, and coupling of radical species. In a similar process, addition of $[IPh_2][PF_6]$ to $[PPN][Mn(CO)_5]$ (PPN = bis-triphenylphosphoranylideneammonium) results in complete conversion to $Mn_2(CO)_{10}$ and $MnPh(CO)_5$.[200]

The coordination complex $Fe^{II}(TPP)$ (TPP = tetraphenylporphyrinate dianion) undergoes a $1e^-$ oxidation by $[IPh_2]X$ (X = Cl, I) or 4-chlorophenyl(2-thienyl)-iodine(III) chloride to form $Fe^{III}(TPP)X$ but, in the presence of an excess of iron(0) as a reducing agent, the phenyl- or 4-chlorophenyliron(III) complexes $Fe^{III}(TPP)Ar$ are obtained.[201] Evidence was also obtained for transfer to hepatic cytochromes P-450 in the presence of NADPH or sodium dithionite. Radical processes occurring in these reactions may be represented in general terms by Scheme 35, showing oxidation in the first sequence, and the role of the reducing agent in the second

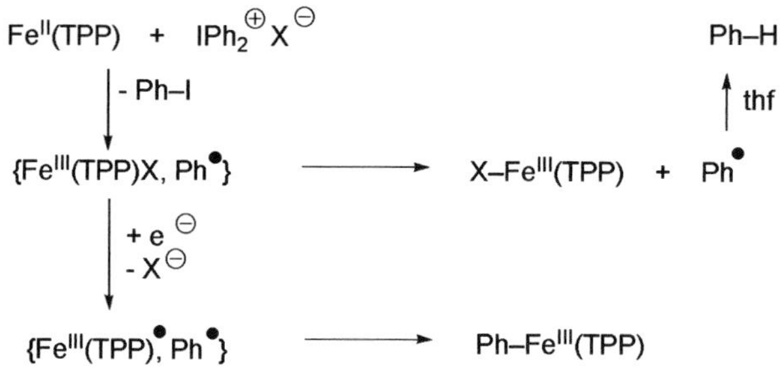

$Fe^{II}(TPP) + IPh_2^{\oplus}X^{\ominus}$ → (− Ph–I) → $\{Fe^{III}(TPP)X, Ph^{\bullet}\}$ → $X–Fe^{III}(TPP) + Ph^{\bullet}$ → (thf) → Ph–H

$\{Fe^{III}(TPP)X, Ph^{\bullet}\}$ → ($+ e^{\ominus}$, $- X^{\ominus}$) → $\{Fe^{III}(TPP)^{\bullet}, Ph^{\bullet}\}$ → $Ph–Fe^{III}(TPP)$

SCHEME 35.

sequence. In a similar procedure, but with excess $[IPh_2]X$, a product for which a nitrogen atom is phenylated was also obtained, Fe^{II}(N-Ph-TPP)X.

Diaryliodonium(III) triflates undergo complex radical reactions with low oxidation state ytterbium and samarium reagents.[202] Thus, IPh_2^+ reacts with metallic ytterbium, but not samarium, to form benzene in tetrahydrofuran, and iodobenzene is formed together with benzene in the reaction with YbI_2 or SmI_2. The presence of SiH_2MePh leads to formation of $SiHMePh_2$, and it is proposed that this results from trapping of undetected "MIPh", formed from the metal and PhI after decomposition of $IPh_2\bullet$ to PhI and Ph• (following transfer of an electron from Yb or YbI_2 to IPh_2^+), and that benzene is formed on interaction of Ph• with tetrahydrofuran. In reactions of unsymmetrical reagents $I(Ar)Ph^+$, the least electronegative group is transferred to silicon, indicating fragmentation to form the radical of the least electronegative group, as discussed in Section II.A.

Square-planar d^8 diorganoplatinum(II) and palladium(II) complexes, with a 2,2′-bipyridine (bpy) or 4,4′-di(*tert*-butyl)-2,2′-bipyridine (Bu^t_2bpy) ligand environment known to facilitate nucleophilic behaviour in reactions involving $2e^-$ oxidation of the metal centre, react readily with diaryliodine(III) cations, forming Pt(IV) and Pd(IV) complexes (Schemes 36 and 37).[17,203] In this system, iodide or chloride ions are added to form Pt(IV) and Pd(IV) complexes that are more stable than the initially formed triflates. Mixtures of isomers are formed relating to the orientation of organyl groups trans to triflato, iodo or chloro ligands (**57a**/**57b**, **59a**/**59b**, **61a**/**61b**, **62a**/**62b**); a solution of isomer **60b** isolated at −50°C converts to the *cis* isomer on warming to room temperature. Complexes **57b** and **60b** have been characterised by X-ray crystallography. In the case of palladium, these reactions represent the first examples of formal transfer of Ar^+ groups to Pd(II) to form Pd(IV) species, but none of the Pd(IV) complexes were sufficiently stable to allow isolation as solids. For platinum, the iodonium reagents provide a convenient route to dimethyl(aryl)platinum(IV) chemistry. This has allowed the first example of a bridging 1,4-benzenediyl group between Pt(IV) centres (**64b**) obtained using a di-iodonium reagent, and the archetypal triarylplatinum(IV) group (**63**, **64**) isolated 97 years after the report of the first trialkylplatinum(IV) complex.[17] Isomers **57b** and **63b** are formed in 1:1 ratio, indicating a lack of selectivity in transfer of aryl groups from the unsymmetrical iodonium reagent $[I(C_6H_4\text{-}4\text{-}I)Ph]^+$.

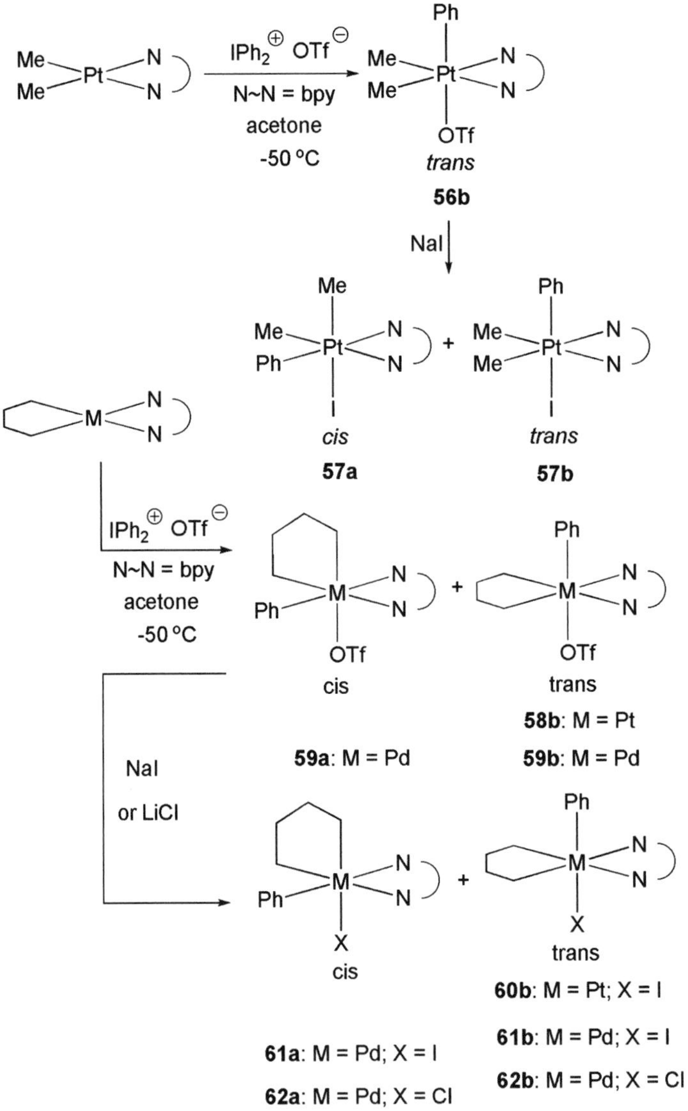

SCHEME 36.

B. *Alkynyl(aryl)iodine(III) Reagents*

As noted above for reactions of $[IPh_2]^+$ salts with metal carbonyl anions, reactions do not always result in formation of new metal-carbon bonds. Thus, as shown in Scheme 38, $[M(CN)(CO)_5]^-$ reacts with $[I(C{\equiv}CR)Ph]^+$ in a manner consistent with nucleophilic attack by the nitrogen atom of the cyanide ligand.[204] Complex **67** (R = Ph) has been characterised by X-ray crystallography. For formation of the cyclobutenyl products **68** (M = Cr also characterised by X-ray

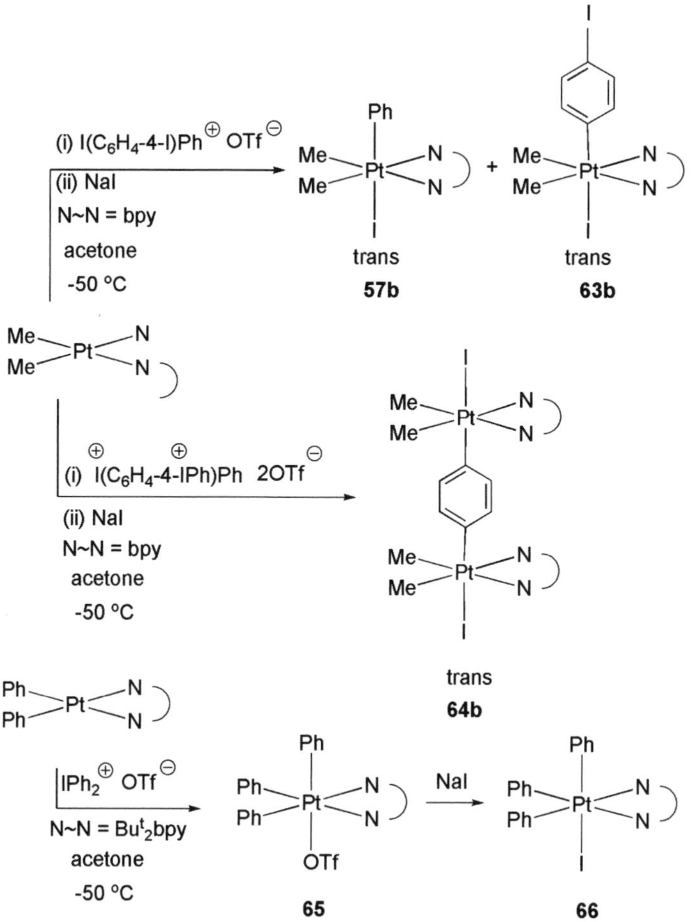

SCHEME 37.

diffraction), a mechanism involving insertion by a carbene intermediate at the β-C–H bond, as for the cyclisation shown in Scheme 14, is clearly implicated. The syntheses of Scheme 38 employ the 4-toluenesulphonate salt of $[I(C{\equiv}CR)Ph]^+$, and the reaction of the triflate $[I(C{\equiv}CH)Ph][OTf]$ also gives **67** (R = H), but reaction of $[I(C{\equiv}CSiMe_3)Ph][BF_4]$ followed by column chromatography gives **67** ($R = SiMe_3$) together with the product of hydrodesilylation of **67** (R = H). Preliminary results indicate that the synthetic approach to alkynyl isocyanide complexes reported here can be extended to the bis(diphenylphosphino)ethane reagents M(CN)Cp(dppe) (M = Fe, Ru).[204]

In the case of formation of the rhenium(I) complex **69** (Scheme 38), the alkynyl reagent $[I(C{\equiv}CSiMe_3)Ph][OTf]$ has been used to add an additional alkyne functionality to convert a diyne to a triyne.[123]

The reaction of $Pt(CH_2{=}CH_2)(PPh_3)_2$ with $[I(C{\equiv}CR)Ph][OTf]$ is complex, giving **70** and/or **71** depending upon reaction conditions and R (Scheme 39). The

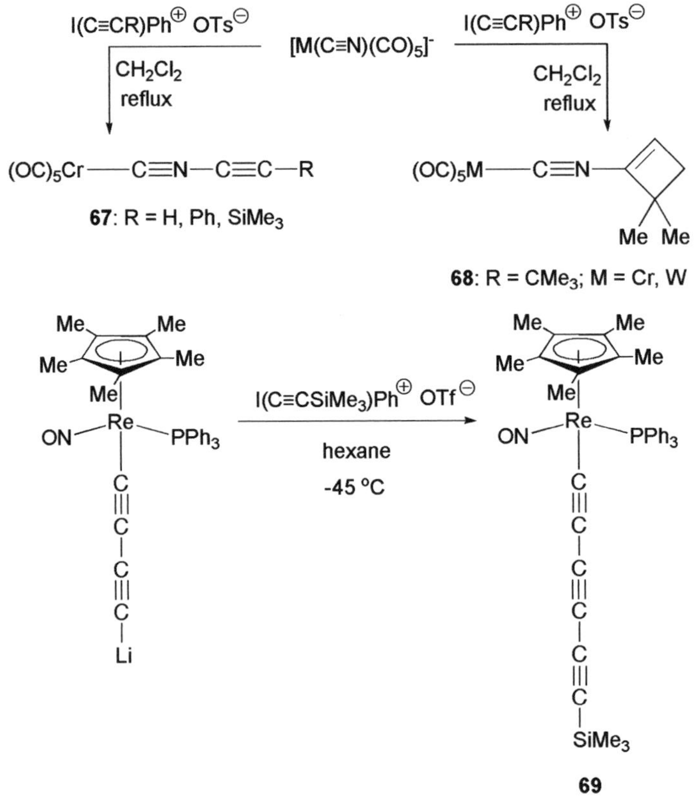

I(C≡CR)Ph$^{\oplus}$ OTs$^{\ominus}$

[M(C≡N)(CO)$_5$]$^-$

I(C≡CR)Ph$^{\oplus}$ OTs$^{\ominus}$

CH$_2$Cl$_2$ reflux

CH$_2$Cl$_2$ reflux

(OC)$_5$Cr—C≡N—C≡C—R

67: R = H, Ph, SiMe$_3$

(OC)$_5$M—C≡N—

Me Me

68: R = CMe$_3$; M = Cr, W

Me Me Me Me Me

ON—Re—PPh$_3$

C≡C—C≡C—Li

I(C≡CSiMe$_3$)Ph$^{\oplus}$ OTf$^{\ominus}$

hexane

-45 °C

Me Me Me Me Me

ON—Re—PPh$_3$

C≡C—C≡C—C≡C—SiMe$_3$

69

SCHEME 38.

‖—Pt(PPh$_3$)$_2$ ⇌ Pt(PPh$_3$)$_2$ + ‖

I(C≡CR)Ph $^{\oplus}$ OTf $^{\ominus}$

CD$_2$Cl$_2$ RT

TfO—Pt(PPh$_3$)(Ph$_3$P)—C≡C—R

+

Me R

Pt

Ph$_3$P PPh$_3$

$\oplus$ OTf$^{\ominus}$

70

71

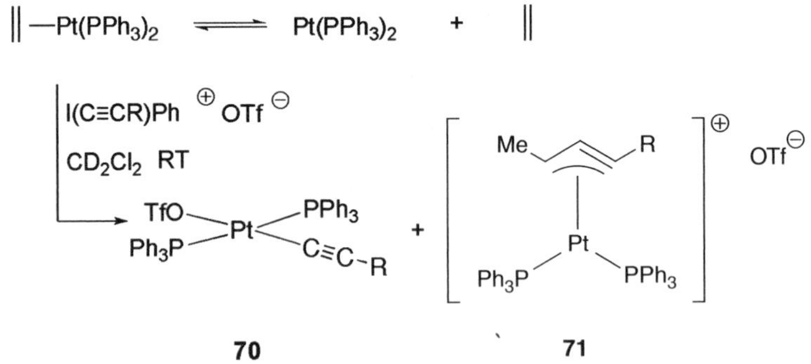

SCHEME 39.

η^3-propargyl/allenyl complex **71** (R = Bu^t) has been characterised by X-ray crystallography.[205] Both forms are detected for R = Me, Bu^n, Bu^t and SiMe$_3$, with increasing size of R favouring η^3-species, and the species are not interconverted on either heating or passing ethene through solutions of the σ-species. Passing argon through solutions during the reaction favours the formation of the σ-complexes, whereas passing ethene favours the formation of the η^3-complexes. These observations are most readily explained in terms of the equilibrium shown in Scheme 39, where loss of ethene leads to formation of the σ-complexes.[205] It is proposed that, in the formation of the η^3-complexes, the Pt(0) nucleophile $Pt(CH_2{=}CH_2)(PPh_3)_2$ interacts with the electrophilic iodine centre to form an alkynylplatinum(II) complex $[Pt(C{\equiv}CR)(CH_2{=}CH_2)(PPh_3)_2]^+$, followed by ethene insertion to form a $Pt^{II}CH_2CH_2C{\equiv}CR$ moiety which rearranges to form the product η^3-species.[205]

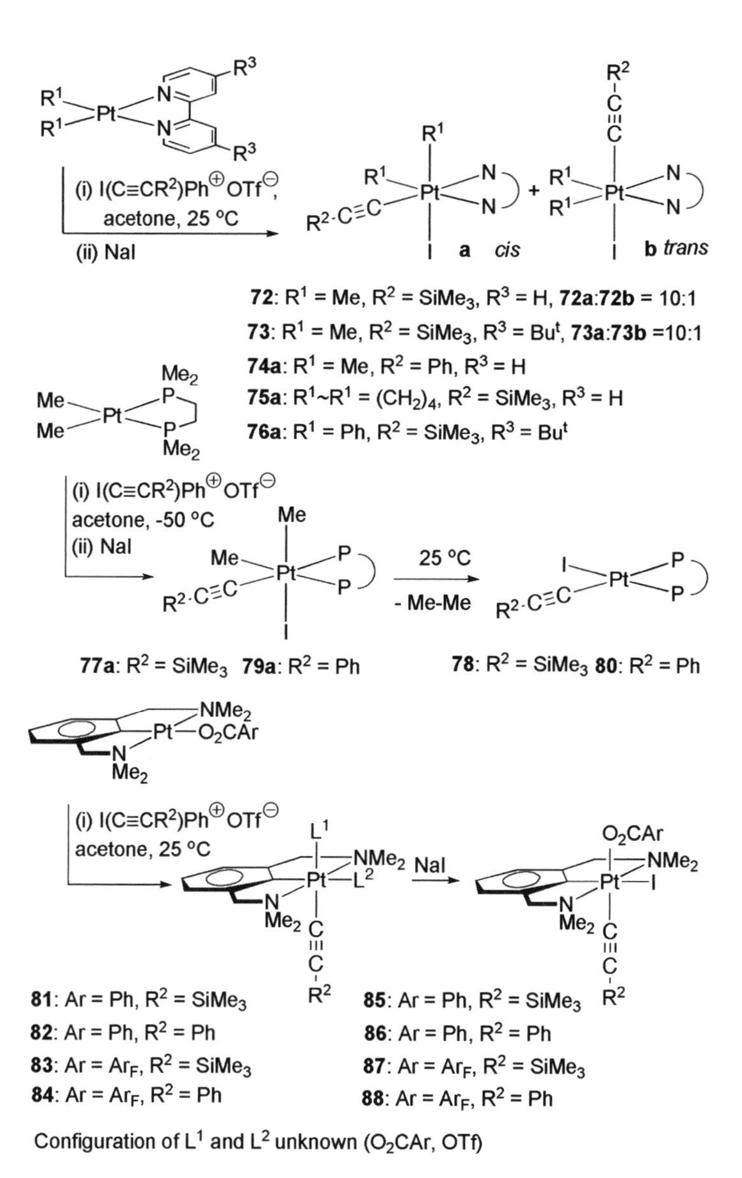

SCHEME 40.

This report provides new routes to transition metal propargyl/allenyl chemistry, an area of current interest in organometallic chemistry.

Organoplatinum(II) and palladium(II) nucleophiles also react cleanly under very mild conditions with alkynyliodonium reagents (Schemes 40 and 41), providing the first examples of dialkyl(alkynyl)platinum(IV) (**72–77a**, **79a**), alkynyl(pincer)platinum(IV) (**81–88**), and of alkynylpalladium(IV) species (**89**, **90**).[206–208] In addition, complex **89** is the first example of an organopalladium(IV) complex containing two phosphine donor atoms. Both **89** and **90** were too unstable for isolation, being formed, detected by NMR and decomposing at −50 °C. As for the syntheses involving $[IPh_2]^+$, iodo groups are added to give more stable metal(IV) species and isomeric mixtures are formed in some cases (**72a**/**72b**, **73a**/**73b**). Complexes **77a** and **79a** were not isolated as they reductively eliminate ethane at room temperature to form **78** and **80**. The aryl/triflato complexes **81–84** were not isolated, but the aryl/iodo complexes **85–88** were readily obtained and the structure of **87** confirmed by X-ray crystallography.

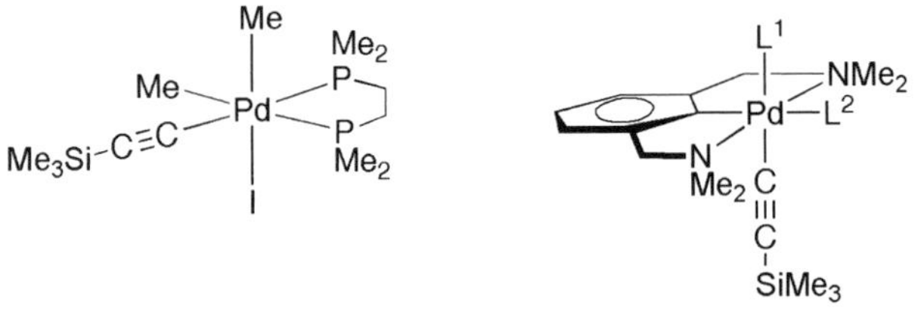

89

90: configuration of L^1 and L^2 unknown (O_2CPh, OTf)

SCHEME 41.

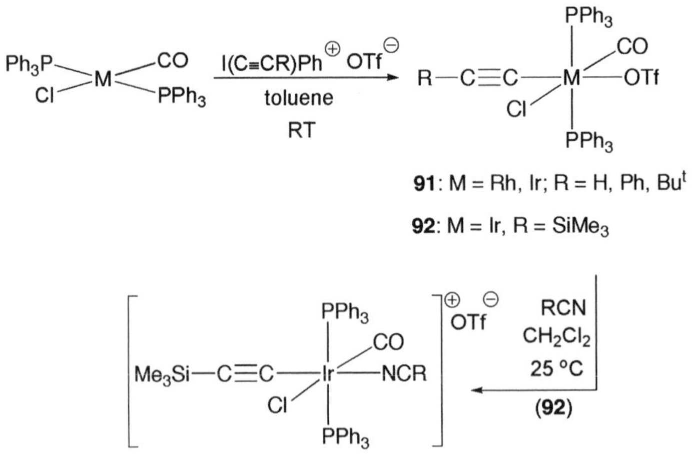

$Ph_3P(Cl)M(CO)(PPh_3)$ —[$I(C{\equiv}CR)Ph^{\oplus}$ $OTf^{\ominus}$, toluene, RT]→ $R{-}C{\equiv}C{-}M(PPh_3)_2(CO)(OTf)(Cl)$

91: M = Rh, Ir; R = H, Ph, Bu^t

92: M = Ir, R = $SiMe_3$

(**92**) —[RCN, CH_2Cl_2, 25 °C]→ $[Me_3Si{-}C{\equiv}C{-}Ir(PPh_3)_2(CO)(NCR)(Cl)]^{\oplus}$ $OTf^{\ominus}$

93: R = Me, Ph, 4-$MeOC_6H_4$, 4$CF_3C_6H_4$

SCHEME 42.

Stang and coworkers,[208–210] and Tykwinski,[18] have reported an extensive synthetic chemistry of d^6 rhodium(III) and iridium(III) alkynyls from alkynyliodonium reagents (Schemes 42–44), providing initial guidance for the development of d^6 Group 8 chemistry described above. Alkynyl(phenyl)iodonium(III) triflates add directly to Vaska's complex and the Rh(I) analogue in toluene or dichloromethane to form alkynyl(triflato)metal(III) complexes **91** and **92**,[18,208–210] but the most widely used protocol has involved reactions in acetonitrile to yield cationic acetonitrile complexes **95**–**100**. In a recent development, the triflate complex **92** ($R = Me_3Si$, $M = Ir$) has been used to synthesise a range of nitrile complexes **93** in dichloromethane, allowing crystallisation and X-ray structural analysis of two

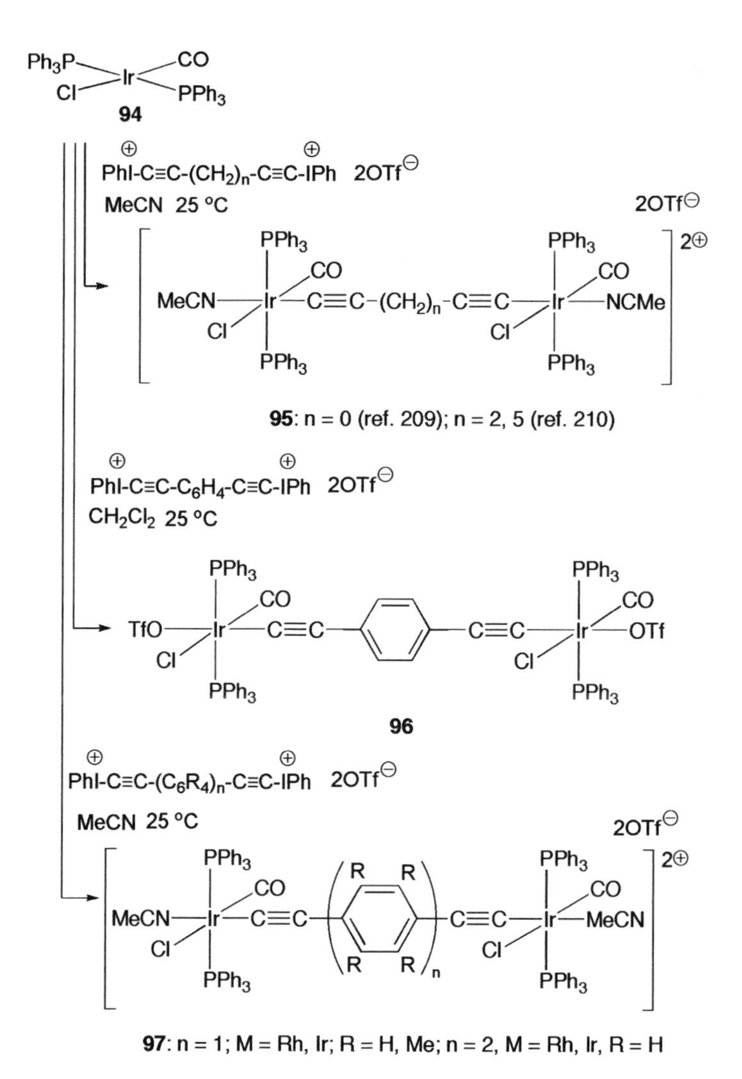

Scheme 43.

complexes, $[IrCl(C{\equiv}CSiMe_3)(PPh_3)_2(CO)(NCR)][OTf]$ (R = Me, 4-$CF_3C_6H_4$).[18] The syntheses in acetonitrile have led to isolation of bimetallic (**95**–**99**)[209,210] and a trimetallic species (**100**)[210] of considerable interest for the development of advanced materials, including those with potentially interesting electronic and optical properties, and examples containing moieties of interest in understanding the role of organometallic species in catalysis.

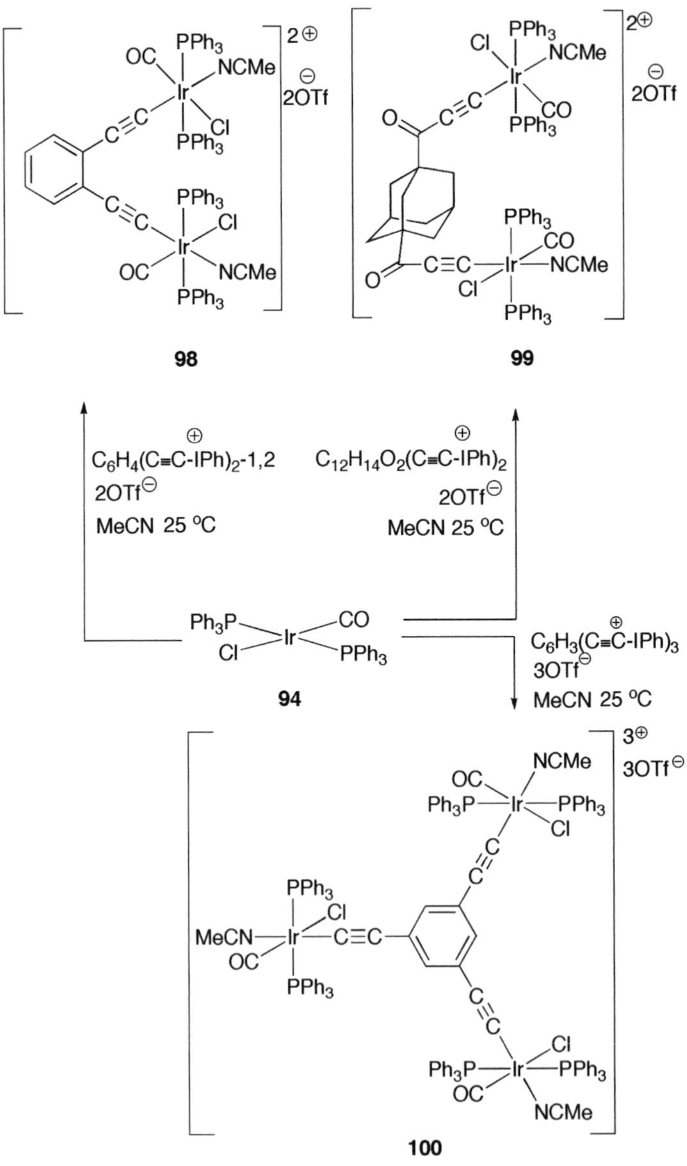

98 **99**

94

100

Scheme 44.

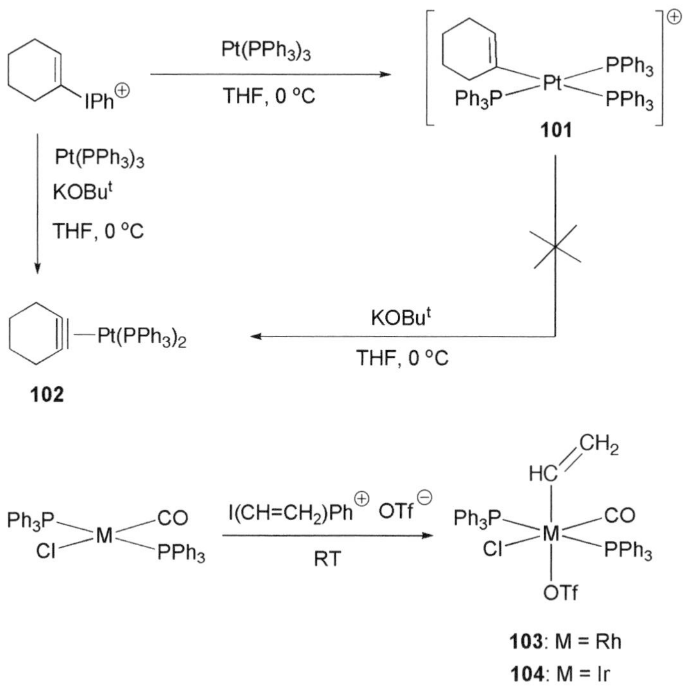

SCHEME 45.

C. *Alkenyl(aryl)iodine(III) Reagents*

Cyclohex-1-enyl(phenyl)iodonium tetrafluoroborate reacts with nucleophilic d^{10} $Pt^0(PPh_3)_2$ to form the σ-cyclohexenylplatinum(II) complex **101**, characterised by X-ray crystallography, but in the presence of a base forms the cyclohexyne complex **102** (Scheme 45).[211] Since **101** is not converted to **102** on addition of base, it is concluded that **102** is formed by trapping of transient cyclohexyne formed by deprotonation of the iodonium reagent by the base.

In similar reactions, iridium(I) and rhodium(I) complexes react with the vinyliodonium salt $[I(CH{=}CH_2)Ph][OTf]$ to form octahedral complexes **103** and **104**.[212]

IV

SUMMARY AND EMERGING OPPORTUNITIES IN ORGANOMETALLIC CHEMISTRY

For the very limited range of systems studied to date, diaryliodine(III) reagents react with metal carbonyl anions and low oxidation state lanthanides in a manner indicative of the occurrence of radical processes.

Readily synthesised and stable diaryliodine(III) and alkynyl(aryl)iodine(III) reagents react under mild conditions with representative d^8 square-planar complexes to form a range of new classes of d^6 octahedral complexes. For these sterically and electronically unsaturated d^8 substrates, the reactions most likely occur *via* polar mechanism, consistent with the known reactivity of these nucleophiles toward other electrophiles. Transfer of alkenyl groups from $[I(alkenyl)Ph]^+$ to d^{10} and d^8 reagents appear to proceed similarly. However, the range of transition metals studied to date is limited: Pd(II) and Pt(II) for aryl transfer; Pt(0), Pd(II), Pt(II), Rh(I) and Ir(I) for alkynyl transfer; and Pt(0), Rh(I) and Ir(I) for alkenyl transfer.

For alkynyl transfer, no direct evidence has been presented to determine whether organometallic nucleophiles interact with the electrophilic iodine centre or the β-carbon although, in the case of transfer to the nitrogen nucleophile of a coordinated cyano ligand, the β-interaction mechanism is clearly implicated in the formation of **68** (Scheme 38, related to the mechanism in Scheme 14). This example indicates, for the synthesis of alkynylmetal species, the attractiveness of $RC{\equiv}C(Ph)I^+$ reagents where the R group cannot participate in such reactivity, e.g. R = $SiMe_3$, Ph. Alternatively, this type of reaction, where the metal centre may act as a nucleophile at the β-carbon, is worthy of exploration as a mode of organometallic synthesis; as also is electrophilic attack at the α-carbon of ylides (illustrated by the mechanism in Scheme 14).

It is remarkable how many new organometallic systems have been discovered from the small number of investigations to date, indicating considerable promise for future development. For example, there appear to be no reports of studies with dialkynyliodine(III) reagents, although some of these reagents are stable at room temperature.[213] In addition to the potential for synthesis, the use of organometallic nucleophiles, for which steric and electronic factors can be so readily altered and for which various spectroscopic techniques are applicable, should provide useful probes for determining mechanism of reactions, in particular for attack of the nucleophile at iodine or the β-carbon of alkynyliodine(III) systems.

Reactions demonstrating the transfer of, formally, Ph^+ groups to Pd(II) to form $PhPd^{IV}$ species (Scheme 36)[17,203] provide model reactions that support the proposed role of Pd(IV) species in catalysis systems involving $[IPh_2]^+$ as a reagent (Scheme 13),[114,115] and the potential for related catalysis (Pd(II)/Pd(IV))) involving alkynyl- and alkenyl(phenyl)iodine(III) reagents is apparently unexplored. Indeed, even model reactions of Pd(0) substrates to support the proposed, and highly probable, transfer of Ar^+ and related species to form $ArPd^{II}$ and $RC{\equiv}CPd^{II}$ intermediates in more conventional Pd(0)/Pd(II) catalysis do not appear to have been reported to date.

Acknowledgments

We thank the Australian Research Council for financial support.

References

(1) For a summary of reviews and books on this subject up to 2002, see Wirth, T. in: *Topics in Current Chemistry, Vol. 224, Hypervalent Iodine Chemistry-/- Modern Developments in Organic Synthesis*, Wirth, T. Ed.; Springer-Verlag Publishers: Berlin Heidelberg 2003, Chapter 1, pp. 1–4.

(2) Varvoglis, A. *Synthesis* **1984**, 709.
(3) Varvoglis, A. *The Organic Chemistry of Polycoordinated Iodine*, VCH Publishers, Inc., New York, 1992.
(4) Varvoglis, A. *Hypervalent Iodine in Organic Synthesis*, Academic Press, London, 1997.
(5) Varvoglis, A.; Spyroudis S., *Syn. Lett.* **1998**, 221.
(6) Stang, P. J.; Zhdankin, V. V. *Chem. Rev.* **1996**, *96*, 1123.
(7) Zhdankin, V. V.; Stang, P. J. *Chem. Rev.* **2002**, *102*, 2523.
(8) Stang, P. J. J. *Org. Chem.* **2003**, *68*, 2997.
(9) Koser, G. F. In: *The Chemistry of Functional Groups, Suppl. D*; Patai, S. Rappaport, Z., Eds; Wiley-Interscience: Chichester, 1983; Chapters 18 and 25, pp. 721–811 and 1265–1351.
(10) Koser, G. F. In: *The Chemistry of Halides, Pseudo-Halides and Azides, Suppl. D2*, Patai, S. Rappaport, Z., Eds.; Wiley-Interscience: Chichester, 1995; Chapter 21, pp. 1173–1274.
(11) Akiba, K., Ed., *Chemistry of Hypervalent Compounds*, VCH Publishers, New York, 1999.
(12) Wirth, T., Ed., Topics in Current Chemistry, Vol. 224, **2002** *Hypervalent Iodine Chemistry-/-Modern Developments in Organic Synthesis*; Springer-Verlag Publishers: Berlin Heidelberg, 2003.
(13) Nesmeyanov, A. N.; Chapovskii, Yu. A.; Lokshin, B. V.; Kisin, A. V.; Makarova, L. G. *Dokl. Akad. Nauk SSSR* **1966**, *171*, 637.
(14) Nesmeyanov, A. N.; Chapovsky, Yu. A.; Polovyanuk, I. V.; Makarova, L. G. J. *Organomet. Chem.* **1967**, *7*, 329.
(15) Ochiai, M. J. *Organomet. Chem.* **2000**, *611*, 494.
(16) Ochiai, M., Kida, M., Sato, K., Takino, T., Goto, S., Donkai, N., Okuyama, T. *Tetrahedron Lett.* **1999**, 1559.
(17) Canty, A. J.; Patel, J.; Rodemann, T.; Ryan, J. H.; Skelton, B. W.; White, A. H. *Organometallics* **2004**, *23*, 3466.
(18) Bykowski, D., McDonald, R. Tykwinski, R. R. *ARKIVOC*, **2003**, 21.
(19) Musher, J. I. *Angew. Chem Int. Ed. Engl.* **1969**, *8*, 54.
(20) Martin, J. C. *Science* **1983**, *221*, 509.
(21) Reed, A. E.; Schleyer, P. v. R. *J. Am. Chem. Soc.* **1990**, *112*, 1434.
(22) Ochiai, M. (K. Akiba, Ed.), *Chemistry of Hypervalent Compounds* Chapter 13, **1999**, Wiley-VCH Publishers, New York, pp. 359–387.
(23) Moriarty, R. M., Vaid, R. K. *Synthesis*, **1990**, 431.
(24) Banks, D. F. *Chem. Rev.* **1966**, *66*, 243.
(25) Beringer, M. F.; Brierley, A.; Drexler, M.; Grindler, E. M.; Lumpkin, C. C. *J. Am. Chem. Soc.* **1953**, *75*, 2708.
(26) Chen, Z.-C.; Jin, Y.-Y.; Stang, P. J. *J. Org. Chem.* **1987**, *52*, 4115.
(27) Ochiai, M.; Kitagawa, Y.; Takayama, N.; Takaoka, Y.; Shiro, M. *J. Am. Chem. Soc.* **1999**, *121*, 9233.
(28) Oh, C. H.; Kim, J. S.; Jung, H. H. *J. Org. Chem.* **1999**, *64*, 1338.
(29) Chen, K.; Koser, G. F. *J. Org. Chem.* **1991**, *56*, 5764.
(30) Iwama, T.; Birman, V. B.; Kozmin, S. A.; Rawal, V. H. *Org. Lett.* **1999**, *1*, 673.
(31) Kornblum, N.; Taylor, H. J. *J. Org. Chem.* **1963**, *28*, 1424.
(32) Lubinowski, J. J.; Gomez, M.; Calderon, J. L.; McEwan, W. E. *J. Org. Chem.* **1978**, *43*, 2432.
(33) Nesmeyanov, A. N.; Makarova, L. G.; Tostaya, T. P. *Tetrahedron* **1957**, *1*, 145.
(34) Ziegler, H.; Marr, C. *J. Org. Chem.* **1962**, *27*, 3335.
(35) Taylor, E. C.; Kienzle, F. *J. Org. Chem.* **1971**, *36*, 233.
(36) Cadogan, J. I. G.; Rowlex, A. G. *Synth. Commun.* **1977**, *7*, 365.
(37) Sheradsky, T.; Nov, E. *J. Chem. Soc. Perkin Trans.* **1980**, *1*, 2781.
(38) McKillop, A.; Kobylecki, R. J. *J. Org. Chem.* **1974**, *39*, 2710.
(39) Crivello, J. V.; Lam, J. H. W. *Synth. Commun.* **1979**, *9*, 151.
(40) Grushin, V. V., Kantor, M. M., Tolstaya, T. P., Shcherbina, T. M. *Izv. Akad. Nauk SSSR, Ser. Khim.* **1984**, 2332.
(41) Kotali, E., Varvoglis, A. *J. Chem. Res. (S)* **1989**, 142.
(42) Kotali, E.; Varvoglis, A. *J. Chem. Soc. Perkin Trans.* **1987**, *1*, 2759.
(43) Chen, Z.-C.; Jin, Y.-Y.; Stang, P. J. *J. Org. Chem.* **1987**, *52*, 4117.
(44) Chen, D.-W.; Zhang, Y.-D.; Chen, Z.-C. *Synth. Commun.* **1995**, *25*, 1627.

(45) Nesmeyanov, A. N., Tolstaya, T. P., Grib, A. V., Kirgizbaeva, S. R., *Izv. Akad. Nauk SSSR, Ser. Khim.* **1973**, 1678.
(46) Gronowitz, S.; Holm, B. *J. Heterocyclic Chem.* **1977**, *14*, 281.
(47) You, J.-Z., Chen, Z.-C. *Synthesis* **1992**, 521.
(48) Liu, D.-D.; Chen, D.-W.; Chen, Z.-C. *Synth. Commun.* **1992**, *22*, 2903.
(49) Xia, M.; Chen, Z.-C. *Synth. Commun.* **1997**, *27*, 1309.
(50) Liu, D.-D., Chen, Z.-C. *Synthesis* **1993**, 373.
(51) You, J.-Z.; Chen, Z.-C. *Synth. Commun.* **1992**, *22*, 1441.
(52) Liu, D.-D.; Zeng, H.; Chen, Z.-C. *Synth. Commun.* **1994**, *24*, 475 Check Liu.
(53) You, J.-Z., Chen, Z.-C. *Synthesis* **1992**, 663.
(54) Davidson, J. M., Dyer, G. *J. Chem. Soc. [A]* **1968**, 1616.
(55) Uchiyama, M., Suzuki, T., Yamazaki, Y., *Nippon Kagaku Kaishi* **1982**, 236; *C. A.* **1982**, 96, 217396.
(56) Abramovitch, R. A.; Alvernhe, G.; Bartnik, R.; Dassanayake, N. L.; Inbasekaran, M. N.; Kato, S. *J. Am. Chem. Soc.* **1981**, *103*, 4558.
(57) Kang, J.; Ku, B. C. *Bull. Korean Chem. Soc.* **1985**, *6*, 375.
(58) (a) Ptitsyna, O. A., Pudeeva, M. E., Reutov, O. A., *Dokl. Akad. Nauk SSSRI* **1965**, 165, 582; *ibid*, 838. (b) Ptitsyna, O. A.; Gurskii, M. E.; Reutov, O. A. J. *Org. Chem. USSR* **1974**, *10*, 2262.
(59) Varvoglis, A. *Tetrahedron Lett.* **1972**, 31.
(60) Kitamura, T., Yamane, M. *J. Chem. Soc. Chem. Commun.* **1995**, 983.
(61) Finet, J.-P. *Ligand Coupling Reactions with Heteroatomic Compounds*, Pergamon Press, Oxford, UK, 1998.
(62) (a) Beringer, F. M.; Forgione, P. S.; Yudis, M. D. *Tetrahedron* **1960**, *8*, 49. (b) Beringer, F. M.; Galton, S. A.; Huang, S. J. *J. Am. Chem. Soc.* **1962**, *84*, 2819. (c) Beringer, F. M.; Forgione, P. S. *Tetrahedron* **1963**, *19*, 739.
(63) Tanner, D. D.; Reed, D. W.; Setiloane, B. P. *J. Am. Chem. Soc.* **1982**, *104*, 3917.
(64) (a) Lubinowski, J. J.; Knapczyk, J. W.; Calderon, J. L.; Petit, L. R.; McEwen, W. E. *J. Org. Chem.* **1975**, *40*, 3010. (b) Lubinowski, J. J.; Arrieche, C. G.; McEwen, W. E. *J. Org. Chem.* **1980**, *45*, 2076.
(65) Barton, D. H. R.; Finet, J.-P.; Giannotti, C.; Halley, F. *J. Chem. Soc. Perkin Trans.* **1987**, *1*, 241.
(66) Budylin, V. A., Ermolenko, M. S., Chugtai, F. A., Kost, A. N., *Khim. Geterotsikl. Soedin.* **1981**, 1494; *Chem. Abstr.* **1982**, *96*, 142617n.
(67) Grushin, V. V. *Acc. Chem. Res.* **1992**, *25*, 529. Grushin, V. V.; Demkina, I. I.; Tolstaya, T. P. *J. Chem. Soc. Perkin Trans.* **1992**, *1*, 505.
(68) Yamada, Y.; Okawara, M. *Bull. Chem. Soc. Jpn.* **1972**, *45*, 1860.
(69) Lancer, K. M.; Wiegand, G. H. *J. Org. Chem.* **1976**, *41*, 3360.
(70) Olah, G. A.; Sakakibara, T.; Asensio, G. *J. Org. Chem.* **1978**, *43*, 463.
(71) (a) Beringer, F. M.; Dehn, J. W. Jr.; Winicov, M. *J. Am. Chem. Soc.* **1960**, *82*, 2948. (b) Beringer, F. M.; Chang, L. L. *J. Org. Chem.* **1971**, *36*, 4055.
(72) Wittig, G.; Clauss, K. *Liebigs Ann. Chem.* **1949**, *562*, 187.
(73) Moriarty, R. M.; Ku, Y. Y.; Sultana, M.; Tuncay, A. *Tetrahedron Lett.* **1987**, *28*, 3071.
(74) Cookson, R. C., Farquharson, B. K. *Tetrahedron Lett.* **1979**, 1255.
(75) Stang, P. J.; Kitamura, T. *J. Am. Chem. Soc.* **1987**, *109*, 7561.
(76) Ryan, J. H.; Stang, P. J. *Tetrahedron Lett.* **1997**, *38*, 5061.
(77) Kang, S.-K.; Yoon, S.-K.; Kim, Y.-M. *Org. Lett.* **2001**, *3*, 2697.
(78) Kang, S.-K.; Yamaguchi, T.; Kim, T.-H.; Ho, P.-S. *J. Org. Chem.* **1996**, *61*, 9082.
(79) Crivello, J. V.; Lam, J. H. W. *J. Org. Chem.* **1978**, *43*, 3055.
(80) Kataoka, T.; Tomoto, A.; Shimizu, H.; Hori, M. *J. Chem. Soc. Perkin Trans.* **1983**, *1*, 2913.
(81) Bumagin, N. A.; Sukhomlinova, L. I.; Banchikov, A. N.; Tolstaya, T. P.; Beletskaya, I. P. *Russ. Chem. Bull.* **1992**, *41*, 2130.
(82) (a) Kang, S.-K.; Jung, K.-Y.; Park, C.-H.; Jang, S.-B. *Tetrahedron Lett.* **1995**, *36*, 8047. (b) Kang, S.-K.; Lee, H.-W.; Jang, S.-B.; Kim, T.-H.; Pyun, S.-J. *J. Org. Chem.* **1996**, *61*, 2604.
(83) Liang, Y.; Luo, S.; Liu, C.; Wu, X.; Yongxiang, M. *Tetrahedron* **2000**, *56*, 2961.
(84) (a) Kang, S.-K.; Yamaguchi, T.; Pyun, S.-J.; Lee, Y.-T.; Baik, T.-G. *Tetrahedron Lett.* **1998**, *39*, 2127. (b) Kang, S.-K.; Baik, T.-G.; Hur, Y. *Tetrahedron* **1999**, *55*, 6863.
(85) Zhu, Q.; Wu, J.; Fathi, R.; Yang, Z. *Org. Lett.* **2002**, *4*, 3333.

(86) Xia, M.; Chen, Z.-C. *Synth. Commun.* **1999**, *29*, 2457.
(87) Kang, S.-K.; Lee, H.-W.; Jang, S.-B.; Ho, P.-S. *J. Org. Chem.* **1996**, *61*, 4720.
(88) (a) Bumagin, N. A.; Sukhomlinova, L. I.; Igushkina, S. O.; Banchikov, A. N.; Tolstaya, T. P.; Beletskaya, I. P. *Russ. Chem. Bull.* **1992**, *41*, 2128. (b) Bumagin, N. A.; Sukhomlinova, L. I.; Tolstaya, T. P.; Beletskaya, I. P. *Zh. Org. Khim.* **1996**, *32*, 1724.
(89) Kang, S.-K.; Lee, Y. T.; Lee, S.-H. *Tetrahedron Lett.* **1999**, *40*, 3573.
(90) Kang, S.-K.; Yamaguchi, T.; Hong, R.-K.; Kim, T.-H.; Pyun, S.-J. *Tetrahedron* **1997**, *53*, 3027.
(91) Kang, S.-K.; Choi, S.-C.; Baik, T.-G. *Synth. Commun.* **1999**, *29*, 2493.
(92) Kang, S.-K.; Ryu, H.-C.; Kim, J.-W. *Synth. Commun.* **2001**, *31*, 1021.
(93) Huang, X.; Sun, A.-M. *Synth. Commun.* **1998**, *28*, 773.
(94) Wang, L.; Chen, Z.-C. *Synth. Commun.* **2000**, *30*, 3607.
(95) Luzikova, E. V.; Sukhomlinova, L. I.; Tolstaya, T. P.; Bumagin, N. A.; Beletskaya, I. P. *Russ. Chem. Bull.* **1995**, *44*, 556.
(96) Kang, S.-K., Lee, H.-W., Jang, S.-B., Ho, P.-S. *Chem. Commun.* **1996**, 835.
(97) Zhou, T.; Chen, Z.-C. *Synth. Commun.* **2001**, *31*, 3289.
(98) Xia, M., Chen, Z.-C. *J. Chem. Res. (S)* **1999**, 328.
(99) Kang, S.-K.; Yamaguchi, T.; Ho, P.-S.; Kim, W.-Y.; Ryu, H.-C. *J. Chem. Soc. Perkin Trans.* **1998**, *1*, 841.
(100) Ryu, H.-C.; Hong, Y.-T.; Kang, S.-K. *Heterocycles* **2001**, *54*, 985.
(101) Wang, L.; Chen, Z.-C. *Synth. Commun.* **2001**, *31*, 1633.
(102) Kang, S.-K.; Lim, K.-H.; Ho, P.-S.; Yoon, S.-K.; Son, H.-J. *Synth. Commun.* **1998**, *28*, 1481.
(103) Xia, M., Chen, Z. *J. Chem. Res. (S)* **1999**, 400–401.
(104) Al-Qahtani, M. H.; Pike, V. W. *J. Chem. Soc. Perkin Trans.* **2000**, *1*, 1033.
(105) Kang, S.-K.; Lee, H.-W.; Choi, W.-K.; Hong, R.-K.; Kim, J.-S. *Synth. Commun.* **1996**, *26*, 4219.
(106) Bumagin, N. A.; Sukhomlinova, L. I.; Luzikova, E. V.; Tolstaya, T. P.; Beletskaya, I. P. *Tetrahedron Lett.* **1996**, *37*, 897.
(107) Radhakrishnan, U.; Stang, P. J. *Org. Lett.* **2001**, *3*, 859.
(108) Beletskaya, I. P.; Davydov, D. V.; Moreno-Mañas, M. *Tetrahedron Lett.* **1998**, *39*, 5621.
(109) Kang, S.-K., Lim, K. H., Ho, P.-S., Kim, W. Y. *Synthesis* **1997**, 874.
(110) Zhou, T., Chen, Z.-C. *J. Chem. Res. (S)* **2001**, 116.
(111) Kang, S.-K.; Kim, W.-Y.; Lee, Y.-T.; Ahn, S.-K.; Kim, J.-C. *Tetrahedron Lett.* **1998**, *39*, 2131.
(112) Kang, S.-K.; Ryu, H.-C.; Lee, S.-W. *J. Chem. Soc. Perkin Trans.* **1999**, *1*, 2661.
(113) J. Tsuji, Ch 1 *Palladium Reagents and Catalysts: Innovations in Organic Synthesis*, J. Wiley and Sons Ltd, Chichester, England, 1995.
(114) Kalyani, D.; Deprez, N. R.; Desai, L. V.; Sanford, M. S. *J. Am. Chem. Soc.* **2005**, *127*, 7330.
(115) Deprez, N. R.; Kalyani, D.; Krause, A.; Sanford, M. S. *J. Am. Chem. Soc.* **2006**, *128*, 4972.
(116) Koser, G. F. in: *The Chemistry of Functional Groups, Supplement D*; Patai, S., Rappaport, Z., Eds.; Wiley: 1983; Chapter 25, pp. 1323–24.
(117) Stang, P. J. *Angew. Chem. Int. Ed. Engl.* **1992**, *31*, 274.
(118) Zhdankin, V. V.; Stang, P. J. *Tetrahedron* **1998**, *54*, 10927.
(119) Beringer, F. M.; Galton, S. A. *J. Org. Chem.* **1965**, *30*, 1930.
(120) Ochiai, M., Ito, T., Takaoka, Y., Masaki, Y., Kunishima, M., Tani, S., Nagao, Y. *J. Chem. Soc., Chem. Commun.* **1990**, 118.
(121) Bachi, M. D.; Bar-Ner, N.; Critell, C. M.; Stang, P. J.; Williamson, B. L. *J. Org. Chem.* **1991**, *56*, 3912.
(122) Kitamura, T.; Nagata, K.; Taniguchi, H. *Tetrahedron Lett.* **1995**, *36*, 1081.
(123) Dembinski, R.; Lis, T.; Szafert, S.; Mayne, C. L.; Bartik, T.; Gladysz, J. A. *J. Organomet. Chem.* **1999**, *578*, 229.
(124) Murch, P., Williamson, B. L., Stang, P. J. *Synthesis* **1994**, 1255.
(125) (a) Kitamura, T.; Tashi, N.; Tsuda, K.; Fujiwara, Y. *Tetrahedron Lett.* **1999**, *39*, 3787. (b) Kitamura, T.; Tashi, N.; Tsuda, K.; Chen, H.; Fujiwara, Y. *Heterocycles* **2000**, *52*, 303.
(126) Witulski, B., Gossmann, M. *J. Chem. Soc., Chem. Commun.* **1999**, 1879.
(127) Stang, P. J.; Zhdankin, V. V. *J. Am. Chem. Soc.* **1991**, *113*, 4571.
(128) Stang, P. J.; Boehshar, M.; Wingert, H.; Kitamura, T. *J. Am. Chem. Soc.* **1988**, *110*, 3272.
(129) Tykwinski, R. R.; Stang, P. J. *Tetrahedron* **1993**, *49*, 3043.

(130) Stang, P. J.; Kitamura, T.; Boehshar, M.; Wingert, H. *J. Am. Chem. Soc.* **1989**, *111*, 2225.
(131) Zhang, J.-L.; Chen, Z.-C. *Synth. Commun.* **1998**, *28*, 175.
(132) Liu, Z. D.; Chen, Z. C. *Synth. Commun.* **1992**, *22*, 1997. Tykwinski, R. R.; Williamson, B. L.; Fischer, D. R.; Stang, P. J.; Arif, A. M. *J. Org. Chem.* **1993**, *58*, 5235.
(133) Tykwinski, R. R.; Williamson, B. L.; Fischer, D. R.; Stang, P. J.; Arif, A. M. *J. Org. Chem.* **1993**, *58*, 5235.
(134) Williamson, B. L., Murch, P., Fischer, D. R., Stang, P. J. *Synlett.* **1993**, 858.
(135) Liu, Z.-D.; Chen, Z.-C. *J. Org. Chem.* **1993**, *58*, 1924.
(136) Stang, P. J., Murch, P. J. *Synthesis* **1997**, 1378.
(137) Zhang, J.-L.; Chen, Z.-C. *Synth. Commun.* **1997**, *27*, 3757.
(138) Zhang, J.-L.; Chen, Z.-C. *Synth. Commun.* **1997**, *27*, 3881.
(139) Ochiai, M., Kunishima, M., Nagao, Y., Fuji, K., Fujita, E. *J. Chem. Soc. Chem. Commun.* **1987**, 1708.
(140) Stang, P. J.; Crittell, C. M. *J. Org. Chem.* **1992**, *57*, 4305.
(141) Laali, K. K.; Regitz, M.; Birkel, M.; Stang, P. J.; Crittell, C. M. *J. Org. Chem.* **1993**, *58*, 4105.
(142) Lodaya, J. S.; Koser, G. F. *J. Org. Chem.* **1990**, *55*, 1513.
(143) Nagaoka, T.; Sueda, T.; Ochiai, M. *Tetrahedron Lett.* **1995**, *36*, 261.
(144) Ochiai, M.; Kunishima, M.; Nagao, Y.; Fuji, K.; Shiro, M.; Fujita, E. *J. Am. Chem. Soc.* **1986**, *108*, 8281.
(145) Kitamura, T.; Stang, P. J. *Tetrahedron Lett.* **1988**, *29*, 1887.
(146) Yoshida, M.; Hara, S. *Org. Lett.* **2003**, *5*, 573.
(147) Stang, P. J.; Kitamura, T. *J. Am. Chem. Soc.* **1987**, *109*, 7561.
(148) Yang, D.-Y.; He, J.; Miao, S. *Synth. Commun.* **2003**, *33*, 2695.
(149) (a) Kitamura, T.; Tanaka, T.; Taniguchi, H.; Stang, P. J. *J. Chem. Soc. Perkin Trans.* **1991**, *1*, 2892. (b) Kitamura, T.; Lee, C. H.; Taniguchi, H.; Matsumoto, M.; Sano, Y. *J. Org. Chem.* **1994**, *59*, 8053. (c) Kitamura, T., Lee, C. H., Taniguchi, Y., Fujiwara, Y., Matsumoto, M., Sano, Y. F. *J. Am. Chem. Soc.* **1997**, 119
(150) Eaton, P. A.; Gallopini, E.; Gilardi, R. *J. Am. Chem. Soc.* **1994**, *116*, 7588.
(151) Stang, P. J.; Surber, B. W.; Chen, Z.-C.; Roberts, K. A.; Anderson, A. G. *J. Am. Chem. Soc.* **1987**, *109*, 228.
(152) (a) Huang, X.; Zhong, P. *J. Chem. Soc. Perkin Trans.* **1999**, *1*, 1543. (b) Sun, A.-M., Huang, X. *J. Chem. Res. Synop.* **1998**, 616.
(153) Kitamura, T., Mihara, I., Taniguchi, H., Stang, P. J. *J. Chem. Soc., Chem. Commun.* **1990**, 614.
(154) Sun, A.-M.; Huang, X. *Tetrahedron* **1999**, *55*, 13201.
(155) Sh. Pirkuliev, S.; Brel, V. K.; Zefirov, N. S. *Russ. Chem. Rev.* **2000**, *69*, 105.
(156) (a) Ochiai, M.; Sumi, K.; Nagao, Y.; Fujita, E. *Tetrahedron Lett.* **1985**, *26*, 2351. (b) Ochiai, M.; Sumi, K.; Takaoka, Y.; Kunishima, M.; Nagao, Y.; Shiro, M.; Fujita, E. *Tetrahedron* **1988**, *44*, 4095.
(157) Ochiai, M.; Shu, T.; Nagaoka, T.; Kitagawa, Y. *J. Org. Chem.* **1997**, *62*, 2130.
(158) Papoutsis, I.; Spyroudis, S.; Varvoglis, A. *Tetrahedron* **1998**, *54*, 1005.
(159) Zefirov, N. S.; Koz'min, A. S.; Kasumov, T.; Potekhin, K. A.; Sorokin, V. D.; Brel, V. K.; Abramkin, E. V.; Struchkov, Yu. T.; Zhdankin, V. V.; Stang, P. J. *J. Org. Chem.* **1992**, *57*, 2433.
(160) Nesmeyanov, A. N.; Tolstaya, T. P.; Sokolova, N. F.; Varfolomeeva, V. N.; Petrakov, A. V. *Dokl. Akad. Nauk. SSSR* **1971**, *198*, 115.
(161) Nesmeyanov, A. N.; Tolstaya, T. P.; Petrakov, A. V. *Dokl. Akad. Nauk. SSSR* **1971**, *197*, 1337.
(162) Stang, P. J.; Schwarz, A.; Blume, T.; Zhdankin, V. V. *Tetrahedron Lett.* **1992**, *33*, 6759.
(163) (a) Ochiai, M.; Oshima, K.; Masaki, Y.; Kunishima, M.; Tani, S. *Tetrahedron Lett.* **1993**, *34*, 4829. (b) Ochiai, M.; Kitagawa, Y.; Toyonari, M.; Uemura, K. *Tetrahedron Lett.* **1994**, *35*, 9407. (c) Ochiai, M.; Kitagawa, Y.; Toyonari, M.; Uemura, K.; Oshima, K.; Shiro, M. *J. Org. Chem.* **1997**, *62*, 8001.
(164) (a) Nesmeyanov, A. N.; Tolstaya, T. P.; Petrakov, A. V.; Gol'tsev, A. N. *Dokl. Akad. Nauk. SSSR* **1977**, *235*, 591. (b) Nesmeyanov, A. N.; Tolstaya, T. P.; Petrakov, A. V.; Leshcheva, I. F. *Dokl. Akad. Nauk. SSSR* **1978**, *238*, 1109.
(165) Yan, J.; Chen, Z.-C. *Synth. Commun.* **1999**, *29*, 2867.
(166) Yan, J.; Chen, Z.-C. *Synth. Commun.* **2000**, *30*, 3897.

(167) Yan, J.; Chen, Z.-C. *Synth. Commun.* **2000**, *30*, 1009.
(168) Yan, J.; Chen, Z.-C. *Tetrahedron Lett.* **1999**, *40*, 5757.
(169) Yan, J.; Chen, Z.-C. *Synth. Commun.* **2000**, *30*, 2359.
(170) Ochiai, M.; Oshima, K.; Masaki, Y. *J. Am. Chem. Soc.* **1991**, *113*, 7059.
(171) Sueda, T.; Nagaoka, T.; Goto, S.; Ochiai, M. *J. Am. Chem. Soc.* **1996**, *118*, 10141.
(172) Ochiai, M.; Sueda, T.; Noda, R.; Shiro, M. *J. Org. Chem.* **1999**, *64*, 8563.
(173) Okuyama, T.; Ochiai, M. *J. Am. Chem. Soc.* **1997**, *119*, 4785.
(174) Ochiai, M., Yamamoto, S., Sato, K. *Chem. Commun.* **1999**, 1363.
(175) Ochiai, M.; Yamamoto, S.; Suefuji, T.; Chen, D.-W. *Org. Lett.* **2001**, *3*, 2753.
(176) Okuyama, T. *Acc. Chem. Res.* **2002**, *35*, 12.
(177) Ochiai, M.; Shu, T.; Nagaoka, T.; Kitagawa, Y. *J. Org. Chem.* **1997**, *62*, 2130.
(178) Ochiai, M., Oshima, K., Masaki, Y. *Tetrahedron Lett.* **1991**, 7711.
(179) (a) Okuyama, T.; Fujita, M. *Acc. Chem. Res.* **2005**, *38*, 679. (b) Fujita, M., Sakanishi, Y., Kim, W. H., Okuyama, T. *Chem. Lett.* **2002**, 908. (c) Fujita, M.; Kim, W. H.; Sakanishi, Y.; Fujiwara, K.; Hirayama, S.; Okuyama, T.; Ohki, Y.; Tatsumi, K.; Yoshioka, Y. *J. Am. Chem. Soc.* **2004**, *126*, 7548.
(180) Okuyama, T.; Ishida, Y.; Ochiai, M. *Bull. Chem. Soc. Jpn.* **1999**, *72*, 163.
(181) Okuyama, T.; Takino, T.; Sueda, T.; Ochiai, M. *J. Am. Chem. Soc.* **1995**, *117*, 3360.
(182) Ochiai, M., Takaoka, Y., Sumi, K., Nagao, Y. *J. Chem. Soc. Chem. Commun.* **1986**, 1382.
(183) Huang, X.; Wang, J.-H.; Yang, D. Y.; Chem. Soc., J. *Perkin Trans.* **1999**, *1*, 673.
(184) Ishikura, M., Terashima, M., *J. Chem. Soc. Chem. Commun.* **1989**, 727.
(185) Ishikura, M.; Tershima, M. *Heterocycles* **1988**, *27*, 2619.
(186) Stang, P. J., Blume, T., Zhdankin, V. V. *Synthesis* **1993**, 35.
(187) Hinkle, R. J.; Leri, A. C.; David, G. A.; Erwin, W. M. *Org. Lett.* **2000**, *2*, 1521.
(188) Huang, X.; Xu, X.-H. *J. Chem. Soc. Perkin Trans.* **1998**, *1*, 3321.
(189) Yan, J.; Chen, Z.-C. *Synth. Commun.* **1999**, *29*, 3275.
(190) Yan, J.; Chen, Z.-C. *Synth. Commun.* **1999**, *29*, 3605.
(191) Moriarty, R. M.; Epa, W. R.; Awasthi, A. K. *J. Am. Chem. Soc.* **1991**, *113*, 6315.
(192) Kurihara, Y.; Sodeoka, M.; Shibasaki *Chem. Pharm. Bull.* **1994**, *42*, 2357.
(193) Roh, K. R.; Kim, J. Y.; Kim, Y. H. *Tetrahedron Lett.* **1999**, *40*, 1903.
(194) Moriarty, R. M.; Epa, W. R. *Tetrahedron Lett.* **1992**, *33*, 4095.
(195) Hinkle, R. J.; Poulter, G. T.; Stang, P. J. *J. Am. Chem. Soc.* **1993**, *115*, 11626.
(196) Sh. Pirguliyev, N.; Brel, V. K.; Zefirov, N. S.; Stang, P. J. *Tetrahedron* **1999**, *55*, 12377.
(197) Ryan, J. H.; Stang, P. J. *J. Org. Chem.* **1996**, *61*, 6162.
(198) Ustynyuk, N. A. *Thesis*, Moscow, INEOS AN SSSR, **1973**.
(199) Nesmeyanov, A. N.; Ustynyuk, N. A.; Novikova, L. N.; Rybina, T. N. *J. Organomet. Chem.* **1980**, *184*, 63.
(200) Wei, C.-H.; Bockmann, T. M.; Kochi, J. K. *J. Organomet. Chem.* **1992**, *428*, 85.
(201) Battioni, J.-P.; Dupre, D.; Delaforge, M.; Jaouen, M.; Mansuy, D. *J. Organomet. Chem.* **1988**, *358*, 389.
(202) Makioka, Y.; Fujiwara, Y.; Kitamura, T. *J. Organomet. Chem.* **2000**, *611*, 509.
(203) Bayler, A.; Canty, A. J.; Ryan, J. H.; Skelton, B. W.; White, A. H. *Inorg. Chem. Commun.* **2000**, *3*, 575.
(204) Kunz, R.; Fehlhammer, W. P. *Angew. Chem. Int. Edn. Engl.* **1994**, *33*, 330.
(205) Stang, P. J.; Crittell, C. M.; Arif, A. M. *Organometallics* **1993**, *12*, 4799.
(206) Canty, A. J.; Rodemann, T. *Inorg. Chem. Commun.* **2003**, *6*, 1382.
(207) Canty, A. J.; Rodemann, T.; Skelton, B. W.; White, A. H. *Inorg. Chem. Commun.* **2005**, *8*, 55.
(208) Stang, P. J.; Crittell, C. M. *Organometallics* **1990**, *9*, 3191.
(209) Stang, P. J.; Tykwinski, R. *J. Am. Chem. Soc.* **1992**, *114*, 4411.
(210) Tykwinski, R. R.; Stang, P. J. *Organometallics* **1994**, *13*, 3203.
(211) Fujita, M.; Kim, W. H.; Sakanishi, Y.; Fujiwara, K.; Hirayama, S.; Okuyama, T.; Ohki, Y.; Tatsumi, K.; Yoshioka, Y. *J. Am. Chem. Soc.* **2004**, *726*, 7548.
(212) Stang, P.; Ullmann, J. *Angew. Chem. Int. Edn. Engl.* **1991**, *30*, 1469.
(213) Stang, P. J.; Zhdankin, V. V.; Arif, A. M. *J. Am. Chem. Soc.* **1991**, *113*, 8997.

Index

Cumulative List of Contributors for Volumes 1–36

Cumulative Index for Volumes 37–55